Water Resources Management:
Principles and Practice

Water Resources Management: Principles and Practice

Edited by Sean Hart

SYRAWOOD
PUBLISHING HOUSE
New York

Published by Syrawood Publishing House,
750 Third Avenue, 9th Floor,
New York, NY 10017, USA
www.syrawoodpublishinghouse.com

Water Resources Management: Principles and Practice
Edited by Sean Hart

International Standard Book Number: 978-1-64740-148-1 (Hardback)

Cataloging-in-Publication Data

Water resources management : principles and practice / edited by Sean Hart.
     p. cm.
Includes bibliographical references and index.
ISBN 978-1-64740-148-1
1. Water resources development. 2. Water-supply--Management. 3. Water. I. Hart, Sean.
GB665 .W38 2022
333.91--dc23

TABLE OF CONTENTS

PREFACE

The main aim of this book is to educate learners and enhance their research focus by presenting diverse topics covering this vast field. This is an advanced book which compiles significant studies by distinguished experts in the area of analysis. This book addresses successive solutions to the challenges arising in the area of application, along with it; the book provides scope for future developments.

Water is the most important resource for all life on Earth. Freshwater resources are being exploited by the increasing demand for drinking, manufacturing, agriculture and sanitation. The need to optimize the use of water and minimize the environmental effects of water use on the natural environment gave rise to water resource management. It is a sub-field of water cycle management which includes the developing, distributing, planning and managing the water resources efficiently. The biggest concern of water resource management is the sustainability of the allocation of water-based resources. The topics included in this book on water resource management are of utmost significance and bound to provide incredible insights to readers. From theories to research to practical applications, case studies related to all contemporary topics of relevance to this field have been included in this book. It will provide comprehensive knowledge to the readers.

It was a great honour to edit this book, though there were challenges, as it involved a lot of communication and networking between me and the editorial team. However, the end result was this all-inclusive book covering diverse themes in the field.

Finally, it is important to acknowledge the efforts of the contributors for their excellent chapters, through which a wide variety of issues have been addressed. I would also like to thank my colleagues for their valuable feedback during the making of this book.

Editor

A dual-inexact fuzzy stochastic model for water resources management and non-point source pollution mitigation under multiple uncertainties

C. Dong[1], Q. Tan[1,3], G.-H. Huang[1], and Y.-P. Cai[2,3]

[1]MOE Key Laboratory of Regional Energy and Environmental Systems Optimization, Resources and Environmental Research Academy, North China Electric Power University, Beijing 102206, China

[2]State Key Laboratory of Water Environment Simulation, School of Environment, Beijing Normal University, Beijing 100875, China

[3]Institute for Energy, Environment and Sustainable Communities, University of Regina, Regina S4S 7H9, Canada

Correspondence to: Y.-P. Cai (yanpeng.cai@bnu.edu.cn) and Q. Tan (tanqian@iseis.org)

Abstract. In this research, a dual-inexact fuzzy stochastic programming (DIFSP) method was developed for supporting the planning of water and farmland use management system considering the non-point source pollution mitigation under uncertainty. The random boundary interval (RBI) was incorporated into DIFSP through integrating fuzzy linear programming (FLP) and chance-constrained programming (CCP) approaches within an interval linear programming (ILP) framework. This developed method could effectively tackle the uncertainties expressed as intervals and fuzzy sets. Moreover, the lower and upper bounds of RBI are continuous random variables, and the correlation existing between the lower and upper bounds can be tackled in RBI through the joint probability distribution function. And thus the subjectivity of decision making is greatly reduced, enhancing the stability and robustness of obtained solutions. The proposed method was then applied to solve a water and farmland use planning model (WFUPM) with non-point source pollution mitigation. The generated results could provide decision makers with detailed water supply–demand schemes involving diversified water-related activities under preferred satisfaction degrees. These useful solutions could allow more in-depth analyses of the trade-offs between humans and environment, as well as those between system optimality and reliability. In addition, comparative analyses on the solutions obtained from ICCP (Interval chance-constraints programming) and DIFSP demonstrated the higher application of this developed approach for supporting the water and farmland use system planning.

1 Introduction

Due to population growth, ongoing urbanization, industrialization and the intensification of agriculture, water and land demands are increasing globally, while the availability and quality of water resources and farmland are decreasing and non-point source pollution sharpening. These phenomena often cause a reduction in environmental quality and endanger sustainable development (Sessa, 2007). Thus, effective water resource and farmland use planning with non-point source pollution mitigation is necessary for ensuring economic and environmental welfare of regional populations (Ray et al., 2012; Mehta et al., 2013).

Previously, plenty of modeling technologies were applied to water resources and farmland use system planning with non-point source pollution mitigation (Satti et al., 2004; Chen et al., 2005; Riquelme and Ramos, 2005; Victoria et al., 2005; Kondilia and Kaldellis, 2006; Gregory et al., 2006; Khare et al., 2007; Castelletti et al., 2008; Qin et al., 2011; Mahmoud et al., 2011; Deviney Jr. et al., 2012; Zarghami and Hajykazemian, 2013; Canter et al., 2014). For example, Satti et al. (2004) used the GIS-based water resources and agricultural permitting and planning system to simulate the effect of climate, soil, and crop parameters on crop irrigation requirements. Chen et al. (2005) established force-state-response (DSR) dynamic strategy planning procedure to assist responsible authorities in obtaining alternatives of sustainable top river basin land use management. Riquelme and Ramos (2005) built up a geographic information system (GIS) on vine growing for supporting decision-making processes related to land and water management in Castilla–La Mancha, Spain. Victoria et al. (2005) adopted modeling tools, ISAREG model and SAGBAH model, to solve multi-scale problems with irrigation water uses and non-point source pollution in basins. Qin et al. (2011) proposed a system dynamics and water environmental model to operate the integrated socio-economic and water management system in a rapidly urbanizing catchment. Mehta et al. (2013) developed integrated water resources management models using the water evaluation and planning decision support system, for three towns in the Lake Victoria region. Zarghami and Hajykazemian (2013) proposed a new optimization algorithm by coupling the mutation process to the particle swarm optimization, which was successfully applied to the urban water resources management with a non-point source pollution problem for Tabriz, Iran.

However, effective planning for water resource and farmland use management with non-point source pollution mitigation is actually complicated with a variety of uncertainties and dynamics. For example, intricate interactions exist between various subsystems (such as economy, eco-environment, society, administration, etc.), which will inevitably produce a variety of uncertainties. Moreover, subjective judgments obtained from experts and stakeholders also exert significant impacts on data acquisition and system reliability. These complexities lead to the difficulties in solving the resulted uncertain optimization problems (Azaiez et al., 2005; Sethi et al., 2006; Tan et al., 2011; Bender and Simonovic, 2000; Guo et al., 2010; Qin et al., 2007; Lu et al., 2012; Huang et al., 2012; Zhang et al., 2009; Gu et al., 2013; Dessai and Hulme, 2007; Cai et al., 2011, 2012; Li et al., 2013). Nowadays, the stochastic linear programming (SLP) and interval linear programming (ILP) have become two of the most effective optimization approaches, especially the chance-constraints programming (CCP). For instance, Azaiez et al. (2005) tackled the uncertainties in inflows through adopting chance constraints and penalties of failure for optimal multi-period operation of a multi-reservoir system. And in 2006, Sethi et al. (2006) developed deterministic linear programming (DLP) and chance-constrained linear programming (CCLP) models to allocate available land and water resources optimally on seasonal basis. Tan et al. (2011) developed a radial interval chance-constrained programming (RICCP) approach for supporting source-oriented

non-point source pollution control under uncertainty. Another useful method handling uncertainties existing in water resources and farmland use management is based on fuzzy set theory. Bender and Simonovic (2000) applied a fuzzy compromise approach into water resource systems planning of the Tisza River. Qin et al. (2007) developed an interval-fuzzy nonlinear programming (IFNP) model for water quality management under uncertainty.

In reality, a high degree of uncertainty may exist among some parameters and coefficients of related water resources and farmland use management models. For example, the availabilities of various water resources are sensitive to geographical conditions and climate, technology selection and utilization efficiency, as well as water-saving consciousness, causing difficulties to related data acquisition, even the determination of interval numbers, when the lower and upper bounds are correlated. In the past decades, little work has been conducted to handle this type of uncertainty (dual uncertainty) existing in the processes of water and farmland use planning, which might result in missed information and thus impractical decision support (Cao et al., 2010). Therefore, in this study, a concept of random boundary interval (RBI) will be introduced to reflect such dual uncertainty. Specifically, the lower and upper bounds of RBI are continuous random variables, and the distribution information can be incorporated into the model. And, moreover, correlation existing between the lower and upper bounds can be tackled in RBI through the joint probability distribution function.

Finally, the proposed RBI theory and joint probability distribution will be integrated with ILP, CCP, and FLP technologies, leading to a dual-inexact fuzzy stochastic programming (DIFSP) method. Such an approach can tackle uncertainties expressed as interval numbers with known upper and lower bounds, fuzzy sets, as well as RBI. Due to the consideration of the intersection between lower and upper bounds, the robustness of the developed model could be enhanced. Then, this method will be applied to a water and farmland use planning model (WFUPM) with non-point source pollution mitigation for solving practical problems. And then many useful results will be generated, covering farmland use arrangement, water allocations among various consumers, resources supplies, and water pollution control under various water supply conditions and system reliabilities. The trade-offs between system benefit and failure risks can be balanced through the use of probability of constraint violations and satisfaction degrees. In addition, the solutions obtained through the existing ICCP method and the DIFSP approach proposed in this study will be compared to demonstrate how DIFSP would become improved upon ICCP in the planning of a water and farmland use system.

2　Methodology

2.1　The concept of RBI

In many practical problems, the lower and upper bounds of the right-hand sides can rarely be acquired as deterministic values. Instead, the obtained data can be presented as a random boundary interval (RBI) with its lower and upper bounds being random variables (Cao et al., 2010). Specifically, the parameters $\tilde{B}_i^-$ and $\tilde{B}_i^+$ on the right-hand side of constraints can be formulated as follows: (u_1, v_1), $(u_2, v_2), \ldots, (u_n, v_n)$, where $u_1, u_2, \ldots, u_n$ and $v_1, v_2, \ldots, v_n$ are the random numbers of lower and upper bounds of $\tilde{B}_i^{\pm}$ ($\tilde{B}_i^-$ and $\tilde{B}_i^+$). And $f(s, t)$ can be defined as the joint probability distribution function of $\left(\tilde{B}_i^-, \tilde{B}_i^+ \right)$, where s is the variable referring to $\tilde{B}_i^-$, and t is the variable referring to $\tilde{B}_i^+$. Then, RBI can be incorporated into the interval fuzzy linear programming (IFLP) model through the introduction of membership grade λ (Cao et al., 2010):

$$\text{Max } \lambda \tag{1}$$

subject to

$$C^{\pm} X^{\pm} \geq f^+ + (1 - \lambda)\left(f^+ - f^- \right), \tag{1a}$$

$$A^{\pm} X^{\pm} \leq \tilde{B}^- + (1 - \lambda)\left(\tilde{B}^+ - \tilde{B}^- \right), \tag{1b}$$

$$X^{\pm} \geq 0, \tag{1c}$$

$$0 \leq \lambda \leq 1. \tag{1d}$$

Let $Z_i = \tilde{B}^- + (1 - \lambda)\left(\tilde{B}^+ - \tilde{B}^- \right) = \lambda \tilde{B}_i^- + (1 - \lambda) \tilde{B}_i^+$. Given the joint probability distribution function of $\left(\tilde{B}_i^-, \tilde{B}_i^+ \right)$ is available, the distribution function of its linear combination $Z_i = \lambda \tilde{B}_i^- + (1 - \lambda) \tilde{B}_i^+$ can be calculated as follows:

$$G_i(z, \lambda) = \Pr\{Z_i \leq z\} = \Pr\left\{ \lambda \tilde{B}_i^- + (1 - \lambda) \tilde{B}_i^+ \leq z \right\}$$

$$= \int \int_{\lambda \tilde{B}_i^- + (1-\lambda)\tilde{B}_i^+} f(s, t)\,ds\,dt. \tag{1e}$$

In this model, λ represents not only the level of satisfying the objectives and constrains but also a linear combination parameter of the lower and upper bounds of RBI. Then RBI can be converted into a new random variable $\left[Z_i = \lambda \tilde{B}_i^- + (1 - \lambda) \tilde{B}_i^+ \right]$ in IFLP, and the distribution function $[G_i(z, \lambda)]$ of Z_i can be generated (Cao et al., 2010).

2.2　Dual inexact fuzzy stochastic programming

When a right-hand-side parameter is random, the CCP method should be adopted. Since parameters on the left-hand side are intervals, an interval chance-constrained linear programming (ICCP) can be developed (Huang et al., 1992,

1995). As Z_i is a random variable with known distribution function, model (1) can be converted into the following (Cao et al., 2010; Cai et al., 2007, 2009a, b):

$$\text{Max } \lambda \tag{2}$$

subject to

$$C^{\pm} X^{\pm} \geq f^+ + (1 - \lambda)\left(f^+ - f^- \right), \tag{2a}$$

$$\Pr\left\{ A^{\pm} X^{\pm} \leq Z_i \right\} \geq 1 - p_i, i = 1, 2, \ldots, m, \tag{2b}$$

$$X^{\pm} \geq 0, \tag{2c}$$

$$0 \leq \lambda \leq 1. \tag{2d}$$

Model (2) can be converted into an "equivalent" deterministic version as follows:

$$\text{Max } \lambda \tag{3}$$

subject to

$$C^{\pm} X^{\pm} \geq f^+ + (1 - \lambda)\left(f^+ - f^- \right), \tag{3a}$$

$$A_i^{\pm} X^{\pm} \leq Z_i^{(p_i)}, A_i^{\pm} \in A^{\pm}, i = 1, 2, \ldots, m, \tag{3b}$$

$$X^{\pm} \geq 0, \tag{3c}$$

$$0 \leq \lambda \leq 1, \tag{3d}$$

where $Z_i^{(p_i)} = G_i^{-1}(p_i, \lambda)$. In CCP, the random variable on the right-hand side can be handled as several deterministic numbers corresponding to different violation probabilities (p_i). However, $Z_i^{(p_i)}$ in this part is a function of λ corresponding to p_i, because the distribution function of $Z_i\left[G_i^{-1}(p_i, \lambda) \right]$ is a function of z and λ. When p_i is a constant, then $G_i^{-1}(p_i, \lambda)$ is a function of λ (Huang et al., 1992, 1995). Thus the solution method of the developed model will be different from the conventional CCP.

If $Z_i^{(p_i)}$ is a linear function of λ, according to solution algorithm developed by Huang et al. (1992, 1995), this model can be divided into two deterministic sub-models:

$$\text{Max } \lambda \tag{4}$$

subject to

$$\sum_{j=1}^{k} c_j^+ x_j^+ + \sum_{j=k+1}^{n} c_j^+ x_j^- \geq f^- + (1 - \lambda)\left(f^+ - f^- \right), \tag{4a}$$

$$\sum_{j=1}^{k_1} |a_{ij}|^- \text{sign}(a_{ij}^-) x_j^+ + \sum_{j=k_1+1}^{n} |a_{ij}|^+ \text{sign}(a_{ij}^+) x_j^-$$

$$\leq Z_i^{(p_i)}, i = 1, 2, \ldots, m, \tag{4b}$$

$$x_j^+ \geq 0, j = 1, 2, \ldots, k, \tag{4c}$$

$$x_j^- \geq 0, j = k + 1, k + 2, \ldots, n, \tag{4d}$$

$$0 \leq \lambda \leq 1, \tag{4e}$$

and

$$\text{Max} \quad \lambda \tag{5}$$

subject to

$$\sum_{j=1}^{k} c_j^- x_j^- + \sum_{j=k+1}^{n} c_j^- x_j^+ \geq f^- + (1-\lambda)\left(f^+ - f^-\right), \tag{5a}$$

$$\sum_{j=1}^{k_1} |a_{ij}|^+ \text{sign}(a_{ij}^+) x_j^- + \sum_{j=k_1+1}^{n} |a_{ij}|^- \text{sign}(a_{ij}^-) x_j^+$$
$$\leq Z_i^{(p_i)}, i = 1, 2, \ldots, m, \tag{5b}$$

$$x_j^{\pm} \geq 0, \forall j, \tag{5c}$$

$$x_j^- \leq x_{j\text{opt}}^+, j = 1, 2, \ldots, k, \tag{5d}$$

$$x_j^+ \geq x_{j\text{opt}}^-, \ldots j = k+1, k+2, \ldots, n. \tag{5e}$$

Among these two submodels, f^+ and f^- correspond to the lower and upper bounds of the objective function values. When the objective function is to be minimized, sub-model corresponding to f^- is firstly formulated. And then the sub-model corresponding to f^+ can be obtained based on the solution of the first sub-model. Through solving these two submodels, the final interval solutions can be acquired as λ, $x_{j\text{opt}} = [x_{j\text{opt}}^-, x_{j\text{opt}}^+]$, and $f_{\text{opt}} = [f_{\text{opt}}^-, f_{\text{opt}}^+]$.

However, the two-step method encounters difficulties if $Z_i^{(p_i)}$ is not a linear function of λ. In this case, it should be converted to linear or stepwise linear functions of λ. The particular flow of this developed optimization method is displayed in Fig. 1.

3 Application to water and farmland use management system with non-point source pollution mitigation

3.1 Overview of water and farmland use management system with non-point source pollution mitigation

Increasing population, diminishing supplies and changing climatic conditions amplify difficulties in resolving the conflicts between human activities and environment. Since agriculture is one of the most important water users, the farmland use arrangement can directly or indirectly influence the water resources utilization and environment. Specifically, the abuse of fertilizer and pesticide can cause extensive anthropogenic non-point source pollution. Conversely, the water pollution control can also exert an impact on associated human activities, such as water allocation and cultivation. These all call for the need to integrate pollution mitigation efforts into the framework of water resources management. However, in real-world problems, various components in the water resource and agricultural land use management system impact each other, which inevitably leads to complexities and dynamics. For example, the interactions between population

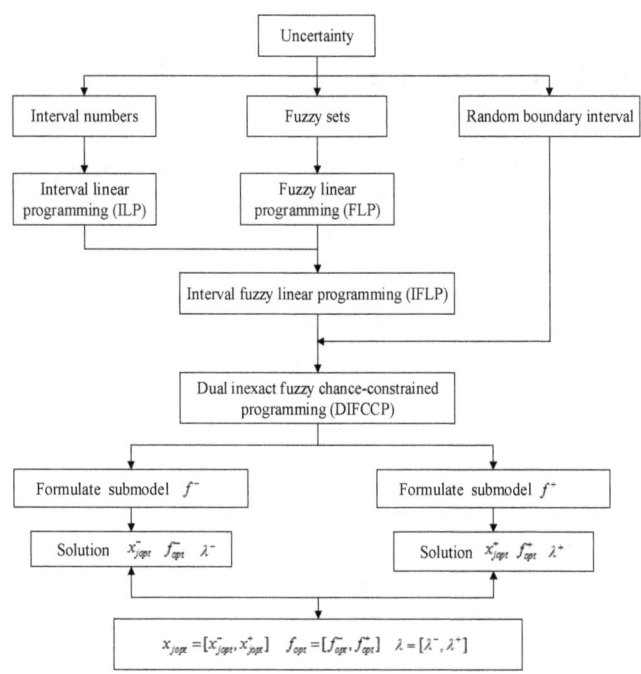

Fig. 1. Optimization method.

and water supplies can directly cause complex water utilization among various end users. This makes it critical to clarify the interactions among system factors and those intimately involved in the planning process (Chung et al., 2008).

Figure 2 presents a general water and farmland use management system, including internal and external factors. Specifically, resources availability, distribution, utilization technology, policy, security and other internal factors of water and agricultural land comprise the microscopic system, directly affecting the related planning processes. Besides, external factors such as social, economy, natural conditions, institutions, eco-environment (i.e., non-point source pollution), and population can exert indirect impacts on the entire system operation. Given the complexity of this system and the interactions among various components, uncertainty is a necessary consideration in the process of modeling. In addition, the uncertainties existing in this system can be also generated from the errors in data collection and parameter settings, subjective judgments of experts or stakeholders, as well as uncertainty due to the structure of adopted model (Lindenschmidt et al., 2007; Dong et al., 2013).

Therefore, several optimization technologies will be introduced to handle these uncertainties in this system. For example, economic coefficients (e.g., unit benefit of water supply and pollutants treatment cost), technological efficiencies, and continuous variables can be expressed as interval numbers. Given the random and dynamic features of water resources availabilities (i.e., surface drainage water, groundwater and river water), it is rather hard to accurately determine their two bounds. And the random boundary interval (RBI) will

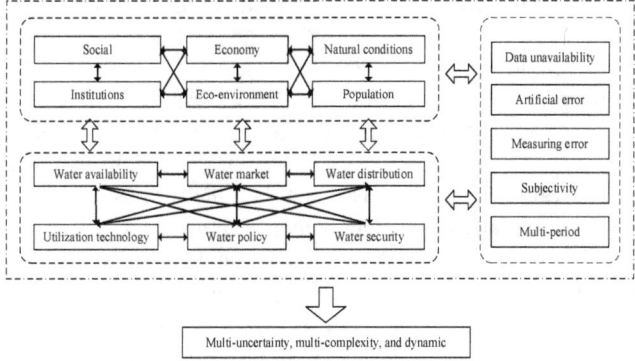

Fig. 2. Water and farmland use management system.

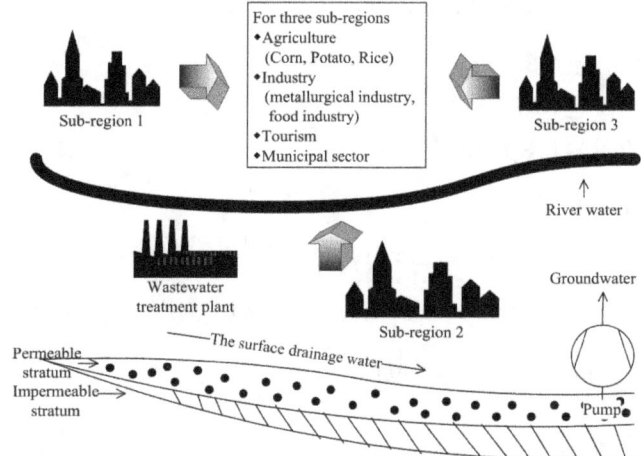

Fig. 3. The typical regional farmland use and water resources system.

be adapted to reflect their dual uncertainty, with the lower and upper bounds of RBI being continuous random variables. Then, the developed dual-inexact fuzzy stochastic programming (DIFSP) method will be applied into a water and farmland use planning model (WFUPM) with non-point source pollution mitigation.

3.2 Modeling formulation

In the study case, three types of water resources (i.e., surface drainage water, groundwater, and river water) are major water supplies, meeting the regional water demands of various end users (i.e., agriculture, industry, tourism, residents, and municipal sector). Depending upon different intended uses for end users, surface water can be sent directly for industrial production and irrigation, or should be treated prior to drinking and other uses. Pumped groundwater can be delivered directly to all users before disinfection. River water can be provided to agricultural irrigation and industry production. Water is transferred between end users by pipes with limited capacities. In this study region, corn, potato, and rice are selected as the major crops, and metallurgical and food industries constitute local industry. Figure 3 gives an overview of the components and factors that need to be taken into account in this model (Dong et al., 2013). The study time horizon is 15 years and is further divided into three planning periods. With the consideration of these elements, water and farmland use planning model (WFUPM) with non-point source pollution mitigation can be formulated. Its objective is to maximize the total system benefit, covering benefit for agriculture irrigation, water supply benefits for industry, tourism, residents, and minus the costs for water pumping and delivering, as well as wastewater treatment, specifically as follows:

$$\text{Max} f^{\pm} = f_{\text{BC}}^{\pm} + f_{\text{BI}}^{\pm} + f_{\text{BT}}^{\pm} + f_{\text{BT}}^{\pm} - f_{\text{CW}}^{\pm} - f_{\text{CE}}^{\pm} \tag{6}$$

1. Benefit for agriculture irrigation

$$f_{\text{BC}}^{\pm} = \sum_{i=1}^{3} \sum_{t=1}^{3} \left(\text{PC}_{it}^{\pm} \cdot Y_{it}^{\pm} - \text{CC}_{it}^{\pm} \right) A_{it}^{\pm} \tag{6a}$$

2. Water supply benefit for industry

$$f_{\text{BI}}^{\pm} = \sum_{k=1}^{2} \sum_{t=1}^{3} \text{QI}_{kt}^{\pm} \cdot \text{ZI}_{kt}^{\pm} \tag{6b}$$

3. Water supply benefit for tourism

$$f_{\text{BT}}^{\pm} = \sum_{t=1}^{3} \text{QT}_{t}^{\pm} \cdot \text{ZT}_{t}^{\pm} \tag{6c}$$

4. Water supply benefit for residents

$$f_{\text{BT}}^{\pm} = \sum_{t=1}^{3} \text{QR}_{t}^{\pm} \cdot \text{ZR}_{t}^{\pm} \tag{6d}$$

5. Cost of water pumping and delivering

$$f_{\text{CW}}^{\pm} = \sum_{t=1}^{3} \left(\text{QS}_{t}^{\pm} \cdot \text{WS}_{t}^{\pm} + \text{QG}_{t}^{\pm} \cdot \text{WG}_{t}^{\pm} + \text{QR}_{t}^{\pm} \cdot \text{WR}_{t}^{\pm} \right)$$

$$\tag{6e}$$

6. Cost of wastewater treatment

$$f_{\text{CE}}^{\pm} = \sum_{t=1}^{3} \left(\text{QWT}_{t}^{\pm} \cdot \text{DWT}_{t}^{\pm} \cdot \text{ZT}_{t}^{\pm} + \text{QWM}_{t}^{\pm} \cdot \text{DWM}_{t}^{\pm} \right.$$

$$\left. \cdot \text{ZM}_{t}^{\pm} + \text{QWR}_{t}^{\pm} \cdot \text{DWM}_{t}^{\pm} \cdot \text{ZR}_{t}^{\pm} \right)$$

$$+ \sum_{k=1}^{2} \sum_{t=1}^{3} \left(\text{QWI}_{kt}^{\pm} \cdot \text{DWI}_{kt}^{\pm} \cdot \text{ZI}_{kt}^{\pm} \right) \tag{6f}$$

Constraints:

1. Balance for farmland use

$$\text{MINA}_{t} \leq \sum_{i=1}^{3} A_{it}^{\pm} \leq \text{MAXA}_{t}, \forall t \tag{6g}$$

2. Balance for water resource availability

$$WS_t^{\pm} \leq MS_t^{\pm}, \forall t \tag{6h}$$

$$WG_t^{\pm} \leq MG_t^{\pm}, \forall t \tag{6i}$$

$$WR_t^{\pm} \leq MR_t^{\pm}, \forall t \tag{6j}$$

$$\sum_{i=1}^{2} RWC_{it}^{\pm} \cdot A_{it}^{\pm} + \sum_{k=1}^{2} ZI_{kt}^{\pm} + ZT_t^{\pm} + ZM_t^{\pm} + ZR_t^{\pm}$$
$$+ RWG_t^{\pm} \cdot GA_t^{\pm} \leq WS_t^{\pm} + WG_t^{\pm} + WR_t^{\pm}, \forall t \tag{6k}$$

3. Balance for water supply

$$\sum_{i=1}^{2} RWC_{it}^{\pm} \cdot A_{it}^{\pm} \geq MA_t^{\pm}, \forall t \tag{6l}$$

$$\sum_{k=1}^{2} ZI_{kt}^{\pm} \geq MI_t^{\pm}, \forall t \tag{6m}$$

$$ZT_t^{\pm} \geq MT_t^{\pm}, \forall t \tag{6n}$$

$$ZM_t^{\pm} \geq MM_t^{\pm}, \forall t \tag{6o}$$

$$ZR_t^{\pm} \geq MR_t^{\pm}, \forall t \tag{6p}$$

4. Balance for wastewater treatment

$$DWI_{kt}^{\pm} \cdot ZI_{kt}^{\pm} + DWT_t^{\pm} \cdot ZT_t^{\pm} + DWM_t^{\pm} \cdot ZM_t^{\pm}$$
$$+ DWR_t^{\pm} \cdot ZR_t^{\pm} \leq TWC_t, \forall t \tag{6q}$$

5. Non-point pollution control constraints

$$\sum_{i=1}^{3} NA_t^{\pm} \cdot A_{it}^{\pm} +$$
$$\left(\sum_{k=1}^{2} NI_{kt}^{\pm} \cdot ZI_{kt}^{\pm} + NT_t^{\pm} \cdot ZT_t^{\pm} + NM_t^{\pm} \cdot ZM_t^{\pm} + NR_t^{\pm} \cdot ZR_t^{\pm} \right)$$
$$\left(1 - NRE_t^{\pm} \right) \leq TN_t^{\pm}, \forall t \tag{6r}$$

$$\sum_{i=1}^{3} PA_t^{\pm} \cdot A_{it}^{\pm} +$$
$$\left(\sum_{k=1}^{2} PI_{kt}^{\pm} \cdot ZI_{kt}^{\pm} + PT_t^{\pm} \cdot ZT_t^{\pm} + PM_t^{\pm} \cdot ZM_t^{\pm} + PR_t^{\pm} \cdot ZR_t^{\pm} \right)$$
$$\left(1 - PRE_t^{\pm} \right) \leq TP_t^{\pm}, \forall t, \tag{6s}$$

where f = expected net system benefit (USD); t = time period, $t = 1, 2, 3$; i = type of crop, $i = 1, 2, 3$ (where $i = 1$ for corn, 2 for potato, 3 for rice); k = type of industry, $k = 1, 2$ (where $k = 1$ for metallurgical industry, 2 for food industry); $PC_{it}^{\pm}$ = price of crop i in period t (USD kg^{-1}); $Y_{it}^{\pm}$ = yield of crop i in period t (kg km^{-2}); $CC_{it}^{\pm}$ = cost of cultivating crop i in period t (USD km^{-2}); $QI_{kt}^{\pm}$ = unit benefit of water allocated to industry k in period t (USD m^{-3});

$QT_t^{\pm}$ = unit benefit of water allocated to tourism in period t (USD m^{-3}); $QR_t^{\pm}$ = unit benefit of water allocated to household in period t (USD m^{-3}); $QG_t^{\pm}$ = cost of cultivating green field in period t (USD km^{-2}); $QS_t^{\pm}$ = cost of pumping and delivering the surface drainage water in period t (USD m^{-3}); $QG_t^{\pm}$ = cost of pumping and delivering the groundwater in period t (USD m^{-3}); $QR_t^{\pm}$ = cost of pumping and delivering the river water in period t (USD m^{-3}); $QWI_{kt}^{\pm}$ = treatment cost of wastewater from industry k in period t (USD t^{-1}); $QWT_t^{\pm}$ = treatment cost of wastewater from tourism in period t (USD t^{-1}); $QWM_t^{\pm}$ = treatment cost of wastewater from municipal sector in period t (USD t^{-1}); $QWR_t^{\pm}$ = treatment cost of wastewater from household in period t (USD t^{-1}); $DWI_{kt}^{\pm}$ = unit wastewater discharge by industry k in period t (t m^{-3}); $DWT_t^{\pm}$ = unit wastewater discharge by tourism industry in period t (t m^{-3}); $DWM_t^{\pm}$ = unit wastewater discharge by municipal sector in period t (t m^{-3}); $DWR_t^{\pm}$ = unit wastewater discharge by household in period t (t m^{-3}); $MAXA_t$ = the maximum area allocated to crop i in period t (km^2); $MINA_t$ = the minimum area allocated to crop i in period t(km^2); $MS_t^{\pm}$ = the maximum allocated amount of surface drainage water in period t (m^3); $MG_t^{\pm}$ = the maximum allocated amount of groundwater in period t (m^3); $MR_t^{\pm}$ = the maximum allocated amount of river water in period t (m^3); $RWC_{it}^{\pm}$ = unit irrigation demand for crop i in period t (m^3 km^{-2}); $RWG_t^{\pm}$ = unit irrigation demand for green field in period t (m^3 km^{-2}); $MA_t^{\pm}$ = water demand of agriculture in period t (m^3); $MI_t^{\pm}$ = water demand of industry in period t (m^3); $MT_t^{\pm}$ = water demand of tourism in period t (m^3); $MM_t^{\pm}$ = water demand of municipal sector in period t (m^3); $MR_t^{\pm}$ = water demand of household in period t (m^3); TWC_t = total wastewater treatment capacity in period t (t); $NA_t^{\pm}$ = nitrogen percent content of the soil in period t (%); $PA_t^{\pm}$ = phosphorus percent content of the soil in period t (%); $NI_{kt}^{\pm}$ = unit nitrogen discharge by industry k in period t (t m^{-3}); $PI_{kt}^{\pm}$ = unit phosphor discharge by industry k in period t (t m^{-3}); $NT_t^{\pm}$ = unit nitrogen discharge by tourism in period t (t m^{-3}); $PT_t^{\pm}$ = unit phosphor discharge by tourism in period t (t m^{-3}); $NM_t^{\pm}$ = unit nitrogen discharge by municipal sector in period t (t m^{-3}); $PM_t^{\pm}$ = unit phosphor discharge by municipal sector in period t (t m^{-3}); $NR_t^{\pm}$ = unit nitrogen discharge by household in period t (t m^{-3}); $PR_t^{\pm}$ = unit phosphor discharge by household in period t (t m^{-3}); $NRE_t^{\pm}$ = nitrogen removal efficiency in period t (%); $PRE_t^{\pm}$ = phosphor removal efficiency in period t (%); $TN_t^{\pm}$ = the maximum allowed amount of nitrogen discharge in period t (kg); $TP_t^{\pm}$ = the maximum allowed amount of phosphor discharge in period t (kg); $A_{it}^{\pm}$ = area allocated to crop i in period t (km^2); $ZI_{kt}^{\pm}$ = water allocated to industry k in period t (m^3); $ZT_t^{\pm}$ = water allocated to tourism in period t (m^3); $ZM_t^{\pm}$ = water allocated to municipal sector in period t (m^3); $ZR_t^{\pm}$ = water allocated to household in period t (m^3); $WS_t^{\pm}$ = allocated amount of surface drainage water in period t (m^3); $WG_t^{\pm}$ = allocated

Table 1. Benefits of water supply for end users (USD m^{-3}).

End user	Period		
	$t = 1$	$t = 2$	$t = 3$
Metallurgical industry	[27.57, 29.56]	[25.53, 27.31]	[23.77, 24.67]
Food industry	[14.86, 15.09]	[14.29, 14.45]	[13.64, 13.77]
Tourism	[9.11, 9.25]	[8.76, 8.86]	[8.36, 8.44]
Household	[14.26, 25.3]	[31.95, 43.42]	[44.77, 45.48]

Data source: Dong et al. (2013, 2014).

Table 2. Costs for pumping and delivering water resources (USD m^{-3}).

Water resource type	Period		
	$t = 1$	$t = 2$	$t = 3$
Surface drainage water groundwater	[0.0033, 0.0034]	[0.0032, 0.0033]	[0.0031, 0.0032]
Groundwater	[0.0056, 0.0062]	[0.0054, 0.0059]	[0.0052, 0.0057]
River water	[0.0062, 0.0063]	[0.0060, 0.0061]	[0.0058, 0.0059]

Data source: Dong et al. (2013, 2014).

amount of groundwater in period t (m^3); WR$_t^{\pm}$ = allocated amount of river water in period t (m^3).

In order to generate optimal system solutions, several effective constraints are formulated to restrain the entire model for the purpose of maximizing system benefit. Particularly, farmland use, including the planting areas of corn, potato, and rice, should be limited to available farmland resources. All kinds of water resources have their own availabilities every period due to the natural and policy limitations. Water supplies to each end user should satisfy their operational demands. The wastewater discharged to the central treatment plant should not excess its fixed capacity. Finally, the total quantity control should be applied to control the discharge amount of non-point source pollution (i.e., nitrogen and phosphorus). The benefits of water supply for end users and costs for pumping and delivering water resources are displayed in Tables 1 and 2. In this model, the RBIs are combined with the water resources availabilities, and Table 3 presents the linearization results of surface water availability.

4 Analysis of results

Through computing a developed water and farmland use planning model (WFUPM) with non-point source pollution mitigation, a series of related schemes were generated. Particularly, they can provide useful plans of planting areas for crops, water allocations to each end user, water resources supplies, and non-point source pollution control under various water supply conditions and system reliabilities. In addition, the solutions obtained from ICCP and DIFSP were compared to demonstrate the efficiency of new developed optimization method for tackling uncertainties in water and farmland use system.

According to the climate of northern China, corn, potato, and rice are chosen as the staple crops. The crop planting areas under $p_i = 0.01$ are presented in Table 4. Obviously, potato would be the major crop in this region, encompassing planting areas of [138.31, 146.85], [131.98, 140.62], and [125.66, 132.61] km^2 in periods 1 to 3, respectively. Then corn would occupy the second position of crop planning, which would require [7.65, 9.9], [7.3, 9.48], and [6.95, 8.94] km^2 in these three periods. Finally, [7.04, 8.25], [6.72, 7.9], and [6.39, 7.45] km^2 of areas would be used to plant rice in periods 1 to 3, respectively. Mainly due to imported vegetables and food, the planting areas of these crops decrease each period.

Table 5 shows the solutions of water allocations to various end users under $p_i = 0.01$, mainly including agriculture, metallurgical industry, food industry, tourism, residents, and municipal sector. Among them, residents are still the biggest consumer, utilizing [32.61, 33.69], [31.34, 32.27], and [29.93, 30.76] million m^3 in periods 1, 2, and 3, respectively. For providing vegetables and rice to local residents, [27.57, 29.56], [25.53, 27.31], and [23.77, 24.67] million m^3 of water would be allocated to agricultural production in these three periods. Municipal sector is another important water user, which would require [16.78, 19.47], [16.13, 18.64], and [15.41, 17.77] million m^3. Furthermore, local industries also need adequate water to ensure their normal operations, such as metallurgical industry consuming [14.86, 15.09], [14.29, 14.45], and [13.64, 13.77] million m^3 of water resources in periods 1 to 3, respectively. As the tourism develops, it would consume a rather large proportion of water usage, increasing from [14.26, 25.3] million m^3 in period 1, through [31.95, 43.42] million m^3 in period 2, to [44.77,

Table 3. Linearization results of surface water availability.

Period	p_i value	$G_i^{-1} p_i, \lambda/Z_i^{p_i}$	$\lambda = [0, 0.10]$	$\lambda = [0.81, 0.90]$
$t = 1$	$p_i = 0.01$	$-2.23\sqrt{0.34\lambda^2 - 0.51\lambda + 0.56} + 27.85 - 1.43\lambda$	$-0.0068\lambda + 26.107$	$-0.0157\lambda + 25.291$
	$p_i = 0.05$	$-1.64\sqrt{0.34\lambda^2 - 0.51\lambda + 0.56} + 27.85 - 1.43\lambda$	$-0.0090\lambda + 26.623$	$-0.0153\lambda + 25.710$
	$p_i = 0.10$	$-1.28\sqrt{0.34\lambda^2 - 0.51\lambda + 0.56} + 27.85 - 1.43\lambda$	$-0.0102\lambda + 26.893$	$-0.0150\lambda + 25.929$
	$p_i = 0.15$	$-1.03\sqrt{0.34\lambda^2 - 0.51\lambda + 0.56} + 27.85 - 1.43\lambda$	$-0.0110\lambda + 27.080$	$-0.0149\lambda + 26.080$
$t = 2$	$p_i = 0.01$	$-2.23\sqrt{0.27\lambda^2 - 0.67\lambda + 0.94} + 26.91 - 1.76\lambda$	$-0.0098\lambda + 24.651$	$-0.0152\lambda + 23.590$
	$p_i = 0.05$	$-1.64\sqrt{0.27\lambda^2 - 0.67\lambda + 0.94} + 26.91 - 1.76\lambda$	$-0.0121\lambda + 25.320$	$-0.0153\lambda + 24.257$
	$p_i = 0.10$	$-1.28\sqrt{0.27\lambda^2 - 0.67\lambda + 0.94} + 26.91 - 1.76\lambda$	$-0.0133\lambda + 25.669$	$-0.0158\lambda + 24.530$
	$p_i = 0.15$	$-1.03\sqrt{0.27\lambda^2 - 0.67\lambda + 0.94} + 26.91 - 1.76\lambda$	$-0.0141\lambda + 25.912$	$-0.0162\lambda + 24.720$
$t = 3$	$p_i = 0.01$	$-2.23\sqrt{0.14\lambda^2 - 0.22\lambda + 0.72} + 26.32 - 1.33\lambda$	$-0.0105\lambda + 24.343$	$-0.0136\lambda + 23.402$
	$p_i = 0.05$	$-1.64\sqrt{0.14\lambda^2 - 0.22\lambda + 0.72} + 26.32 - 1.33\lambda$	$-0.0113\lambda + 24.929$	$-0.0135\lambda + 23.951$
	$p_i = 0.10$	$-1.28\sqrt{0.14\lambda^2 - 0.22\lambda + 0.72} + 26.32 - 1.33\lambda$	$-0.0117\lambda + 25.234$	$-0.0135\lambda + 24.237$
	$p_i = 0.15$	$-1.03\sqrt{0.14\lambda^2 - 0.22\lambda + 0.72} + 26.32 - 1.33\lambda$	$-0.0120\lambda + 25.446$	$-0.0134\lambda + 24.436$

Table 4. Crop planting (km^2).

End user	Period		
	$t = 1$	$t = 2$	$t = 3$
Corn	[7.65, 9.9]	[7.3, 9.48]	[6.95, 8.94]
Potato	[138.31, 146.85]	[131.98, 140.62]	[125.66, 132.61]
Rice	[7.04, 8.25]	[6.72, 7.9]	[6.39, 7.45]

Table 6. Water pollution control (10^3 t).

End user	Period		
	$t = 1$	$t = 2$	$t = 3$
Wastewater	[28.12, 29.75]	[30.64, 31.29]	[28.22, 31.15]
Total nitrogen	[1.27, 1.34]	[1.18, 1.24]	[0.98, 1]
Total phosphorus	[0.49, 0.50]	[0.44, 0.45]	[0.34, 0.36]

Table 5. Water allocations to end users (million m^3).

End user	Period		
	$t = 1$	$t = 2$	$t = 3$
Agriculture	[27.57, 29.56]	[25.53, 27.31]	[23.77, 24.67]
Metallurgical industry	[14.86, 15.09]	[14.29, 14.45]	[13.64, 13.77]
Food industry	[9.11, 9.25]	[8.76, 8.86]	[8.36, 8.44]
Tourism	[14.26, 25.3]	[31.95, 43.42]	[44.77, 45.48]
Residents	[32.61, 33.69]	[31.34, 32.27]	[29.93, 30.76]
Municipal sector	[16.78, 19.47]	[16.13, 18.64]	[15.41, 17.77]

45.48] million m^3 in period 3, which should arouse the general concern of relevant department.

Table 6 presents the solutions of water pollution control. In this study, the model mainly considers wastewater and nonpoint source pollution (i.e., total nitrogen and phosphorus). Restrained by the capacities of pollution treatment, [28.12, 29.75], [30.64, 31.29], and [28.22, 31.15] × 10^3 t of wastewater would be allowed to discharge to the sewage treatment facilities in periods 1, 2, and 3. For total nitrogen, [1.27, 1.34], [1.18, 1.24], and [0.98, 1] × 10^3 t would be the nitrogen allowances in these three periods. In addition, [0.49, 0.50], [0.44, 0.45], and [0.34, 0.36] × 10^3 t of total phosphorus could be disposed to local treatment system. In order to protect the water environment, the discharge quantities of

water pollutants should be controlled within local capacities according to current technological development situation.

In this research, four p_i values are defined, including 0.01, 0.05, 0.10, and 0.15. Generally, a higher p_i value indicates a higher probability of constraint violation, resulting in a larger volume of water supplies and a higher system benefit. As shown in Table 7, the quantity from surface drainage water in period 1 would be [20.73, 21.54], [21.07, 21.96], [21.25, 22.18], and [21.37, 22.24] million m^3 under a p_i level of 0.01, 0.05, 0.10, and 0.15, respectively. The corresponding volume of groundwater would be [28.33, 29.14], [28.69, 29.62], [28.87, 29.88], and [29, 30.05] million m^3. Similarly, when the p_i value changes from 0.01 to 0.15, the amount of river water would increase from [102.79, 107.79] to [104.61, 109.38] million m^3. From periods 1 to 3, a downward trend would be observed for the amounts of water supplied. For example, under a p_i level of 0.01, the amount of groundwater would be [28.33, 29.14], [26.52, 27.61], and [21.91, 24.16] million m^3 in periods 1 to 3, respectively. Such a decrease is probably contributed by the advancement of water-saving techniques and the improved efficiency in water utilization.

Since p_i value represents the probability of the constraint being violated, higher p_i values mean higher probability of constraint violations, presenting higher system failure risks and leading to a decreased reliability in fulfilling the system

Table 7. Water resources supplies under different p_i values (million m^3).

Water resource type	Period	p_i value			
		$p_i = 0.01$	$p_i = 0.05$	$p_i = 0.10$	$p_i = 0.15$
Surface drainage water	$t = 1$	[20.73, 21.54]	[21.96, 21.07]	[22.18, 21.25]	[22.24, 21.37]
	$t = 2$	[19.45, 20.34]	[20.89, 19.88]	[21.17, 20.1]	[21.25, 20.26]
	$t = 3$	[19.18, 20.08]	[20.57, 19.63]	[20.81, 19.86]	[20.88, 20.03]
Groundwater	$t = 1$	[28.33, 29.14]	[29.62, 28.69]	[29.88, 28.87]	[30.05, 29]
	$t = 2$	[26.52, 27.61]	[27.97, 26.77]	[28.16, 26.91]	[28.29, 27]
	$t = 3$	[21.91, 24.16]	[24.94, 22.71]	[25.34, 23.13]	[25.62, 23.42]
River water	$t = 1$	[102.79, 107.79]	[108.63, 103.75]	[109.07, 104.26]	[109.38, 104.61]
	$t = 2$	[98.26, 103.88]	[105.24, 99.14]	[105.95, 99.6]	[106.44, 99.92]
	$t = 3$	[96.37, 102.06]	[103.2, 97.15]	[103.79, 97.56]	[104.2, 97.84]

Table 8. System benefit from ICCP and DIFSP (USD million).

Optimization method	p_i value			
	$p_i = 0.01$	$p_i = 0.05$	$p_i = 0.10$	$p_i = 0.15$
ICCP	[314.95, 370.04]	[314.97, 371.82]	[314.95, 372.76]	[314.94, 373.40]
DIFSP	[316.02, 367.71]	[317.54, 370.16]	[317.71, 371.09]	[317.70, 371.74]

requirements, but generating higher benefits. Conversely, lower p_i values correspond to lower system risks and lower benefits. Figure 4 shows the effects of varied p_i values on the system benefit under upper bound. Specifically, the system benefits would increase from USD [316.02, 367.71], [317.54, 370.16], and [317.71, 371.09], to [317.70, 371.74] million USD under $p_i = 0.01, 0.05, 0.10$, and 0.15, respectively.

Moreover, the satisfaction degree $\lambda^{\pm}$ presents the flexibility in the constraints and fuzziness in the objective, which indicates the decision makers' preferences regarding the trade-offs between environment and economy, as well as system reliability and benefit. Generally, higher $\lambda^{\pm}$ level means decreased system reliability, with a higher benefit, being consistent with higher p_i value; in comparison, lower $\lambda^{\pm}$ level presents decreased system reliability with a lower system benefit, corresponding to a lower p_i value. Figure 5 represents the satisfaction degrees under different p_i values under upper bound. Particularly, the satisfaction degrees would increase with p_i values, rising from [0.017, 0.829], [0.041, 0.868], and [0.043, 0.882], to [0.043, 0.892] under $p_i = 0.01, 0.05, 0.10$, and 0.15, respectively.

In order to further demonstrate that the method DIFSP is more applicable than ICCP in dealing with water resources and farmland use management problems under uncertainty, a comparable study was conducted between the generated solutions from these two optimization methods. Letting the lower and upper bounds of RBIs equal their mid-values, the model would be simplified into a conventional inexact chance-constrained programming (ICCP) problem (Cao et al., 2010). The system benefits obtained through ICCP and

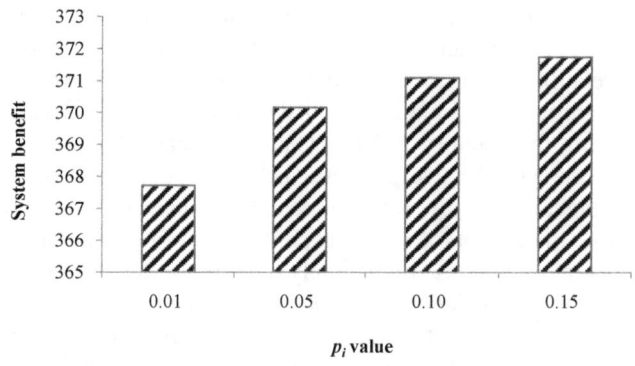

Fig. 4. System benefit under different p_i values (upper bound).

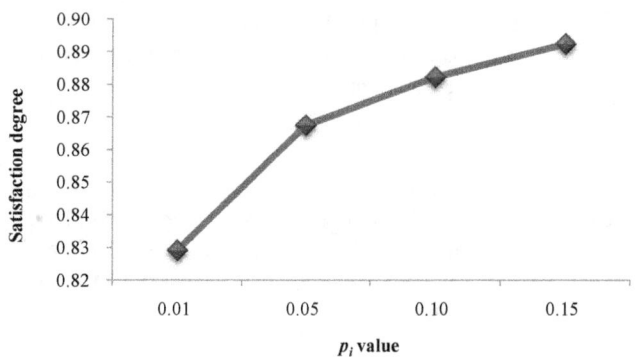

Fig. 5. Satisfaction degrees under different p_i values (upper bound).

DIFSP are presented in Table 8, which indicates that the results of DIFSP are much more robust than those of ICCP, meaning that the solution width of DIFSP is tighter and less uncertain. For instance, the system benefits computed from ICCP and DIFSP would be [314.95, 370.04] and [316.02, 367.71] million USD under $p_i = 0.01$, obviously becoming tightened. This comparison convincingly certifies the effectiveness of DIFSP (introducing of RBI) in tackling the uncertainty (dual uncertainty) existing in the water resources and farmland use management system. Obtained solutions could provide decision makers with the desired schemes under preferable system reliability and economic benefit.

5   Conclusions

In this research, a dual-inexact fuzzy stochastic programming (DIFSP) method was proposed through incorporating the random boundary interval (RBI) with fuzzy programming (FP), chance-constrained programming (CCP), and interval linear programming (ILP) techniques. And then the developed method was applied to the water and farmland use planning model (WFUPM) with non-point source pollution mitigation. Overall, this study can (1) conduct comprehensive analysis of water and farmland use management system; (2) tackle multiple uncertainties presented as interval numbers, fuzzy sets, and probability distributions; (3) tackle the correlation exiting between the lower and upper bounds of RBI through the joint probability distribution function, enhancing the stability and robustness of obtained solutions; (4) generate effective schemes including planting area arrangement of crops, water allocations among various end users, water resources supplies, and water pollution control plans under various water supply conditions and system reliabilities; (5) balance the trade-offs between system benefit and failure risks through utilizing the probability of constraint violations and satisfaction degrees; and (6) compare the solutions obtained from ICCP and DIFCCP to demonstrate the application of this developed method for supporting water and farmland use system planning. In the future research, this developed DIFSP method can be applied to other environmental planning problems, and can be incorporated with other optimization technologies to handle various practical issues under uncertainty.

Acknowledgements. This research was supported by the National Science Foundation for Innovative Research Group (no. 51121003), the National Natural Science Foundation of China (no. 51209087), and the special fund of State Key Lab of Water Environment Simulation (11Z01ESPCN).

Edited by: Y. Cai

References

Azaiez, M. N., Hariga, M., and Al-Harkan, I.: A chance-constrained multi-period model for a special multi-reservoir system, Comput. Oper. Res., 32, 1337–1351, 2005.

Bender, M. J. and Simonovic, S. P.: A fuzzy compromise approach to water resource systems planning under uncertainty, Fuzzy Sets Syst., 115, 35–44, 2000.

Cai, Y. P., Huang, G. H., Nie, X. H., Li, Y. P., and Tan, Q.: Municipal Solid Waste Management Under Uncertainty: A Mixed Interval Parameter Fuzzy-Stochastic Robust Programming Approach, Environ. Eng. Sci., 24, 338–352, 2007.

Cai, Y. P., Huang, G. H., Yang, Z. F., Lin, Q. G., and Tan, Q.: Community–scale renewable energy systems planning under uncertainty – An interval chance–constrained programming approach, Renew. Sustain. Energy Rev., 13, 721–735, 2009a.

Cai, Y. P., Huang, G. H., Tan, Q., and Yang, Z. F.: Planning of community-scale renewable energy management systems in a mixed stochastic and fuzzy environment, Renew. Energy, 34, 1833–1847, 2009b.

Cai, Y. P., Huang, G. H., Tan, Q., and Chen, B.: Identification of optimal strategies for improving eco-resilience to floods in ecologically vulnerable regions of a wetland, Ecol. Model., 222, 360–369, 2011.

Cai, Y. P., Huang, G. H., Wang, X., Li, G. C., and Tan, Q.: An inexact programming approach for supporting ecologically sustainable water supply with the consideration of uncertain water demand by ecosystems, Stochast. Environ. Res. Risk Assess., 25, 721–735, 2012.

Canter, L. W., Chawla, M. K., and Swo, C. T.: Addressing trend-related changes within cumulative effects studies in water resources planning, Environ. Impact Assess. Rev., 44, 58–66, 2014.

Cao, M. F., Huang, G. H., Sun, Y., Xu, Y., and Yao, Y.: Dual inexact fuzzy chance-constrained programming for planning waste management systems, Stochast. Environ. Res. Risk Assess., 24, 1163–1174, 2010.

Castelletti, A., Pianosi, F., and Sessa, R. S.: Integration, participation and optimal control in water resources planning and management, Appl. Math. Comput., 206, 21–33, 2008.

Chen, C. H., Liu, W. L., Liaw, S. L., and Yu, C. H.: Development of a dynamic strategy planning theory and system for sustainable river basin land use management, Sci. Total Environ., 346, 17–37, 2005.

Chung, G., Lansey, K., Blowers, P., Brooks, P., Ela, W., Stewart, S., and Wilson P.: A general water supply planning model: Evaluation of decentralized treatment, Environ. Model. Softw., 23, 893–905, 2008.

Dessai, S. and Hulme, M.: Assessing the robustness of adaptation decisions to climate change uncertainties: A case study on water resources management in the East of England, Global Environ. Change, 17, 59–72, 2007.

Deviney Jr., F. A., Brown, D. E., and Rice, K. C.: Evaluation of Bayesian Estimation of a Hidden Continuous-Time Markov Chain Model with Application to Threshold Violation in Water-Quality Indicators, J. Environ. Inf., 19, 70–78, 2012.

Dong, C. L., Schoups, G., and van de Giesen, N.: Scenario development for water resource planning and management: A review, Tech. Forecast. Soc. Change, 80, 749–761, 2013.

Dong, C., Huang, G. H, Tan, Q., and Cai Y. P.: Coupled planning of water resources and agricultural land-use based on an inexact-stochastic programming model, Front. Earth Sci., 8, 70–80, 2014.

Gregory, R., Failing, L., and Higgins, P.: Adaptive management and environmental decision making: A case study application to water use planning, Ecol. Econ., 58, 434–447, 2006.

Gu, J. J., Huang, G. H., Guo, P., and Shen, N.: Interval multistage joint-probabilistic integer programming approach for water resources allocation and management, J. Environ. Manage., 128, 615–624, 2013.

Guo, P., Huang, G. H., and Li, Y. P.: An inexact fuzzy-chance-constrained two-stage mixed-integer linear programming approach for flood diversion planning under multiple uncertainties, Adv. Water Resour., 33, 81–91, 2010.

Huang, G. H., Batez, B. W., and Patry, G. G.: A grey linear programming approach for municipal solid waste management planning under uncertainty, Civ. Eng. Syst., 9, 319–335, 1992.

Huang, G. H., Batez, B. W., and Patry, G. G.: Grey integer programming: an application to waste management planning under uncertainty, Eur. J. Oper. Res., 83, 594–620, 1995.

Huang, Y., Li, Y. P., Chen, X., and Ma, Y. G.: Optimization of the irrigation water resources for agricultural sustainability in Tarim River Basin, China, Agric. Water Manage., 107, 74–85, 2012.

Khare, D., Jat, M. K., and Deva, J.: Sunder Assessment of water resources allocation options: Conjunctive use planning in a link canal command, Resour. Conserv. Recycling, 51, 487–506, 2007.

Kondilia, E. and Kaldellis, J. K.: Model development for the optimal water systems planning, Comput. Aided Chem. Eng., 21, 1851–1856, 2006.

Li, Z., Huang, G., Zhang, Y. M., and Li, Y.P.: Inexact two-stage stochastic credibility constrained programming for water quality management, Resour. Conserv. Recycling, 73, 122–132, 2013.

Lindenschmidt, K. E., Fleischbein, K., and Baborowski, M.: Structural uncertainty in a river water quality modelling system, Ecol. Model., 204, 289–300, 2007.

Lu, H .W., Huang, G. H., Zhang, Y. M., and He, L.: Strategic agricultural land-use planning in response to water-supplier variation in a China's rural region, Agric. Syst., 108, 19–28, 2012.

Mahmoud, M. I., Gupta, H. V., and Rajagopal, S.: Scenario development for water resources planning and watershed management: Methodology and semi-arid region case study, Environ. Model. Softw., 26, 873–885, 2011.

Mehta, V. K., Aslam, O., Dale, L., Miller, N., and Purkey, D. R.: Scenario-based water resources planning for utilities in the Lake Victoria region, Phys. Chem. Earth, 61–62, 22–31, 2013.

Qin, H. P., Su, Q., and Khu, S. T.: An integrated model for water management in a rapidly urbanizing catchment, Environ. Model. Softw., 26, 1502–1514, 2011.

Qin, X. S., Huang, G. H., Zeng, G. M., Chakma, A., and Huang, Y. F.: An interval-parameter fuzzy nonlinear optimization model for stream water quality management under uncertainty, Eur. J. Oper. Res., 180, 1331–1357, 2007.

Ray, D. K., Pijanowski, B. C., Kendall, A. D., and Hyndman, D. W.: Coupling land use and groundwater models to map land use legacies: Assessment of model uncertainties relevant to land use planning, Appl. Geogr., 34, 356–370, 2012.

Riquelme, F. J. M. and Ramos, A. B.: Land and water use management in vine growing by using geographic information systems in Castilla-La Mancha, Spain, Agric. Water Manage., 77, 82–95, 2005.

Satti, S. R., Jacobs, J. M., and Irmak, S.: Agricultural water management in a humid region: sensitivity to climate, soil and crop parameters, Agric. Water Manage., 70, 51–65, 2004.

Sessa, R. S.: Modelling and control for participatory planning and managing water systems, Control Eng. Pract., 15, p. 985, 2007.

Sethi, L. N., Panda, S. N., and Nayak, M. K.: Optimal crop planning and water resources allocation in a coastal groundwater basin, Orissa, India, Agric. Water Manage., 83, 209–220, 2006.

Tan, Q., Huang, G. H., and Cai, Y. P.: Radial interval chance-constrained programming for agricultural non-point source water pollution control under uncertainty, Agric. Water Manage., 98, 1595–1606, 2011.

Victoria, F. B., Filho, J. S. V., Pereira, L. S., Teixeira, J. L., and Lanna, A. E.: Multi-scale modeling for water resources planning and management in rural basins, Agric. Water Manage., 77, 4–20, 2005.

Zarghami, M. and Hajykazemian, H.: Urban water resources planning by using a modified particle swarm optimization algorithm, Resour. Conserv. Recycling, 70, 1–8, 2013

Zhang, Y. M., Huang, G. H., and Zhang, X. D.: Inexact de Novo programming for water resources systems planning, Eur. J. Oper. Res., 199, 531–541, 2009.

Minimum forest cover required for sustainable water flow regulation of a watershed: a case study in Jambi Province, Indonesia

Suria Tarigan[1], Kerstin Wiegand[2], Sunarti[3], and Bejo Slamet[4]

[1]Department of Soil Sciences and Natural Resource Management, Bogor Agricultural University, Bogor, Indonesia

[2]Department of Ecosystem Modeling, University of Göttingen, Büsgenweg 4, 37077 Göttingen, Germany

[3]Faculty of Agriculture, University of Jambi, Jambi, Indonesia

[4]Faculty of Agriculture, North Sumatra University, Medan, Indonesia

Correspondence: Suria Tarigan (sdtarigan@apps.ipb.ac.id)

Abstract. In many tropical regions, the rapid expansion of monoculture plantations has led to a sharp decline in forest cover, potentially degrading the ability of watersheds to regulate water flow. Therefore, regional planners need to determine the minimum proportion of forest cover that is required to support adequate ecosystem services in these watersheds. However, to date, there has been little research on this issue, particularly in tropical areas where monoculture plantations are expanding at an alarming rate. Therefore, in this study, we investigated the influence of forest cover and oil palm (*Elaeis guineensis*) and rubber (*Hevea brasiliensis*) plantations on the partitioning of rainfall into direct runoff and subsurface flow in a humid, tropical watershed in Jambi Province, Indonesia. To do this, we simulated streamflow with a calibrated Soil and Water Assessment Tool (SWAT) model and observed several watersheds to derive the direct runoff coefficient (C) and baseflow index (BFI). The model had a strong performance, with Nash–Sutcliffe efficiency values of 0.80–0.88 (calibration) and 0.80–0.85 (validation) and percent bias values of -2.9–1.2 (calibration) and 7.0–11.9 (validation). We found that the percentage of forest cover in a watershed was significantly negatively correlated with C and significantly positively correlated with BFI, whereas the rubber and oil palm plantation cover showed the opposite pattern. Our findings also suggested that at least 30 % of the forest cover was required in the study area for sustainable ecosystem services. This study provides new adjusted crop parameter values for monoculture plantations, particularly those that control surface runoff and baseflow processes, and it also describes the quantitative association between forest cover and flow indicators in a watershed, which will help regional planners in determining the minimum proportion of forest and the maximum proportion of plantation to ensure that a watershed can provide adequate ecosystem services.

1 Introduction

In recent years, monoculture plantations have rapidly expanded in Southeast Asia, and the areas under oil palm (*Elaeis guineensis*) and rubber (*Hevea brasiliensis*) plantations are expected to increase further (Fox et al., 2012; Van der Laan et al., 2016). In Indonesia, which is currently the largest palm oil producer worldwide, the oil palm plantation area increased from 7000 km^2 in 1990 to 110 000 km^2 in 2015 (Ditjenbun, 2015; Tarigan et al., 2016b), and a further 170 000–200 000 km^2 is projected for future oil palm development (Colchester et al., 2006; Wicke et al., 2011; Afriyanti et al., 2016). This rapid expansion of oil palm plantations has been partly triggered by an increased demand for biofuel production (Mukherjee and Sovacoo, 2014). In addition, rubber plantations, which are also prevalent in Southeast Asia (Ziegler et al., 2009), currently cover 35 000 km^2 of land in Indonesia (Ditjenbun, 2015).

Although oil palm is of economic value to farmers and the local regions in which it is grown, it has received environmental and social criticism, often being held responsible for deforestation (Wicke et al., 2011; Vijay et al., 2016; Gatto et al., 2017), biodiversity loss (Fitzherbert et al., 2008; Koh and Wilcove, 2008; Wilcove and Koh, 2010; Carlson et al., 2012; Krashevska et al., 2015), decreased soil carbon stocks (Guillaume et al., 2015, 2016; Pransiska et al., 2016), and increased greenhouse gas emissions (Allen et al., 2015; Hassler et al., 2017). Similarly, rubber plantations have environmental impacts such as reducing the soil infiltration capacity, accelerating soil erosion, increasing stream sediment loads (Ziegler et al., 2009; Tarigan et al., 2016b), and decreasing soil carbon stocks (Ziegler et al., 2011). Furthermore, the conversion of tropical rainforest into oil palm and rubber plantations affects the local hydrological cycle by increasing transpiration (Ziegler et al., 2009; Sterling et al., 2012; Röll et al., 2015; Hardanto et al., 2017), increasing evapotranspiration (ET) (Meijide et al., 2017), decreasing infiltration (Banabas et al., 2008; Tarigan et al., 2016b), increasing the flooding frequency (Tarigan, 2016a), and decreasing low flow levels (Yusop et al., 2007; Adnan and Atkinson, 2011; Comte et al., 2012; Merten et al., 2016). These climatic impacts that occur due to land use change are expected to be stronger under maritime conditions, such as those in Indonesia, than under continental conditions because 40 % of the global tropical latent heating of the upper troposphere occurs over the maritime continent (Van der Molen et al., 2006).

The forests in Jambi Province, Indonesia, have been largely transformed into plantations (Drescher et al., 2016), resulting in inhabitants experiencing water shortages during the dry season and a dramatic increase in flooding frequency during the wet season (Merten et al., 2016; Tarigan, 2016a) because plantations promote higher levels of direct runoff than forested lands (Bruijnzeel, 1989, 2004; Tarigan et al., 2016b; Dislich et al., 2017). However, this negative impact of plantation expansion could be minimized by maintaining an adequate proportion of forested land as a watershed, which raises the following question: what is the minimum proportion of forest cover that is required in a watershed to support adequate water flow regulation?

The water flow regulation function of watersheds represents their ability to retain rainwater and is one of the most important soil hydrological processes in tropical regions where rainfall is highly seasonal (Lele, 2009). Functional water flow regulation by a watershed reduces flood peaks by moderating direct runoff (Le Maitre et al., 2014; Ellison et al., 2017) via soil water infiltration through the soil surface and percolation through the soil profile. This vertical movement of water through the soil determines how much water flows as direct runoffs and how much reaches the water table where it is sustained as baseflow or groundwater (Hewlett and Hibbert, 1967; Bruijnzeel, 1990; Le Maitre et al., 2014; Tarigan et al., 2016b). Forest vegetation provides organic matter and habitat for soil organisms, thereby facilitating higher levels of infiltration than other land uses (Hewlett and Hibbert, 1967).

A number of empirically based and process-based approaches can be used for assessing the impacts of expanding rubber and oil palm plantations on hydrological characteristics in the Southeast Asia region. Empirically based approaches use long-term historical data to correlate land use changes with corresponding streamflow data (Adnan and Atkinson, 2011; Rientjes et al., 2011; Mwangi et al., 2016) or paired catchment studies (Bosch and Hewlett, 1982; Brown et al., 2005), whereas process-based approaches use physically based hydrological models in which the impact of land use changes is determined by varying the land use/cover settings (Khoi and Suetsugi, 2014; Guo et al., 2016; Zhang et al., 2016; Marhaento et al., 2017; Wangpimool et al., 2017). Process-based approaches have the drawback of requiring more data to be input and having high uncertainty in parameter estimation (Xu et al., 2014; Zhang et al., 2016). However, there is currently an absence of long-term historical data for Jambi Province, precluding the use of an empirically based approach.

Distributed hydrological models are useful for understanding the effects of land use changes on watershed flow regulation. One such model is the Soil and Water Assessment Tool (SWAT) ecohydrological model (2012), which quantifies the water balance of a watershed on a daily basis (Neitsch et al., 2011) and has been recommended for evaluating the hydrological ecosystem services of a watershed (Vigerstol and Aukema, 2011). The SWAT model approach is one of the most widely used and scientifically accepted tools for assessing water management in a watershed (Gassman et al., 2007). Consequently, its popularity has also increased in Southeast Asia. Marhaento et al. (2017) recently used the SWAT model to analyze the impact of forest cover and agriculture land use on the runoff coefficient (C) and baseflow index (BFI) on Java Island, Indonesia, and found that a decrease in forest cover from 48.7 to 16.9 % and an increase in agriculture area from 39.2 to 45.4 % increased C from 35.7 to 44.6 % and decreased BFI from 40 to 31.1 %. Meanwhile, Wangpimool et al. (2017) found that the expansion of rubber plantations in Thailand between 2002 and 2009 led to an annual reduction of approximately 3 % in the average water yield of the basin, whereas Babel et al. (2011) found that the expansion of oil palm plantations in Thailand increased nitrate loading (1.3–51.7 %) in the surface water based on SWAT simulations. Tarigan et al. (2016b) also used the SWAT model to simulate the impact of soil and water conservation practices on low flow levels in oil-palm-dominated watersheds in Jambi Province, Indonesia. This study aimed to quantify the minimum proportion of forest cover that is required to allow a watershed to provide adequate ecosystem services. We selected Jambi Province as our study area because of the rapid expansion of oil palm and rubber plantations in that area. The study findings provide new adjusted values for crop parameters of monoculture plantations, particularly those that control surface runoff and baseflow processes, and they describe the quantitative association between forest cover and flow in-

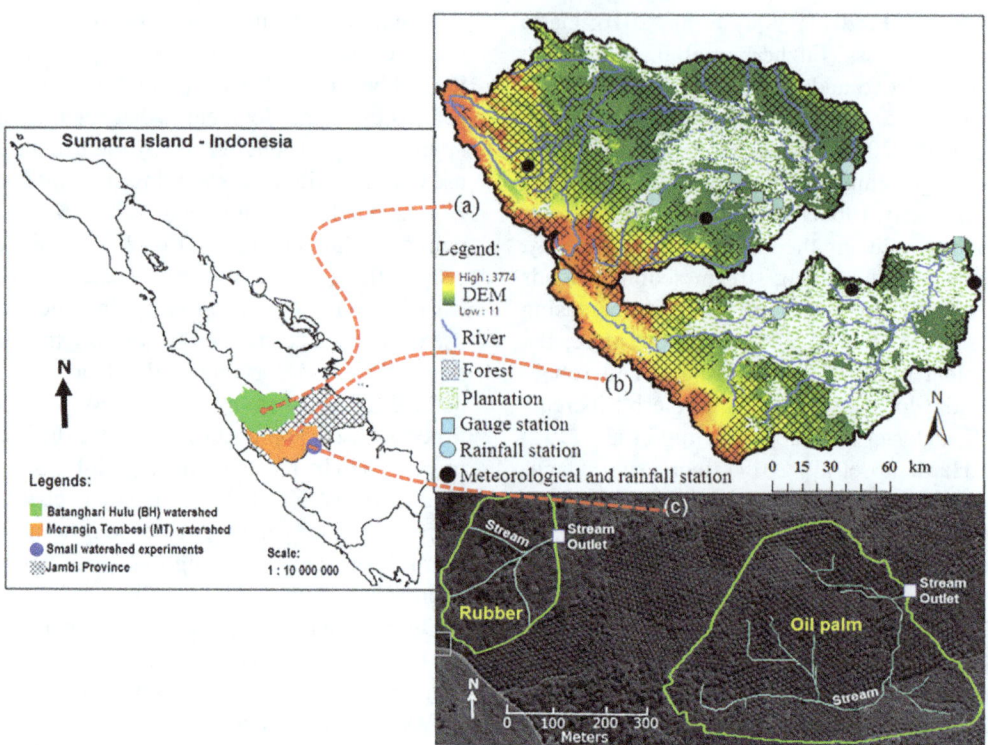

Figure 1. Locations of the **(a, b)** macro and **(c)** small watershed experiments in Jambi Province, Sumatra, Indonesia.

dicators in a watershed, which will help regional planners in determining the minimum proportion of forest that needs to be conserved to ensure that a watershed can provide adequate ecosystem services.

2   Methods

2.1   Study area

The study area is located in Jambi Province, Sumatra (1°54′31.4″ S, 103°16′7.9″ E; Fig. 1). There has been a rapid expansion of plantations in this area, particularly oil palm and rubber (Drescher et al., 2016). The area has a tropical humid climate, with an average temperature of 27 °C and an average rainfall of 2700 mm yr^{-1}. The rainy season occurs from October to March. The area under oil palm plantation in the study area (Jambi Province) increased from 1500 km^2 in 1996 to 6000 km^2 in 2011, representing an almost 400 % increase (Setiadi et al., 2011), whereas the area under rubber plantation increased from 5000 to 6500 km^2 over the same period (Ditjenbun, 2015). In 2013, only 30 % of Jambi Province was covered with rainforest (mainly located in mountainous regions), with 55 % of the land having been converted into agricultural land, of which 10 % was degraded/fallow and will potentially be converted into monoculture plantations (Drescher et al., 2016).

The study area consists of two macro watersheds for the simulation of the C and BFI values with the SWAT model, namely Batanghari Hulu (BH; Fig. 1a) and Merangin Tembesi (MT; Fig. 1b) watersheds, which cover areas of 18 415 and 13 452 km^2, respectively. The dominant land uses in both watersheds are forest (BH, 50 %; MT, 30 %) and plantation (BH, 18 %; MT, 48 %). The dominate soil types in the study area are classified as Tropodult and Dystropept, which are characterized as consisting of medium to heavy texture (Allen et al., 2015).

To ensure that the C and BFI values obtained from the macro watershed simulations (particularly those subwatersheds that were dominated by oil palm or rubber) reflected the real observed values in the field, we carried out the direct C measurements in two small watersheds, with sizes of 14 and 9 ha, respectively, in the study area (Fig. 1c). These small watersheds were covered with 90 % oil palm and 80 % rubber plantations, respectively.

Oil palm and rubber are perennial crops that have a life cycle of 25 years. Both crops are planted in rows at planting distances of 8 and 4 m, respectively. In oil palm plantations, there are two types of paths between the planting rows: the harvest path, which is used to transport freshly harvested fruit bunches, and the so-called death path, which is used for piling pruned leaf fronds, which occupy approximately 2 m or one-quarter of this path. Both oil palm and rubber require very intensive harvesting activities, which occur twice per month for oil palm and almost daily for rubber; thus, soils

Table 1. Water storage capacity of the leaf axils along the trunks of oil palm trees.

Replicate	Water storage (mm)			
	Tree 1	Tree 2	Tree 3	Tree 4
1	14.2	10.8	4.4	6.2
2	10.6	10.2	4.2	5.9
3	9.4	10.5	5.9	7.9
4	8.1	10.4	5.4	7.4
5	8.8	11.5	3.8	9.8
6	9.4	10.9	6.0	8.0
7	9.3	10.6	5.9	7.5
8	10.1	11.0	5.2	7.3
9	8.9	11.2	4.7	10.5
10	9.5	11.3	4.9	7.2

Average = 8.4 mm

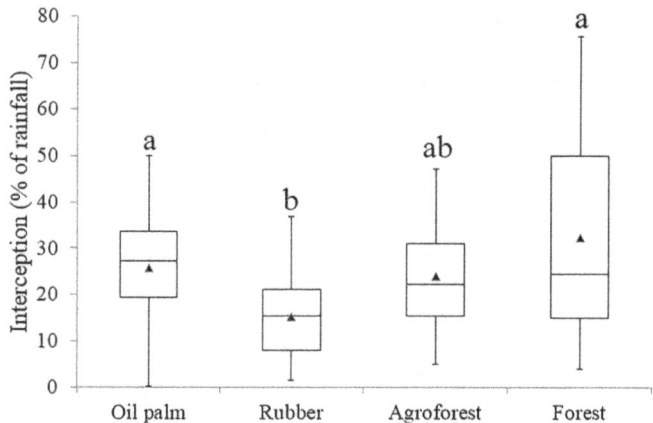

Figure 2. Canopy interception of rainfall under different land uses. Different letters indicate significant differences between the means (Bonferroni-corrected post hoc t-test based on analysis of variance (ANOVA); $p < 0.05$).

along the harvest path and part of the death path are very compacted. The soils under oil palm and rubber plantations remain unploughed for the entire growing period. Weeds in the oil palm plantations are regularly eradicated using herbicides or mechanical equipment. Intensive inorganic fertilization ($1000 \, \mathrm{kg \, ha^{-1} \, yr^{-1}}$) also occurs, contributing to the degradation of the soil structure and fauna.

2.2 SWAT model

The SWAT model is a continuous long-term yield model that was developed to simulate the impact of different land cover/management practices on streamflow in complex watersheds with varying soil, land use, and management conditions over long time periods. The major model components include weather, hydrology, soil temperature, soil properties, plant growth, nutrients, and land management (Neitsch et al., 2011; Arnold et al., 2012).

During the modeling process, a watershed is subdivided into several sub-watersheds, which are then further partitioned into hydrological response units (HRUs) that are defined by their topography, soil, and land use characteristics, which are not spatially referenced in the model. The hydrological outputs of HRUs are calculated using the water balance equation and include total streamflow, surface flow, and baseflow. These output components can then be used to calculate indicators of the water flow regulation functions of a watershed, namely C, which is the ratio of direct runoff to rainfall, and BFI, which is the proportion of baseflow in the streamflow.

Because the SWAT model was designed for temperate regions, adapting crop parameter inputs for use in tropical regions is necessary (Strauch et al., 2013; Van Griensven et al., 2014; Alemayehu et al., 2017). To avoid incorrect parameterization of sensitive values, we carried out field measurements for interception, infiltration, and surface runoff to adapt the parameter values, particularly those that control

surface runoff and baseflow processes. We then performed SWAT model simulation in two study watersheds and conducted small watershed experiments to compare the observed C values with those obtained from simulations.

2.2.1 Model setup

Delineation of watersheds and their sub-watersheds in our study area was automatically performed by the SWAT model and was based on a digital elevation model (DEM) with a 30 m resolution. During this automatic delineation, we predefined 50 000 ha as a threshold for the minimum sub-watershed area, based on subdivision of the BH and MT watersheds into 25 and 23 sub-watersheds, respectively.

Crop parameters

Oil palm plantations exhibit specific characteristics, particularly with respect to rainfall partitioning. These characteristics include high interception, high ET, low soil infiltration, high proportion of surface runoff, and absence of leaf litter, of which the first four can potentially reduce baseflow. Therefore, to consider these specific characteristics, we conducted field measurements and adjusted several crop parameters that are related to flow components, including canopy storage (CANMX), plant uptake compensation factor (EPCO), hydrologic soil group (HSG), and Soil Conservation Service (SCS) curve number (CN).

a. *Interception.* CANMX is the maximum amount of water that can be stored in the canopy and trunks of fully developed trees. Thus, an increase in this parameter reflects a reduction in the amount of rainfall that reaches the ground. In oil palms, rainfall is intercepted not only by leaves and branches but also by water reservoirs in leaf axils along the trunk. Therefore, we measured the

Table 2. Adapted parameter inputs for the study area.

SWAT parameter	Definition	Oil palm	Rubber	Agroforest	Forest
CANMX (mm)	Maximum trunk storage	8.4	0	0	0
	Maximum canopy storage	4.7	2.7	4.3	5.8
HYDGRP	Hydrologic soil group	D	D	B	C
CN2	Curve number	83	83	65	45
SOL_BD ($g\,cm^{-3}$)	Soil bulk density	1.2–1.3	1.2–1.3	1	0.9
EPCO	Plant uptake compensation factor	1	1	1	1
OV_N	Manning's n value for overland flow	0.07	0.14	0.4	0.5
SOL_K ($mm\,h^{-1}$)	Saturated hydraulic conductivity	30	78	400	470
SOL_AWC ($mm\,mm^{-1}$)	Available water capacity	0.1	0.1	0.2	0.2
BLAI ($m^2\,m^{-2}$)	Maximum potential leaf area index	3.6*	2.6	5	5
CHTMX (m)	Maximum canopy height	12	13	14	20
T_BASE	Base temperature	20	20	20	20
ALPHA_BF	Baseflow recession constant	0.90–0.95	0.90–0.95	0.90–0.95	0.90–0.95
T_OPT	Optimal temperature	28	28	30	30

* Fan et al. (2015).

Table 3. Model input data sources.

Data type	Resolution	Description	Source
Topography	30 m	DEM with a resolution of 30 m per pixel	SRTM
Soil map	1 : 250 000	Additional soil data were collected from the field and previous studies	Soil Research Institute, Ministry of Agriculture
Land use	1 : 100 000	Land use map with intensive ground check	Regional Planning Office (BAPPEDA[a])
Rainfall and climate	Daily	Rainfall and meteorological stations at Rantau Pandan, Siulak Deras, Muara Hemat, Padang Aro, Depati Parbo, Bangko, Bungo, Pematang Kabau, and Bungku	BMKG[b] office and CRC990[c]
Streamflow	Daily discharge data	Stations at Muara Tembesi, Air Gemuruh, Batang Tabir, Batang Pelepat, and Muara Kilis	Ministry of Public Works (BBWS[d])

[a] BAPPEDA, Regional Planning Office (*Badan Perencanaan Daerah*). [b] BMKG, Meteorology, Climatology and Geophysics Agency (*Badan Meteorologi, Klimatologi dan Geofisika*). [c] CRC990, Collaborative Research Centre 990. [d] BBWS, Ministry of Public Works (*Balai Besar Wilayah Sungai*).

water storage capacity of leaf axils along the trunks of four 10–12-year-old oil palm trees with 10 replications per tree. We found that the leaf axils along the trunk can store up to 20 L or 8.4 mm of water (Table 1), which matches previous reports that the leaf axils along oil palm trunks have a high water storage capacity (Merten et al., 2016; Meijide et al., 2017).

We also measured the canopy interception by oil palm, rubber, agroforest, and forest canopies between November 2012 and February 2013. Rainfall interception was assessed by measuring throughfall and stemflow and subtracting these from the incident rainfall. In total, there were 30 rainfall events during this time, representing light to heavy rain. We found that oil palm plantations tended to exhibit higher levels of canopy interception (Fig. 2), with our estimates falling within the range of values that were previously reported for tropical forests in Southeast Asia (commonly 10–30 %; Ku-

magai et al., 2005; Dietz et al., 2006). These interception assessments were used to adjust the CANMX parameter of the SWAT model (Table 2).

b. *ET.* Actual ET was determined by measuring the daily depletion of soil moisture content at a distance of 2 m from the trunks of oil palm trees. Soil moisture measurements were made on consecutive no-rain days over the 16-day period from 25 July to 10 August 2012. On average, soil moisture decreased by 6 % (vol) over this period, which is equivalent to 72 mm or 4.5 mm day^{-1} and is relatively high compared with the average land use in the study area. Similarly, Meijide et al. (2017) also reported a yearly oil palm ET of 1216 mm (4.7 mm day^{-1}) using Eddy covariance measurements in the study area. This ET rate is similar to or even higher than ET rates for forests in Southeast Asia (Kumagai et al., 2005), despite oil palm having a much lower stand density and

Table 4. The percentage of land use types, C, and BFI in each sub-watershed within the BH and MT watersheds.

Sub-wat. no.	Percentage of land use types					C	BFI
	F	AF	RP	OP	S		
BH watershed							
1	56	0	17	0	26	0.21	0.71
2	46	0	26	0	26	0.27	0.63
3	90	0	0	0	0	0.00	0.99
4	76	0	0	0	0	0.18	0.75
5	16	0	67	0	17	0.40	0.32
6	0	0	67	0	34	0.41	0.23
7	68	0	0	0	19	0.17	0.76
8	100	0	0	0	0	0.01	0.98
9	80	0	0	0	0	0.15	0.79
11	55	0	0	0	0	0.31	0.57
12	61	0	23	0	0	0.23	0.62
13	30	0	41	0	18	0.32	0.42
14	84	0	0	0	16	0.04	0.91
15	49	0	0	0	0	0.31	0.58
16	0	20	39	0	24	0.36	0.35
17	59	0	0	0	24	0.16	0.71
18	0	0	82	19	0	0.48	0.03
19	19	30	0	0	52	0.13	0.75
20	69	0	0	0	0	0.22	0.69
21	45	24	0	0	32	0.08	0.87
22	0	26	16	58	0	0.35	0.28
23	0	48	33	0	20	0.23	0.61
24	66	5	25	5	0	0.14	0.76
25	44	32	0	7	17	0.04	0.93
MT watershed							
1	0	13	0	85	0	0.59	0.11
2	56	0	0	44	0	0.36	0.45
3	0	16	0	82	0	0.58	0.13
4	21	20	12	48	0	0.45	0.32
5	32	0	19	38	12	0.40	0.29
6	0	46	0	52	0	0.43	0.37
8	0	0	15	86	0	0.54	0.01
9	48	0	0	53	0	0.59	0.11
10	0	0	0	96	0	0.34	0.37
11	0	0	19	69	0	0.67	0.00
12	0	0	13	88	0	0.66	0.01
14	0	0	20	80	0	0.54	0.01
15	0	27	14	57	0	0.65	0.01
16	70	0	0	0	31	0.53	0.22
17	0	28	11	57	0	0.28	0.65
19	0	12	31	57	0	0.53	0.23
20	0	0	63	37	0	0.60	0.10
21	48	0	0	36	16	0.65	0.01
22	36	0	17	17	16	0.32	0.42
23	100	0	0	0	0	0.40	0.32

Sub-wat. no., sub-watershed number (see Fig. 3a and b); F, forest; AF, agroforest; RP, rubber plantation; OP, oil palm; S, shrubland.

biomass per hectare. Therefore, we adjusted EPCO accordingly, which is related to ET (Table 2).

c. *Infiltration and surface runoff.* One important parameter of the SWAT model that is related to surface runoff is CN (Arnold et al., 2012), which determines the proportion of rainfall that becomes surface runoff (range, 0–100, with a higher value reflecting a higher level of surface runoff). The CN value is grouped into four HSGs (i.e., A, B, C, and D) according to the soil infiltration capacity. To adjust the CN value, we measured soil infiltration and surface runoff in each land use type (i.e., oil palm, rubber, agroforest, and forest) using a double-ring infiltrometer and multidivisor runoff collectors mounted at the lower end of each plot, respectively. The infiltration rate in different land use types increases in the following order: oil palm harvest path ($3\,\mathrm{cm\,h^{-1}}$) < oil palm circle ($3\,\mathrm{cm\,h^{-1}}$) < rubber harvest path ($7\,\mathrm{cm\,h^{-1}}$) < between rubber trees ($7.8\,\mathrm{cm\,h^{-1}}$) < forest ($47\,\mathrm{cm\,h^{-1}}$). The infiltration rates in the oil palm and rubber plantations were markedly lower than those in the forest. Low infiltration rate in the oil palm is associated with the soil compaction due to the intensive harvest activities.

For all HRUs with oil palm and rubber land uses, we selected HSG category D (Table 2) owing to its high surface runoff and low infiltration rate. We assumed that CN values of the forest and agroforest were similar to those of the evergreen and mixed forest, respectively, in the SWAT crop database.

d. *Litter fall.* In the oil palm plantations, negligible litter was found outside the frond piles. Litter fall in oil palm plantations does not naturally occur, but leaves are cut during fruit harvest and piled up in a frond pile, which occupies only 12 % of the entire oil palm plantation area. Consequently, the ground surface of oil palm plantations is managed mostly without litter, leading to higher surface runoffs. There was also negligible understory vegetation (grasses) because herbicides were routinely sprayed. The absence of the litter fall affected Manning's n value for overland flow (OV_N).

e. *Baseflow.* The baseflow recession constant (ALPHA_BF) was calculated by plotting the selected daily streamflow hydrograph on semi-log paper and determining the average values from several individual rainfall events. A previous study (Tarigan et al., 2016b) showed a similar range of ALPHA-BF values in the study area.

General input data

The SWAT model requires considerable other types of input data in addition to the crop parameters described above, such

Table 5. Observed C values obtained from field experiments in two small watersheds that were dominated by plantation cover.

Event no.	Rainfall (cm h^{-1})	Rainfall volume (m^3)		Runoff (m^3)		C runoff coefficient	
		Small wat. 1	Small wat. 2	Small wat. 1	Small wat. 2	Small wat.1	Small wat. 2
1	6.0	8960	3136	4500	1320	0.50	0.42
2	3.0	5180	1813	2625	840	0.51	0.46
3	1.4	4095	1433	3000	810	0.73	0.57
4	0.6	1456	509.6	1080	255	0.74	0.50
5	10.7	14923	5223	11250	3900	0.75	0.75
6	3.1	6006	2102	4050	1020	0.67	0.49
7	2.3	8188	2885	6150	1584	0.75	0.55
8	2.2	4416	1465	2400	780	0.54	0.53
9	9.6	8916	3121	5400	1650	0.61	0.53
Average						0.65	0.53
Total average						0.59	

Table 6. Simulated C values for the sub-watersheds used in the SWAT simulation (Table 4) that had similar percentages of plantation cover to the two small watersheds used in the field experiments.

Sub-wat. code (see Table 4)*	Percentage of plantation cover		C
	Oil palm	Rubber	
BH-18	82	19	0.49
MT-1	85	0	0.59
MT-8	86	15	0.54
MT-10	96	0	0.67
MT-11	69	19	0.67
MT-12	88	13	0.54
MT-14	80	20	0.65
MT-20	37	63	0.65

* Sub-watersheds were selected that had > 80 % plantation (oil palm and rubber) cover, allowing a comparison to be made with the small watersheds used in the field experiment.

as the climate, topography, soil type, and land use for each sub-watershed (Table 3).

A DEM with a resolution of 30 m per pixel was derived from the NASA Shuttle Radar Topography Mission (SRTM). The soil map was obtained from the Soil Research Institute at a scale of 1 : 250000. Some soil parameters such as soil hydraulic conductivity, bulk density, available water content, and texture were derived from a previous study (Sunarti et al., 2008). Additional soil data were collected from Batang Tabir sub-watershed and from CRC990 plots in Bukit Duabelas and Hutan Harapan landscape (Drescher et al., 2016). Daily rainfall and climate data between 2000 and 2014 were sourced from the rainfall and meteorological stations at Rantau Pandan, Siulak Deras, Muara Hemat, Padang Aro, Depati Parbo, Bangko, Bungo, Pematang Kabau, and Bungku. Daily streamflow data between 2000 and 2014 were

provided by the Ministry of Public Works (BBWS). All these data are freely available for research purposes upon official request to the corresponding institutions. The streamflow time series and rainfall records for the small catchments and the soil data have been deposited by the first author at Bogor Agricultural University and in the EFForTS Database (https://efforts-is.uni-goettingen.de). The land use for the study area was obtained from Jambi Province Regional Planning and Agricultural Plantation offices (Ditjenbun, 2015).

2.2.2 Model validation and calibration

The first step in the calibration and validation process in SWAT is the determination of the most sensitive parameters for a given watershed (Van Griensven et al., 2006; Arnold et al., 2012). The sensitivity analysis was performed using the SWAT Calibration and Uncertainty Procedure (SWAT-CUP) package, which is an interface for autocalibration that was specifically developed for SWAT and which links any calibration/uncertainty or sensitivity program to SWAT (Abbaspour, 2015).

Following the sensitivity analysis, we calibrated the SWAT model using the Latin hypercube sampling approach of the SWAT-CUP software. We first determined parameter ranges based on the minimum and maximum values allowed in SWAT. We then performed calibration and validation of the SWAT model by comparing the simulated monthly streamflows with observed data at the Muara Kilis and Muara Tembesi gauging stations from 2007 to 2009 for calibration and 2012 to 2014 for validation. Moriasi et al. (2007, 2015) recommended the use of three quantitative statistics for model evaluation: the Nash–Sutcliffe efficiency (NSE), percent bias (PBIAS), and the ratio of the root mean square error to the standard deviation of measured data. In this study, we used NSE and PBIAS to evaluate the model performance, which

Table 7. Sensitivity rank and initial and final values of the calibration parameters that were used in the study for the BH and MT watersheds.

Parameter	Description	Sensitivity rank	Initial value range	Best-fit values	
			BH and MT	BH	MT
ALPHA_BF	Baseflow recession constant	1	0.0 to 1.0	0.94	0.91
CN2	SCS runoff curve number for moisture condition II	2	−0.2 to 0.2 (V)*	0.14	0.12
GW_DELAY	Groundwater delay time (days)	3	30 to 450	62.5	57.2
CANMX	Maximum canopy storage (mm)	4	−0.2 to 1.0 (V)*	0.95	0.76
SOL_BD	Soil bulk density	5	−0.5 to 0.6 (V)*	0.46	0.47
GWQMN	Water depth in a shallow aquifer for a return flow (mm H_2O)	6	0.0 to 2.0	0.99	0.95
SOL_K	Saturated hydraulic conductivity (mm h^{-1})	7	−0.8 to 0.8 (V)*	0.71	0.62
CH_N2	Manning's n value for the main channel	8	0.0 to 0.3	0.05	0.15
SOL_AWC	Available water capacity of the soil (mm H_2O mm^{-1} soil)	9	−0.2 to 0.4 (V)*	0.09	0.04
OV_N	Manning's n value for overland flow	10	−0.2 to 1.0 (V)*	0.51	0.3

* (V) = Variable depends on land use and soil, and so changes in calibration were expressed as a fraction.

is consistent with the majority of the existing SWAT literature (Gassman et al., 2007, 2014; Douglas-Mankin et al., 2010; Tuppad et al., 2011; Bressiani et al., 2015). NSE is a normalized statistic that determines the relative magnitude of the residual variance ("noise") compared with the measured data variance ("information") (Nash and Sutcliffe, 1970). PBIAS measures the average tendency of simulated data to be larger or smaller than that of observational data (Gupta et al., 1999), with an optimum value of zero and lower values that indicate better simulations. Positive values of PBIAS indicate model underestimation, whereas negative values indicate model overestimation.

2.3 Simulated C and BFI values and the proportion of land use types in a watershed

The output of the validated SWAT model consisting of flow components for each sub-watershed was used to calculate indicators of the water flow regulation functions of a watershed, namely C, which is the ratio of direct runoff to rainfall, and BFI, which is the proportion of baseflow in the streamflow. To analyze the association between the C and BFI values and the proportion of each land use type in a watershed, we derived data vectors from the BH and MT watersheds. Each of these vectors corresponded to the percentage of land use types, C and BFI, in each of the 25 sub-watersheds from the BH watershed and 23 sub-watersheds from the MT watershed (Fig. 3a and b; Table 4).

2.4 Measured C values

We carried out the direct C measurements from the two small watersheds (Fig. 1c) between 2013 and 2015 using rectangular weirs and water level recorders for comparison with sim-

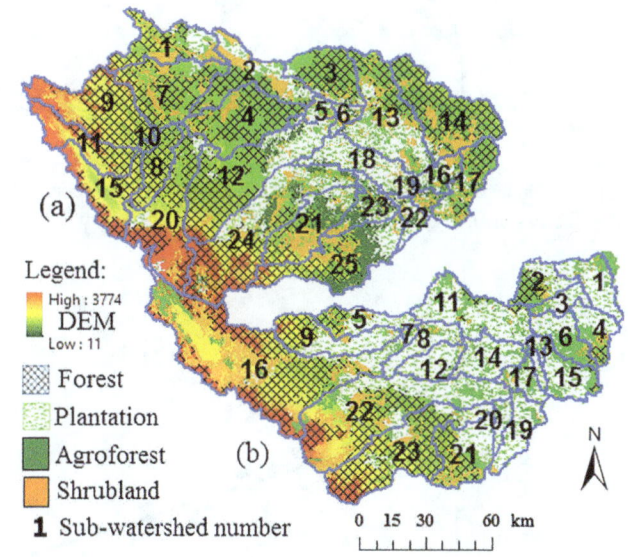

Figure 3. Land use types and sub-watershed number of the **(a)** BH and **(b)** MT watersheds.

ulated C values. The land use in the small watersheds is similar to the proportions found in several of the sub-watersheds shown in Table 4 (e.g., BH 18, MT 1, MT 8, MT 10, MT 11, MT 12, MT 14, and MT 20). The direct runoff components of the hydrographs were separated using the straight-line method described by Blume et al. (2007), following which C was calculated. We did not calculate BFI values along with C values in the small watershed experiments because BFI calculation requires hydrograph records over a longer period.

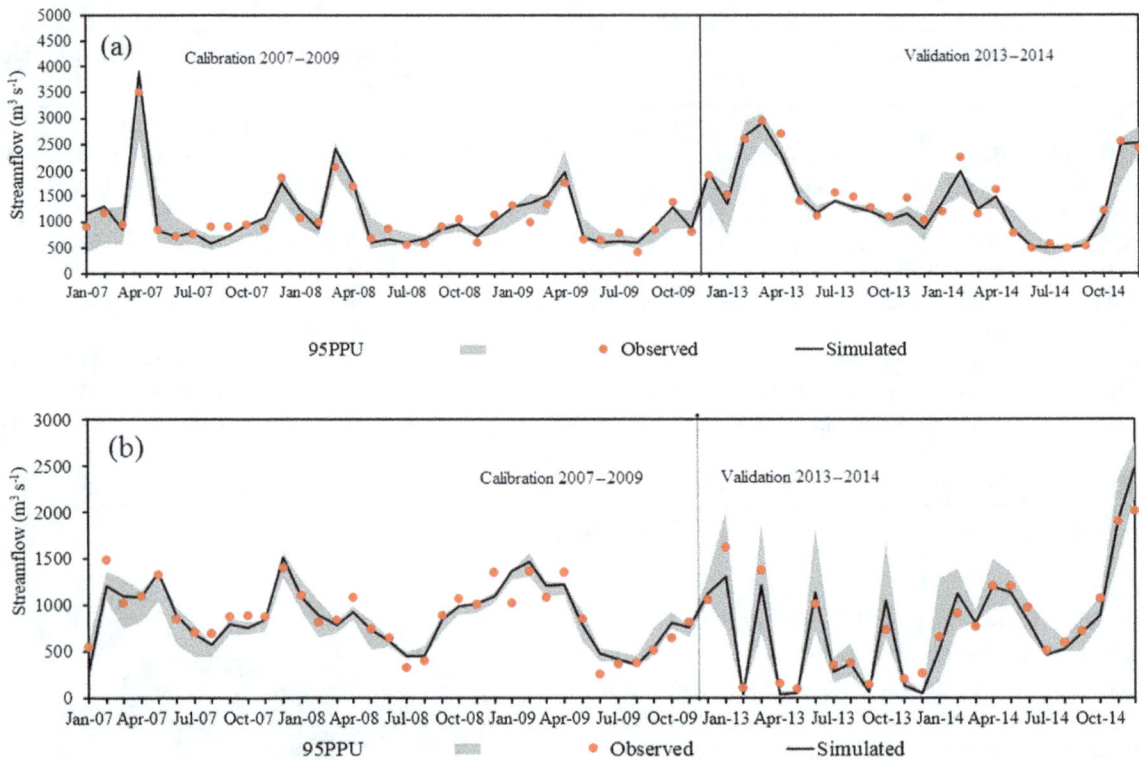

Figure 4. Observed vs. simulated streamflow and 95 % uncertainty interval (95PPU; see Abbaspour, 2015) with P factors of 0.83 and 0.76 and R factors of 0.65 and 0.69 for the **(a)** BH and **(b)** MT watersheds, respectively.

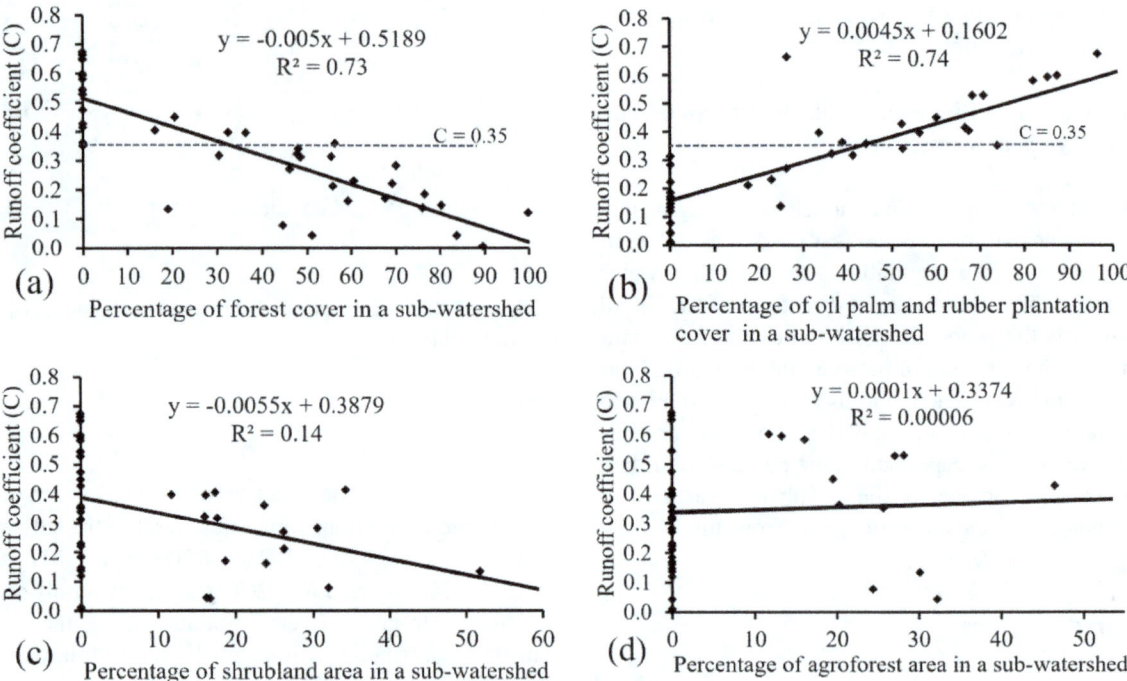

Figure 5. Association between simulated C values and the percentage of each land use type in a particular sub-watershed. Dotted lines indicate the maximum acceptable C value according to the Ministry of Forestry Decree (2013).

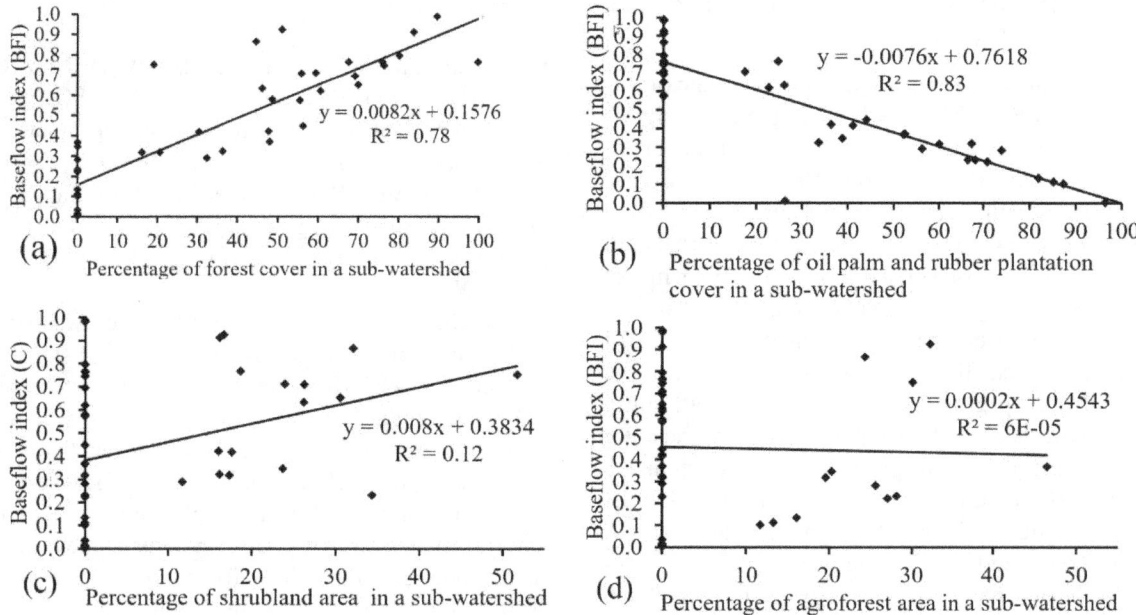

Figure 6. Association between simulated BFI values and the percentage of each land use type in a given sub-watershed.

3　Results and discussion

3.1　Measured C values

The average C value was based on nine individual rainfall events during the field experiment. The observed C values that were measured during two small watershed experiments was 0.59 (Table 5). This value was comparable with the averaged simulated values of 0.60 for sub-watersheds with comparable proportions of land use type with those of the small watershed experiment (Table 6).

3.2　SWAT model performance

The sensitive parameters that were included in the calibration of the SWAT model are ranked in Table 7. Some of these parameters play an important role in controlling the initial abstraction of rainfall (e.g., CANMX), rainfall partitioning into surface runoff (e.g., CN2 and OV_N), and vertical movement of water through the soil (e.g., SOL_BD, SOL_K, and SOL_AWC).

A visual comparison of best-fit simulations and observed data is shown in Fig. 4, with NSE values of 0.80–0.88 (calibration) and 0.84–0.85 (validation) and PBIAS values of −2.9 to 1.2 (calibration) and 7.0–11.9 (validation) for BH and MT watersheds, respectively. Based on the criterion proposed by Moriasi et al. (2007, 2015), the model performance was considered very good and satisfactory for calibration and validation, respectively.

3.3　Simulated C and BFI values and the proportion of land use types in a watershed

The proportion of a particular land use type in a sub-watershed was significantly correlated with C and BFI values obtained from 48 data vectors (Table 4). C values significantly decreased as the percentage of forest cover increased ($R^2 = 0.73$; $p < 0.05$; Fig. 5a) and significantly increased as the percentage of plantation cover increased ($R^2 = 0.74$; $p < 0.05$; Fig. 5b). Low infiltration capacity in oil palm and rubber plantations was the reason for higher C values in the sub-watersheds with high proportions of plantation land use. There were no significant associations between C values and any of the other land use types such as shrubland (Fig. 5c), agroforest (Fig. 5d), and dryland farming (data not shown). Some sub-watersheds had low C values despite having low levels of forest cover (e.g., BH 19). This can be explained by the fact that BH 19 had no oil palm or rubber plantations but consisted of 52 % of shrubland, which will have helped in reducing the C value. Furthermore, some watersheds had 100 % forest cover but a low BFI value (e.g., MT 23). This is the only sub-watershed in the MT watershed to have a high proportion of steeper slopes (76 % of the sub-watershed), which will increase C and decrease BFI. Among the 48 sub-watersheds we considered, only two had these slope characteristics.

The Ministry of Forestry of Indonesia considers C values of < 0.35 to be adequate for supporting the required ecosystem services of Indonesian watersheds (Ministry of Forestry Decree, 2013). Based on our findings, $\geq 30\,\%$ forest cover (Fig. 5a) and $\leq 40\,\%$ plantation cover (Fig. 5b) are required in a given sub-watershed with rapid expansion of plantation to achieve the desired C value.

BFI values significantly increased as the percentage of forest cover increased ($R^2 = 0.78$; $p < 0.05$; Fig. 6a) and significantly decreased as the percentage of plantation cover increased ($R^2 = 0.83$; $p < 0.05$; Fig. 6b). BFI was not significantly related to any other land use types such as shrubland (Fig. 6c), agroforest (Fig. 6d), and dryland farming (data not shown). According to Neitsch et al. (2009), the SWAT model considers only shallow groundwater in stream flow simulation. Therefore, we expected that SWAT underestimated BFI values in our study area. To improve the performance of the SWAT model for deep groundwater flow (low flow) simulation, Pfannerstill et al. (2014) modified the groundwater module by splitting the active groundwater storage into a fast and a slow contributing aquifer. Similar studies that focused on modifications of the SWAT groundwater component to obtain improved baseflow and overall streamflow results have also been reported by Luo et al. (2012) and Wang and Brubaker (2015). Similar modifications are needed in the standard SWAT model in order to more accurately simulate the conditions such as those encountered in this study.

3.4 Application of the research findings

The conversion of tropical rainforest into oil palm and rubber plantations affects the local hydrological cycle by increasing ET, decreasing infiltration, decreasing low flow levels, and increasing flooding frequency. In Jambi Province, Indonesia, forested areas have been largely transformed into plantations, resulting in inhabitants experiencing water shortages during the dry season and dramatic increases in flooding frequency during the wet season. One way in which this problem could be mitigated is by maintaining an adequate proportion of forested and plantation areas in a particular watershed, but this raises the question about what the minimum percentage of forest area and the maximum proportion of plantation area in a watershed is that will allow the maintenance of adequate water flow regulation. This study is the first to describe the quantitative association between forest and plantation areas and the flow indicators C and BFI; this understanding is required by spatial planners if they are to balance the ecology and socioeconomic functions of a landscape with the rapid expansion of plantation crops. In addition, our study provides data regarding how SWAT input parameters related to tropical plantations, such as oil palm and rubber, should be adjusted, particularly those that play an important role in controlling rainfall initial abstraction (e.g., CANMX), rainfall partitioning into surface runoffs (e.g., CN2, OV_N), and vertical movement of water through (e.g., SOL_BD, SOL_K, and SOL_AWC).

4 Summary

We found that ALPHA_BF, CN2, GW_DELAY, CANMX, SOL_BD, GWQMN, SOL_K, CH_N2, SOL_AWC, and OV_N were sensitive parameters in our model, some of which play an important role in controlling the initial abstraction of rainfall (e.g., CANMX), rainfall partitioning into surface runoff (e.g., CN2, OV_N), and vertical movement of water through the soil (e.g., SOL_BD, SOL_K, and SOL_AWC).

Overall, the SWAT model performance was strong, with NSE values of 0.80–0.88 (calibration) and 0.80–0.85 (validation) and PBIAS values of -2.9–1.2 (calibration) and 7.0–11.9 (validation). We found that the percentage of forest cover in a watershed was significantly negatively correlated with C and positively correlated with BFI, whereas the percentage of rubber and oil palm plantation cover showed the opposite pattern. Finally, our findings suggest that a watershed should contain $\geq 30\%$ forest cover and a maximum of 40 % plantation cover for maintaining sustainable water flow regulation ecosystem services.

The quantitative association between forest cover and flow indicators, which was derived in this study, will help regional planners in determining the minimum proportion of forest cover that needs to be maintained to ensure effective water flow regulation in a watershed.

Data availability. The land use data are freely available for research purposes upon official request to the corresponding institutions: the rainfall and climate data can be obtained from the Meteorology and Geophysics Agency; the streamflow data of the macro watersheds can be obtained from the Ministry of Public Works; and the land use data can be obtained from the Regional Planning Office. The streamflow time series and rainfall records for the small watersheds and data for the resampled soil hydraulic conductivity, bulk density, available water content, and texture have been deposited by the first author at Bogor Agricultural University and in the EFForTS database (https://efforts-is.uni-goettingen.de).

Competing interests. The authors declare that they have no conflict of interest.

Special issue statement. This article is part of the special issue "Coupled terrestrial-aquatic approaches to watershed-scale water resource sustainability". It is not associated with a conference.

Acknowledgements. This study was performed in the framework of the joint Indonesian–German research project EFForTS-CRC 990 (http://www.uni-goettingen.de/crc990) and was funded by the Directorate General of Higher Education (DIKTI), Indonesia.

Edited by: Ann van Griensven

References

Abbaspour, K. C.: SWAT-CUP 2012: SWAT Calibration and Uncertainty Programs – A User Manual, 2012, Eawag–Swiss Federal Institute of Aquatic Science and Technology, 2015.

Adnan, N. A. and Atkinson, P. M.: Exploring the impact of climate and land use changes on streamflow trends in a monsoon catchment, Int. J. Climatol., 31, 815–831, 2011.

Afriyanti, D., Kroeze, C., and Saad, A.: Indonesia palm oil production without deforestation and peat conversion by 2050, Sci. Total. Environ., 557/558, 562–570, 2016.

Alemayehu, T., van Griensven, A., Woldegiorgis, B. T., and Bauwens, W.: An improved SWAT vegetation growth module and its evaluation for four tropical ecosystems, Hydrol. Earth Syst. Sci., 21,9, 4449–4467, https://doi.org/10.5194/hess-21-4449-2017, 2017.

Allen, K., Corre, M., Tjoa, A., and Veldkamp, E.: Soil nitrogen-cycling response to conversion of lowland forests to oil palm and rubber plantations in Sumatra, Indonesia, PLoS One, 10, e0133325, https://doi.org/10.1371/journal.pone.0133325, 2015.

Arnold, J. G., Moriasi, D. N., Gassman, P. W., Abbaspour, K. C., White, M. J., Srinivasan, R., Santhi, R. C., Harmel, R. D., van Griensven, A., Van Liew, M. W., Kannan, N., and Jha, M. K.: SWAT: Model use, calibration, and validation, T. ASABE., 55, 1491–1508, 2012.

Babel, M. S., Shrestha, B., and Perret, S. R.: Hydrological impact of biofuel production: a case study of the Khlong Phlo Watershed in Thailand, Agric. Water. Manag., 101, 8–26, https://doi.org/10.1016/j.agwat.2011.08.019, 2011.

Banabas, M., Turner, M. A., Scotter, D. R., and Nelson, P. N.: Losses of nitrogen fertilizer under oil palm in Papua New Guinea: 1. Water balance, and nitrogen in soil solution and runoff, Aust. J. Soil Res., 46, 332–339, 2008.

Blume, T., Zehe, E., and Bronstert, A.: Rainfall–runoff response, event-based runoff coefficients and hydrograph separation, Hydrol. Sci. J., 52, https://doi.org/10.1623/hysj.52.5.843, 2007.

Bosch, J. M. and Hewlett, J. D.: A review of catchment experiments to determine the effect of vegetation changes on water yield and evapotranspiration, J. Hydrol., 55, 3–23, 1982.

Bressiani, D. D. A., Gasman, P., Fernades J. G., and Mediondo, E. M.: A review of Soil and Water Assessment Tool (SWAT) applications in Brazil: Challenges and prospects, IJABE, 8, 9–35, https://doi.org/10.3965/j.ijabe.20150803.1765, 2015.

Brown, A. E., Zhang, L., McMahon, T. A., Western, A. W., and Vertessy, R. A.: A review of paired catchment studies for determining changes in water yield resulting from alterations in vegetation, J. Hydrol., 310, 28–61, 2005.

Bruijnzeel, L. A.: (De)forestation and dry season flow in the tropics: a closer look, J. Trop. For. Sci., 1, 229–243, 1989.

Bruijnzeel, L. A.: Hydrology of moist tropical forests and effects of conversion: a state of knowledge review, UNESCO International Hydrological Programme, A publication of the Humid Tropics Programme, UNESCO, Paris, 1990.

Bruijnzeel, L. A.: Hydrological functions of tropical forests: not seeing the soil for the trees?, Agr. Ecosyst. Environ., 104, 185–228, 2004.

Carlson, K. M., Curran, L. M., Ratnasari, D., Pittman, A. M., Soares-Filho, B. S., and Asner, G. P.: Committed carbon emissions, deforestation, and community land conversion from oil palm plantation expansion in West Kalimantan, Indonesia, Proc. Natl. Acad. Sci. USA, 109, 7559–7564, 2012.

Colchester, M., Jiwan, N., Andiko, Sirait, M., Firdaus, A. Y., Surambo, A., and Pane, H.: Promised Land: Palm Oil and Land Acquisition in Indonesia - Implications for Local Communities and Indigenous Peoples, ISBN: 979-15188-0-7, Forest Peoples Programme, Perkumpulan Sawit Watch, HuMA and the World Agroforestry Centre, Bogor, Indonesia, 2006.

Comte, I., Colin, F., Whalen, J. K., Gruenberger, O., and Caliman, J. P.: Agricultural practices in oil palm plantations and their impact on hydrological changes, nutrient fluxes and water quality in Indonesia: a review, Adv. Agron., 116, 71–124, 2012.

Dietz, J., Hölscher, D., and Leuschner, C.: Rainfall partitioning in relation to forest structure in differently managed montane forest stands in Central Sulawesi, Indonesia For. Ecol. Manage., 237, 170–178, 2006.

Dislich, C., Faust, H., Kisel, Y., Knohl, A., Otten, F., Meyer. K., Pe'er, G., Salecker, J., Steinebach, S., Tarigan, S., Tölle, M., and Wiegand, K.: A review of the ecosystem functions in oil palm plantations, using forests as a reference system, Biol. Rev., 92, 1539–1569, https://doi.org/10.1111/brv.12295, 2017.

Ditjenbun: Indonesian plantation statistics, Directorate General for Plantations, Ministry of Agriculture Indonesia, 2015.

Douglas-Mankin, K. R., Srinivasan, R., and Arnold, A. J.: Soil and Water Assessment Tool (SWAT) model: Current developments and applications, T. ASABE, 53, 1423–1431, https://doi.org/10.13031/2013.34915, 2010.

Drescher, J., Rembold, K., Allen, K., Beckschäfer, P., Buchori, D., Clough, Y., Faust, H., Fauzi, A. M., Gunawan, D., Hertel, D., Irawan, B., Jaya, I. N. S., Klarner, B., Kleinn, C., Knohl, A., Kotowska, M. M., Krashevska, V., Krishna, V., Leuschner, C., Lorenz, W., Meijide, A., Melati, D., Nomura, M., Pérez-Cruzado, C., Qaim, M., Siregar, I. Z., Steinebach, S., Tjoa, A., Tscharntke, T., Wick, B., Wiegand, K., Kreft, H., and Scheu, S.: Ecological and socioeconomic functions across tropical land-use systems after rainforest conversion, Phil. T. R. Soc. B, 371, 20150275, https://doi.org/10.1098/rstb.2015.0275, 2016.

Ellison, D., Morris, C. E., Locatelli, B., Sheil, D., Cohen, J., Murdiyarso, D., Gutierrez, V., van Noordwijk. M., Creed, I. F., Pokorny, J., Gaveau, D., Spracklen, D. V., Tobella, A. B., Ilstedt, U., Teuling, A. J., Gebrehiwot, S. G., Sands, D. C., Muys, B., Verbist, B., Springgay, E., Sugandi, G., Sullivan, C. A.: Trees, forests and water: Cool insights for a hot world, Glob. Environ. Change, 43, 51–61, https://doi.org/10.1016/j.gloenvcha.2017.01.002, 2017.

Fan, Y., Roupsard, O., Bernoux, M., Le Maire, G., Panferov, O., Kotowska, M. M., and Knohl, A.: A sub-canopy structure for simulating oil palm in the Community Land Model (CLM-Palm): phenology, allocation and yield, Geosci. Model Dev., 8, 3785–3800, https://doi.org/10.5194/gmd-8-3785-2015, 2015.

Fitzherbert, E. B., Struebig, M. J., Morel, A., Danielsen, F., Brühl, C. A., Donald, P. F., and Phalan, B.: How will oil palm expansion affect biodiversity?, Trends. Ecol. Evol., 23, 538–545, 2008.

Fox, J., Vogler, J. B., Sen, O. L., Giambelluca, T. W., and Ziegler, A. D.: Simulating land-cover change in Montane mainland Southeast Asia, Environ. Manage., 49, 968–979, https://doi.org/10.1007/s00267-012-9828-3, 2012.

Gassman, P. W., Reyes, M. R., Green, C. H., and Arnold, J. G.: The soil and water assessment tool: historical development, applications, and future research direction, Amer. Soc. Agric. Bio. Eng., 50, 1211–1250, 2007.

Gassman, P. W., Sadeghi, A. M., and Srinivasan, R.: Applications of the SWAT Model Special Section: Overview and Insights, J. Environ. Qual., 43, 1–8, https://doi.org/10.2134/jeq2013.11.0466, 2014.

Gatto, M., Wollni, M., Asnawi, R., and Qaim, M.: Oil palm boom, contract farming, and rural economic development: village-level evidence from Indonesia, World Dev., 95, 127–140, 2017.

Guillaume, T., Damris, M., and Kuzyakov, Y.: Losses of soil carbon by converting tropical forest to plantations: erosion and decomposition estimated by δ^{13}C, Glob. Chang. Biol., 21, 3548–3560, 2015.

Guillaume, T., Holtkamp, A. M., Damris, M., Brümmer, B., and Kuzyakov, Y.: Soil degradation in oil palm and rubber plantations under land resource scarcity, Agr. Ecosyst. Environ., 232, 110–118, 2016.

Guo, J., Su, X., Singh, V. P., and Jin, J.: Impacts of climate and land use/cover change on streamflow using SWAT and a separation method for the Xiying River basin in Northwestern China, Water, 8, 1–14, https://doi.org/10.3390/w8050192, 2016.

Gupta, H. V., Sorooshian, S., and Yapo, P. O.: Status of automatic calibration for hydrologic models: Comparison with multilevel expert calibration, J. Hydrol. Eng., 4, 135–143, 1999.

Hardanto, A., Röll, A., Niu, F., and Meijide, A.: Oil palm and rubber tree water use patterns – effects of topography and flooding, Front. Plant. Sci., 8, 1–12, 2017.

Hassler, E., Corre, M. D., Kurniawan, S., and Veldkamp, E.: Soil nitrogen oxide fluxes from lowland forests converted to smallholder rubber and oil palm plantations in Sumatra, Indonesia, Biogeosciences, 14, 2781–2798, https://doi.org/10.5194/bg-14-2781-2017, 2017.

Hewlett, J. D. and Hibbert, A. R.: Factors affecting response of small watersheds to precipitation in humid areas, in: Forest Hydrology, edited by: Sopper, W. E. and Lull, H. W., Pergamon Press, New York, 1967.

Khoi, D. N. and Suetsugi, T.: Impact of climate and land-use changes on hydrological processes and sediment yield – a case study of the Be River catchment, Vietnam, Hydrol. Sci. J., 59, 1095–1108, https://doi.org/10.1080/02626667.2013.819433, 2014.

Koh, L. P. and Wilcove, D. S.: Is oil palm agriculture really destroying tropical biodiversity?, Conserv. Lett., 1, 60–64, 2008.

Krashevska, V., Klarner, B., Widyastuti, R., Maraun, M., and Scheu, S.: Impact of tropical lowland rainforest conversion into rubber and oil palm plantations on soil microbial communities, Biol. Fertil. Soil., 51, 697–705, 2015.

Kumagai, T., Saitoh, T. M., Sato, Y., Takahashi, H., Manfroi, O. J., Morooka, T., Kuraji, K., Suzuki, M., Yasunari, T., and Komatsu, H.: Annual water balance and seasonality of evapotranspiration in a Bornean tropical rainforest, Agr. Forest Meteorol., 128, 81–92, 2005.

Lele, S.: Watershed services of tropical forests: from hydrology to economic valuation to integrated analysis, Curr. Opin. Env. Sust., 1, 148–155, 2009.

Le Maitre, D. C., Kotzee, I. M., and O'Farrell, P. J.: Impacts of land-cover change on the water flow regulation ecosystem service: Invasive alien plants, fire and their policy implications, Land Use Policy, 36, 171–181, 2014.

Luo, Y., Arnold, J., Allen, P., and Chen, X.: Baseflow simulation using SWAT model in an inland river basin in Tianshan Mountains, Northwest China, Hydrol. Earth Syst. Sci., 16, 1259–1267, https://doi.org/10.5194/hess-16-1259-2012, 2012.

Marhaento, H., Booij, M. J., Rientjes, T. H. M., and Hoekstra, A. Y.: Attribution of changes in the water balance of a tropical catchment to land use change using the SWAT model, Hydrol. Process., 31, 2029–2040, https://doi.org/10.1002/hyp.11167, 2017.

Meijide, A., Röll, A., Fan, Y., Herbst, M., Niu, F., Tiedemann, F., June, T., Rauf, A., Hölscher, D., and Knohl, A.: Controls of water and energy fluxes in oil palm plantations: environmental variables and oil palm age, Agr. Forest Meteorol., 239, 71–85, 2017.

Merten, J., Röll, A., Guillaume, T., Meijide, A., Tarigan, S., Agusta, H., Dislich, C., Dittrich, C., Faust, H., Gunawan, D., and Hein, J.: Water scarcity and oil palm expansion: social views and environmental processes, Ecol. Soc., 21, 5–16, 2016.

Ministry of Forestry Decree – Permenhut/Menhut-V/2013.: Criteria for watershed classification (Kriteria Penetapan Klassifikasi Daerah Aliran Sungai), Ministry of Forestry, Republic of Indonesia, 2013.

Moriasi, D. N., Arnold, J. G., van Liew, M. W., Binger, R. L. Harmel, R. D., and Veith, T.: Model evaluation guidelines for systematic quantification of accuracy in watershed simulations, T. ASABE, 50, 885–900, 2007.

Moriasi, D. N., Gitau, M. W., Pai, N., and Daggupati, P.: Hydrologic and water quality models: performance measures and evaluation criteria, T. ASABE, 58, 1763–1785, https://doi.org/10.13031/trans.58.10715, 2015.

Mukherjee, I. and Sovacool, B. K.: Palm oil-based biofuels and sustainability in southeast Asia: a review of Indonesia, Malaysia, and Thailand, Renew. Sust. Energ. Rev., 37, 1–12, https://doi.org/10.1016/j.rser.2014.05.001, 2014.

Mwangi, H. M., Julich, S., Pati, S. D., McDonald, M. A., and Feger, K. H.: Relative contribution of land use change and climate variability on discharge of upper Mara River, Kenya, J. Hydrol., 5, 244–260, 2016.

Nash, J. and Sutcliffe, J. V.: River flow forecasting through conceptual models, 1, A discussion of principles, J. Hydrol., 10, 282–290, 1970.

Neitsch, S. L., Arnold, S. G., Kiniry, J. R., and Williams, J. R.: Soil and water assessment tool theoretical documentation, Version 2009, Texas Water Resources Institute Technical Report No. 406, Texas A&M University System, College Station, Texas, 1–8, 2011.

Pfannerstill, M., Guse, B., and Fohrer, N.: A multi-storage groundwater concept for the swat model to emphasize non- linear groundwater dynamics in lowland catchments, Hydrol. Process., 28, 5599–5612, https://doi.org/10.1002/hyp.10062, 2014.

Pransiska, Y., Triadiati, T., Tjitrosoedirjo, S., Hertel, D., and Kotowska, M. M.: Forest conversion impacts on the fine and coarse root system, and soil organic matter in tropical lowlands of Sumatera (Indonesia), Forest Ecol. Manag., 379, 288–298, 2016.

Rientjes, T. H. M., Haile, A. T., Kebede, E., Mannaerts, C. M. M., Habib, E., and Steenhuis, T. S.: Changes in land cover, rainfall and stream flow in Upper Gilgel Abbay catchment, Blue Nile basin – Ethiopia, Hydrol. Earth Syst. Sci., 15, 1979–1989, https://doi.org/10.5194/hess-15-1979-2011, 2011.

Röll, A., Niu, F., Meijide, A., Hardanto, A., Hendrayanto, Knohl, A., and Hölscher, D.: Transpiration in an oil palm landscape: effects of palm age, Biogeosciences, 12, 5619–5633, https://doi.org/10.5194/bg-12-5619-2015, 2015.

Setiadi,B., Diwyanto, K., Puastuti, W., Mahendri, I. G. A. P., and Tiesnamurti, B.: Peta Potensi dan Sebaran Areal Perkebunan Kelapa Sawit di Indonesia, Pusat Penelitian dan Pengembangan Peternakan, BadanLitbang Pertanian, Kementerian Pertanian (Area distribution of oil palm plantation in Indonesia, Center for Research and Development, Ministry of Agriculture Indonesia), ISBN 978-602-8475-45-7, 2011.

Sterling, S. M., Ducharne, A., and Polcher, J.: The impact of global land cover change on the terrestrial water cycle, Nature Climate Change, 3, 385–390, https://doi.org/10.1038/nclimate1690, 2012.

Strauch, M. and Volk, M.: SWAT plant growth modification for improved modeling of perennial vegetation in the tropics, Ecol. Modell., 269, 98–112, https://doi.org/10.1016/j.ecolmodel.2013.08.013, 2013.

Sunarti, Sinukaban, N., Sanim, B., and Tarigan, S. D.: Forest conversion to rubber and oil palm plantation and its effect on runoff and soil erosion in Batang Pelepat watershed, Jambi, J. Tanah Tropika, Vol. 13. No.3, ISSN 0852-257X, Lampung, 2008.

Tarigan, S. D.: Land cover change and its impact on flooding frequency of Batanghari Watershed, Jambi Province, Indonesia, Procedia, Environ. Sci., 33, 386–392, 2016a.

Tarigan, S. D., Wiegand, K., Dislich, C., Slamet, B., Heinonen, J., and Meyer K.: Mitigation options for improving the ecosystem function of water flow regulation in a watershed with rapid expansion of oil palm plantations, Sustainability of Water Quality and Ecology, 8, 4–13, 2016b.

Tuppad, P., Douglas-Mankin, K. R., Lee, T., Srinivasan, R., and Arnold, J. G.: Soil and Water Assessment Tool (SWAT) hydrologic/water quality model: Extended capability and wider adoption, T. ASABE, 54, 1677–1684, https://doi.org/10.13031/2013.34915, 2011.

Van der Laan, C., Wicke, B., Verweij, P. A., and Faaij, A. P. C.: Mitigation of unwanted direct and indirect land-use change-an integrated approach illustrated for palm oil, pulpwood, rubber and rice production in North and East Kalimantan, Indonesia, GCB Bioenergy, 9, 429–444, https://doi.org/10.1111/gcbb.12353, 2016.

Van der Molen, M. K., Dolman, A. J., Waterloo, M. J., and Bruijnzeel, L. A.: Climate is affected more by maritime than by continental land use change: a multiple scale analysis, Global Planet. Change, 54, 128–149, 2006.

Van Griensven, A., Meixner, T., Grunwald, S., Bishop, T., Diluzio, M., and Srinivasan, R. A.: Global sensitivity analysis tool for the parameters of multi-variable catchment models, J. Hydrol., 324, 10–23, 2006.

Van Griensven, A., Maharjan, S., and Alemayehu, T.: Improved simulation of evapotranspiration for land use and climate change impact analysis at catchment scale, International Environmental Modelling and Software Society (iEMSs) 7th International Congress on Environmental Modelling and Software, 2014.

Vigerstol, K. L., Aukema, J. E.: A comparison of tools for modeling freshwater ecosystem services, J. Environ. Manage., 92, 2403–2409, 2011.

Vijay, V., Pimm, S. L., Jenkins, C. N., and Smith, S. J.: The impacts of oil palm on recent deforestation and biodiversity loss, PLoS One, 11, e0159668, https://doi.org/10.1371/journal.pone.0159668, 2016.

Wang, Y. and Brubaker, K.: Implementing a nonlinear groundwater module in the soil and water assessment tool (SWAT), Hydrol. Process., 28, 3388–3403, https://doi.org/10.1002/hyp.9893, 2014

Wangpimool, W., Pongput, K., Tangtham, N., Prachansri, S., and Gassman, P. W.: The impact of para rubber expansion on streamflow and other water balance components of the Nam Loei River Basin, Thailand, Water, 9, 1–20, https://doi.org/10.3390/w9010001, 2017.

Wicke, B., Sikkema, R., Dornburg, V., and Faaij, A.: Exploring land use changes and the role of palm oil production in Indonesia and Malaysia, Land Use Policy, 28, 193–206, 2011.

Wilcove, D. S. and Koh, L. P.: Addressing the threats to biodiversity from oil-palm agriculture, Biodivers. Conserv., 19, 999–1007, 2010.

Xu, X., Yang D., Yang, H., and Lei, H.: Attribution analysis based on the Budyko hypothesis for detecting the dominant cause of runoff decline in Haihe basin, J. Hydrol., 510, 530–540, 2014.

Yusop, Z., Chan, C. H., and Katimon, A.: Runoff characteristics and application of HEC-HMS for modelling stormflow hydrograph in an oil palm catchment, Water Sci. Technol., 56, 41–48, 2007.

Zhang, L., Nan, Z., Xu, Y., and Li, S.: Hydrological impacts of land use change and climate variability in the headwater region of the Heihe river basin, Northwest China, PLoS One, 11, e0158394, https://doi.org/10.1371/journal.pone.0158394, 2016.

Ziegler, A. D., Fox, J. M., and Xu, J.: The rubber juggernaut, Science, 324, 1024–1025, https://doi.org/10.1126/science.1173833, 2009.

Ziegler, A. D., Fox, J. M., Webb, E. L., Padoch, C., Leisz, S. J., Cramb, R. A., Mertz, O., Bruun, T. B., and Vien, T. D.: Recognizing contemporary roles of swidden agriculture in transforming landscapes of Southeast Asia, Conserv. Biol., 25, 846–848, 2011.

3

On inclusion of water resource management in Earth system models – Problem definition and representation of water demand

A. Nazemi and H. S. Wheater

Global Institute for Water Security, University of Saskatchewan, 11 Innovation Boulevard, Saskatoon, SK, S7N 3H5, Canada

Correspondence to: A. Nazemi (ali.nazemi@usask.ca)

Abstract. Human activities have caused various changes to the Earth system, and hence the interconnections between human activities and the Earth system should be recognized and reflected in models that simulate Earth system processes. One key anthropogenic activity is water resource management, which determines the dynamics of human–water interactions in time and space and controls human livelihoods and economy, including energy and food production. There are immediate needs to include water resource management in Earth system models. First, the extent of human water requirements is increasing rapidly at the global scale and it is crucial to analyze the possible imbalance between water demands and supply under various scenarios of climate change and across various temporal and spatial scales. Second, recent observations show that human–water interactions, manifested through water resource management, can substantially alter the terrestrial water cycle, affect land–atmospheric feedbacks and may further interact with climate and contribute to sea-level change. Due to the importance of water resource management in determining the future of the global water and climate cycles, the World Climate Research Program's Global Energy and Water Exchanges project (WRCP-GEWEX) has recently identified gaps in describing human–water interactions as one of the grand challenges in Earth system modeling (GEWEX, 2012). Here, we divide water resource management into two interdependent elements, related firstly to water demand and secondly to water supply and allocation. In this paper, we survey the current literature on how various components of water demand have been included in large-scale models, in particular land surface and global hydrological models. Issues of water supply and allocation are addressed in a companion paper. The available algorithms to represent the dominant demands are classified based on the demand type, mode of simulation and underlying modeling assumptions. We discuss the pros and cons of available algorithms, address various sources of uncertainty and highlight limitations in current applications. We conclude that current capability of large-scale models to represent human water demands is rather limited, particularly with respect to future projections and coupled land–atmospheric simulations. To fill these gaps, the available models, algorithms and data for representing various water demands should be systematically tested, intercompared and improved. In particular, human water demands should be considered in conjunction with water supply and allocation, particularly in the face of water scarcity and unknown future climate.

1 Background and scope

1.1 Large-scale modeling – an introduction to land-surface and global hydrological models

The Earth system is an integrated system that unifies the physical processes at the Earth's surface. These processes include a wide range of feedbacks and interactions between and within the atmosphere, land and oceans and cover the global cycles of climate, water and carbon that support planetary life (e.g., Schellnhuber, 1999; Kump et al., 2010). From the advent of digital computers, Earth system models have been a key tool to identify past changes and to predict the future of planet Earth. These models normally include sub-models that represent various functions of the land, atmosphere and oceans (Claussen, 2001; Schlosser et al., 2007). A crucial sub-model in Earth system models is the land-surface model (LSM) that represents the land portion of the

Earth system. LSMs contain interconnected computational modules that characterize physical processes related to soil, vegetation and water over a gridded mesh, and account for their influences on water, energy and, increasingly, carbon exchanges. A wide range of LSMs is currently available, and these can be differentiated based on how, and to what extent, different land-surface processes are represented; nonetheless, a LSM should explicitly or implicitly include the dynamics of these processes, and account for their drivers at various temporal and spatial scales (see Trenberth, 1992; Sellers, 1992).

The importance of representing the terrestrial water cycle in LSMs is well-established (see Pitman, 2003, and references therein), and there has been progressive development of LSMs in representing various components of the hydrologic cycle, such as soil moisture, vegetation, snowmelt and evaporation. In early LSMs, hydrology was conceptualized as a simple lumped bucket model (Manabe, 1969), but this representation has progressively been improved by including more complexity and explicit physics in canopy, soil moisture and runoff calculations (see Deardorff, 1978; Dickinson, 1983, 1984; Sellers et al., 1986, 1994, 1996a; Nicholson, 1988; Pitman et al., 1990). Despite these improvements, major limitations and uncertainties remain in the hydrological simulations, causing systematic bias in water and energy balance calculations. These deficiencies have been attributed (in part) to unrealistic assumptions and incomplete parameterizations of catchment response in LSMs (Soulis et al., 2000; Music and Caya, 2007; Sulis et al., 2011). Further attempts, therefore, have focused on including catchment-scale runoff generation and routing processes (e.g., Miller et al., 1994; Hagemann and Dümenil, 1997; Oki and Sud, 1998; Oleson et al., 2008; Lawrence et al., 2011). These components determine the hydrological response at the larger scales and have been frequently used in large-scale hydrological models, so-called global hydrologic models (GHMs). Similar to LSMs, GHMs are gridded large-scale models; however, they are typically simpler in structure and focus on representing the water cycle rather than other land-surface processes (such as the energy and carbon cycles). LSMs have been applied frequently in regional and global modeling (e.g., Liang et al., 1994; Pietroniro et al., 2007; Adam et al., 2007; Livneh et al., 2011) and compared to GHMs (see Haddeland et al., 2011). At this stage of research, however, both LSMs and GHMs are still imperfect and incomplete, as current simulations cannot match recent hydrological observations (see Lawrence et al., 2012).

1.2 Modeling human–water interactions

While external forcing, mainly the energy flux from the Sun, is the main driver of the Earth system, internal disturbances such as volcanic eruptions, wildfires and human activities can substantially affect the natural Earth system cycles (Vitousek et al., 1997; Trenberth and Dai, 2007; Bowman et al.,

2009). In particular, post-industrial human activities, from the mid-20th century onwards, have severely perturbed the Earth system (Crutzen and Steffen, 2003; Crutzen, 2006). This has initiated a new geological epoch, informally termed the "Anthropocene", in which it is recognized that the natural processes within the land surface system are highly controlled and regulated by humans (see McNeil, 2000; Steffen et al., 2007, 2011). Accordingly, Earth system models should address feedbacks and interactions between the natural Earth system and the anthroposphere, which includes human cultural and socio-economic activities (Schellnhuber, 1998, 1999; Claussen, 2001). The terrestrial water cycle is one set of Earth system processes that is greatly perturbed by human activities; it also is of critical importance in determining human health, safety and livelihoods, as well as local, regional and global economies (e.g., Nilsson et al., 2005). However, although some anthropogenic effects, such as the emission of greenhouse gases and land-use change, have been incorporated in LSMs (e.g., Lenton, 2000; Zhao et al., 2001; Karl and Trenberth, 2003; Brovkin et al., 2006; Solomon et al., 2009), less effort has been made to represent human–water interactions (e.g., Trenberth and Asrar, 2012; Lawrence et al., 2012; Oki et al., 2013). This can be a major reason for current deficiencies in hydrological performance of large-scale modes (i.e., LSMs and/or GHMs). In fact, large-scale models still widely assume that human effects on the terrestrial water cycle can be ignored. This assumption is highly questionable and can result in the neglect of important hydrologic processes (see Gleick et al., 2013).

Human–water interactions include a wide spectrum of anthropogenic interventions, including land-use change and water resource management. During the past century, human water consumption has increased more than 6-fold, with around 5, 18 and 10 times increase in agricultural, industrial and municipal consumption, respectively (see Shiklomanov, 1993, 1997, 2000). Supplying such intensive demands has required large changes in the natural water cycle – which can be even more than the effects of warming climate (see Haddeland et al., 2014), and is associated with major environmental water stress at the global scale. Smakhtin et al. (2004) concluded that over 1.4 billion people currently live in river basins with high environmental water stress and this number will increase as water withdrawals grow. For instance, surface-water withdrawals for supplying human needs decrease downstream flows, often substantially, and result in seasonal decline in flows of major rivers such as the Colorado River (e.g., Cayan et al., 2010). Similarly, dam operations considerably change the timing, volume, peak and the age of natural streamflow and reduce inputs to wetlands, lakes and seas (e.g., Vörösmarty et al., 1997, 2005; Vörösmarty and Sahagian, 2000; Meybeck, 2003; Tang et al., 2010). This is associated with some extreme effects, such as the death of the Aral Sea (e.g., Precoda, 1991; Small et al., 2001) and aggressive decline in the area of Lake Urmia in Iran (Aghakouchak et al., 2014). In parallel, groundwater abstractions are associated with declining groundwater levels, reduced baseflow contributions and loss of wetlands. For instance, current assessments reveal significant groundwater depletion in some areas of the globe, such as Indian peninsula, the US Midwest, and Iran (Giordano, 2009; Rodell et al., 2009; Gleeson et al., 2012; Döll et al., 2014). Without considering human withdrawals, these changes in surface-water and groundwater availability cannot be captured by large-scale models. It should be noted that human activities have large effects on water quality as well. For instance, extensive groundwater pumping is also associated with potential long-term contamination, for example by salt-water intrusion (Sophocleous, 2002; Antonellini et al., 2008), and nutrient pollution of surface and groundwater is an outstanding global challenge. These water quality impacts, however, remain beyond the scope of this survey.

As human life and water availability are tightly interconnected (see Sivapalan et al., 2012), current and future changes in the water availability are not only important for Earth system modeling, but are also of major importance to human society, and these issues can be explored to a large extent with large-scale models. Although human water use still accounts for a small proportion of total water on and below the surface (see Oki and Kanae, 2006), total human withdrawals currently include around 26 % of terrestrial evaporation and 54 % of the accessible surface runoff that is geographically and temporally available (Postel et al., 1996). There are already major water scarcity issues across highly populated regions of the globe (e.g., Falkenmark, 2013; Schiermeier, 2014), which raise fundamental concerns about how future demand should be supplied, particularly considering climate change (e.g., Arnell, 1999, 2004; Tao et al., 2003; Döll, 2009; Taylor et al., 2013; Hanasaki et al., 2013a, b; Wada et al., 2013; Milano et al., 2013; Mehta et al., 2013; Schewe et al., 2014). Such important threats to water security necessitate a detailed understanding of water availability and demand in time and space; and therefore large-scale models are required for impact assessments.

Apart from the hydrologic and water security relevance discussed above, human–water interactions can have broader implications for the water cycle and affect climate, although these issues are yet to be fully explored, and remain in some cases controversial. For instance, irrigation can disturb the "natural" atmospheric boundary conditions (e.g., Sacks et al., 2009; Destouni et al., 2010; Gerten et al., 2011; Pokhrel et al., 2012; Hossain et al., 2012; Guimberteau et al., 2012; Dadson et al., 2013). At this stage of model development, the available quantitative understanding of these land–atmospheric implications is limited. To explore these issues it is necessary to include these processes in coupled land–atmospheric models, and this requires explicit representation of relevant human–water interactions within LSM computational schemes. Moreover, the return flows from human usage, entering the seas and oceans, can affect salinity and temperature and consequently impact their circulation patterns

(e.g., Rohling and Bryden, 1992; Skliris and Lascaratos, 2004; Vargas-Yàñez et al., 2010). This is of particular concern for closed oceans and the polar environment, where a change in freshwater input can modify the oceanic circulations and thus feedback on continental rainfall (Polcher, 2014). However as noted above, issues related to water quality remain beyond the scope of our survey.

1.3   Aim and scope of this survey

The aim of our survey is to consider the associated scientific and data challenges, the state of current practice, and directions for future research around including human effects on the terrestrial water cycle. In this paper and a companion paper (hereafter Nazemi and Wheater, 2015), we focus on human–water activities manifested through water resource management and note that this is subject to operational and policy constraints. We only consider water quantity aspects of water resource management, which we define as a suite of anthropogenic activities related to storage, abstraction and redistribution of available water sources for various human demands. Although a fully coupled representation of water resource management in Earth system models is not currently available, important progress is being made, and more generally a body of literature is gradually shaping around describing different aspects of water resource management in large-scale models, in particular within the context of GHMs. Nonetheless, there are still fundamental obstacles in including water resource management within large-scale models.

First, a fundamental principle in Earth system models as well as LSMs and GHMs is the conservation of water. To represent water resource management, therefore, it is necessary to fully capture water in a coupled human–natural system. To achieve this (i) modeling complexity should be increased, (ii) process representations related to both natural and anthropogenic systems should be improved and (iii) modeling capability should be extended to new domains (see Polcher, 2014, for an in-depth discussion). For instance, a large proportion of human demand is supplied by groundwater, which is often absent or crudely represented in both LSMs and GHMs and is widely considered disjoint from other elements of the Earth system such as climate.

Second, multiple factors affect water resource management at the larger scales, such as climate, hydrology, land-cover and socio-economy as well as land and environment management. Moreover, real-world management decisions often include cultural values and political concerns (Gober and Wheater, 2014). These various influences are so far considered in isolation and the interactions among them are widely unseen (e.g., Beddington, 2013).

Third, there is considerable lack of regional and global data concerning the actual use and operation of water resources systems, and therefore large-scale models cannot be properly tuned or validated. This major limitation, for instance, has led the research community to use estimated demand as a surrogate for actual use. Lack of data about human operations can also introduce large uncertainty into simulations of terrestrial storage and runoff. For instance Gao et al. (2012) noted that the "[...] results from global reservoir simulations are questionable" as "there are no direct observations of reservoir storage".

Fourth, there is a major gap between the scope of local operational water resource models and large-scale applications and research needs. Essentially, the scale at which local water resource management takes place is often within the sub-grid resolution of current large-scale models, which requires narrowing the resolution in large-scale models for explicit representation (see Wood et al., 2011) or adding more sub-grid heterogeneity into grid calculations for implicit parameterization. In addition, there is (and will increasingly be) competition between various water demands which requires allocation decisions. At this stage of model development, however, it is still unclear how operational policies should best be reflected at larger scales. At the local scale, detailed information on physical and operational systems as well as climate and water supply conditions are available (or can be generated as scenarios; see, e.g., Nazemi et al., 2002, 2013; Nazemi and Wheater, 2014a, b) and the competition between demands is often reflected as an optimization problem. As the simulation scale moves from local and small basin scales to regional and global scales, the data availability degrades considerably and the high level of calculations within optimization algorithms cannot be maintained, due to computational barrier.

Conceptually, water resource management at larger scales can be seen as an integration of two fully interactive elements, related to water demand as well as water supply and allocation: water demand is constrained by water availability and drives water allocation, which results in extraction from water sources and determines the extent of change in hydrological elements of the land surface. Moreover, as noted briefly above, perturbations in the terrestrial water cycle due to water resource management can further interact with other elements of the Earth system, particularly with climate (see Fig. 1). To assess the impacts of water resource management on land-surface processes and associated feedbacks with climate, the elements of water demand and water allocation should be described using computational algorithms and included in large-scale models. For the purpose of our survey, and reflecting the state of algorithm development and data availability, we focus in this paper only on the representation of water demand, and in the Nazemi and Wheater (2015) on water supply and allocation. Here, we classify human water demands under two general categories, namely irrigative and non-irrigative, and further divide non-irrigative demands into municipal, industrial, environmental, energy-related, and livestock water needs. This is useful to put current algorithms and modeling applications into context. Accordingly, we discuss how these demands are characterized using various computational algorithms. As will be

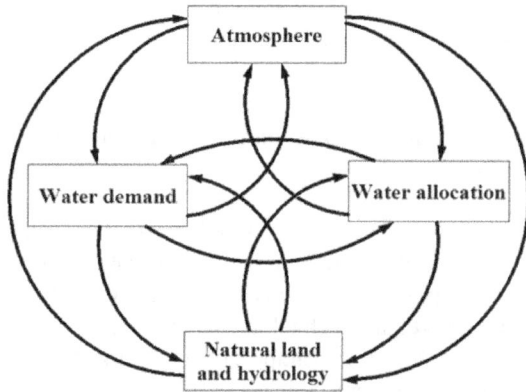

Figure 1. Water resource management as an integration of water demand and water allocation and its interactions with natural land-surface and climate.

shown later in this paper, human demands are mainly quantified either using downscaling (i.e., top-down approaches) or through direct modeling at the grid scale (i.e., bottom-up approaches). Depending on the type of application, the algorithms can be included in a wide range of large-scale models. Throughout our review, we consider both offline and online implications of water demand. Offline simulations assess the effects of water demand on land-surface processes without considering the associated feedbacks to the climate system, but can be linked to atmospheric driving variables to simulate land-surface and/or hydrological responses to climate and water resource management. Online models also account for the effects of water demand on land–atmospheric feedbacks and are further coupled with climate models. This is done by considering the effects of water demand on the dynamics of land-surface variables and updating the surface boundary conditions in climate models (Verseghy, 1991, 2000; Verseghy et al., 1993). Online applications are also termed in the LSM community as coupled land–atmospheric simulations (e.g., Entekhabi and Eagleson, 1989; Noilhan and Planton, 1989) and are more computationally demanding compared to offline simulations. While offline models include both LSMs and GHMs, it should be noted that GHMs cannot be used for online applications as they do not account for the energy balance and therefore cannot fully represent land–atmosphere feedbacks.

The structure of this paper is as follows: in Sect. 2 we highlight the impacts of irrigative and non-irrigative water demands on the terrestrial water cycle and land–atmospheric feedbacks. Sections 3 and 4 provide an overview of available representations of irrigative and non-irrigative demands at larger scales, respectively. In Sect. 5, we briefly explore state-of-the-art applications and highlight current limitations and uncertainties in estimating current and future water demand and associated online and offline impacts. We further discuss current gaps in Sect. 6 and provide some suggestions for future developments. Finally, Sect. 7 summarizes this first part of our survey and outlines our main findings with respect to representing human water demand.

2 Types of human demand and their impacts on the water cycle

Human water demands can be divided into irrigative and non-irrigative categories. Irrigation is the dominant human water use and has significantly intensified since the 1950s, due to population growth and technological development (Steffen et al., 2011). This has major importance for global food security, as it produces approximately 40 % of the world's food (Abdullah, 2006). Currently, around 25 % of harvested crop area is irrigated (Portmann et al., 2010). This accounts for some 90 % of water consumption at the global scale (Döll et al., 2009; Siebert et al., 2010), which is around 70 % of the total water withdrawals from surface and groundwater resources (Wisser et al., 2008; Gerten and Rost, 2010). Clearly supplying such a large water demand can severely disturb the "natural condition" by decreasing streamflow volume (e.g., Meybeck, 2003; Gaybullaev et al., 2012; Lai et al., 2014) and groundwater levels (e.g., Rodell et al., 2009; Gleeson et al., 2012; Wada et al., 2010, 2012, 2014; Döll et al., 2014). Currently, surface water is the main supplier of global irrigative needs, accounting for 57 % of the total consumptive irrigation use at the global scale (Siebert et al., 2010).

Apart from driving hydrological changes, irrigation-induced changes in soil moisture can affect land surface–atmosphere feedbacks (see Eltahir, 1998). Pokhrel et al. (2012) showed that increased soil water content through irrigation substantially enhances evapotranspiration, and therefore transforms the surface energy balance. Evapotranspiration due to irrigation leads to cooling of the land surface (e.g., Haddeland et al., 2006; Saeed et al., 2009; Destouni et al., 2010), as well as enhanced cloud cover and chance of convective precipitation (e.g., Moore and Rojstaczer, 2001; Douglas et al., 2009; Harding and Snyder, 2012a, b; Qian et al., 2013). Irrigation may also alter regional circulation patterns due to temperature difference between irrigated areas and neighboring regions (e.g., DeAngelis et al., 2010; Wei et al., 2013). Over highly irrigated regions, this can mask important climate change signals. Gerten et al. (2011), for instance, showed that the irrigation in South Asia has offset the increasing temperature in the region.

Non-irrigative water demands include municipal and industrial uses, energy-related withdrawals, other agricultural uses, such as livestock, as well as designated environmental water uses, which can be an important constraint on water management. Non-irrigative demands contribute a lesser proportion to total human water use at the global scale. This proportion, however, has significant spatial variability (Vassolo

and Döll, 2005; Flörke et al., 2013) as regional differences in population, income, life style and technological developments can alter the extent of non-irrigative demand significantly (e.g., Alcamo et al., 2003; Flörke and Alcamo, 2004; Hejazi et al., 2013a). However, while irrigation is predominantly a consumptive water use, only a small portion of the non-irrigative withdrawal is consumptive (e.g., Hanasaki et al., 2013a). Non-irrigative withdrawals, therefore, partially or totally return to surface-water or groundwater systems with varying degrees of time lag. Still, this can considerably perturb the streamflow regime (e.g., Maybeck, 2003; Förster and Lilliestam, 2010). Non-irrigative water demands are currently on a rapid incline due to growing population and industrial development. This can increase water stress in both time and space (Hejazi et al. 2013a, b, c, d). As non-irrigative demands are mainly non-consumptive, they are less likely to change the energy balance and/or perturb the atmospheric moisture condition significantly and therefore they are less relevant to land–atmospheric interactions. However, changing timing of flows can have significant local effects, for example on wetland inundation. Similarly, for some large-scale mining activities in which the extent of water withdrawals is considerable, the associated changes in soil moisture and land cover can be potentially relevant to land–atmospheric feedbacks. To the best of our knowledge, such online considerations for non-irrigative withdrawals have not yet been explored in the literature.

3 Available representations of irrigative demand in large-scale models

Irrigation is an important element of water resource management and has been explored more in depth than non-irrigative demands. To simplify our presentation, we classify current representations with respect to the scale (regional vs. global) and/or mode of simulation (offline vs. online). Tables 1 and 2 summarize representative examples of offline simulations at both regional (Table 1) and global (Table 2) scales. Table 3 presents some online examples. In brief, current online applications have mainly been performed at rather fine temporal and spatial resolutions with shorter simulation periods than offline representations. In contrast, a wide spectrum of host models (i.e., large-scale models in which the irrigation algorithm is embedded), as well as forcing and land-use data, has been used in current offline examples (see Tables 1 and 2). Model resolutions in offline applications can vary in time from 1 h (e.g., Leng et al., 2013) to 1 day (e.g., Haddeland et al., 2007) with a grid size ranging from a few kilometers (e.g., Siebert and Döll, 2010; Nakayama and Shankman, 2013) to a few hundred kilometers (e.g., Gueneau et al., 2012) in space. Moreover, offline irrigation demand calculations have already been performed globally under future climate conditions.

Table 1. Representative examples including regional irrigation in large-scale models (offline mode).

Reference	Irrigation data	Irrigation demand	Region	Host model	Forcing	Temporal resolution	Spatial resolution
Haddeland et al. (2006)	Döll and Siebert (2002)	Difference between current soil moisture content and minimum of FAO Penman–Monteith crop-specific evapotranspiration and soil moisture content at field capacity.	Colorado (USA) and Mekong (east Asia)	VIC (Liang et al., 1994)	Adam and Lettenmaier (2003); Maurer et al. (2002)	3 h	$0.5° \times 0.5°$
Haddeland et al. (2007)	Siebert et al. (2005)	Haddeland et al. (2006)	North America and Asia	VIC (Liang et al., 1994)	Maurer et al. (2002)	24 h	$0.5° \times 0.5°$
Gueneau et al. (2012)	GAEZ (IIASA/FAO, 2012); FRIS (USDA, 2008)	Difference between actual and potential evapotranspiration based on Farmer et al. (2011). Crop growth and irrigation losses included.	USA	CLM3.5 (Oleson et al., 2004, 2008)	NCC (Ngo-Duc et al., 2005b)	6 h	$2.5° \times 2.5°$
Leng et al. (2013)	MODIS (Ozdogan and Gutman, 2008); NASS (USDA, 2002)	Difference between current and ideal soil moisture content based on CLM4CNcrop crop growth model of CLM4 (Levis and Sacks, 2011; Levis et al., 2012).	Contermi-nous USA	CLM4 (Lawrence et al., 2011)	NLDAS (Cosgrove et al., 2003)	1 h	$0.125° \times 0.125°$
Nakayama and Shankman (2013)	Liu et al. (2010)	Difference between current soil moisture content and soil moisture at the field capacity.	Changjing, Yellow River basins (China)	NICE (Nakayama, 2011)	ECMWF (http://www.ecmwf.int/en/forecasts/datasets)	6 h	10km × 10km
Voisin et al. (2013)	Crop area projections in Chaturvedi et al. (2013a, b).	Downscaling GCAM model estimations (Wise and Calvin, 2011; Wise et al., 2009a) using methods of Hejazi et al. (2013a), Siebert and Döll (2008) and Hanasaki et al. (2013a, b).	US Mid-west	SCLM-MOSART (Lawrence et al., 2011; Li et al., 2013); Tesfa et al. (2014)	CASCaDE (http://cascade.wr.usgs.gov)	1 h	$0.125° \times 0.125°$

Table 2. Representative examples including global irrigation in large-scale models (offline mode).

Reference	Irrigation data	Irrigation demand	Host model	Forcing	Temporal resolution	Spatial resolution
Döll and Siebert (2002)	Döll and Siebert (2000)	Difference between Smith (1992) effective rainfall and Priestley and Taylor (1972) crop-specific potential evapotranspiration and Allen et al. (1998) multipliers.	WaterGAP (Alcamo et al., 2003)	CRU TS 1.0 (New et al., 1999, 2000)	24 h	$0.5° × 0.5°$
de Rosnay et al. (2003)*	Döll and Siebert (2002)	Difference between effective rainfall and FAO potential evapotranspiration (Allen et al., 1998) without considering irrigation efficiency.	ORCHIDEE (Ducoudré et al., 1993)	ISLSCP-I (Sellers et al., 1996b)	24 h	$1° × 1°$
Hanasaki et al. (2006)	Döll and Siebert (2000)	Similar to Döll and Siebert (2002). Reference evaporation is based on FAO Penman–Monteith.	TRIP (Oki and Sud, 1998)	ISLSCP-I (Sellers et al., 1996b)	24 h	$0.5° × 0.5°$
Wisser et al. (2008)	Siebert et al. (2005, 2007); GIAM (Thenkabail et al., 2009)	Similar to Haddeland et al. (2006) using Allen et al. (1998) procedure.	WBM (Vörösmarty et al., 1998)	CRU TS 2.1 (Mitchell and Jones, 2005); NCEP (Kalnay et al., 1996)	24 h	$0.5° × 0.5°$
Rost et al. (2008, 2009)	Siebert et al. (2007)	Difference between available plant moisture and an updated Priestley and Taylor (1972) potential evaporation based on potential canopy conductance of carbon and water (Sitch et al., 2003).	LPJmL (Bondeau et al., 2007)	CRU TS 2.1 (Mitchell and Jones, 2005)	24 h	$0.5° × 0.5°$
Hanasaki et al. (2008a, b)	Döll and Siebert (2000)	Difference between current and 75 % of field capacity. Irrigation applied 30 days prior to planting. Detailed crop growth representation based on SWIM (Krysanova et al., 1998).	H08 (Hanasaki et al., 2008a, b)	NCEP-DOE (Kanamitsu et al., 2002); GSWP-2 (Zhao and Dirmeyer, 2003)	24 h	$1° × 1°$
Siebert and Döll (2010)	MIRCA2000 (Portmann et al., 2010)	Difference between actual and crop-dependent reference evapotranspiration computed according to Priestley and Taylor (1972). Crop coefficients obtained from Allen et al. (1998).	GCWM (Siebert and Döll, 2008)	CRU TS 2.1 (Mitchell and Jones, 2005)	24 h	$0.08° × 0.08°$
Wada et al. (2011, 2012)	MIRCA2000 (Portmann et al., 2010)	Difference between actual and potential transpiration according to van Beek et al. (2011), using Priestley and Taylor (1972) for calculating crop-specific and transpiration (Allen et al., 1998).	PCR-GLOBWB (van Beek et al., 2011)	CRU TS 1.0 (New et al., 1999, 2000)	24 h	$0.5° × 0.5°$
Pokhrel et al. (2012)	Siebert et al. (2007)	Procedure of Hanasaki et al. (2008a, b). Crop calendar is based on potential evapotranspiration (Allen et al., 1998).	MASTIRO (Takata et al., 2003)	Kim et al. (2009); GPCC (Rudolf et al., 2005)	6 h	$1° × 1°$
Wada et al. (2014)	MIRCA2000 (Portmann et al., 2010)	Constant 50 mm surface-water depth for paddy irrigation until 20 days before harvesting. For non-paddy areas, the difference between current and ideal plant available moisture at field capacity with dynamic root zone.	PCR-GLOBWB (van Beek et al., 2011)	ERA-Interim (Dee et al., 2011); MERRA (http://gmao.gsfc.nasa.gov/merra/)	24 h	$0.5° × 0.5°$

* The simulation is performed globally but the results are analyzed only over the Indian peninsula.

Table 3. Representative examples including irrigation in coupled land-surface models (online mode).

Reference	Irrigation data	Irrigation demand	Region	Host LSM	Climate model	Temporal resolution	Spatial resolution
Adegoke et al. (2003)	LandSat (http://landsat.gsfc.nasa.gov/)	Target soil moisture deficit (difference between actual and saturated soil moisture).	High Plains (USA)	LEAF-2 (Walko et al., 2000)	RAMS (Pielke et al., 1992)	30 s nested in 1 min	10 km × 10 km nested in 40 km × 40 km
Sacks et al. (2009)	FAO-AQUASTAT (http://www.fao.org/nr/water/aquastat/main/index.stm)	AQUASTAT irrigated water uses applied at constant rate when LAI exceeds 80 % of the maximum annual value.	Global	CLM3.5 (Oleson et al., 2008)	CAM (Collins et al., 2004, 2006)	20 min	2.8° × 2.8°
Sorooshian et al. (2011)	CIMIS-MODIS (http://wwwcimis.water.ca.gov/)	Target soil moisture deficit (irrigation starts when the soil moisture drops below a maximum depletion threshold beyond which the plant in stressed (a percentage of field capacity, depending on the crop) and continues to field capacity)	California Central Valley (USA)	Noah (Ek et al., 2003)	NCAR-MM5 (Chen and Dudhia, 2001a, b)	30 min / 1 h	4 km × 4 km / 12 km × 12 km / 36 km × 36 km
Harding and Snyder (2012a, b)	MODIS (Friedl et al., 2002; Ozdogan and Gutman, 2008; NASS (USDA, 2002)	Target soil moisture deficit (difference between actual and saturated soil moisture to depth of 2 m).	Great Plains (USA)	Noah (Ek et al., 2003)	WRF (Skamarock et al., 2005)	30 and 25 s	10 km × 10 km
Guimberteau et al. (2012)	Döll and Siebert (2002)	Difference between potential transpiration and the net water amount kept by the soil (i.e., the difference between precipitation reaching the soil and total runoff).	Global	ORCHIDEE (Ducoudré et al., 1993)	LMDZ4 (Hourdin et al., 2006)	30 min	2.5° × 1.25°
Qian et al. (2013)	MODIS (Ozdogan and Gutman, 2008; Ozdogan et al., 2010)	Similar to Sorooshian et al. (2011). Based on Ozdogan et al. (2010), moisture threshold is fixed at 50 % of filed capacity. Roots grow based on the greenness index.	Southern Great Plains (USA)	Noah (Ek et al., 2003)	WRF (Skamarock et al., 2005)	3 h	12 km × 12 km

3.1   Framework and general procedure

Irrigated lands normally introduce heterogeneity into the computational grids of LSMs and GHMs. Such sub-grid heterogeneity can be represented as an additional "tile", similar to forested land, bare soil and snow cover (Polcher et al., 2011). Essentially, irrigation algorithms are required to estimate the irrigation demand, and accordingly irrigative water use, at the grid scale. Here we refer to the irrigation demand as the water required for ideal crop growth in addition to the available water from precipitation. To simulate the grid-based irrigation demand, crop type and the extent of irrigated regions and growing seasons should be first identified. The location and area of irrigation districts and the associated crop types can be extracted from regional and global data sets (e.g., USDA, 2002, 2008; Siebert et al., 2005, 2007; Portmann et al., 2010) and/or remotely sensed data (e.g., Adegoke et al., 2003; Qian et al., 2013). There are two general approaches for identifying growing seasons. The choice of these options depends on the level of detail in the host model. In simpler models, where no energy-balance calculation is available (i.e., GHMs), crops can grow when and where simple temperature- and precipitation-based criteria are met (e.g., Döll and Siebert, 2002). In more detailed models (i.e., LSMs) the optimal growing season can be identified based on biophysical conditions of crop growth and/or soil water, canopy and energy balance conditions to estimate the cropping period that is necessary to obtain mature and optimal plant biomass (e.g., Rost et al., 2008; Pokhrel et al., 2012). Both approaches are subject to uncertainty. On one hand, models with fixed crop calendars ignore inter-annual variability in growing seasons. On the other hand, even models with fully dynamic crop growth algorithms may misrepresent the seasonality. After the growing season is identified, the irrigation demands (and under some assumptions, actual irrigation withdrawals) at each simulation time step can be calculated. A variety of top-down and bottom-up procedures are available for calculating the irrigation demand in large-scale models and are reviewed further below. If the irrigation demand is completely fulfilled, then the actual evapo(transpi)ration would be equal to crop-specific evapo(transpi)ration under standard conditions (see Allen et al., 1998). In offline applications, the irrigation rate can perturb soil moisture content, evaporation, deep percolation and runoff in irrigated tiles (e.g., Hanasaki et al., 2008a, b; Wada et al., 2011, 2012, 2014). In online applications, the vertical vapor and heat fluxes need also to be considered. The total fluxes for each grid can be then calculated as the sum of the flux contributions from irrigated and non-irrigated portions of the grid (e.g., Haddeland et al., 2006; Pokhrel et al., 2012), and can be further introduced to climate models as coupled surface boundary conditions (e.g., Sorooshian et al., 2011; Harding and Snyder, 2012a, b).

3.2   Top-down algorithms for calculating irrigation demand

In top-down approaches, the irrigation demand is not directly calculated, but estimated based on downscaling information available at coarser scales, often at national or geopolitical scales. Such information is based on census-based inventories (e.g., Sacks et al., 2009) or socio-economic model outputs (e.g., Voisin et al., 2013). Top-down approaches are highly influenced by the availability of global data on water use, such as FAO's Information System on Water and Agriculture (AQUASTAT; http://www.fao.org/nr/water/aquastat/main/index.stm), which provides annual inventory data on national (and in some cases also sub-national) scales, and has been further extended to include socio-economic model outputs. An example of such a model is the Global Change Assessment Model (GCAM; Wise et al., 2009a, b; Wise and Calvin 2011), which estimates agricultural production based on socio-economic variables, from which the irrigation water use is indirectly calculated using the water required for each crop per unit of land. Downscaling is performed mainly using land-use, technological and/or socio-economic proxies. There are various sources of uncertainty associated with top-down algorithms. First, both inventory and model-based products have major limitations due to their spatial and temporal scales as irrigation practices are highly variable within a country and a typical year. Moreover, the quality of both census and model-based products is poor. For instance, there are inconsistencies between census data and data quality varies from country to country (see Portman et al., 2010, for a detailed discussion). Also, socio-economic models widely ignore water availability constraints (Hejazi et al., 2013d). As a result, calculation of irrigation demand is mainly pursued through bottom-up schemes.

3.3   Bottom-up algorithms for calculating irrigation demand

In contrast to top-down schemes, bottom-up approaches estimate the irrigation demand directly at the grid scale by mimicking the optimal crop growth for irrigated tiles. Despite major limitations due to the heterogeneity in soil and crops, bottom-up algorithms have been widely used in the literature. These algorithms include a range of modeling assumptions; however, they are all centered around estimation of an ideal crop water requirement, i.e., where there is no water deficit. This requirement is based on estimation of "potential evapo(transpi)ration", which characterizes the atmospheric moisture deficit (Hobbins et al., 2008). There are multiple approaches to estimate the potential evapo(transpi)ration, and the estimates obtained may vary considerably. LSMs typically include detailed energy balance calculations and resolve the diurnal cycle; therefore, they can directly calculate potential and actual evaporations (see Milly, 1992; Barella-Ortiz et al., 2013, for a detailed description). Alternative approaches adopt a variety of methods, and are heavily influenced by FAO's guidelines for calculating irrigation water requirements (see Allen et al., 1998). These approaches are mainly used in GHMs, where the evapotranspiration is calculated for a reference crop and corrected as a function of crop type and development stage using a set of empirical coefficients. Various methods are used to characterize the reference evapotranspiration, such as FAO Penman–Monteith (Allen et al., 1998), Priestley and Taylor (1972) and modified Hargreaves (Farmer et al., 2011) to name a few (see McKenney and Rosenberg, 1993, for more examples). The choice of appropriate formulation for reference evapotranspiration is rather arbitrary and depends largely on the data availability as well as the level of detail supported in the host model. It should be noted that, due to the difference in estimation of evaporation, incorporating FAO's guidelines for estimation irrigation demand in LSMs can introduce inconsistencies with the evaporation estimated by the model at various timescales, particularly over dry regions where the irrigation is likely to occur (Polcher, 2014).

Here we briefly explain the currently available bottom-up algorithms, from the more simple to the more comprehensive algorithms, and highlight their strengths and weaknesses.

In the most simple bottom-up representations, the irrigation demand at every time step is the water required to bring the soil moisture at the root zone to saturation (e.g., Lobell et al., 2006; Harding and Snyder, 2012a, b), which describes an extreme demand condition and clearly overestimates the actual irrigation water requirement (Sacks et al., 2009). In a more realistic but still naïve representation, the soil moisture requirement during the growing season is considered to be the field capacity (e.g., Nakayama and Shankman, 2013); therefore, the irrigation water need is the water required to bring the soil moisture to field capacity. The description of the irrigation demand based on the field capacity can also overestimate the actual water requirements, as the evaporation often reaches potential level before the soil reaches field capacity. The threshold at which the evaporation reaches potential evaporation is crop-dependent, but often considered as a constant value in large-scale models. As an offline example, Hanasaki et al. (2008a) assumed that paddy and non-paddy crops require soil moisture content of 100 or 75 % of the field capacity at the root zone with constant depth at the global scale. Yoshikawa et al. (2014) later updated the assumption for non-paddy soil moisture requirement and used 60 % of field capacity, referring to the requirement for wheat. This is again rather unrealistic as (1) by assuming a constant percentage of the field capacity for all crop types, the diversity in crop water requirement is ignored; and (2) a constant root zone depth at the global scale can result in misestimating the irrigation demand. There are attempts to address these limitations. For instance, Sorooshian et al. (2011) assumed that the required soil moisture content can change for each grid based on the dominant crop. Leng et al. (2013) and Qian et al. (2013) implemented root growth in their irrigation demand algorithm to avoid overestimation of demand due to a constant root zone. It should be noted that calculating the root growth is also subject to uncertainty; however, associated limitations remain beyond the scope of this paper.

More realistic definitions of irrigation water demand are based on the difference between the crop-dependent potential evapotranspiration and available crop water. This definition has been widely used in global irrigation demand projections (see Table 2). In earlier examples (e.g., Döll and Siebert, 2002; de Rosnay et al., 2003), crop development is described by constant monthly multipliers for potential evapotranspiration and the effective rainfall is used as a surrogate for available crop water. In more advanced algorithms, the correction factors are considered as functions of daily climate, stage of vegetation and root growth. Moreover, actual evapotranspiration or soil moisture content can be used instead of effective rainfall (Haddeland et al., 2006, 2007; Gueneau et al., 2012). There are two key limitations associated with this approach to simulation of irrigation demands. First, FAO's definition of irrigation water requirement considers both transpiration from crop and evaporation from soil. It has been noted that this quantification may result in overestimating the irrigation demand and may not properly represent the dynamics of vegetation (Polcher et al., 2011). Second, it is assumed that crop growth is a function of water availability only; therefore, the effects of other drivers such as CO_2 on photosynthesis are wholly ignored.

Some efforts try to overcome these limitations by defining irrigation demand based on potential transpiration instead of potential evapotranspiration (e.g., Wada et al., 2011, 2012), in conjunction with models that have more comprehensive vegetation schemes. Potential transpiration is the transpiration that would occur if the crop is not water stressed. Potential transpiration takes into account CO_2 fertilization effects and can represent the adaptation of the plants to climatic conditions and/or crop growth cycles, if the host model is equipped with relevant calculations (Guimberteau et al., 2012); therefore, this approach is mainly used in LSMs with detailed consideration of vegetation growth. As an example, Rost et al. (2008) coupled a transpiration deficit algorithm with the Lund-Potsdam-Jena managed Land scheme (LPJmL; Bondeau et al., 2007), which has a detailed vegetation growth module based on carbon and water availability (see Sitch et al., 2003; Gerten et al., 2004). The crop water limitation was calculated based on the atmospheric water deficit, soil moisture, plant hydraulic states as well as the CO_2 effects. Considering the effects of both carbon and water in vegetation can provide a basis for explicit linkage between CO_2 emission, crop growth and irrigation water requirement. This would be important for future predictions under increasing CO_2 effects. Moreover, some recent simulations showed that the irrigation requirement changes if a dynamic growth model is used; and this can improve the partitioning of latent heat flux, which is relevant to online applications (e.g., Lu, 2013). Nonetheless, it should be noted that the success of potential transpiration algorithm depends strongly on the way various tiles are treated at the grid scale. Normally, LSMs can define multiple crops at the grid scale and can distinguish the various water needs across different tiles within a grid. If potential transpiration is implemented consistently with sub-grid soil moisture divisions, then the water taken from the irrigated tiles optimizes photosynthesis and is only evaporated by the crops and not used by other surface types (e.g., bare soil, non-irrigated crops). In contrast, if all tiles share the same soil moisture reservoir at the grid scale, irrigation will increase the soil moisture and evaporation and therefore reduce water stress over the whole grid.

3.4 Projection of irrigative demand

From water and food security perspectives, particularly under various global change scenarios, it is crucial to investigate future irrigation demand and assess various possibilities for irrigation deficit. Climate model projections under IPCC emission scenarios (IPCC, 2000) have been widely used to force bottom-up irrigation demand algorithms (e.g., Arnell, 1999; Wada et al., 2013; Rosenzweig et al., 2014). Efforts have been also made to include intermediate socio-economic scenarios that can be matched to current climate change scenarios (see, e.g., Arnell, 2004; Fischer et al., 2007; Alcamo et al., 2007). For irrigation, intermediate scenarios describe changes in irrigated areas, irrigation efficiency and crop type, using empirical approaches. For example, Hanasaki et al. (2013a) recently proposed intermediate scenarios based on newly developed Shared Socio-economic Pathways (SSPs; Kriegler et al., 2012; see also Moss et al., 2010), which are consistent with Representative Concentration Pathways (RCPs; Meinshausen et al., 2011; K. E. Taylor et al., 2012). Constructing intermediate scenarios using empirical procedures, however, is uncertain as mechanisms that link irrigation expansion to socio-economic factors are not fully known and current empirical relationships can contain large uncertainties. More dynamic linkage between irrigation expansion and socio-economic drivers can be provided by coupled socio-economy–energy–carbon models. One emerging model of such a kind is GCAM, which has been recently implemented for simulating the future expansions in irrigation areas and demands (Hejazi et al., 2013b, c, d) as well as policy implications for irrigation water requirements (e.g., Chaturvedi et al., 2013a, b). Although these models can represent the dynamic effects of various drivers on irrigation, they remain uncertain as their simulations are rather coarse and do not incorporate water availability constraints. There are emerging efforts to avoid this limitation by linking the irrigation demand to climate, economy and water management constraints. This can result in prediction of regions in which irrigation can be developed and sustained considering changing climate, water availability, water price and water management infrastructure (see Nassopoulos et al., 2008,

2012). Such approaches however have not been applied at larger regional and global scales.

4 Available representations of non-irrigative demand

4.1 Forms and drivers of non-irrigative demand

Non-irrigative water demands relate to a wide range of environmental, municipal, industrial and energy-related uses, as well as other agricultural water needs (e.g., livestock), and include both consumptive and non-consumptive withdrawals. Among these, livestock water demand is assumed fully consumptive, and can be estimated by livestock number and demand per livestock head (e.g., Wada et al., 2011; Strzepek et al., 2012b; Hejazi et al., 2013d). Wada et al. (2014) made a further improvement by estimating daily livestock requirements at $0.5° \times 0.5°$ spatial resolution using livestock data of Steinfeld et al. (2006). Daily demand was considered as a function of daily temperature.

In contrast to livestock water demand, environmental flow needs can be considered as a fully non-consumptive need, required to protect rivers' health and aquatic life. Considering the extent of environmental degradation at the global scale, accounting for environmental flow needs becomes more and more relevant and should be considered as an integral part of water resource management at larger scales. Tharme (2003) made an extensive review of available methodologies for estimating environmental flow needs and identified more than 200 methodologies based on various hydrological, hydraulic rating, habitat simulation and holistic guidelines at the river basin scale. There are also some recent trends to involve scientists, water-resource managers and stakeholders to analyze available hydrological information and convert them into ecologically based and socially acceptable goals for estimating the environmental flow needs (see Poff et al., 2009). Such procedures however are widely dependent on the availability of relevant information and, therefore, cannot be easily implemented in large-scale models. Currently, implementation of environmental flow needs in large-scale models remains rather limited and simplistic and these needs are often calculated based on generic rules. For instance, Smakhtin et al. (2004) assigned thresholds for fair (Q_{90}), natural (Q_{50}) and good (Q_{75}) natural flow conditions. Shirakawa (2004, 2005, referenced from Hanasaki et al., 2008a) distinguished between two factors, i.e., minimum and perturbation flow requirements, which can also accommodate transient streamflow conditions. Currently, the perturbation flow requirements are often ignored in large-scale models and the environmental needs are estimated as a minimum flow threshold (often Q_{90} or 10 % of mean annual), which should be maintained in the river reaches (e.g., Hanasaki et al., 2008a; Döll et al., 2009; Strzepek et al., 2010, 2012b; Blanc et al., 2013). Other rules have been also suggested. For instance, Haddeland et al. (2006) considered a 7-day consecutive low flow with a 10-year recurrence period as the environmental flow requirement. Although these rules are easily implementable for larger regions and global scales, they widely ignore natural system complexity and the local policy context and can contribute to misunderstanding of the extent of environmental water stress (Arthington et al., 2006).

At this stage of model development, municipal, industrial and energy-related water demands are considered as the most dominant forms of non-irrigative uses. These demands are estimated using complex functions of socio-economic and technological factors, with high variability in time and space. Population is the most significant factor driving these withdrawals (e.g., Alcamo et al., 2003; Hanasaki et al., 2008a; Wada et al., 2014). National gross domestic product (GDP) is also a strong factor (e.g., Gleick, 1996; Cole, 2004; Wada et al., 2011). Although higher GDP may trigger more municipal water use per capita (Alcamo et al., 2007), Hughes et al. (2010) showed that, in general, water uses per capita are greater in developing than developed countries due to low-tech water delivery and industrialization. Strzepek et al. (2010) argued that industrial water use increases with the level of resource industry and decreases when a country moves toward the service sector. Industrial technology is another important factor for non-irrigative use as the extent of both consumptive and non-consumptive uses can significantly change based on the type of technology. Macknick et al. (2011), for instance, provided estimates of total water withdrawals and consumption for most electricity generation technologies within the US. Comparing to recirculating cooling technology, they noted that once-through cooling requires 10 to 100 times more water withdrawal per unit of electric generation. However, the latter consumes less than half of the water consumed by recirculating cooling technology. Climate can be another important factor controlling both consumptive and non-consumptive withdrawals (e.g., Wada et al., 2011, 2014; Hejazi et al., 2013a; Voisin et al., 2013), but has often been ignored as an explicit driver of non-irrigative water demand.

4.2 Top-down algorithms for estimation of grid-based non-irrigative withdrawals

Unlike irrigation demand, top-down approaches have been widely used for non-irrigative withdrawals to transfer national or geopolitical data to basin or grid scales. Various downscaling procedures have been suggested, based on different proxies (see Table 4). These top-down schemes are heavily influenced by the availability of national and global data sets and the downscaling algorithms within the Water – Global Assessment and Prognosis scheme, which is a global water budget and use model (WaterGAP; Alcamo et al., 1997, 2003, 2007). Currently, the availability of different global information sources has provided the opportunity to generate gridded products from different sources. As an example, Hanasaki et al. (2008a) merged the FAO-AQUASTAT

data with population distributions and national boundary information from Columbia University (CIAT, 2005) and the consumptive ratios of Shiklomanov (2000) to come up with gridded industrial and municipal water withdrawals and uses at the global scale. More detailed information on various industrial uses resulted in breaking down the industrial withdrawals into their components. For instance, Vassolo and Döll (2005) distinguished between industrial water uses related to thermoelectric power generation and manufacturing production. Temporal disaggregation of annual withdrawals, however, has received much less attention. Recently Wada et al. (2011, 2014) and Voisin et al. (2013) developed simple algorithms to disaggregate annual data to monthly and daily estimates (see Table 5).

4.3 Projection of non-irrigative demand

Characterizing the past and future evolution of non-irrigative demands is required to understand the mechanisms controlling water use and water allocation. Current projections have coarse temporal and spatial resolution and describe non-irrigative demands as functions of socio-economic and technological developments (e.g., Davies et al., 2013; Blanc et al., 2013; Hejazi et al., 2013b, d; Voisin et al., 2013). These changes can be characterized by intermediate socio-economic and technological scenarios, as briefly explained above for irrigation expansion (see Sect. 3.4). The projected demands can be further downscaled using various proxy variables, as explained in Sect. 4.2. Table 6 summarizes some representative efforts, which can be classified as explicit and implicit algorithms. In explicit algorithms, changes in water withdrawals are directly described as functions of changes in socio-economy, technology and water price using simple parametric structures (e.g., Strzepek et al., 2012b; Flörke et al., 2013; Hanasaki et al., 2013a; Hejazi et al., 2013a). The parameters can be assigned using the available global and regional data. In implicit procedures, first the production (or population) is estimated based on integrated economy and population models or prescribed scenarios. By considering the amount of water withdrawal per unit of production (or population) and accounting for technological and/or socio-economic shifts, water withdrawals are consequently projected.

5 State of large-scale modeling applications

The algorithms reviewed in Sects. 3 and 4 have had a wide range of online and offline applications. In comparison to offline applications, online simulations are still at a relatively early stage of development; they typically only include irrigation, mainly implemented at regional scale and under current conditions, and present rather contradictory results. Offline applications in contrast include both irrigative and non-irrigative demands, performed under current and future con-

Table 4. Representative examples calculating grid-based non-irrigative demands using downscaling of coarse-scale estimates.

Reference	Estimated demand	Downscaling procedure	Data support	Targeted resolution
Alcamo et al. (2003)	Domestic	Distributing country-level withdrawals based on population, ratio of rural to urban population (constant for each country) and percentage of population with access to drinking water	Population (van Woerden et al., 1995); access to drinking water (WRI, 1998)	$0.5° \times 0.5°$ (global)
	Industrial	Downscaling countywide industrial withdrawals based on proportion of urban population	Population (van Woerden et al., 1995)	
Vassolo and Döll (2005)	Thermoelectric cooling	Calculating the gridded data for power production based on downscaling global estimates. Allocating constant flow to each unit of production according to type of cooling system.	World Electric Power Plants Data Set (http://www.platts.com)	$0.5° \times 0.5°$ (global)
	Manufacturing	Estimating countrywide sectoral production volumes along with water intensity for each unit of production in each sector. Downscaling total demand to the grid scale based on city nighttime light.	Industrial production volumes (UN, 1997; CIA, 2001); sectoral intensity (Shiklomanov, 2000; WRI, 2000); night city light pollution (US Air Force, www.ngdc.noaa.gov/dmsp)	
Hanaskai et al. (2008a)	Domestic and industrial	Countrywide data downscaled to grid scale by weighting population and national boundary information, further converted to water consumption estimates.	AQUASTAT countrywide withdrawals, population and national boundaries (CIAT, 2005); ratio of consumption to withdrawal (Shiklomanov, 2000).	$1° \times 1°$ (global)
Hejazi et al. (2013b)	Municipal and industrial	Demand estimates of GCAM model (http://wiki.umd.edu/gcam) downscaled as a function of population. Population density assumed static in time.	Global population density data based on WWDR-II and methodology of Wada et al. (2011, 2013a)	$0.5° \times 0.5°$ (global)

Table 5. Representative examples for disaggregating annual non-irrigative demand into monthly estimates.

Reference	Estimated demand	Disaggregation procedure	Data support
Wada et al. (2011, 2014)	Municipal and livestock	Downscaling annual demand to monthly fluctuations as a function of temperature	CRU (New et al., 1999, 2000)
Voisin et al. (2013)	Electrical	Dividing electrical use into industry, transportation and building sectors. Assuming uniform distribution for industry and transportation uses and capturing the monthly fluctuations in building use based on heating/cooling degree days.	CASCaDE (http://cascade.wr.usgs.gov)

Table 6. Representative examples for projection of non-irrigative water demands using socio-economic variables.

Reference	Simulated demands	Simulation procedure	Temporal resolution	Spatial resolution
Alcamo et al. (2003)	Domestic and industrial	Explicit simulation of change in industrial and domestic withdrawal as functions of usage intensity and technological change. Usage intensities are functions of GDP.	Annual	Countrywide
Strzepek et al. (2012b)	Municipal and industrial	Explicit simulation of change in municipal water use as a function of population and per capita income. Industrial water use considered as a function of water use per capita and GDP considering growth rate and climatic and water availability factors.	Annual	Assessment sub-regions (global)
Flörke et al. (2013)	Domestic and industrial	Explicit simulation of domestic demand using Alcamo et al. (2003) with parameterization based on HYDE (http://themasites.pbl.nl/tridion/en/themasites/hyde/) and UNEP (http://www.unep.org/) data sets. Technological change influenced electrical demand. Manufacturing water use computed as a function of baseline structural intensity and rates of manufacturing gross value and technological change.	Annual	Countrywide (global)
Davies et al. (2013)	Electrical	Implicit simulation – changes in regional cooling system shares estimated based on shift from wet to dry cooling technologies. Reductions in water withdrawal and consumptions estimated based on level of technological change.	Annual	Geopolitical regions (global)
Hanasaki et al. (2013a)	Industrial and municipal	Explicit simulation of industrial withdrawal as a function of electricity production and water intensity which decreases linearly in time. Municipal water use calculated as a function of population and change in municipal intensity, varying based on GDP.	Five-year interval	Countrywide
Blanc et al. (2013)	Electrical, domestic, industrial and mining	Electrical demand projected implicitly using ReEDS (Short et al., 2009) and integration with USREP model (Rausch and Mowers, 2013). Water withdrawal and consumption to meet electrical demand estimated using Strzepek et al. (2012a). Other demands categorized into three groups: public supply, self-supply and mining supply and simulated explicitly. Public supply considered as a function of population and GDP per capita. Self-supply considered as function of sectoral GDP. Mining supply considered as a function of mining's GDP.	Annual	Assessment sub-regions (US)
Hejazi et al. (2013a)	Municipal	Withdrawal per capita explicitly determined as a function of GDP per capita, water price and technological development. Technological development considered as a function of operational efficiency, which further determines extent of water use.	Annual	Geopolitical regions (global)
Hejazi et al. (2013b, d)	Industrial	Manufacturing water demand is explicitly simulated based on population and GDP. Water demand for primary energy scaled by amount of fuel production and water demand for secondary energy.	Annual	Geopolitical regions (global)
Wada et al. (2014)	Industrial and municipal	Industrial and municipal withdrawal taken from WWDR-II data set (Shiklomanov, 1997; Vörösmarty et al., 2005) and backcasted explicitly using economic and technological proxies. Net municipal water demand calculated as a function of fraction of urban to total population and recycling ratio.	Annual	Countrywide (global)

ditions, and provide relatively more consistent results. Here, we briefly summarize recent applications and highlight the limitations in current simulations.

5.1 Online representation

Recent studies have shown that including irrigation in coupled land-surface schemes can generally improve climate simulations. With respect to regional temperature, for instance, Saeed et al. (2009) showed that representing irrigation activities over northwestern India and Pakistan can reduce climate model simulation bias by $5°$ K. It should be noted, however, that there are still large disagreements in quantifying the effects of irrigation on regional and global temperature (see, e.g., Boucher et al., 2004 vs. Lobell et al., 2006), mainly attributed to the difference in the implemented irrigation demand calculations. Sacks et al. (2009) tried to overcome the limitations in demand algorithms by downscaling the AQUASTAT irrigative water use data to the grid scale. They concluded that irrigation has significant importance for regional temperature, but at global scale the temperature cooling in some regions due to irrigation is canceled by temperature warming in some other areas due to climate, land-cover and circulation changes. There are, however, some limitations in their study, as the irrigation demand did not vary between years and they applied irrigation only when the LAI is around 80 % of the annual LAI. These assumptions can result in large uncertainty.

Irrigation-induced precipitation has been studied for quite some time and irrigation has been shown to have a significant effect on local and regional precipitation patterns (e.g., Barnston and Schickedanz, 1984; Moore and Rojstaczer, 2001). For instance despite regional decline, Tuinenberg et al. (2011) found a positive precipitation trend in climate stations located in the irrigated regions of the southern Asia. Lucas-Picher et al. (2011) tested four climate models and argued that lack of representation of irrigation is the main reason for precipitation bias over the Indian monsoon area. Guimberteau et al. (2012) showed that irrigation can also affect the onset of mean monsoon date over the Indian peninsula, leading to a significant decrease in precipitation during May to July. Nonetheless, there are still large disagreements in (1) identifying the dominant mechanisms that drive the irrigation-induced precipitation; and (2) estimating the amount and spatial extension of change in precipitation. DeAngelis et al. (2010) noted that the growing season precipitation increased in the Great Plains of the US during the 20th century as a result of intensive irrigation. Using vapor tracking analysis, they indicated that evaporation from irrigated lands adds to downwind precipitation, which increases as the evaporation increases. Harding and Snyder (2012a, b), however, noted that the extent of effects on precipitation also depend on the antecedent soil moisture. They argued that, in low soil moisture conditions, further irrigation can result in suppression of regional precipitation. Guimberteau

et al. (2012) argued that these contrasting results might be due to differences in local moisture, where the irrigation is applied. Based on a 30-year simulation, they showed an increase in summer precipitation over the arid western region of the Mississippi River basin in association with enhanced evapotranspiration. However, a decrease in precipitation was identified over the wet eastern part of the basin. These results, however, are based on only one set of models and the coarse grid resolution might degrade the quality of simulations – see the discussion below. With respect to the scale of disturbance, Sorooshian et al. (2011) showed that irrigation over California's Central Valley significantly decreases local temperature and increases local precipitation; however, they argued that the effects of irrigation do not expand far from the place where irrigation takes place. In contrast, Lo and Famiglietti (2013) argued that irrigation in California's Central Valley intensifies the water cycle in the southwestern US and can increase the flow in the Colorado River.

There are two main limitations associated with available simulations of irrigation-induced rainfall discussed above. First, in most of the online studies, water availability is not a constraint. As a result, the water balance is not closed and they simply analyze whether evaporation increase can enhance atmospheric moisture convergence or not. This can be considered as a major limitation as the available water can control the extent of irrigation (and consequently evaporation) and stabilize the associated feedback processes. Second, it is known that sharp landscape contrasts (i.e., transitions between wet and cool as well as dry and hot areas) critically affect rainfall formation (e.g., Taylor 2009; C. M. Taylor et al., 2012). Although irrigation can create such transitions due to enhanced evaporation and decreased surface temperature, current LSMs are generally unable to generate the atmospheric perturbations due to these transitions (Polcher, 2014). Due to these limitations, the results of current sensitivity analyses should be considered with caution.

Online simulations under future climate change are limited and have been performed mainly at regional scales. Gerten et al. (2011) used a nested regional climate model to dynamically downscale the future simulations of a global climate model over southern Asia and considered two modes of simulation: with or without irrigation. They concluded that including irrigation can result in roughly half of the temperature increase predicted without representing irrigation. With respect to future precipitation, simulation with and without irrigation both showed a decrease in precipitation over northern India and increase in precipitation over the southern peninsula; the latter was enhanced with irrigation. They noted that the increase in precipitation cannot be seen if the global-scale simulations are not dynamically downscaled. This highlights the importance of including irrigation schemes in regional climate models for dynamic downscaling of future climate change scenarios.

In summary, despite current limitations and differences in the host climate and LSM models, irrigation demand algorithms and simulation settings, significant feedback effects are associated with irrigation. Large uncertainties, however, exist in current coupled irrigation–land-surface–climate modeling, which emphasizes the need for more research in this area.

5.2 Offline representation

Offline representation of water demands is more common, and a wide variety of GHMs and LSMs in conjunction with different demand algorithms have been used to simulate the dynamics of water demand under both current and future conditions. The available global simulations under current conditions are compared and summarized in Wada et al. (2014) and Chaturvedi et al. (2013a, b) for irrigative demands and in Alcamo et al. (2003) and Hejazi et al. (2013b) for total water consumption. Although incorporating water demand calculations can generally result in more realistic river discharge simulations (see Ngo-Duc et al., 2005a, b, 2007), current simulations exhibit large differences in estimates of water demand and use at countrywide, continental and global scales. This can be referred to the differences in data support, demand calculation schemes and host models – see the discussion of Sect. 6.

Normally, future projections of water demands include more uncertainty than simulation of current conditions as they are also conditioned on uncertain climate futures and/or socio-economic and technological scenarios. Considering future climate projections, with or without considering irrigation expansion, irrigation demand algorithms have mainly projected increase in irrigation demand under climate change scenarios. As an early example, Fischer et al. (2007) estimated irrigation water requirement as a function of both projected irrigated land and climate change from 1990 to 2080. They showed that the impact of climate change on increasing irrigation water requirement could be nearly as large as the changes initiated by socio-economic developments. There are, however, two sets of uncertainty associated with future projections of irrigation demand. First, gridded climate products have significant deficiencies in representing current and future climate, particularly with respect to precipitation (e.g., Lorenz and Kunstmann, 2012; Grey et al., 2013). This can further propagate to estimation of irrigation demand at the sub-grid scale. Second, there are large disagreements between irrigation demand projections with respect to different climate model simulation, irrigation algorithms and host large-scale models. One possible approach to account for these uncertainties would be using a multi-model approach, as recommended by Gosling et al. (2011) and Haddeland et al. (2011, 2014) and implemented to some extent by Wada et al. (2013) and Rosenzweig et al. (2014). Based on the latest IPCC climate scenarios (K. E. Taylor et al., 2012), these studies generally concluded that a significant increase in future demand is likely, with possibly 1-month or more shift in the peak irrigation demand in mid-latitude regions

(Wada et al., 2013), but large uncertainties are associated with the predictions (see Rosenzweig et al., 2014). Moreover, both studies noted that CO_2 increases might have beneficial effects on crop transpiration efficiency, if other factors are not limiting (see also Gerten et al., 2011; Konzmann et al., 2013). Nonetheless, it still remains unclear whether increased transpiration efficiency is canceled out by increased transpiration due to increasing biomass and plant growth. More studies, therefore, are required in this direction (see Gerten, 2013). This is a context for which LSMs can offer an ideal platform as they have the explicit modules required for considering dynamic interactions of carbon, vegetation and water – see the discussion of Sect. 6.

Similar conclusions were obtained with respect to non-irrigative demands. Alcamo et al. (2007) and Hejazi et al. (2013d) showed that increasing domestic and industrial water uses, if not controlled, can be a major threat for water supply. There are, however, large discrepancies between different projections of non-irrigative demands (Gleick, 2003), in which the divergence between modeling results becomes more highlighted as the projection horizon increases (see Davis et al. (2013) for electrical demand and associated water use). These uncertainties can be referred to limitations in current data availability for supporting robust and reliable projections, differences in socio-economic and technological scenarios, as well as some underlying assumptions in demand calculation algorithms, which can limit their efficiency in future simulations.

As the current global potential for expanding water demand is rather limited (Rost et al., 2009; Gerten and Rost, 2010), adaptation and mitigation strategies are required to moderate human water demands. In such cases prescribed "policy" scenarios can be introduced into large-scale models for impact assessment. Using this approach, it has been shown that mitigation can significantly decrease future global water demand. For example, Hanasaki et al. (2013a) showed approximately 7-fold and 2.5-fold variation in industrial and municipal demands, depending on the SSP considered. The effects of mitigation, however, have large regional variation. For irrigative demands, Fischer et al. (2007) showed that some regions may be negatively affected by mitigation actions, which depend on specific combinations of CO_2 changes that affect crop water requirement and projected precipitation and temperature changes. Kyle et al. (2013) showed that applying CO_2 mitigation policies can result in high deployment of other high-tech solutions for electrical generation (e.g., solar power) that have low water requirements. Hejazi et al. (2013c) further showed that taxation can be an important factor in mitigating the effect of water scarcity by regulating more water efficient options for irrigation. Hejazi et al. (2013a) further showed the possibility of a slight decrease in municipal withdrawals in the year 2100 under a high-tech scenario, despite significant population growth. Davies et al. (2013) showed similar results for electricity water withdrawals if high-tech solutions are employed. Large-scale models also showed that promoting international trade can be a strong adaptation option for controlling regional demand, in which water-limited regions can import water-expensive products from other areas (e.g., Siebert and Döll, 2010; Hanasaki et al., 2010; Konar et al., 2013). Assessment of trade scenarios and water footprinting, however, needs detailed tracking of the water cycle (see Chenoweth et al., 2014) and is highly dependent on how reasonable the human demands and production, as well as water availability and water allocation, are described in time and space. Such a level of accuracy is currently not available and therefore the assessments remain widely uncertain.

In summary, current offline projections agree on large impacts of future change in climate, socio-economy and technology on water demands and the importance of adaptation and mitigation strategies for managing future water security threats. Available projections, however, are rather limited and suffer from major sources of uncertainty, which is revealed by large discrepancies between different simulation products under current and future conditions. We now turn to discuss these gaps in more detail and identify the research needs and priorities.

6 Discussion

Major gaps remain in the current capability in modeling water demands and understanding their online and offline impacts on the terrestrial water cycle and human livelihoods. These gaps are partially due to inherent complexity in modeling Earth system processes, which is more significant in coupled simulation modes. Apart from various computational barriers, one main challenge in online simulations is the uncertainty associated with coupling land and atmospheric models, as given a unique land-surface boundary condition, the simulations obtained by different climate models can be divergent (Koster et al., 2004; Pitman et al., 2009; Dadson et al., 2013). Another major challenge for coupled irrigation–land-surface–climate simulations is the choice of appropriate temporal and spatial resolutions, at which the relevant physical processes and feedbacks between land and atmosphere should be represented and described. Ideally, the optimal modeling resolution should be identified based on physical realism; nonetheless, the choice of resolution in coupled simulations is mainly constrained by computational resources, data availability and the complexity supported by the LSMs. If these are not limiting factors, it has been shown that finer temporal and spatial resolutions can improve online representation of irrigation. For instance, using six different combinations of temporal/spatial resolutions, Sorooshian et al. (2011) concluded that spatial and temporal resolution in coupled irrigation–land–climate models can significantly change both temperature and precipitation simulations over irrigated grids and a fine level of detail is required for representing the physical processes controlling the feedbacks

between irrigation and atmosphere. However, these findings remain regionally and seasonally dependent and are closely linked to the level of complexity supported in the considered irrigation parameterization and host model. It should be noted that, by increasing the spatial resolution, more processes need to be included in order to ensure water conservation within the model and that can further complicate the issues related to water availability – see the discussion below. The effects of fine modeling resolution seem to be in general less significant in offline runs, as far as the evaporation calculation is consistent with estimation of crop water requirements and each crop is supplied by a unique moisture reservoir. Compton and Best (2011) conducted offline global simulations and showed that fine spatial resolution has little importance on long-term modeling of evaporation and runoff; however, the temporal resolution does change the mean evaporation/runoff balance. The issues around modeling resolution are explored further in Nazemi and Wheater (2015).

Large uncertainties are also associated with offline human water demand simulations under current and future conditions. Lissner et al. (2012), for instance, noticed significant difference in terms of water demand per capita between the simulated products of WaterGAP and reported AQUASTAT data. These uncertainties are mainly related to (i) available data support, (ii) demand calculation algorithms and (iii) host models. These sources are widely connected and cannot be easily addressed and quantified independently. Here we briefly discuss these sources and propose few directions for future developments.

1. Uncertainty in current data support: primarily, there are considerable uncertainties across the input and forcing data required for executing large-scale models. Generally, large-scale models discussed in this paper are forced and initialized using various data sources that are developed and maintained independently. This results in major inconsistencies, particularly at the grid scale, where it is often the case that information coming from different sources does not match each other (e.g., soil properties do not fit to land use). Typically, modelers fix these issues by applying simple rules or assumptions; however, inconstancies in personal judgments can highly affect the quality of simulations at the local and regional scales (see Bormann et al., 2011, for a local study). Major uncertainties are also associated with the data required for executing demand calculation algorithms. Siebert et al. (2005) noted that even the locations of irrigation districts are uncertain in many regions and sub-grid variability of crops within irrigated areas is not generally available. Wisser et al. (2008) argued that major uncertainties are associated with forcing, irrigation and crop maps, and this can result in large differences between simulations of irrigation water requirement. Another source of data uncertainty is the generally sparse information on irrigation techniques.

This can be important for understanding the amount of water loss and thus estimating the actual irrigation use and evaporation (see, e.g., Evan and Zaitchik, 2008). The issues around data support apply to non-irrigative demands as well. For the case of water use for electricity generation in the US, Macknick et al. (2011) noted that "federal data sets on water use in power plants have numerous gaps and methodological inconsistencies." Data uncertainty can propagate into structural and parametric identification during model development and can further extend to future projections. The availability of different sources of global and regional data has resulted in emergence of various data sets, with varying degrees of quality, which can potentially support demand calculation algorithms. At this stage of research, the various data sets have not been systematically compared with respect to their uncertainty and the associated effects on demand simulations. This is a major need for future exploration.

2. Uncertainty in demand calculation algorithms: this includes both irrigative and non-irrigative demands.

    a. Irrigative demand: limitations in current algorithms mainly include the uncertainty in describing the crop moisture requirements in time and space and constraining the irrigation to water availability. If the irrigation is limited by the water available at the grid scale, then the quality of simulation is hindered by the ability of the host model to describe water allocation from surface and groundwater resources (see Nazemi and Wheater, 2015). In addition, current bottom-up algorithms do not appropriately consider plant-specific water requirements at the sub-grid scale due to missing soil and crop diversity. This can result in misestimating the irrigation demand. In the best situation, where the same assumption is used for the calculation of the crop evaporation and the irrigation demand, the uncertainty of the irrigative demand is the same as evaporation, but this can still vary greatly across various host models. Considering future simulations, widely used irrigation demand estimates based on FAO guidelines often require several input variables (see, e.g., Farmer et al., 2011 and Hejazi et al., 2013b, for simplifications), and given the need for downscaling of climate variables for future simulations, these can be outperformed by simpler models (e.g., Vörösmarty et al., 1998; Wisser et al., 2010). At the current stage of research, different methods for calculating irrigative demand have not yet been fully intercompared to identify appropriate algorithms with respect to region, climate and type of crops. This can be considered as an important need for further research. Another avenue for future development is improving the demand simulations us-

ing data assimilation and model calibration. These opportunities will be discussed further in Nazemi and Wheater (2015).

    b. Non-irrigative demand: the current offline modeling capability is generally temporally coarse, and available downscaling and projection algorithms mainly do not account for seasonal variations in water demand. There are also parametric and structural uncertainties in functional mappings that link water demand to socio-economic and technological proxies due to limitations in available data as well as the diversity and spatiotemporal variability in non-irrigative demands. At this stage, it is not fully understood how these uncertainties propagate into future projections considering additional uncertainty in future climate and socio-economic scenarios. Developing robust downscaling and projection algorithms for estimation of non-irrigative demands therefore is an important need for future development.

3. Uncertainty in host models: host models can add substantial uncertainty to demand simulations, particularly for irrigation. As noted in Sect. 3, the calculation of irrigation demand involves solving the soil water balance at every simulation time step and this is determined by how the relevant natural processes, such as actual evapotranspiration and soil moisture, are parameterized in the host model. Haddeland et al. (2011) showed major differences in the global simulations obtained from six LSMs and five GHMs due to differences in underlying assumptions, process representations, and related parameterizations. It is also shown that considering feedback effects between irrigation and atmosphere can considerably change potential evaporation (e.g., Blyth and Jacobs, 2011; Lu, 2013); therefore offline irrigation demand simulations based on GHMs might be biased as they inherently ignore climate feedbacks. Moreover, GHMs often cannot represent important processes such as the effects of increased carbon concentration on irrigation demand. This limitation may result in major deficiencies in simulating climate change scenarios as CO_2 increases can significantly change vegetation dynamics (e.g., Prudhomme et al., 2014), which can further alter the evaporation and runoff regimes (Gerten et al., 2004). From this perspective, it can be concluded that online LSMs are superior to GHMs with respect to simulations under increasing CO_2 concentration and future water stress, as they often include many of the required computational components for investigating interactions between climate, carbon, vegetation and water cycles. Efforts are however needed to transfer recent demand calculation algorithms developed in the context of GHMs into LSMs. In addition, although it has been argued that the uncertainties in host models are more

significant than in climate forcing (e.g., Wada et al., 2013), uncertainties in irrigation algorithms and large-scale host models have not been fully disaggregated and distinguished. This requires a "mix and match" of multiple demand algorithms with multiple host models to conduct a systematic intercomparison and sensitivity analysis. This can be considered as an important research direction – see Nazemi and Wheater (2015).

7   Summary and concluding remarks

The terrestrial water cycle has been greatly affected in time and space by human activities during the recent past, to the extent that the current geological era has been named the "Anthropocene". Anthropogenic activities, therefore, are required to be represented in models that are used for impact assessments, large-scale hydrological modeling and land–atmosphere feedback representations. Current human–water interactions are mainly manifested through water resource management, which can be further broken down into two interacting components, related to water demand as well as water supply and allocation. In this paper we considered the representation of water demand in large-scale models. Water demand was further divided into irrigative and non-irrigative categories. We summarized current demand calculation algorithms based on type of demand, modeling procedure and underlying assumptions. Current applications were overviewed; and limitations in knowledge were identified and discussed. Considering current gaps in representing the anthropogenic demands in large-scale models, three main directions are suggested for future developments. These include (1) systematic intercomparisons between different data sets, demand algorithms and host models and associated uncertainties with respect to different geographic regions as well as various socio-economic and climate conditions; (2) developing improved algorithms for calculating both irrigative and non-irrigative demands in time and space considering data limitations as well as diversity and spatiotemporal variability in human demand; and finally (3) transferring the algorithms developed in the context of GHMs to LSMs for (a) improved irrigation demand calculation under increasing CO_2 effects; and (b) further coupled studies with climate models to address various scientific questions with respect to interactions between carbon, irrigation and climate under climate change conditions. Apart from these immediate research needs, efforts are also required to link with socio-economic and energy models to have a full understanding of the dynamic interactions between natural and anthropogenic drivers of human water availability, demand and consumption (Calvin et al., 2013). This seems to be more of a long-term development due to the limitations in current demand algorithms, LSMs as well as socio-economic and energy models.

As a final remark, it must be noted that the effects of water demand on both the terrestrial water cycle and water security cannot be fully studied unless considered in conjunction with water supply and allocation, which determine the extent of human intervention in water cycle. This is particularly important for future predictions, as increasing water scarcity is a major limiting factor for water demand and can substantially increase competition over available water sources. In Nazemi and Wheater (2015), we review how water supply and allocation have been represented at larger scales and been integrated with various water demands and natural land-surface processes at grid and sub-grid scales.

Acknowledgements. The first author would like to thank Amir Aghakouchak for his valuable inputs during the early stages of this survey. The authors gratefully acknowledge the constructive comments from the editor, Jan Polcher, Ingjerd Haddeland, Jason Evans and two anonymous reviewers, which have enabled us to make significant improvements to this paper. Financial support for this study was provided by the Canada Excellence Research Chair in Water Security at the University of Saskatchewan.

Edited by: W. Buytaert

References

Abdullah, K. B.: Use of water and land for food security and environmental sustainability, Irrig. Drain., 55, 219–222, doi:10.1002/ird.254, 2006.

Adam, J. C. and Lettenmaier, D. P.: Adjustment of global gridded precipitation for systematic bias, J. Geophys. Res., 108, 4257, doi:10.1029/2002JD002499, 2003.

Adam, J. C., Haddeland, I., Su, F., and Lettenmaier, D. P.: Simulation of reservoir influences on annual and seasonal streamflow changes for the Lena, Yenisei and Ob' rivers, J. Geophys. Res.-Atmos., 112, D24114, doi:10.1029/2007JD008525, 2007.

Adegoke, J. O., Pielke Sr., R. A., Eastman, J., Mahmood, R., and Hubbard, K. G.: Impact of irrigation on midsummer surface fluxes and temperature under dry synoptic conditions: A regional atmospheric model study of the US High Plains, Mon. Weather Rev., 131, 556–564, 2003.

AghaKouchak, A., Norouzi, H.-R., Madani, K., Mirchi, A., Azarderakhsh, M., Nazemi, A., Nasrollahi, N., Farahmand, A.-R., Mehran, A., and Hasanzadeh, E.: Aral Sea syndrome desiccates Lake Urmia: Call for action, Journal of Great Lakes Research, doi:10.1016/j.jglr.2014.12.007, in press, 2014.

Alcamo, J., Döll, P., Kaspar, F., and Siebert, S.: Global change and global scenarios of water use and availability: an application of WaterGAP 1.0, Center for Environmental Systems Research (CESR), University of Kassel, Germany, available at: http://www.usf.uni-kassel.de/usf/archiv/dokumente/projekte/watergap.teil1.pdf (last access: 6 May 2014), 1997.

Alcamo, J., Döll, P., Henrichs, T., Kaspar, F., Lehner, B., Rösch, T., and Siebert, S.: Development and testing of the WaterGAP 2 global model of water use and availability, Hydrolog. Sci. J., 48, 317–337, 2003.

Alcamo, J., Flörke, M., and Märker, M.: Future long-term changes in global water resources driven by socio-economic and climatic changes, Hydrolog. Sci. J., 52, 247–275, 2007.

Allen, R. G., Pereira, L. S., Raes, D., and Smith, M.: Crop evapotranspiration-Guidelines for computing crop water requirements-FAO Irrigation and drainage paper 56, FAO, Rome, http://www.engr.scu.edu/~emaurer/classes/ceng140_watres/handouts/FAO_56_Evapotranspiration.pdf (last access: 6 May 2014), 1998.

Antonellini, M., Mollema, P., Giambastiani, B., Bishop, K., Caruso, L., Minchio, A., Pellegrini, L., Sabia, M., Ulazzi, E., and Gabbianelli, G.: Salt water intrusion in the coastal aquifer of the southern Po Plain, Italy, Hydrogeol. J., 16, 1541–1556, 2008.

Arnell, N. W.: Climate change and global water resources, Global Environ. Change, 9, 31–49, 1999.

Arnell, N. W.: Climate change and global water resources: SRES emissions and socio-economic scenarios, Global Environ. Change, 14, 31–52, 2004.

Arthington, A. H., Bunn, S. E., Poff, N. L., and Naiman, R. J.: The challenge of providing environmental flow rules to sustain river ecosystems, Ecol. Appl., 16, 1311–1318, 2006.

Barella-Ortiz, A., Polcher, J., Tuzet, A., and Laval, K.: Potential evaporation estimation through an unstressed surface-energy balance and its sensitivity to climate change, Hydrol. Earth Syst. Sci., 17, 4625–4639, doi:10.5194/hess-17-4625-2013, 2013.

Barnston, A. G. and Schickedanz, P. T.: The effect of irrigation on warm season precipitation in the southern Great Plains, J. Clim. Appl. Meteorol., 23, 865–888, 1984.

Beddington, J.: Catalysing sustainable water security: role of science, innovation and partnerships, Philos. T. Roy. Soc. A, 371, 414, doi:10.1098/rsta.2012.0414, 2013.

Blanc, E., Strzepek, K., Schlosser, A., Jacoby, H. D., Gueneau, A., Fant, C., Rausch, S., and Reilly, J.: Analysis of U.S. water resources under climate change, MIT Joint Program on the Science and Policy of Global Change, Report No. 239, http://globalchange.mit.edu/files/document/MITJPSPGC_Rpt239.pdf (last access: 6 May 2014), 2013.

Blyth, E. and Jacobs, C.: Including climate feedbacks in regional water resource assessments, WATCH Water and global change, Report No. 38, http://www.eu-watch.org/publications/technical-reports (last access: 6 May 2014), 2011.

Bondeau, A., Smith, P. C., Zaehle, S., Schaphoff, S., Lucht, W., Cramer, W., Gerten, D., Lotze-Campen, H., Müller, C., Reichstein, M., and Smith, B.: Modelling the role of agriculture for the 20th century global terrestrial carbon balance, Global Change Biol., 13, 679–706, doi:10.1111/j.1365-2486.2006.01305.x, 2007.

Bormann, H., Holländer, H. M., Blume, T., Buytaert, W., Chirico, G. B., Exbrayat, J.-F., Gustafsson, D., Hölzel, H., Kraft, P., Krauße, T., Nazemi, A., Stamm, C., Stoll, S., Blöschl, G., and Flühler, H.: Comparative discharge prediction from a small artificial catchment without model calibration: Representation of initial hydrological catchment development, Bodenkultur, 62, 23–29, 2011.

Boucher, O., Myhre, G., and Myhre, A.: Direct human influence of irrigation on atmospheric water vapour and climate, Clim. Dynam., 22), 597–603, 2004.

Bowman, D. M. J. S., Balch, J. K., Artaxo, P., Bond, W. J., Carlson, J. M., Cochrane, M. A., D'Antonio, C. M., DeFries, R. S., Doyle, J. C., Harrison, S. P., Johnston, F. H., Keeley, J. E., Krawchuk, M. A., Kull1, C. A., Marston, J. B., Moritz, M. A., Prentice, I. C., Roos, C. I., Scott, A. C., Swetnam, T. W., van der Werf, G. R., and Pyne, S. J.: Fire in the Earth system, Science, 324, 481–484, 2009.

Brovkin, V., Claussen, M., Driesschaert, E., Fichefet, T., Kicklighter, D., Loutre, M.-F., Matthews, H. D., Ramankutty, N., Schaeffer, M., and Sokolov, A.: Biogeophysical effects of historical land cover changes simulated by six Earth system models of intermediate complexity, Clim. Dynam., 26, 587–600, 2006.

Calvin, K., Wise, M., Clarke, L., Edmonds, J., Kyle, P., Luckow, P., and Thomson, A.: Implications of simultaneously mitigating and adapting to climate change: initial experiments using GCAM, Climatic Change, 117, 545–560, 2013.

Cayan, D. R., Das, T., Pierce, D. W., Barnett, T. P., Tyree, M., and Gershunov, A.: Future dryness in the southwest US and the hydrology of the early 21st century drought, P. Natl. Acad. Sci., 107, 21271–21276, 2010.

Chaturvedi, V., Hejazi, M., Edmonds, J., Clarke, L., Kyle, P., Davies, E., and Wise, M.: Climate mitigation policy implications for global irrigation water demand, Mitig. Adapt. Strat. Global Change, 18, 1–19, 2013a.

Chaturvedi, V., Hejazi, M., Edmonds, J., Clarke, L., Kyle, P., Davies, E., Wise, M., and Calvin, K. V.: Climate Policy Implications for Agricultural 5 Water Demand, Pacific Northwest National Laboratory, Richland, WA, available at: http://www.globalchange.umd.edu/wp-content/uploads/projects/PNNL-22356.pdf (last access: 6 May 2014), 2013b.

Chen, F. and Dudhia, J.: Coupling an advanced land surface-hydrology model with the Penn State-NCAR MM5 modeling system Part I: Model implementation and sensitivity, Mon. Weather Rev., 129, 569–585, 2001a.

Chen, F. and Dudhia, J.: Coupling an advanced land surface-hydrology model with the Penn State-NCAR MM5 modeling system Part II: Preliminary model validation, Mon. Weather Rev., 129, 587–604, 2001b.

Chenoweth, J., Hadjikakou, M., and Zoumides, C.: Quantifying the human impact on water resources: a critical review of the water footprint concept, Hydrol. Earth Syst. Sci., 18, 2325–2342, doi:10.5194/hess-18-2325-2014, 2014.

CIA: CIA World Factbook [CD-ROM], Washington, D.C., https://www.cia.gov/library/publications/the-world-factbook (last access: 6 May 2014), 2001.

CIAT: Gridded Population of the World, Version 3 (GPWv3): Population Density Grid, NASA Socioeconomic Data and Applications Center (SEDAC), http://sedac.ciesin.columbia.edu/data/set/gpw-v3-population-density (last access: 6 May 2014), 2005.

Claussen, M.: Earth system models, in: Understanding the Earth System: Compartments, Processes and Interactions, edited by: Ehlers, E., and Krafft, T., Springer-Verlag, Heidelberg, 145–162, 2001.

Cole, M. A.: Economic growth and water use, Appl. Econ. Lett., 11, 1–4, 2004.

Collins, W. D., Rasch, P. J., Boville, B. A., Hack, J. J., McCaa, J. R., Williamson, D. L., Briegleb, B., Bitz, C., Lin, S.-J., Zhang, M., and Dai, Y.: Description of the NCAR community atmosphere model (CAM 3.0), NCAR Tech. Note NCAR/TN-464+STR, 226, available at: http://hanson.geog.udel.edu/hanson/hanson/

CLD_GCMExperimentS11_files/description.pdf (last access: 6 May 2014), 2004.

Collins, W. D., Rasch, P. J., Boville, B. A., Hack, J. J., McCaa, J. R., Williamson, D. L., Briegleb, B., Bitz, C., Lin, S.-J., and Zhang, M.: The formulation and atmospheric simulation of the Community Atmosphere Model version 3 (CAM3), J. Climate, 19, 2144–2161, 2006.

Compton, E. and Best, M.: Impact of spatial and temporal resolution on modelled terrestrial hydrological cyce components, WATCH Water and global change. Report No. 44, http://www.eu-watch.org/publications/technical-reports (last access: 6 May 2014), 2011.

Cosgrove, B. A., Lohmann, D., Mitchell, K. E., Houser, P. R., Wood, E. F., Schaake J. C., Robock, A., Marshall, C., Sheffield, J., Duan, Q., Luo, L., Wayne Higgins, R., Pinker R. T., Dan Tarpley, J., and Meng, J.: Real-time and retrospective forcing in the North American Land Data Assimilation System (NLDAS) project, J. Geophys. Res., 108, 8842, doi:10.1029/2002JD003118, 2003.

Crutzen, P. J.: The "anthropocene", in: Earth System Science in the Anthropocene, edited by: Ehlers, E., Krafft, T., and Moss, C., Springer, Berlin, Heidelberg, 13–18, 2006.

Crutzen, P. J. and Steffen, W.: How long have we been in the Anthropocene era?, Climatic Change, 61, 251–257, 2003.

Dadson, S., Acreman, M., and Harding, R.: Water security, global change and land–atmosphere feedbacks, Philos. T. Roy. Soc. A, 371, 2002, doi:10.1098/rsta.2012.0412, 2013.

Davies, E. G., Kyle, P., and Edmonds, J. A.: An integrated assessment of global and regional water demands for electricity generation to 2095, Adv. Water Resour., 52, 296–313, 2013.

DeAngelis, A., Dominguez, F., Fan, Y., Robock, A., Kustu, M. D., and Robinson, D.: Evidence of enhanced precipitation due to irrigation over the Great Plains of the United States, J. Geophys. Res., 115, D15115, doi:10.1029/2010JD013892, 2010.

Deardorff, J. W.: Efficient prediction of ground surface temperature and moisture, with inclusion of a layer of vegetation, J. Geophys. Res.-Oceans, 83, 1889–1903, 1978.

Dee, D. P., Uppala, S. M., Simmons, A. J., Berrisford, P., Poli, P., Kobayashi, S., Andrae, U., Balmaseda, M. A., Balsamo, G., Bauer, P., Bechtold, P., Beljaars, A. C. M., van de Berg, L., Bidlot, J., Bormann, N., Delsol, C., Dragani, R., Fuentes, M., Geer, A. J., Haimberger, L., Healy, S. B., Hersbach, H., Hólm, E. V., Isaksen, L., Kållberg, P., Köhler, M., Matricardi, M., McNally, A. P., Monge-Sanz, B. M., Morcrette, J.-J., Park, B.-K., Peubey, C., de Rosnay, P., Tavolato, C., Thépaut, J.-N., and Vitart, F.: The ERA-interim reanalysis: configuration and performance of the data assimilation system, Q. J. Roy. Meteorol. Soc., 137, 553–597, doi:10.1002/qj.828, 2011.

de Rosnay, P., Polcher, J., Laval, K., and Sabre, M.: Integrated parameterization of irrigation in the land surface model ORCHIDEE: Validation over Indian Peninsula, Geophys. Res. Lett., 30, 1986, doi:10.1029/2003GL018024, 2003.

Destouni, G., Asokan, S. M., and Jarsjö, J.: Inland hydro-climatic interaction: Effects of human water use on regional climate, Geophys. Res. Lett., 37, L18402, doi:10.1029/2010GL044153, 2010.

Dickinson, R. E.: Land surface processes and climate-surface albedos and energy balance, Adv. Geophys., 25, 305–353, 1983.

Dickinson, R. E.: Modeling evapotranspiration for three-dimensional global climate models, Geophys. Monogr. Ser., 29, 58–72, 1984.

Döll, P.: Vulnerability to the impact of climate change on renewable groundwater resources: a global-scale assessment, Environ. Res. Lett., 4, 035006, doi:10.1088/1748-9326/4/3/035006, 2009.

Döll, P. and Siebert, S.: A digital global map of irrigated areas, ICID J., 49, 55–66, 2000.

Döll, P. and Siebert, S.: Global modeling of irrigation water requirements, Water Resour. Res., 38, 8-1–8-10, doi:10.1029/2001WR000355, 2002.

Döll, P., Fiedler, K., and Zhang, J.: Global-scale analysis of river flow alterations due to water withdrawals and reservoirs, Hydrol. Earth Syst. Sci., 13, 2413–2432, doi:10.5194/hess-13-2413-2009, 2009.

Döll, P., Müller Schmied, H., Schuh, C., Portmann, F. T., and Eicker, A.: Global-scale assessment of groundwater depletion and related groundwater abstractions: Combining hydrological modeling with information from well observations and GRACE satellites. Water Resour. Res., 50, 5698–5720, doi:10.1002/2014WR015595, 2014.

Douglas, E. M., Beltrán-Przekurat, A., Niyogi, D., Pielke Sr., R. A., and Vörösmarty, C. J.: The impact of agricultural intensification and irrigation on land–atmosphere interactions and Indian monsoon precipitation – A mesoscale modeling perspective, Global Planet. Change, 67, 117–128, 2009.

Ducoudré, N. I., Laval, K., and Perrier, A.: SECHIBA, a new set of parameterizations of the hydrologic exchanges at the land-atmosphere interface within the LMD atmospheric general circulation model, J. Climate, 6, 248–273, 1993.

Ek, M. B., Mitchell, K. E., Lin, Y., Rogers, E., Grunmann, P., Koren, V., Gayno, G., and Tarpley, J. D.: Implementation of Noah land surface model advances in the National Centers for Environmental Prediction operational mesoscale Eta model, J. Geophys. Res., 108, 8851, doi:10.1029/2002JD003296, 2003.

Eltahir, E. A.: A soil moisture–rainfall feedback mechanism: 1. Theory and observations, Water Resour. Res., 34, 765–776, 1998.

Entekhabi, D. and Eagleson, P. S.: Land surface hydrology parameterization for atmospheric general circulation models including subgrid scale spatial variability, J. Climate, 2, 816–831, 1989.

Evans, J. P. and Zaitchik, B. F.: Modeling the large-scale water balance impact of different irrigation systems, Water Resour. Res., 44, W08448, doi:10.1029/2007WR006671, 2008.

Falkenmark, M.: Growing water scarcity in agriculture: future challenge to global water security, Philos. T. Roy. Soc. A, 371, 2002, doi:10.1098/rsta.2012.0410, 2013.

Farmer, W., Strzepek, K., Schlosser, C. A., Droogers, P., and Gao, X.: A Method for Calculating Reference Evapotranspiration on Daily Time Scales, MIT Joint Program on the Science and Policy of Global Change, Report number 195, http://18.7.29.232/handle/1721.1/61773 (last access: 6 May 2014), 2011.

Fischer, G., Tubiello, F. N., Van Velthuizen, H., and Wiberg, D. A.: Climate change impacts on irrigation water requirements: effects of mitigation, 1990–2080, Technol. Forecast. Social Change, 74, 1083–1107, 2007.

Flörke, M. and Alcamo, J.: European outlook on water use, Final Report, EEA/RNC/03/007, Center for Environmental Systems Research, University of Kassel, http://www.improve.novozymes.

com/Documents/European_OutlookonWaterUse.pdf, (last access: 6 May 2014), 2004.

Flörke, M., Kynast, E., Bärlund, I., Eisner, S., Wimmer, F., and Alcamo, J.: Domestic and industrial water uses of the past 60 years as a mirror of socio-economic development: A global simulation study, Global Environ. Change, 23, 144–156, 2013.

Förster, H. and Lilliestam, J.: Modeling thermoelectric power generation in view of climate change, Reg. Environ. Change, 10, 327–338, 2010.

Friedl, M. A., McIver, D. K., Hodges, J. C., Zhanga, X. Y., Muchoneyb, D., Strahlera, A. H., Woodcocka, C. E., Gopala, S., Schneidera, A., Coopera, A., Baccinia, A., Gaoa, F., and Schaafa, C.: Global land cover mapping from MODIS: algorithms and early results, Remote Sens. Environ., 83, 287–302, 2002.

Gao, H., Birkett, C., and Lettenmaier, D. P.: Global monitoring of large reservoir storage from satellite remote sensing, Water Resour. Res., 48, W09504, doi:10.1029/2012WR012063, 2012.

Gaybullaev, B., Chen, S. C., and Kuo, Y. M.: Large-scale desiccation of the Aral Sea due to over exploitation after 1960, J. Mount. Sci., 9, 538–546, 2012.

Gerten, D.: A vital link: water and vegetation in the Anthropocene, Hydrol. Earth Syst. Sci., 17, 3841–3852, doi:10.5194/hess-17-3841-2013, 2013.

Gerten, D. and Rost, S.: Climate change impacts on agricultural water stress and impact mitigation potential, World Bank, Washington, D.C., USA, http://siteresources.worldbank.org/INTWDR2010/Resources/5287678-1255547194560/WDR2010_BGNoteGerten.pdf (last access: 3 November 2014), 2010.

Gerten, D., Schaphoff, S., Haberlandt, U., Lucht, W., and Sitch, S.: Terrestrial vegetation and water balance – hydrological evaluation of a dynamic global vegetation model, J. Hydrol., 286, 249–270, 2004.

Gerten, D., Hagemann, S., Biemans, H., Saeed, F., and Konzmann, M.: Climate Change and Irrigation: Global Impacts and Regional Feedbacks, WATCH Technical Report Number 47, http://www.eu-watch.org/publications/technical-reports (last access: 6 May 2014), 2011.

GEWEX: GEWEX plans for 2013 and beyond - GEWEX science questions (version 1), GEWEX document series No. 2012-2, http://www.gewex.org/pdfs/GEWEXScience_Questionsfinal.pdf, (last access: 6 May 2014), 2012.

Giordano, M.: Global groundwater? Issues and solutions, Annu. Rev. Environ. Resour., 34, 153–178, 2009.

Gleeson, T., Wada Y., Bierkens, M. F., and van Beek, L. P.: Water balance of global aquifers revealed by groundwater footprint, Nature, 488, 197–200, 2012.

Gleick, P. H.: Basic water requirements for human activities: Meeting basic needs, Water Int., 21, 83–92, 1996.

Gleick, P. H.: Water use, Annual Rev. Environ. Resour., 28, 275–314, 2003.

Gleick, P. H., Cooley, H., Famiglietti, J. S., Lettenmaier, D. P., Oki, T., Vörösmarty, C. J., and Wood, E. F.: Improving Understanding of the Global Hydrologic Cycle, in: Climate Science for Serving Society, edited by: Asrar, G. R. and Hurrell, J. W., Springer Netherlands, 151–184, 2013.

Gober, P. and Wheater, H. S.: Socio-hydrology and the science–policy interface: a case study of the Saskatchewan River basin,

Hydrol. Earth Syst. Sci., 18, 1413–1422, doi:10.5194/hess-18-1413-2014, 2014.

Gosling, S. N., Taylor, R. G., Arnell, N. W., and Todd, M. C.: A comparative analysis of projected impacts of climate change on river runoff from global and catchment-scale hydrological models, Hydrol. Earth Syst. Sci., 15, 279–294, doi:10.5194/hess-15-279-2011, 2011.

Grey, D., Garrick, D., Blackmore, D., Kelman, J., Muller, M., and Sadoff, C.: Water security in one blue planet: twenty-first century policy challenges for science, Philos. T. Roy. Soc. A, 371, 2002, doi:10.1098/rsta.2012.0406, 2013.

Gueneau, A., Schlosser, C. A., Strzepek, K. M., Gao, X., and Monier, E.: CLM-AG: An Agriculture Module for the Community Land Model version 3.5, MIT Joint Program on the Science and Policy of Global Change, http://dspace.mit.edu/handle/1721.1/73007 (last access: 6 May 2014), 2012.

Guimberteau, M., Laval, K., Perrier, A., and Polcher, J.: Global effect of irrigation and its impact on the onset of the Indian summer monsoon, Clim. Dynam., 39, 1329–1348, doi:10.1007/s00382-011-1252-5, 2012.

Haddeland, I., Lettenmaier, D. P., and Skaugen T.: Effects of irrigation on the water and energy balances of the Colorado and Mekong river basins, J. Hydrol., 324, 210–223, 2006.

Haddeland, I., Skaugen, T., and Lettenmaier, D. P.: Hydrologic effects of land and water management in North America and Asia: 1700–1992, Hydrol. Earth Syst. Sci., 11, 1035–1045, doi:10.5194/hess-11-1035-2007, 2007.

Haddeland, I., Clark, D. B., Franssen, W., Ludwig, F., Voß, F., Arnell, N. W., Bertrand, N., Best, M., Folwell, S., Gerten, D., Gomes, S., Gosling, S. N., Hagemann, S., Hanasaki, N., Harding, R., Heinke, J., Kabat, P., Koirala, S., Oki, T., Polcher, J., Stacke, T., Viterbo, P., Weedon, G. P., and Yeh, P.: Multimodel estimate of the global terrestrial water balance: setup and first results, J. Hydrometeorol., 12, 869–884, doi:10.1175/2011JHM1324.1, 2011.

Haddeland, I., Biemans, H., Eisner, S., Flörke, M., Hanasaki, N., Konzmann, M., Ludwig, F., Masaki, Y., Schewe, J., Stacke, T., Tessler, Z. D., Wada, Y., and Wisser, D.: Global water resources affected by human interventions and climate change, P. Natl. Acad. Sci., 111, 3251–3256, doi:10.1073/pnas.1222475110, 2014.

Hagemann, S. and Dümenil, L.: A parameterization of the lateral waterflow for the global scale, Clim. Dynam., 14, 17–31, doi:10.1007/s003820050205, 1997.

Hanasaki, N., Kanae, S., and Oki, T.: A reservoir operation scheme for global river routing models, J. Hydrol., 327, 22–41, 2006.

Hanasaki, N., Kanae, S., Oki, T., Masuda, K., Motoya, K., Shirakawa, N., Shen, Y., and Tanaka, K.: An integrated model for the assessment of global water resources – Part 1: Model description and input meteorological forcing, Hydrol. Earth Syst. Sci., 12, 1007–1025, doi:10.5194/hess-12-1007-2008, 2008a.

Hanasaki, N., Kanae, S., Oki, T., Masuda, K., Motoya, K., Shirakawa, N., Shen, Y., and Tanaka, K.: An integrated model for the assessment of global water resources – Part 2: Applications and assessments, Hydrol. Earth Syst. Sci., 12, 1027–1037, doi:10.5194/hess-12-1027-2008, 2008b.

Hanasaki, N., Inuzuka, T., Kanae, S., and Oki, T.: An estimation of global virtual water flow and sources of water withdrawal for major crops and livestock products using a global hydrological model, J. Hydrol., 384, 232–244, 2010.

Hanasaki, N., Fujimori, S., Yamamoto, T., Yoshikawa, S., Masaki, Y., Hijioka, Y., Kainuma, M., Kanamori, Y., Masui, T., Takahashi, K., and Kanae, S.: A global water scarcity assessment under Shared Socio-economic Pathways – Part 1: Water use, Hydrol. Earth Syst. Sci., 17, 2375–2391, doi:10.5194/hess-17-2375-2013, 2013a.

Hanasaki, N., Fujimori, S., Yamamoto, T., Yoshikawa, S., Masaki, Y., Hijioka, Y., Kainuma, M., Kanamori, Y., Masui, T., Takahashi, K., and Kanae, S.: A global water scarcity assessment under Shared Socio-economic Pathways – Part 2: Water availability and scarcity, Hydrol. Earth Syst. Sci., 17, 2393–2413, doi:10.5194/hess-17-2393-2013, 2013b.

Harding, K. J. and Snyder, P. K.: Modeling the Atmospheric Response to Irrigation in the Great Plains, Part I: General Impacts on Precipitation and the Energy Budget, J. Hydrometeorol., 13, 1667–1686, 2012a.

Harding, K. J. and Snyder, P. K.: Modeling the atmospheric response to irrigation in the Great Plains, Part II: The precipitation of irrigated water and changes in precipitation recycling, J. Hydrometeorol., 13, 1687–1703, 2012b.

Hejazi, M. I., Edmonds, J., Chaturvedi, V., Davies, E., and Eom, J.: Scenarios of global municipal water-use demand projections over the 21st century, Hydrolog. Sci. J., 58, 519–538, 2013a.

Hejazi, M. I., Edmonds, J., Clarke, L., Kyle, P., Davies, E., Chaturvedi, V., Wise, M., Patel, P., Eom, J., and Calvin, K.: Integrated assessment of global water scarcity over the 21st century – Part 1: Global water supply and demand under extreme radiative forcing, Hydrol. Earth Syst. Sci. Discuss., 10, 3327–3381, doi:10.5194/hessd-10-3327-2013, 2013b.

Hejazi, M. I., Edmonds, J., Clarke, L., Kyle, P., Davies, E., Chaturvedi, V., Eom, J., Wise, M., Patel, P., and Calvin, K.: Integrated assessment of global water scarcity over the 21st century – Part 2: Climate change mitigation policies, Hydrol. Earth Syst. Sci. Discuss., 10, 3383–3425, doi:10.5194/hessd-10-3383-2013, 2013c.

Hejazi M. I., Edmonds, J. A., Clarke, L. A., Kyle, G. P., Davies, E., Chaturvedi, V., Wise, M. A., Patel, P. L., Eom, J., Calvin, K. V., Moss, R. H., and Kim, S. H.: Long-term global water projections using six socioeconomic scenarios in an integrated assessment modeling framework, Technol. Forecast. Soc., 81, 205–226, 2013d.

Hobbins, M. T., Dai, A., Roderick, M. L., and Farquhar, G. D.: Revisiting the parameterization of potential evaporation as a driver of long-term water balance trends, Geophys. Res. Lett., 35, L12403, doi:10.1029/2008GL033840, 2008.

Hossain, F., Degu, A. M., Yigzaw, W., Burian, S., Niyogi, D., Shepherd, J. M., and Pielke Sr., R.: Climate Feedback–Based Provisions for Dam Design, Operations, and Water Management in the 21st Century, J. Hydrol. Eng., 17, 837–850, 2012.

Hourdin, F., Musat, I., Bony, S., Braconnot, P., Codron, F., Dufresne, J.-L., Fairhead, L., Filiberti, M.-A., Friedlingstein, P., Grandpeix, J.-Y., Krinner, G., LeVan, P., Li, Z.-X.: The LMDZ4 general circulation model: climate performance and sensitivity to parametrized physics with emphasis on tropical convection, Clim. Dynam., 27, 787–813, doi:10.1007/s00382-006-0158-0, 2006.

Hughes, G., Chinowsky, P., and Strzepek, K.: The costs of adaptation to climate change for water infrastructure in OECD countries, Utilities Policy, 18, 142–153, 2010.

IIASA/FAO: Global Agro-ecological Zones (GAEZ v3.0), IIASA, Laxenburg, Austria and FAO, Rome, Italy, available at: http://webarchive.iiasa.ac.at/Research/LUC/GAEZv3.0/docs/GAEZModelDocumentation.pdf (last access: 15 July 2014), 2012.

IPCC: The IPCC Special Report on Emissions Scenarios (SRES), IPCC, Geneva, http://www.ipcc.ch/pdf/special-reports/spm/sres-en.pdf (last access: 6 May 2014), 2000.

Kalnay, E., Kanamitsu, M., Kistler, R., Collins, W., Deaven, D., Gandin, L., Iredell, M., Saha, S., White, G., Woollen, J., Zhu, Y., Leetmaa, A., Reynolds, R., Chelliah, M., Ebisuzaki, W., Higgins, W., Janowiak, J., Mo, K. C., Ropelewski, C., Wang, J., Jenne, R., and Joseph, D.: The NCEP/NCAR 40-year reanalysis project, B. Am. Meteorol. Soc., 77, 437–471, 1996.

Kanamitsu, M., Ebisuzaki, W., Woollen, J., Yang, S.-K., Hnilo, J. J., Fiorino, M., and Potter, G. L.: NCEP-DOE AMIP-II Reanalysis (R-2), B. Am. Meteorol. Soc., 83, 1631–1644, 2002.

Karl, T. R. and Trenberth K. E.: Modern global climate change, Science, 302, 1719–1723, 2003.

Kim, H., Yeh, P. J.-F., Oki, T., and Kanae, S.: Role of rivers in the seasonal variations of terrestrial water storage over global basins, Geophys. Res. Lett., 36, L17402, doi:10.1029/2009GL039006, 2009.

Konar, M., Hussein, Z., Hanasaki, N., Mauzerall, D. L., and Rodriguez-Iturbe, I.: Virtual water trade flows and savings under climate change, Hydrol. Earth Syst. Sci., 17, 3219–3234, doi:10.5194/hess-17-3219-2013, 2013.

Konzmann, M., Gerten, D., and Heinke, J.: Climate impacts on global irrigation requirements under 19 GCMs, simulated with a vegetation and hydrology model, Hydrolog. Sci. J., 58, 88–105, 2013.

Koster, R. D., Dirmeyer, P. A., Guo, Z., Bonan, G., Chan, E., Cox, P., Gordon, C. T., Kanae, S., Kowalczyk, E., Lawrence, D., Liu, P., Lu, C.-H., Malyshev, S., McAvaney, B., Mitchell, K., Mocko, D., Oki, T., Oleson, K., Pitman, A., Sud, Y. C., Taylor, C. M., Verseghy, D., Vasic, R., Xue, Y., and Yamada, T.: Regions of strong coupling between soil moisture and precipitation, Science, 305, 1138–1140, 2004.

Kriegler, E., O'Neill, B. C., Hallegatte, S., Kram, T., Lempert, R. J., Moss, R. H., and Wilbanks, T.: The need for and use of socio-economic scenarios for climate change analysis: a new approach based on shared socio-economic pathways, Global Environ. Change, 22, 807–822, 2012.

Krysanova, V., Müller-Wohlfeil, D. I., and Becker, A.: Development and test of a spatially distributed hydrological/water quality model for mesoscale watersheds, Ecol. Model., 106, 261–289, 1998.

Kump, L. R., Kasting, J. F., and Crane, R. G.: The earth system, Prentice Hall, San Francisco, 2010.

Kyle, P., Davies, E. G., Dooley, J. J., Smith, S. J., Clarke, L. E., Edmonds, J. A., and Hejazi, M.: Influence of climate change mitigation technology on global demands of water for electricity generation, Int. J. Greenh. Gas Con., 13, 112–123, 2013.

Lai, X., Jiang, J., Yang, G., and Lu, X. X.: Should the Three Gorges Dam be blamed for the extremely low water levels in the middle–lower Yangtze River?, Hydrol. Process., 28, 150–160, doi:10.1002/hyp.10077, 2014.

Lawrence, D. M., Oleson, K. W., Flanner, M. G., Thornton, P. E., Swenson, S. C., Lawrence, P. J., Zeng, X., Yang, Z.-L., Levis, S., Sakaguchi, K., Bonan, G. B., and Slater, A. G.: Parameterization improvements and functional and structural advances in Version 4 of the Community Land Model, J. Adv. Model. Earth Syst., 3, M03001, doi:10.1029/2011MS00045, 2011.

Lawrence, D., Maxwell, R., Swenson, S., Lopez, S., and Famiglietti, J.: Challenges of Representing and Predicting Multi-Scale Human–Water Cycle Interactions in Terrestrial Systems, http://climatemodeling.science.energy.gov/sites/default/files/Topic_3_final.pdf (last access: 6 May 2014), 2012.

Leng, G., Huang, M., Tang, Q., Sacks, W. J., Lei, H., and Leung, L. R.: Modeling the effects of irrigation on land surface fluxes and states over the conterminous United States: Sensitivity to input data and model parameters, J. Geophys. Res.-Atmos., 118, 9789–9803, doi:10.1002/jgrd.50792, 2013.

Lenton, T. M.: Land and ocean carbon cycle feedback effects on global warming in a simple Earth system model, Tellus B, 52, 1159–1188, 2000.

Levis, S. and Sacks W.: Technical descriptions of the interactive crop management (CLM4CNcrop) and interactive irrigation models in version 4 of the Community Land Model, http://www.cesm.ucar.edu/models/cesm1.1/clm/CLMcropANDirrigTechDescriptions.pdf (last access: 6 May 2014), 2011.

Levis, S., Bonan, G. B., Kluzek, E., Thornton, P. E., Jones, A., Sacks, W. J., and Kucharik, C. J.: Interactive Crop Management in the Community Earth System Model (CESM1): Seasonal Influences on Land-Atmosphere Fluxes, J. Climate, 25, 4839–4859, 2012.

Li, H., Wigmosta, M. S., Wu, H., Huang, M., Ke, Y., Coleman, A. M., and Leung, L. R.: A physically based runoff routing model for landsurface and earth system models, J. Hydrometeorol., 14, 808–828, 2013.

Liang, X., Lettenmaier, D. P., Wood, E. F., and Burges, S. J.: A simple hydrologically based model of land surface water and energy fluxes for general circulation models, J. Geophys. Res.-Atmos., 99, 14415–14428, 1994.

Lissner, T. K., Sullivan, C. A., Reusser, D. E., and Kropp, J. P.: Water stress and livelihoods: A review of data and knowledge on water needs, use and availability, in: 4th EGU Leonardo Conference: Hydrology and Society – Connections between Hydrology and Population dynamics, Policymaking and Power generation, 14–16 November, Torino, Italy, 2012.

Liu, J., Zhang, Z., Xu, X., Kuang, W., Zhou, W., Zhang, S., Li, R., Yan, C., Yu, D., Wu, S., and Jiang N.: Spatial patterns and driving forces of land use change in China during the early 21st century, J. Geogr. Sci., 20, 483–494, 2010.

Livneh, B., Restrepo, P. J., and Lettenmaier, D. P.: Development of a Unified Land Model for prediction of surface hydrology and land-atmosphere interactions, J. Hydrometeorol., 12, 1299–1320, 2011.

Lo, M.-H. and Famiglietti, J. S.: Irrigation in California's Central Valley strengthens the southwestern U.S. water cycle, Geophys. Res. Lett., 40, 301–306, doi:10.1002/grl.50108, 2013.

Lobell, D. B., Bala, G., and Duffy, P. B.: Biogeophysical impacts of cropland management changes on climate, Geophys. Res. Lett., 33, L06708, doi:10.1029/2005GL025492, 2006.

Lorenz, C. and Kunstmann, H.: The Hydrological Cycle in Three State-of-the-Art Reanalyses: Intercomparison and Performance Analysis, J. Hydrometeorol., 13, 1397–1420, 2012.

Lu, Y.: Development and application of WRF3.3-CLM4crop to study of agriculture-climate interaction, PhD Thesis, University of California, Merced, http://escholarship.org/uc/item/12b6p87z (last access: 6 May 2014), 2013.

Lucas-Picher, P., Christensen, J. H., Saeed, F., Kumar, P., Asharaf, S., Ahrens, B., Wiltshire, A. J., Jacob, D., and Hagemann, S.: Can regional climate models represent the Indian monsoon?, J. Hydrometeorol., 12, 849–868, 2011.

Macknick, J., Newmark, R., Heath, G., and Hallett, K. C.: A review of operational water consumption and withdrawal factors for electricity generating technologies, Technical Report NREL/TP-6A20-5090, http://www.cwatershedalliance.com/pdf/SolarDoc01.pdf (last access: 6 May 2014), 2011.

Manabe, S.: Climate and the ocean circulation part I. The atmospheric circulation and the hydrology of the earth's surface, Mon. Weather Rev., 97, 739–774, 1969.

Maurer, E. P., Wood, A. W., Adam, J. C., Lettenmaier, D. P., and Nijssen, B.: A long-term hydrologically based dataset of land surface fluxes and states for the conterminous United States, J. Climate, 15, 3237–3251, 2002.

McKenney, M. S. and Rosenberg, N. J.: Sensitivity of some potential evapotranspiration estimation methods to climate change, Agr. Forest Meteorol., 64, 81–110, 1993.

McNeill, J. R.: Something New Under the Sun: An Environmental History of the Twentieth-Century World, WW Norton & Company, New York, 2000.

Mehta, V. K., Haden V. R., Joyce, B. A., Purkey, D. R., and Jackson, L. E.: Irrigation demand and supply, given projections of climate and land-use change in Yolo County, California, Agr. Water Manage., 117, 70–82, 2013.

Meinshausen, M., Smith, S. J., Calvin, K., Daniel, J. S., Kainuma, M. L. T., Lamarque, J.-F., Matsumoto, K., Montzka, S. A., Raper, S. C. B., Riahi, K., Thomson, A., Velders, G. J. M., and van Vuuren, D. P. P.: The RCP greenhouse gas concentrations and their extensions from 1765 to 2300, Climatic Change, 109, 213–241, 2011.

Meybeck, M.: Global analysis of river systems: from Earth system controls to Anthropocene syndromes, Philos. T. Roy. Soc. Lond. B, 358, 1935–1955, 2003.

Milano, M., Ruelland, D., Fernandez, S., Dezetter, A., Fabre, J., Servat, E., Fritsch, J.-M., Ardoin-Bardin, S., and Thivet, G.: Current state of Mediterranean water resources and future trends under climatic and anthropogenic changes, Hydrolog. Sci. J., 58, 498–518, 2013.

Miller, J. R., Russell, G. L., and Caliri, G.: Continental-scale river flow in climate models, J. Climate, 7, 914–928, doi:10.1175/1520-0442(1994)007<0914:CSRFIC>2.0.CO;2, 1994.

Milly, P. C. D.: Potential evaporation and soil moisture in general circulation models, J. Climate, 5, 209–226, 1992.

Mitchell, T. D. and Jones, P. D.: An improved method of constructing a database of monthly climate observations and associated high-resolution grids, Int. J. Climatol., 25, 693–712, doi:10.1002/joc.1181, 2005.

Moore, N. and Rojstaczer, S.: Irrigation-induced rainfall and the Great Plains, J. Appl. Meteorol., 40, 1297–1309, 2001.

Moss, R. H., Edmonds, J. A., Hibbard, K. A., Manning, M. R., Rose, S. K., van Vuuren, D. P., Carter, T. R., Emori, S., Kainuma, M., Kram, T., Meehl, G. A., Mitchell, J. F. B., Nakicenovic, N., Riahi, K., Smith, S. J., Stouffer, R. J., Thomson, A. M., Weyant, J. P., and Wilbanks, T. J.: The next generation of scenarios for climate change research and assessment, Nature, 463, 747–756, 2010.

Music, B. and Caya, D.: Evaluation of the hydrological cycle over the Mississippi River basin as simulated by the Canadian Regional Climate Model (CRCM), J. Hydrometeorol., 8, 969–988, 2007.

Nakayama, T.: Simulation of the effect of irrigation on the hydrologic cycle in the highly cultivated Yellow River Basin, Agr. Forest Meteorol., 151, 314–327, 2011.

Nakayama, T. and Shankman D.: Evaluation of uneven water resource and relation between anthropogenic water withdrawal and ecosystem degradation in Changjiang and Yellow River basins, Hydrol. Process., 27, 3350–3362, doi:10.1002/hyp.9835, 2013.

Nassopoulos, H., Dumas, P., and Hallegatte, S.: Climate change, precipitation and water management infrastructures, presented at: Water in Africa: Hydro-Pessimism or Hydro-Optimism, 2–3 October 2008, Porto, Portugal, available at:http://www.slideshare.net/water.in.africa/hypatia-nassopoulos-ppt-presentation (last access: 15 October 2014), 2008.

Nassopoulos, H., Dumas, P., and Hallegatte, S.: Adaptation to an uncertain climate change: cost benefit analysis and robust decision making for dam dimensioning, Climatic Change, 114, 497–508, doi:10.1007/s10584-012-0423-7, 2012.

Nazemi, A. and Wheater, H. S.: Assessing the vulnerability of water supply to changing streamflow conditions, Eos Trans. Am. Geophys. Un., 95, 288, doi:10.1002/2014EOS320007, 2014a.

Nazemi, A. and Wheater, H. S.: How can the uncertainty in the natural inflow regime propagate into the assessment of water resource systems?, Adv. Water Resour., 63, 131–142, doi:10.1016/j.advwatres.2013.11.009, 2014b.

Nazemi, A. and H. S. Wheater: On inclusion of water resource management in Earth System models – Part 2: Representation of water supply and allocation and opportunities for improved modeling, Hydrol. Earth Syst. Sci., 19, 63–90, doi:10.5194/hess-19-63-2015, 2015.

Nazemi, A., Akbarzadeh-T, M. R., and Hosseini, S. M.: Fuzzy-stochastic linear programming in water resources engineering, in: Proceedings of IEEE Annual Meeting of Fuzzy Information, IEEE, 227–232, doi:10.1109/NAFIPS.2002.1018060, 2002.

Nazemi, A., Wheater, H. S., Chun, K. P., and Elshorbagy, A.: A stochastic reconstruction framework for analysis of water resource system vulnerability to climate-induced changes in river flow regime, Water Resour. Res., 49, 291–305, doi:10.1029/2012WR012755, 2013.

New, M., Hulme, M., and Jones, P.: Representing Twentieth-Century Space–Time Climate Variability Part I: Development of a 1961–90 Mean Monthly Terrestrial Climatology, J.f Climate, 12, 829–857, 1999.

New, M., Hulme, M., and Jones, P. D.: Representing twentieth century space-time climate variability, part II Development of 1901–

96 monthly grids of terrestrial surface climate, J. Climate, 13, 2217–2238, 2000.

Ngo-Duc, T., Laval, K., Polcher, J., Lombard, A., and Cazenave, A.: Effects of land water storage on global mean sea level over the past half century, Geophys. Res. Lett., 32, L09704, doi:10.1029/2005GL022719, 2005a.

Ngo-Duc, T., Polcher, J., and Laval, K.: A 53-year forcing data set for land surface models, J. Geophys. Res., 110, D06116, doi:10.1029/2004JD005434, 2005b.

Ngo-Duc, T., Laval, K., Ramillien, G., Polcher, J., and Cazenave, A.: Validation of the land water storage simulated by Organising Carbon and Hydrology in Dynamic Ecosystems (ORCHIDEE) with Gravity Recovery and Climate Experiment (GRACE) data, Water Resour. Res., 43, W04427, doi:10.1029/2006WR004941, 2007.

Nicholson, S. E.: Land surface atmosphere interaction, Prog. Phys. Geogr., 12, 36–65, 1988.

Nilsson, C., Reidy, C. A., Dynesius, M., and Revenga, C.: Fragmentation and flow regulation of the world's large river systems, Science, 308, 405–408, 2005.

Noilhan, J. and Planton, S.: A simple parameterization of land surface processes for meteorological models, Mon. Weather Rev., 117, 536–549, 1989.

Oki, T. and Kanae, S.: Global hydrological cycles and world water resources, Science, 313, 1068–1072, 2006.

Oki, T. and Sud, Y. C.: Design of Total Runoff Integrating Pathways (TRIP) – A global river channel network, Earth Interact., 2, 1–37, 1998.

Oki, T., Blyth, E. M., Berbery, E. H., and Alcaraz-Segura, D.: Land Use and Land Cover Changes and Their Impacts on Hydroclimate, Ecosystems and Society, in: Climate Science for Serving Society, edited by: Asrar, G. R. and Hurrell, J. W., Springer Netherlands, 185–203, 2013.

Oleson, K. W., Dai, Y., Bonan, G. B., Bosilovichm, M., Dickinson, R., Dirmeyer, P., Hoffman, F., Houser, P., Levis, S., Niu, G.-Y., Thornton, P., Vertenstein, M., Yang, Z., and Zeng, X.: Technical description of the community land model (CLM), NCAR Tech. Note NCAR/TN-461+STR, 173 pp., doi:10.5065/D6N877R0, http://www.cesm.ucar.edu/models/cesm1.0/clm/CLM4_Tech_Note.pdf (last access: 28 December 2014), 2004.

Oleson, K. W., Niu, G. Y., Yang, Z. L., Lawrence, D. M., Thornton, P. E., Lawrence, P. J., Stöckli, R., Dickinson, R. E., Bonan, G. B., Levis, S., Dai, A., and Qian, T.: Improvements to the Community Land Model and their impact on the hydrological cycle, J. Geophys. Res.-Biogeo., 113, G01021, doi:10.1029/2007JG000563, 2008.

Ozdogan, M. and Gutman, G.: A new methodology to map irrigated areas using multi-temporal MODIS and ancillary data: An application example in the continental US, Remote Sens. Environ., 112, 3520–3537, 2008.

Ozdogan, M., Rodell, M., Beaudoing, H. K., and Toll, D. L.: Simulating the Effects of Irrigation over the United States in a Land Surface Model Based on Satellite-Derived Agricultural Data, J. Hydrometeorol., 11, 171–184, 2010.

Pielke, R. A., Cotton, W. R., Walko, R. L., Tremback, C. J., Lyons, W. A., Grasso, L. D., Nicholls, M. E., Moran, M. D., Wesley, D. A., Lee, T. J., and Copeland, J. H.: A comprehensive meteorological modeling system – RAMS, Meteorol. Atmos. Phys., 49, 69–91, 1992.

Pietroniro, A., Fortin, V., Kouwen, N., Neal, C., Turcotte, R., Davison, B., Verseghy, D., Soulis, E. D., Caldwell, R., Evora, N., and Pellerin, P.: Development of the MESH modelling system for hydrological ensemble forecasting of the Laurentian Great Lakes at the regional scale, Hydrol. Earth Syst. Sci., 11, 1279–1294, doi:10.5194/hess-11-1279-2007, 2007.

Pitman, A. J.: The evolution of, and revolution in, land surface schemes designed for climate models, Int. J. Climatol., 23, 479–510, 2003.

Pitman, A. J., Henderson-Sellers, A., and Yang, Z. L.: Sensitivity of regional climates to localized precipitation in global models, Nature, 346, 734–737, 1990.

Pitman, A. J., de Noblet-Ducoudré, N., Cruz, F. T., Davin, E. L., Bonan, G. B., Brovkin, V., Claussen, M., Delire, C., Ganzeveld, L., Gayler, V., van den Hurk, B. J. J. M., Lawrence, P. J., van der Molen, M. K., Müller, C., Reick, C. H., Seneviratne, S. I., Strengers, B. J., and Voldoire, A.: Uncertainties in climate responses to past land cover change: first results from the LU-CID intercomparison study, Geophys. Res. Lett., 36, L14814, doi:10.1029/2009GL039076, 2009.

Pokhrel, Y., Hanasaki, N., Koirala, S., Cho, J., Yeh, P. J.-F., Kim, H., Kanae, S., and Oki, T.: Incorporating anthropogenic water regulation modules into a land surface model, J. Hydrometeorol., 13, 255–269, 2012.

Polcher, J.: Interactive comment on "On inclusion of water resource management in Earth System models – Part 1: Problem definition and representation of water demand" by A. Nazemi and H. S. Wheater, Hydrol. Earth Syst. Sci. Discuss., 11, C3403–C3410, 2014.

Polcher, J., Bertrand, N., Biemans, H., Clark, D. B., Floerke, M., Gedney, N., Gerten, D., Stacke, T., van Vliet, M., and Voss, F.: Improvements in hydrological processes in general hydrological models and land surface models within WATCH, WATCH Technical Report Number 34, available at: http://www.eu-watch.org/publications/technical-reports (last access: 6 May 2014), 2011.

Poff, N. L., Richter, B. D., Arthington, A. H., Bunn, S. E., Naiman, R. J., Kendy, E., Acreman, M., Apse, C., Bledsoe, B. P., Freeman, M. C., Henriksen, J., Jacobson, R. B., Kennen, J. G., Merritt, D. M., O'Keeffe, J. H., Olden, J. D., Rogers, K., Tharme, R. E., and Warner, A.: The ecological limits of hydrologic alteration (ELOHA): a new framework for developing regional environmental flow standards, Freshwater Biol., 55, 1365–2427, doi:10.1111/j.1365-2427.2009.02204.x, 2009.

Portmann, F. T., Siebert, S., and Döll, P.: MIRCA2000 – Global monthly irrigated and rainfed crop areas around the year 2000: A new high-resolution data set for agricultural and hydrological modeling, Global Biogeochem. Cycl., 24, GB1011, doi:10.1029/2008GB003435, 2010.

Postel, S. L., Daily, G. C., and Ehrlich, P. R.: Human appropriation of renewable fresh water, Science, 271, 785–788, 1996.

Precoda, N.: Requiem for the Aral Sea, Ambio, 20, 109-114, 1991.

Priestley, C. H. B. and Taylor, R. J.: On the assessment of surface heat flux and evaporation using large-scale parameters, Mon. Weather Rev., 100, 81–92, 1972.

Prudhomme, C., Giuntoli, I., Robinson, E. L., Clark, D. B., Arnell, N. W., Dankers, R., Fekete, B. M., Franssen, W., Gerten, D., Gosling, S. N., Hagemann, S., Hannah, D. M., Kim, H., Masaki, Y., Satoh, Y., Stacke, T., Wada, Y., and Wisser, D.: Hydrological droughts in the 21st century, hotspots and uncertainties from a global multimodel ensemble experiment, P. Natl. Acad. Sci. USA, 111, 3262–3267, doi:10.1073/pnas.1222473110, 2014.

Qian, Y., Huang, M., Yang, B., and Berg, L. K.: A Modeling Study of Irrigation Effects on Surface Fluxes and Land-Air-Cloud Interactions in the Southern Great Plains, J. Hydrometeorol., 14, 700–721, 2013.

Rausch, S. and Mowers, M.: Distributional and efficiency impacts of clean and renewable energy standards for electricity, Resour. Energy Econ., 36, 556–585, 2013.

Rodell, M., Velicogna, I., and Famiglietti, J. S.: Satellite-based estimates of groundwater depletion in India, Nature, 460, 999–1002, 2009.

Rohling, E. J. and Bryden, H. L.: Man-induced salinity and temperature increases in western Mediterranean deep water, J. Geophys. Res.-Oceans (1978–2012), 97, 11191–11198, 1992.

Rosenzweig, C., Elliott, J., Deryng, D., Ruane, A. C., Müller, C., Arneth, A., Boote, K. J., Folberth, C., Glotter, M., Khabarov, N., Neumann, K., Piontek, F., Pugh, T. A. M., Schmid, E., Stehfest, E., Yang, H., and Jones, J. W.: Assessing agricultural risks of climate change in the 21st century in a global gridded crop model intercomparison, P. Natl. Acad. Sci. USA, 111, 3268–3273, doi:10.1073/pnas.1222463110, 2014.

Rost, S., Gerten, D., Bondeau, A., Luncht, W., Rohwer, J., and Schaphoff, S.: Agricultural green and blue water consumption and its influence on the global water system, Water Resour. Res., 44, W09405, doi:10.1029/2007WR006331, 2008.

Rost, S., Gerten, D., Hoff, H., Lucht, W., Falkenmark, M., and Rockström, J.: Global potential to increase crop production through water management in rainfed agriculture, Environ.l Res. Lett., 4, 044002, doi:10.1088/1748-9326/4/4/044002, 2009.

Rudolf, B., Beck, C., Grieser, J., and Schneider, U.: Global precipitation analysis products of the GPCC, Climate Monitoring – Tornadoklimatologie – Aktuelle Ergebnisse des Klimamonitorings, available at: http://www.juergen-grieser.de/publications/publications_pdf/GPCC-intro-products-2005.pdf, last access: 16 July 2014, 163–170, 2005.

Sacks, W. J., Cook, B. I., Buenning, N., Levis, S., and Helkowski, J. H.: Effects of global irrigation on the near-surface climate, Clim. Dynam., 33, 159–175, 2009.

Saeed, F., Hagemann, S., and Jacob, D.: Impact of irrigation on the South Asian summer monsoon, Geophys. Res. Lett., 36, L20711, doi:10.1029/2009GL040625, 2009.

Schellnhuber, H. J.: Discourse: Earth System Analysis – The Scope of the Challenge, in: Earth System Analysis – Integrating science for sustainability, edited by: Schellnhuber, H. J. and Wenzel, V., Springer, Heidelberg, 1998.

Schellnhuber, H. J.: Earth system analysis and the second Copernican revolution, Nature, 402, C19–C23, 1999.

Schewe, J., Heinke, J., Gerten, D., Haddeland, I., Arnell, N. W., Clark, D. B., Dankers, R., Eisner, S., Fekete, B. M., Colón-González, F. J., Gosling, S. N., Kim, H., Liu, X., Masaki, Y., Portmann, F. T., Satoh, Y., Stacke, T., Tang, Q., Wada, Y., Wisser, D., Albrecht, T., Frieler, K., Piontek, F., Warszawski, L., and Kabat, P.: Multimodel assessment of water scarcity under climate change, P. Natl. Acad. Sci. USA, 111, 3245–3250, doi:10.1073/pnas.1222460110, 2014.

Schiermeier, Q.: Water risk as world warms, Nature, 505, 7481, doi:10.1038/505010a, 2014.

Schlosser, C. A., Kicklighter, D., and Sokolov, A.: A global land system framework for integrated climate-change assessments, MIT Joint Program on the Science and Policy of Global Change, Report No. 147, http://dspace.mit.edu/handle/1721.1/38461 (last access: 6 May 2014), 2007.

Sellers, P. J.: Biophysical models of land surface processes, in: Climate system modeling, edited by: Trenberth, K. E., Cambridge University Press, Cambridge, UK, 451–490, 1992.

Sellers, P. J., Mintz, Y. C. S. Y., Sud, Y. E. A., and Dalcher, A.: A simple biosphere model (SiB) for use within general circulation models, J. Atmos. Sci., 43, 505–531, 1986.

Sellers, P. J., Tucker, C. J., Collatz, G. J., Los, S. O., Justice, C. O., Dazlich, D. A., and Randall, D. A.: A global 1 by 1 NDVI data set for climate studies – Part 2: The generation of global fields of terrestrial biophysical parameters from the NDVI, Int. J. Remote Sens., 15, 3519–3545, 1994.

Sellers, P. J., Randall, D. A., Collatz, G. J., Berry, J. A., Field, C. B., Dazlich, D. A., Zhang, C., Collelo, G. D., and Bounoua, L.: A revised land surface parameterization (SiB2) for atmospheric GCMs – Part I: Model formulation, J. Climate, 9, 676–705, 1996a.

Sellers, P. J., Meeson, B. W., Closs, J., Collatz, J., Corprew, F., Dazlich, D., Hall, F. G., Kerr, Y., Koster, R., Los, S., Mitchell, K., McManus, J., Myers, D., Sun, K.-J., and Try, P.: The ISLSCP Initiative I global datasets: surface boundary conditions and atmospheric forcings for land–atmosphere studies, B. Am. Meteorol. Soc., 77, 1987–2005, 1996b.

Shiklomanov, I. A.: World water resources, UNESCO, 1998, Paris, http://www.ce.utexas.edu/prof/mckinney/ce385d/Papers/Shiklomanov.pdf (last access: 6 May 2014), 1993.

Shiklomanov, I. A.: Assessment of Water Resources and Water Availability in the World, Comprehensive Assessment of the Freshwater Resources of the World, WMO and SEI, Geneva, 1997.

Shiklomanov, I. A.: World water resources and water use: Present assessment and outlook for 2025, in: World water scenarios, edited by: Rijsberman, F. R., Earthscan, London, 160–203, 2000.

Short, W., Blair, N., Sullivan, P., and Mai, T.: ReEDS model documentation: base case data and model description, Golden, CO: National Renewable Energy Laboratory, http://www.nrel.gov/analysis/reeds/documentation.html (last access: 6 May 2014), 2009.

Siebert, S. and Döll, P.: The Global Crop Water Model (GCWM): Documentation and first results for irrigated crops, https://www2.uni-frankfurt.de/45217788/FHP07_SiebertandDoell__2008.pdf (last access: 6 May 2014), 2008.

Siebert, S. and Döll, P.: Quantifying blue and green virtual water contents in global crop production as well as potential production losses without irrigation, J. Hydrol., 384, 198–217, 2010.

Siebert, S., Döll, P., Hoogeveen, J., Faures, J.-M., Frenken, K., and Feick, S.: Development and validation of the global map of irrigation areas, Hydrol. Earth Syst. Sci., 9, 535–547, doi:10.5194/hess-9-535-2005, 2005.

Siebert, S., Döll, P., Feick, S., Hoogeveen, J., and Frenken, K.: Global map of irrigation areas version 4.0.1, Food and Agriculture Organization of the United Nations, Rome, Italy, https://www2.uni-frankfurt.de/45218039/Global_IrrigationMap (last access: 6 May 2014), 2007.

Siebert, S., Burke, J., Faures, J. M., Frenken, K., Hoogeveen, J., Döll, P., and Portmann, F. T.: Groundwater use for irrigation – a global inventory, Hydrol. Earth Syst. Sci., 14, 1863–1880, doi:10.5194/hess-14-1863-2010, 2010.

Sitch, S., Smith, B., Prentice, I. C., Arneth, A., Bondeau, A., Cramer, W., Kaplan, J. O., Levis, S., Lucht, W., Sykes, M. T., Thonicke, K., and Venevsky, S.: Evaluation of ecosystem dynamics, plant geography and terrestrial carbon cycling in the LPJ dynamic global vegetation model, Global Change Biol., 9, 161–185, 2003.

Sivapalan, M., Savenije, H. H., and Blöschl, G.: Socio-hydrology: A new science of people and water, Hydrol. Process., 26, 1270–1276, 2012.

Skamarock, W. C., Klemp, J. B., Dudhia, J., Gill, D. O., Barker, D. M., Wang, W., and Powers, J. G.: A description of the advanced research WRF version 2 (No. NCAR/TN468+STR), available at: http://oai.dtic.mil/oai/oai?verb=getRecord&metadataPrefix=html&identifier=ADA487419 (last access: 6 May 2014), 2005.

Skliris, N. and Lascaratos, A.: Impacts of the Nile River damming on the thermohaline circulation and water mass characteristics of the Mediterranean Sea, J. Mar. Syst., 52, 121–143, doi:10.1016/j.jmarsys.2004.02.005, 2004.

Smakhtin, V., Revenga, C., and Döll, P.: A pilot global assessment of environmental water requirements and scarcity, Water Int., 29, 307–317, 2004.

Small, I., Van der Meer, J., and Upshur, R. E.: Acting on an environmental health disaster: the case of the Aral Sea, Environ. Health Perspect., 109, 547–549, 2001.

Smith, M.: CROPWAT – A computer program for irrigation planning and management, Irrigation and Drainage, Pap. 46, Food and Agric. Org. of the UN, Rome, http://www.fao.org/nr/water/infores_databasescropwat.html (last access: 6 May 2014), 1992.

Solomon, S., Plattner, G. K., Knutti, R., and Friedlingstein, P.: Irreversible climate change due to carbon dioxide emissions, P. Natl. Acad. Sci., 106, 1704–1709, 2009.

Sophocleous, M.: Interactions between groundwater and surface water: the state of the science, Hydrogeol. J., 10, 52–67, 2002.

Sorooshian, S., Li, J., Hsu, K.-L., and Gao, X.: How significant is the impact of irrigation on the local hydroclimate in California's Central Valley? Comparison of model results with ground and remote-sensing data, J. Geophys. Res., 116, D06102, doi:10.1029/2010JD014775, 2011.

Soulis, E. D., Snelgrove, K. R., Kouwen, N., Seglenieks, F., and Verseghy, D. L.: Towards closing the vertical water balance in Canadian atmospheric models: coupling of the land surface scheme CLASS with the distributed hydrological model WATFLOOD, Atmos.-Ocean, 38, 251–269, 2000.

Steffen, W., Crutzen, P. J., and McNeill, J. R.: The Anthropocene: are humans now overwhelming the great forces of nature, Ambio, 36, 614–621, 2007.

Steffen, W., Grinevald, J., Crutzen, P., and McNeill, J.: The Anthropocene: conceptual and historical perspectives, Philos. T. Roy. Soc. A, 369, 842–867, 2011.

Steinfeld, H., Gerber, P., Wassenaar, T., Castel, V., Rosales, M., and de Haan, C.: Livestock's long shadow: Environmental issues and options, Food and Agriculture Organization – LEAD,

Rome, Italy, http://www.fao.org/docrep/010/a0701e/a0701e00. HTM (last access: 6 May 2014), 2006.

Strzepek, K., Schlosser, A., Farmer, W., Awadalla, S., Baker, J., Rosegrant M., and Gao, X.: Modeling the global water resource system in an integrated assessment modeling framework: IGSM-WRS, MIT Joint Program on the Science and Policy of Global Change, Report No. 189, available at: http://dspace.mit.edu/handle/1721.1/61767 (last access: 6 May 2014), 2010.

Strzepek, K., Baker, J., Farmer, W., and Schlosser, C. A.: Modeling water withdrawal and consumption for electricity generation in the United States, MIT Joint Program on the Science and Policy of Global Change, Report No. 222, http://dspace.mit.edu/handle/1721.1/71168 (last access: 6 May 2014), 2012a.

Strzepek, K., Schlosser, A., Gueneau, A. Gao, X., Blanc, É., Fant, C., Rasheed, B., and Jacoby, H. D.: Modeling water resource system under climate change: IGSM-WRS, MIT Joint Program on the Science and Policy of Global Change, Report No. 236, http://dspace.mit.edu/handle/1721.1/75774 (last access: 6 May 2014), 2012b.

Sulis, M., Paniconi, C., Rivard, C., Harvey, R., and Chaumont, D.: Assessment of climate change impacts at the catchment scale with a detailed hydrological model of surface-subsurface interactions and comparison with a land surface model, Water Resour. Res., 47, W01513, doi:10.1029/2010WR009167, 2011.

Takata, K., Emori, S., and Watanabe, T.: Development of the minimal advanced treatments of surface interaction and runoff, Global Planet. Change, 38, 209–222, 2003.

Tang, Q., Gao, H., Yeh, P., Oki, T., Su, F., and Lettenmaier, D. P.: Dynamics of Terrestrial Water Storage Change from Satellite and Surface Observations and Modeling, J. Hydrometeorol., 11, 156–170, 2010.

Tao, F., Yokozawa, M., Hayashi, Y., and Lin, E.: Terrestrial water cycle and the impact of climate change, Ambio, 32, 295–301, 2003.

Taylor, C. M.: Feedbacks on convection from an African wetland, Geophys. Res. Lett., 37, L05406, doi:10.1029/2009GL041652, 2009.

Taylor, C. M., de Jeu, R. A., Guichard, F., Harris, P. P., and Dorigo, W. A.: Afternoon rain more likely over drier soils, Nature, 489, 423–426, doi:10.1038/nature11377, 2012.

Taylor, K. E., Stouffer, R. J., and Meehl, G. A.: An Overview of CMIP5 and the Experiment Design, B. Am. Meteorol. Soc., 93, 485–498, 2012.

Taylor, R. G., Scanlon, B., Döll, P., Rodell, M., van Beek, R., Wada, Y., Longuevergne, L., Leblanc, M., Famiglietti, J. S., Edmunds, M., Konikow, L., Green, T. R., Chen, J., Taniguchi, M., Bierkens, M. F. P., MacDonald, A., Fan, Y., Maxwell, R. M., Yechieli, Y., Gurdak, J. J., Allen, D. M., Shamsudduha, M., Hiscock, K., Yeh, P. J.-F., Holman, I., and Treidel, H.: Ground water and climate change, Nat. Clim. Change, 3, 322–329, 2013.

Tesfa, T. K., Li, H.-Y., Leung, L. R., Huang, M., Ke, Y., Sun, Y., and Liu, Y.: A subbasin-based framework to represent land surface processes in an Earth system model, Geosci. Model Dev., 7, 947–963, doi:10.5194/gmd-7-947-2014, 2014.

Tharme, R. E.: A global perspective on environmental flow assessment: emerging trends in the development and application of environmental flow methodologies for rivers, River Res. Appl., 19, 397–441, doi:10.1002/rra.736, 2003.

Thenkabail, P. S., Biradar, C. M., Noojipady, P., Dheeravath, V., Li, Y., Velpuri, M., Gumma, M., Gangalakunta, O. R. P., Turral, H., Cai, X., Vithanage, J., Schull, M. A., and Dutta, R.: Global irrigated area map (GIAM), derived from remote sensing, for the end of the last millennium, Int. J. Remote Sens., 30, 3679–3733, 2009.

Trenberth, K. E. (Ed.): Climate Systems Modeling, Cambridge University Press, Cambridge, UK, 1992.

Trenberth, K. E. and Asrar, G. R.: Challenges and opportunities in water cycle research: WCRP contributions, Surv. Geophys., 35, 515–532, 2012.

Trenberth, K. E. and Dai, A.: Effects of Mount Pinatubo volcanic eruption on the hydrological cycle as an analog of geoengineering, Geophys. Res. Lett., 34, L15702, doi:10.1029/2007GL030524, 2007.

Tuinenburg, O. A., Hutjes, R. W. A., Jacobs, C. M. J., and Kabat, P.: Diagnosis of Local Land–Atmosphere Feedbacks in India, J.f Climate, 24, 251–266, 2011.

UN: Statistical Yearbook, Stat. Div., New York, 1997.

USDA: 2002 census of agriculture, National Agricultural Statistics Service, http://www.agcensus.usda.gov/Publications/2002/ (last access: 6 May 2014), 2002.

USDA: 2007 census of agriculture, Farm and Ranch Irrigation Survey, Volume 3, Special studies, part 1, http://www.agcensus.usda.gov/Publications/2007/Online_Highlights/FarmandRanch_IrrigationSurvey/fris08.pdf (last access: 6 May 2014), 2008.

van Beek, L. P. H., Wada, Y., and Bierkens, M. F. P.: Global monthly water stress: 1. Water balance and water availability, Water Resour. Res., 47, W07517, doi:10.1029/2010WR009791, 2011.

van Woerden, J., Diedericks, J., and Klein-Goldewjik, K.: Data management in support of integrated environmental assessment and modelling at RIVM – including the 1995 RIVM Catalogue of International Data Sets, RIVM Report no. 402001006, National Institute of Public Health and the Environment, Bilthoven, the Netherlands, 1995.

Vargas-Yáñez, M., Moya, F., García-Martínez, M. C., Tel, E., Zunino, P., Plaza, F., Salat, J., Pascual, J., López-Jurado, J. L., and Serra, M.: Climate change in the Western Mediterranean sea 1900–2008, J. Mar. Syst., 82, 171–176, doi:10.1016/j.jmarsys.2010.04.013, 2010.

Vassolo, S. and Döll, P.: Global-scale gridded estimates of thermoelectric power and manufacturing water use, Water Resour. Res., 41, W04010, doi:10.1029/2004WR003360, 2005.

Verseghy, D. L.: CLASS – A Canadian land surface scheme for GCMs I. Soil model, Int. J. Climatol., 11, 111–133, 1991.

Verseghy, D. L.: The Canadian land surface scheme (CLASS): Its history and future, Atmos.-Ocean, 38, 1–13, 2000.

Verseghy, D. L., McFarlane, N. A., and Lazare, M.: CLASS – A Canadian land surface scheme for GCMs II. Vegetation model and coupled runs, Int. J. Climatol., 13, 347–370, 1993.

Vitousek, P. M., Mooney, H. A., Lubchenco, J., and Melillo, J. M.: Human domination of Earth's ecosystems, Science, 277, 494–499, 1997.

Voisin, N., Liu, L., Hejazi, M., Tesfa, T., Li, H., Huang, M., Liu, Y., and Leung, L. R.: One-way coupling of an integrated assessment model and a water resources model: evaluation and implications of future changes over the US Midwest, Hydrol. Earth Syst. Sci., 17, 4555–4575, doi:10.5194/hess-17-4555-2013, 2013.

Vörösmarty, C. J. and Sahagian, D.: Anthropogenic disturbance of the terrestrial water cycle, BioScience, 50, 753–765, 2000.

Vörösmarty, C. J., Sharma, K. P., Fekete, B. M., Copeland, A. H., Holden, J., Marble, J., and Lough, J. A.: The storage and aging of continental runoff in large reservoir systems of the world, Ambio, 26, 210–219, 1997.

Vörösmarty, C. J., Federer, C. A., and Schloss A. L.: Potential evaporation functions compared on US watersheds: Possible implications for global-scale water balance and terrestrial ecosystem modeling, J. Hydrol., 207, 147–169, 1998.

Vörösmarty, C. J., Leveque, C., and Revenga, C.: Millennium Ecosystem Assessment Volume 1: Conditions and Trends, chap. 7: Freshwater ecosystems, Island Press, Washington, D.C., USA, 165–207, 2005.

Wada, Y., van Beek, L. P. H., van Kempen, C. M., Reckman, J. W. T. M., Vasak, S., and Bierkens, M. F. P.: Global depletion of groundwater resources, Geophys. Res. Lett., 37, L20402, doi:10.1029/2010GL044571, 2010.

Wada, Y., van Beek, L. P. H., Viviroli, D., Dürr, H. H., Weingartner, R., and Bierkens, M. F. P.: Global monthly water stress: 2. Water demand and severity of water stress, Water Resour. Res., 47, W07518, doi:10.1029/2010WR009792, 2011.

Wada, Y., van Beek, L. P. H., and Bierkens, M. F. P.: Nonsustainable groundwater sustaining irrigation: A global assessment, Water Resour. Res., 48, W00L06, doi:10.1029/2011WR010562, 2012.

Wada, Y., Wisser, D., Eisner, S., Flörke, M., Gerten, D., Haddeland, I., Hanasaki, N., Masaki, Y., Portmann, F. T., Stacke, T., Tessler, Z., Schewe, J.: Multimodel projections and uncertainties of irrigation water demand under climate change, Geophys. Res. Lett., 40, 4626–4632, 2013.

Wada, Y., Wisser, D., and Bierkens, M. F. P.: Global modeling of withdrawal, allocation and consumptive use of surface water and groundwater resources, Earth Syst. Dynam., 5, 15–40, doi:10.5194/esd-5-15-2014, 2014.

Walko, R. L., Band, L. E., Baron, J., Kittel, T. G. F., Lammers, R., Lee, T. J., Ojima, D., Pielke, R. A., Taylor, C., Tague, C., Tremback, C. J., and Vidale, P. L.: Coupled atmosphere-biophysics-hydrology models for environmental modeling, J. Appl. Meteorol., 39, 931–944, 2000.

Wei, J., Dirmeyer, P. A., Wisser, D., Bosilovich, M. G., and Mocko, D. M.: Where does the irrigation water go? An estimate of the contribution of irrigation to precipitation using MERRA, J. Hydrometeorol., 14, 275–289, 2013.

Wise, M. and Calvin, K.: GCAM 3.0 agriculture and land use: technical description of modeling approach, Pacific Northwest National Laboratory, Richland, WA, https://wiki.umd.edu/gcam/images/8/87/GCAM3AGTechDescript12511.pdf (last access: 6 May 2014), 2011.

Wise, M., Calvin, K., Thomson, A., Clarke, L., Bond-Lamberty, B., Sands, R., Smith, S. J., Janetos, A., and Edmonds, J.: The implications of limiting CO_2 concentrations for agriculture, land-use change emissions and bioenergy, Technical report PNNL-17943, available at: http://www.usitc.gov/researchand_analysis/economics_seminars/2009/200902_co2landuse.pdf (last access: 6 May 2014), 2009a.

Wise, M., Calvin, K., Thomson, A., Clarke, L., Bond-Lamberty, B., Sands, R., Smith, S. J., Janetos, A., and Edmonds, J.: Implications of limiting CO_2 concentrations for land use and energy, Science, 324, 1183–1186, 2009b.

Wisser, D., Frolking, S., Douglas, E. M., Fekete, B. M., Vörösmarty, C. J., and Schumann, A. H.: Global irrigation water demand: Variability and uncertainties arising from agricultural and climate data sets, Geophys. Res. Lett., 35, L24408, doi:10.1029/2008GL035296, 2008.

Wisser, D., Fekete, B. M., Vörösmarty, C. J., and Schumann, A. H.: Reconstructing 20th century global hydrography: a contribution to the Global Terrestrial Network-Hydrology (GTN-H), Hydrol. Earth Syst. Sci., 14, 1–24, doi:10.5194/hess-14-1-2010, 2010.

Wood, E. F., Roundy, J. K., Troy, T. J., van Beek, L. P. H., Bierkens, M. F. P., Blyth, E., de Roo, A., Döll, P., Ek, M., Famiglietti, J., Gochis, D., van de Giesen, N., Houser, P., Jaffé, P. R., Kollet, S., Lehner, B., Lettenmaier, D. P., Peters-Lidard, C., Sivapalan, M., Sheffield, J., Wade, A., and Whitehead, P.: Hyperresolution global land surface modeling: meeting a grand challenge for monitoring Earth's terrestrial water, Water Resour. Res., 47, W05301, doi:10.1029/2010WR010090, 2011.

WRI: World Resources 1998–99, Oxford Press, New York, USA, 1998.

WRI: World Resources 2000–01, Oxford Press, New York, USA, 2000.

Yoshikawa, S., Cho, J., Yamada, H. G., Hanasaki, N., and Kanae, S.: An assessment of global net irrigation water requirements from various water supply sources to sustain irrigation: rivers and reservoirs (1960–2050), Hydrol. Earth Syst. Sci., 18, 4289–4310, doi:10.5194/hess-18-4289-2014, 2014.

Zhao, M. and Dirmeyer, P. A.: Production and analysis of GSWP-2 near-surface meteorology data sets (Vol. 159), Center for Ocean-Land-Atmosphere Studies, Calverton, http://ww.w.monsoondata.org/gswp/gswp2data.pdf (last access: 6 May 2014), 2003.

Zhao, M., Pitman, A. J., and Chase, T.: The impact of land cover change on the atmospheric circulation, Clim. Dynam., 17, 467–477, 2001.

Learning about water resource sharing through game play

Tracy Ewen[1,2] **and Jan Seibert**[1,3,4]

[1]Department of Geography, University of Zurich, Zurich, Switzerland
[2]Center for Climate Systems Modeling, ETH Zurich, Zurich, Switzerland
[3]Department of Physical Geography and Quaternary Geology, Stockholm University, Stockholm, Sweden
[4]Department of Earth Sciences, Uppsala University, Uppsala, Sweden

Correspondence to: Tracy Ewen (tracy.ewen@geo.uzh.ch)

Abstract. Games are an optimal way to teach about water resource sharing, as they allow real-world scenarios to be enacted. Both students and professionals learning about water resource management can benefit from playing games, through the process of understanding both the complexity of sharing of resources between different groups and decision outcomes. Here we address how games can be used to teach about water resource sharing, through both playing and developing water games. An evaluation of using the web-based game *Irrigania* in the classroom setting, supported by feedback from several educators who have used *Irrigania* to teach about the sustainable use of water resources, and decision making, at university and high school levels, finds *Irrigania* to be an effective and easy tool to incorporate into a curriculum. The development of two water games in a course for masters students in geography is also presented as a way to teach and communicate about water resource sharing. Through game development, students learned soft skills, including critical thinking, problem solving, team work, and time management, and overall the process was found to be an effective way to learn about water resource decision outcomes. This paper concludes with a discussion of learning outcomes from both playing and developing water games.

1 Introduction

One of the best ways to engage students and instill enthusiasm for hydrology is to expose them to hands-on learning. Using (serious) games in the classroom can engage students, and inspire enthusiasm, while also helping to solidify formal concepts learned in standard curriculum. Learning through games has been shown to increase soft skills, such as critical thinking, creative problem solving, and teamwork (Johnson et al., 2012), skills that are important for future water resource managers. When teaching hydrological concepts, and especially in the context of water resource sharing, where compromises between different interest groups need to be made and conflicts sometimes arise, games can be a good tool to enact different real-world scenarios. Learning through game play can thus be instructive in showing the complexity involved in the management of water resources, for both students and professionals alike (Douven et al., 2012; Rajabu, 2007). The active participation in mock decision making, through to the outcomes of those decisions using games, also allows different learning goals, including critical thinking and problem solving, to be better realized (Wu et al., 2012).

There are several games that focus on water resources, many of which have been used and tested at various levels in educational settings. Some examples include *Aqua Republica* (aquarepublica.com), an online game aimed at promoting sustainable water resource management under growing water demand and scarcity, the *World Water Game* (Deltares, 2015), where the player decides on measures to avoid water shortages in different regions of the world, and *Water: more than just a game*, from the Swiss Federal Office for the Environment (FOEN, 2015), where the player can take different water management actions for a city and rural areas along a stream reach. These types of games focus on the player as a single actor, playing to optimize prosperity for the entire society or system. Although single actor games can have a high degree of realism by trying to simulate a real system as much as possible (Medema et al., 2016), the game can become overly complex, making it more difficult to understand and less attractive in educational settings (Jones, 2011). Additionally, the idea of an individual actor is fundamentally unrealistic; in reality there are almost always many actors involved in water resource decisions.

Multi-player, role-playing games, in contrast to single-player games, allow different actors to interact, and are inherently more realistic as they provoke social learning and collaborative task activity (Hummel et al., 2010), and can thus be very useful in learning about water resource sharing in educational settings. Role-playing games may or may not have limited decision options that are evaluated in a quantitative way. Examples of role-playing games with a focus on water resource sharing where players have limited decision options include board games like the *River Basin Game* and *Globalization of Water Management* (Hoekstra, 2012), which demonstrate issues related to sharing a common resource in upstream and downstream settings, incorporating the concepts of a water footprint and virtual water trade. Other role-playing games based on negotiations between different players include the *Irrigation Management Game* (Burton, 1989, 1994) and the *River Basin Game* (Magombeyi et al., 2008). In a recent review that explores using serious games for social learning and stakeholder collaboration in transboundary watershed management, Medema et al. (2016) found that serious games, including multi-player, role-playing games, provide a promising learning platform for developing partnerships and networks, and help to increase interaction and communication between diverse stakeholder groups. Role-playing games allow players to better understand different player (stakeholder) interests and perspectives, and player dynamics, leading to specific decision outcomes. Medema et al. (2016) summarize different characteristics of serious games that lead to success in supporting social learning and stakeholder collaborations. Among these characteristics, the degree of realism is important, but the multi-player, role-playing aspects are critical in exploring the dynamics and uncertainties involved in water resource sharing over a transboundary watershed, and ultimately lead to a better understanding of how optimal outcomes can be achieved with competing interests.

Building on the idea of better understanding multi-stakeholder decisions and how stakeholders reach an outcome (and not necessarily the optimal one), Madani (2010) suggested that game theory provides a suitable framework to study the behavior and decisions of stakeholders in water resource systems. Unlike conventional systems engineering methods which typically apply optimization methods, game theory offers a more realistic approach to studying water resource systems since people inherently have different interests, and do not always act with the best system-wide outcome in mind, which conventional methods might assume (Madani, 2010). Drawing on this, Seibert and Vis (2012) developed a web-based, multi-player game, which illustrates game theoretical aspects, called *Irrigania*, to teach about wa-

ter resource sharing between several actors (or farmers) in educational settings. In *Irrigania* players act as farmers living in a village and decide how to irrigate their fields over several years, and are thus presented with water sharing situations with other farmers that are typical in real-world water-related conflicts. This game is simple in its rules, and there are few options for making decisions, which means that game outcomes can be more easily understood by students, making it a useful addition to a course on water resource management.

In the following, we address how effective games are in teaching about water resource sharing to different educational levels, through both game play and game development. An evaluation of *Irrigania* in the classroom setting is first presented, supported by feedback from several educators who have used *Irrigania* for teaching about water resource conflicts at both university and high school levels. We then discuss our experiences, together with student feedback, from a course on water games that we facilitated for masters students in geography, where students developed a board and computer game, to be used in secondary school classrooms.

2 *Irrigania* as a teaching tool

Since its inception, *Irrigania* (Seibert and Vis, 2012) has been used in different classroom settings and as an outreach tool, to teach about water resource sharing and to explore the role of cooperation in, and competition for, the use of water as a limited common-pool resource (Seibert and Vis, 2012; Pierce and Madani, 2013; Cuadrado et al., 2014). The game is played between villages made up of several farmers (usually four to six farmers per village). Each farmer has 10 fields, and they can choose to irrigate the fields with a combination of rain water, river water, or groundwater. Each irrigation source has a certain cost and revenue associated with it. Rain water and river water both have a fixed cost, while the revenue for river water depends on the number of farmers using it. For groundwater, the revenue is fixed, but the cost of groundwater increases with increasing depth to groundwater, where for $g < 8 : 20$ and for $g \geq 8 : 20 + (g - 8)^2$, where g is the depth to groundwater (in arbitrary units) and dependent upon the amount of precipitation during a given year (determined by a "precipitation indicator" where a normal year $= 1$; a dry year $= 0$, and a wet year $= 2$) as well as the number of fields irrigated with groundwater. In contrast, the cost of irrigating with river water is fixed at 20, but the revenue depends on the precipitation indicator (0; 1; 2), the number of fields irrigated with river water, and the number of farmers in the village.

The goal of the game is for each farmer to maximize his/her individual income (net of farmer revenue and costs), which to some degree requires considering the total village income. The game is usually played several times with different levels of communication and cooperation during play.

Table 1. Two *Irrigania* game scenarios played with international students during a course at CABI (Centre for Agriculture and Biosciences International), Delemont, Switzerland: Game 1 (top), a cooperative game, and Game 2 (bottom), a non-cooperative game, for farmer Susan in the village of Raintown. Farmer Susan tends to irrigate more heavily in Game 2, acting more selfishly, ending up with a lower individual income and a lower accumulated income for her village, as compared to Game 1 where the other farmers in Raintown are known to her.

Game 1: cooperative	Year	GW level	Irrigation GW	Irrigation river	Rainfed	Income	Accum. income	Accum income village
Village: Raintown	1	7.25	2	3	5	525.00	525.00	
Farmer: Susan	2	9.25	2	2	6	453.54	978.54	
	3	11.25	3	3	4	548.31	1526.85	
	4	11.75	3	2	5	442.81	1969.67	
	5	12.25	2	1	7	378.88	2348.54	
	6	13.5	3	4	3	530.92	2879.46	
	7	15.25	2	1	7	309.88	3189.33	
	8	17.75	3	2	5	239.81	3429.15	
	9	18.25	2	3	5	294.88	3724.02	
	10	19	3	4	3	272.00	3996.02	16 745.02
Game 2: non-cooperative	Year	GW level	Irrigation GW	Irrigation river	Rainfed	Income	Accum. income	Accum. income village
Village: Raintown	1	6.5	4	4	2	650.00	650.00	
Farmer: Susan	2	8	4	4	2	663.33	1313.33	
	3	10.25	4	4	2	669.75	1983.08	
	4	12.25	5	4	1	601.35	2584.44	
	5	14.75	5	4	1	517.19	3101.63	
	6	17	6	6	−2	424.00	3525.63	
	7	19.75	4	5	1	142.75	3668.38	
	8	23	5	5	0	−325.00	3343.38	
	9	22.5	1	5	4	236.42	3579.79	
	10	23	1	5	4	271.67	3851.46	11 012.46

Before play the moderator (teacher) sets the length of the game, rainfall conditions, whether or not communication between farmers and/or villages occurs (making the game either cooperative or non-cooperative), and whether users can see each other's input (information is shared). It is recommended that several rounds be played, and the settings adjusted so that different levels of information and cooperation can be explored. The game can also be played over several days, to give students more time to strategize and discuss results after a certain number of years have occurred, before continuing. The student enters the "farming decisions", i.e., the number of fields irrigated with groundwater and river water, and the number of rainfed fields (for a total of 10 fields), through a simple interface (Fig. 1). The "economical status" with balance (annual income) and accumulated balance (accumulated income) of the farmer is shown, as well as the "current hydrological conditions", on which the current year's farming decisions can be based. The student can also see when all the farmers have made their decisions at the bottom (either "submitted" or "irrigating"). Two game scenarios are shown in Table 1: the columns (from left to right) show the game scenario (Game 1, cooperative, vs. Game 2, non-cooperative), the year (1–10) for the given round, the groundwater level at the start of each year (GW level), the farming decisions taken (how many fields are irrigated with groundwater – Irrigation GW), and river water (Irrigation River) and the number of rainfed fields. The outcomes for each year follow, including the income (net revenue and costs) for each year, the accumulated income for the round, and finally the accumulated income for the entire village.

After playing the game several times, patterns related to the amount of communication and information shared usually emerge (Seibert and Vis, 2012; Pierce and Madani, 2013). In a non-cooperative setting, where no information is shared (farmers are not allowed to discuss and do not see each others' input), villages typically perform worse, whereas when full cooperation occurs, and each farmer knows who the other is, there is less selfishness and more cooperation between farmers, and this high amount of cooperation usually results in a high income for the village. This can be seen in Table 1, where two game scenarios are shown for farmer Susan from the village of Raintown. In Game 1 (top), a cooperative game, where players know who each farmer is, farmer Susan tends to irrigate moderately with both ground-

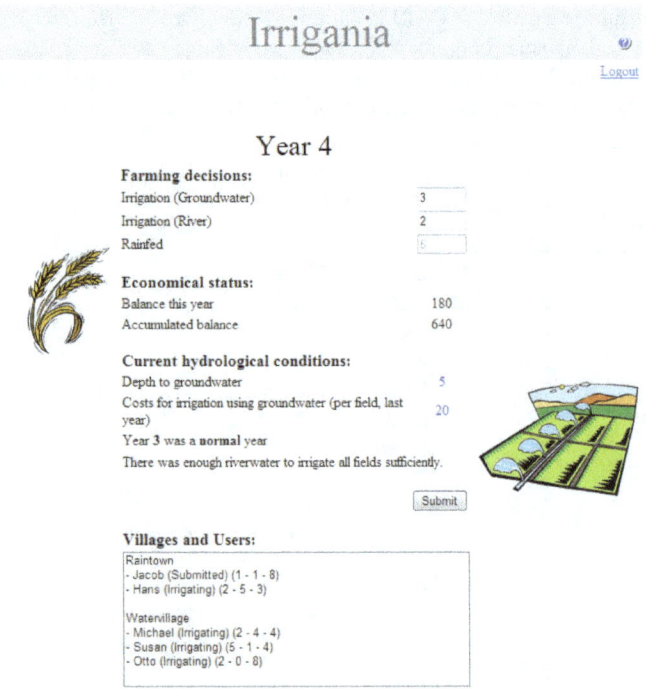

Figure 1. The student (farmer) web-interface during a game of *Irrigania* showing the "Farming decisions" taken for Year 4, the "Economical status" based on Years 1–3, and "Current hydrological conditions" to base the current year's decisions on.

water and river water over all years, reaches a high individual accumulated income, and her village wins with the highest accumulated village income (other villages not shown). Compared to Game 1, in Game 2 (bottom), a non-cooperative game, where players do not know who the other farmers are, farmer Susan tends to irrigate more heavily, reaches a moderate income, and has a lower overall income. The resulting groundwater (GW) level is much lower in Game 2 at the end of the round in year 10, with GW level = 23, compared to 19 in Game 1, reflecting the overall tendency for players to act more selfishly in the non-cooperative game setting. Similar patterns were also found to emerge by others, e.g., Pierce and Madani (2013), who played *Irrigania* as part of a larger study to better understand decision making related to common pool resources. They showed that the most important factors in promoting sustainable resource use were communication and cooperation, followed by trust, information disclosure, and social learning.

When uncertainty is introduced in the weather in the *Irrigania* setting (i.e., random amount of rainfall), decisions become more difficult and differences between farmers in their risk taking also tend to emerge. Between the different water resources, there is also learning as players improve the more they play simply by better understanding the longer-term effects of overuse in groundwater, compared to river water, which, in the game, has no year-to-year memory. In a recent study on sharing common resources among farmers

in Tanzania, Lecoutere et al. (2015) showed that gender and social status were also found to play a role; during times of water scarcity, high-status women shared fairly, whereas rich and powerful men were less worried about being greedy. Low social status (both men and women) tended to distribute water equally when it was abundant, but were more selfish when water was scarce. These different outcomes and aspects that emerge when *Irrigania* is played with different scenarios and groups of players make *Irrigania* a useful tool to both explore and understand the complexities of water resource sharing.

A survey of using *Irrigania*

To evaluate the effectiveness of *Irrigania* in teaching about water resource sharing, we carried out a survey, with an online questionnaire sent out to users (teachers) who had registered to use *Irrigania* (since 2012; 18 in total). We asked these users 15 questions in total and received feedback from nine users (see Table A1). We asked users questions ranging from basic information on how they have used the game in their classrooms, or as an outreach tool, and how they have incorporated playing the game into their curriculum. We then asked for details on the educational level of their class, the type of course it was used in, and how many students played. As responses, teachers have used *Irrigania* mainly at university level, for both bachelor and graduate courses, with one exception of using it for a high school geography course with 30 students. It has mainly been used in courses with a water resource focus (including departments of hydrology, environmental engineering, and natural resource management). One group however, in the department of psychology, played it with students to better understand environmental decision making. Group sizes ranged from 20 students to 110.

This was followed by more detailed questions on the specifics of play (how many times they played with the same group, and with different groups, and duration of play). Although some groups played it only one time, most played it frequently, and some have incorporated it into their regular class curriculum. Most groups played it once during the semester in a block of 2–4 h, but several also played it over several weeks, with up to one full semester for play.

Following the first set of questions, we asked more targeted questions to gauge the effectiveness of *Irrigania* in engaging students (whether the game held students' interest for the duration of play and how enthusiastic students were when playing the game). Teachers' responses depended strongly on the level of study. For bachelor classes that used it, most said that the game held the enthusiasm of the students for the full period, and that the students were quite enthusiastic about playing it. For the graduate level courses however, many said that a 3 h period was sufficient, since after this amount of time, the students understood the mechanics of the game and some lost interest somewhat. The high school students however wanted more graphics and visualizations to make it more interesting, and teachers commented that this would have likely held their attention for longer periods.

Questions to evaluate the effectiveness as a teaching tool were then asked, including how well *Irrigania* taught about collaboration and conflicts with regard to shared water resources and whether there was improved understanding of shared resources like surface/river water and groundwater. All teachers (regardless of level) said that *Irrigania* was moderately (four replies) to very successful (five replies), when asked "how successful" (not; moderately; very) in teaching about collaboration and conflicts with regard to shared water resources. When asked about whether they thought there was an improved understanding of shared resources like surface/river water and groundwater, all answered that there was increased learning about shared water resources, but that a discussion session afterwards was key to solidifying the concepts learnt, especially for the high school and early level bachelor students.

Since *Irrigania* is based on game theory, but is also simple in its rules, it can be a good way to teach about game theoretical considerations related to water resource sharing (Seibert and Vis, 2012). As a follow-up after game play, we asked whether any interesting patterns had evolved and how much discussion the teachers incorporated into the process of playing the game (e.g., whether they had discussions on the topics before and/or after play). We then asked a few questions related to game theory, including whether game theoretical considerations related to water resource sharing were discussed (before and/or after playing) and whether *Irrigania* was successful in teaching students (or other players) about the tragedy of the commons. Almost all teachers discussed game theoretical considerations related to water resource sharing briefly before play, but also in a final discussion after play, and this also helped to solidify learning concepts related to game theory. Almost all teachers also found that students understood, by the end of the session play, that cooperative behavior and communication were both key to succeeding. All teachers said that *Irrigania* was successful in teaching students about the tragedy of the commons and supporting discussion of these concepts (all answered "yes" to this).

Additional questions were asked on whether the teacher had used other educational games and what differences they found in teaching aspects in these games compared to *Irrigania*. Four teachers used other games in the classrooms, and all said that in comparison, *Irrigania* was very easy to use and required little preparation before using it in the class, which made it appealing. In a final question, we asked for general feedback that teachers thought would be useful for evaluating *Irrigania* as an innovative tool for learning about water resource sharing and suggestions for improving the game. Several suggestions were given; e.g., for younger students (high school), it was suggested that it should be more game-like and visually engaging. University level students however seemed to find it engaging enough, but also suggested that a

spatial interface be developed where villages could be represented visually. It was also suggested that more game settings would make it more interesting, allowing students to explore more scenarios and play longer, e.g., by setting different amounts of water from different sources and having rewards or punishments for level of sharing. Two teachers recommended that a more flexible groundwater level evaluation be implemented by allowing the game to be played with different amounts of available water to start. Another commented that allowing the results to easily be exported would be an advantage for follow-up discussion and analysis of game play.

Overall, the feedback from the survey was positive, and all teachers felt that *Irrigania* was a good tool for teaching about both shared water resources and game theory. The results highlight that the use of *Irrigania* for different levels of teaching is quite different, and that it seems to be best suited to higher bachelor level to masters level courses where students were the most engaged, it held their interest for longer, and teachers had less comments for improvements for these groups.

Additional analysis was carried out considering user data collected since July 2013, when user histories began to be saved; this excluded data collected during our own use of *Irrigania*. These data included how often users played *Irrigania* (number of games played), how long their rounds were (average game length), and over what period of time they played. The number of games played varied from only one game (users 8, 9) to 26 games played (user 10), with most users playing games with 10 years (the default setting), although user 10 played consistently shorter games, with an average of 5 years. For the game length, many users played over 1 day, but users 1 and 12 played over a 2-month period, with user 10 (with 26 games played) playing over the full period (July 2013–present). This agrees with some of the user feedback from the online questionnaire, where many teachers had used it once during the semester in a block of 2–4 h, and several also played it over several weeks, with up to one full semester for play.

3   Developing water games in the classroom

An "Integrative Project" course within the masters program at the Department of Geography at the University of Zurich is a six credit point course, corresponding to 180 working hours for the students, running over two semesters. This course has the aim of putting theory learned in the classroom into practice, and is led by different teachers or research groups within the geography department each year. In the "Integrative Project" course on "Water Games" (fall term 2014 and spring term 2015) five students, four female and one male, from the MSc program in geography participated. All students had German as a mother language and the class was taught partly in German and partly in English. In the follow-

Figure 2. Playing the *Wiapuna* board game in the final class.

ing, we first present the course as well as the design and development of two games by students that participated in the course, followed by an evaluation of learning outcomes from the course.

A first goal of the course was for the students to carry out a survey of existing water-related games, including both computer and board games. These games were then played and both positive and negative aspects of each game were discussed, followed by an analysis of what makes a good game. Students also had a couple of lectures, with one on project management followed by two lectures on game theory, given by invited game theory experts, introducing students to game theory (which *Irrigania* is based on). The second part of the course focused on the development of their own games, first through brainstorming ideas for new games, and then forming groups. The students then developed two different games: a board game, *Wiapuna* (Fig. 2), and a computer-based game, *Habitat Ganges* (Fig. 3), over a period of 6 months. Game development began with initial "idea boards" (Fig. 4) where students brainstormed possible game ideas, discussing aspects of each in class, and further in working group sessions, to narrow down their ideas. Most ideas built upon already existing games that the students had reviewed and played in the first part of the course. The games were then developed over 3 months of group work with students organizing their own group time together (including summer). During game development, students also tested (played) the games with a couple of smaller groups of their intended target audiences, to get feedback and make improvements. In a final 3 h class, the games were played by the students in the class and other geographers in the department. Overall, the players enjoyed the games and comments for improvements or changes were discussed amongst the players.

Wiapuna: *Wiapuna* was developed as a multi-player board game (Fig. 2) for both family play or play in schools or as an outreach tool, for ages 10 and older. It is based on the topic of water resource scarcity, and could be incorporated into a regular geography curriculum to supplement and enhance regular lectures. In *Wiapuna*, players build and develop settlements around four central wells (Fig. 5), where water is supplied by buying water pipes, and shared between neighbors using the same well. Natural resources (copper, gravel, wood and food, Fig. 5 right) are used to buy infrastructure. Water supply through wells is slowly depleted as more and larger houses are built around each well. New efficiency measures need to be implemented to reduce the amount of water use (e.g., through buying drip irrigation, harvesting rainwater for agriculture, and increasing efficiency in household appliances). An element of uncertainty is introduced into the game with natural events that include global and regional heavy rainfall, water poisoning, floods, droughts, tornados, or storms. The board design is based on the well-known *Settlers of Catan* board game, where players are also awarded points as their settlements grow, and like *Settlers* is won by the first player to reach a certain number of points. Game play is approximately 70–100 min long, and thus could be incorporated into the regular curriculum, where several sessions could be devoted to game play.

Habitat Ganges: *Habitat Ganges* is an online game (Fig. 3) about the sustainable use and sharing of water resources along the Ganges. This game is aimed at German speaking geography students in secondary schools, ideally for groups of 16–24 students. Time needed is approximately 90 min, which could be played in a classroom where 2×45 min sessions could be planned for play (approx. 15 rounds). The focus of the game is on the development of sustainable water use for communities (the cities of Kanpur, Varanasi, and Calcutta, and the district of Chamoli), and the consequences for the river, the communities relying on it, and the environment, caused by poor river management. Students developed the game based on the sustainability triangle, described by Heins (1994), as a way to show that sustainability needs to be approached by considering ecological, economical, and social aspects equally and all together, in an integrative way. They applied this to the idea of river management and the interaction between upstream and downstream use. The overall objective of the game is to create a sustainable river environment between the different communities (played in teams), with each community's action affecting the others, as in the case of a real river with upstream–downstream consequences for each community. The game is played by buying and trading resources (with the different resources shown in the field; Fig. 5; Table 2), in an attempt to optimize the economy, life quality, and water quality of the Ganges (Table 2, "Effects"), starting with a certain budget. The game is won by achieving the highest overall score from these three indicators, while also taking into account the total population and remaining budget.

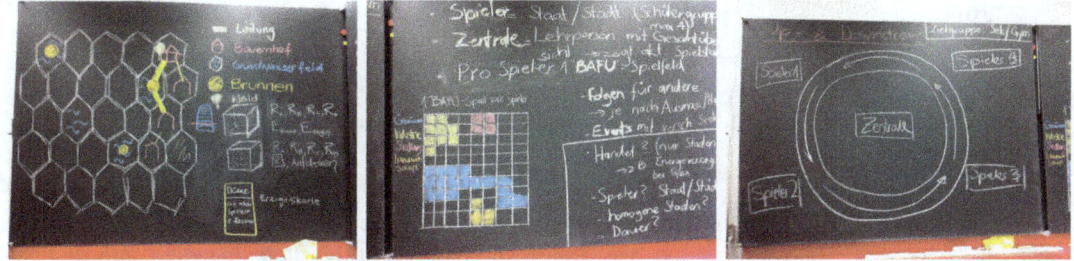

Habitat Ganges – more than just a game

Figure 3. A screenshot of *Habitat Ganges* – more than just a game (Lebensraum Ganges – Mehr als ein Spiel). Shown is the game interface for the district of Chamoli, translated from the German. Note that the resources here can be related to those shown in the resource price list in Table 2.

Figure 4. Initial stages of game development with idea boards. Board 1 (left) shows a hypothetical game board with options for introducing a pipeline (Leitung), farmyard (Bauernhof), groundwater source (Grundwasser Feld), well (Brunnen), forest (Wald), and drought (Dürre) –> event card (Ereigniskarte). Board 2 (middle): game board development based on the FOEN, 2015 game. Board 3 (right): hypothetical game idea for a computer game based on the idea of upstream downstream river use and influence on each player (Spieler).

Evaluation of the "Water Games" course

Based on feedback we received after the course from the students, one of the main comments that most of the students had about this course was that the time (two full semesters) was not enough to get introduced to different games, get into groups, and finally develop, test, and produce their own games. In the end, the rush to complete a final project, and actually produce a game (especially the board game, which required a lot of technical expertise to produce) that could be played during the final session (and used later on as a teaching or outreach tool) meant that the game testing phase was very limited. Since the course was really aimed at getting students to apply theory to practice, there is a goal to produce a product at the end that can be used for either teaching or as a communication tool. This problem in time management likely resulted from a combination of this (not having much experience in turning theory into a practical product in their studies), and from having difficulty getting started with the project (deciding on a group and idea and getting going). The latter could have been improved by giving students more time at the beginning of the course to discuss ideas. The introductory sessions/lectures could have been shorter and pos-

Figure 5. Board set-up for *Wiapuna* centered around four wells (left). Settlements are developed on different land use tiles (right), corresponding to the natural resource cards (copper – Kupfer, gravel – Kies, wood – Holz, food – Nahrung, marsh – Sumpf, and desert – Wüste) that are used to buy infrastructure and energy efficiency measures.

Table 2. Each community in *Habitat Ganges* is given a sheet of paper indicating the list of prices for each resource (in arbitrary monetary units) together with the qualitative outcome (±) for each of the indicators (economy, life quality, and water quality) needed to win the game (here only "Agriculture/fisheries" and "Industry" are shown for Calcutta resource prices, as an example).

Resource	Price/year	Yield/year		Effects		
		Resource	Budget	Economy	Quality of life	Ganges water quality
Agriculture/fisheries						
Tea plantation	60	30	20	+	0	− −
Rice field	60	30	20	+ + +	0	− −
Sugar cane plantation	60	30	20	+ + +	0	− −
Fishery	60	30	20	+ + +	0	−
Industry						
Textile factory	80	50	60	+ + + + +	+	− − −
Leather factory	80	50	60	+ + + + +	0	− − − − −
IT firm	90	60	70	+ + + + + +	+	− − −

sibly more direction while developing ideas and forming the groups given.

Students commented that the lecture on game theory was maybe the least useful part of the course; although they found it interesting, several said that what they learned in the lectures was too theoretical and not useful for them to immediately apply in their game development. Following the lectures, the next part of the course, where students reviewed existing games, worked rather well, and the students all gave positive input about this part and said it was critical for them in developing their own game ideas. This was also clear in the development of the final games, since both of the games were based on existing games that they had reviewed during this part of the course. After this, when students were given time to get into groups, discuss ideas, and get down to work, this proved to be challenging – some students had quite strong ideas about how they wanted to proceed, and what type of game they wanted to develop (based on their skills, interests, and a review of what makes a good game), without wanting to discuss too much with other students. This was intended to be a group activity, and reaching a consensus was rather important for the game development to get started. In the

end it was decided that the two games would be developed, and that one of the students would contribute to both groups. Once this decision was made, game development went reasonably smoothly, and students spent many hours discussing and testing the intricacies and complexities of water resource sharing. In each step of game development, all the possibilities resulting from each player's next move had to be evaluated, and through this process, many scenarios were thought through to the final outcome. This process meant that students learned about water resource sharing in great detail and that soft skills learning, including critical thinking, problem solving, and team work, was reinforced. Several students who did not have a background in either physical geography or hydrology also participated in the course, and although their learning curve for the material was very steep, they had an excellent grasp of the topic after having developed their games.

The overall impression of the course from students was that they had put a lot of work into the course (for the given number of credit points received) – the group project was intense, requiring them to meet and work together frequently. The deadline for the final games to be submitted was also extended into summer and the next fall semester, but they nevertheless scrambled to get the games finished over the summer holiday. As mentioned, this course was meant to emphasize practical aspects of what students learn during their masters curriculum, and students found the transition from theory to practice to be a more challenging step. Although they also had a course on project management, most of them felt that they could not apply the information learnt to their actual project. Indeed, working through the theory of project management is not likely useful without a concrete project to apply those theories to. This lecture could have maybe come later in the course, after they had formed groups, and finalized their project ideas, and then finally applied some of the project management principles to their planning. Given these minor glitches, the students were quite satisfied with having taken the course, and produced their games, and it was definitely a very new (learning) experience for everyone. A next step is to now to get others to play the games, either incorporating the games into teaching curriculum for the age appropriate levels, or possibly during hydrology/water focused outreach events as a communication and teaching tool.

4 Discussion and conclusions

In this paper, we have presented a short evaluation of how both playing games and developing games can be effective ways to learn and communicate about water resource sharing. Using *Irrigania*, a multi-player, web-based game, we presented results from a survey carried out to evaluate the effectiveness of its use in the classroom to teach about water resource sharing. Our survey showed that *Irrigania* is an effective tool for learning about

  i. water resource sharing, and that both cooperation and communication are key factors for sustainable water use;

 ii. different shared resources, including surface/river water and groundwater, and differences between them; and

iii. tragedy of the commons and support discussion of these somewhat theoretical and sometimes difficult concepts for students to grasp.

Overall, teachers found *Irrigania* to be an effective and also easy tool to incorporate into curriculum, ideally for upper level bachelor to masters level students, studying either water resources or decision making.

Learning activated through both playing and developing serious games in the classroom can provide crucial skills for future professionals to solve complex water resource problems. The complex learning through game play and game development emphasizes problem-solving, communication and collaboration, and critical reflection on wicked problems (Hummel et al., 2010), of which water resource management is

one. In a review of learning outcomes of playing serious games, Wouters et al. (2009) found that serious game play improves the acquisition of knowledge and cognitive skills, and that it seems to be promising in accomplishing attitudinal change, likely an important aspect for future water resource professionals as they transition from an educational setting to the workplace, bringing new perspectives with them. In a study on using serious games in acquiring water resource management skills, Hummel et al. (2010) found that the aspect of collaboration within serious games (in the classroom setting) can improve learning about certain problem situations applied in the workplace, according to new modes of more active and experiential learning. The focus on cooperation and communication in *Irrigania*, through its multi-player character and simple game set-up, where communication between farmers is decided before game play, thus also likely leads to improved learning of water resource sharing concepts.

An evaluation of a course on developing water games, based on our experience and student feedback, found that designing and developing their own water games was a positive learning experience for students, although they found it somewhat difficult putting theory into practice to produce their final games. Developing their own games was an active learning exercise, emphasizing what Ruben (1999) describes as "social, collaborative, and peer based" learning. During game development, students had to think through and discuss the intricacies and complexity of water resource sharing, as they enacted players' moves and water resource outcomes, and then had to reevaluate game variables. Through this process, fundamental learning about water resources took place, emphasizing soft skills, including critical thinking, problem solving, collaborative (team) learning, and time management. Several studies that have looked at the effects of collaborative learning in serious game development (Corrigan et al., 2015; Prensky, 2003; Mansour and El-Said, 2008) found that the development of serious games (within the workplace, Corrigan et al., 2015) play a role in fostering the development and improvement of various soft skills, such as communication, collaboration, or negotiation, and enhance overall collaborative learning, similar to learning outcomes from playing serious games. Corrigan et al. (2015) further suggest that "we are at the beginning of a fundamental shift in the way both learning and working is happening in organisations", and that these novel, active learning tools, including both playing and developing serious games, can add a critical collaborative dimension to decision making that cannot be learned otherwise. Our course was a first step in testing serious game development in the classroom and further insight into the learning outcomes as well as carry-on effects into the workplace would be an interesting research question that could shed light on whether just playing games (emphasizing the fun factor) might be enough to achieve similar learning effects to the full process of game development.

Appendix A

Table A1. The *Irrigania* survey questions (16, left column) sent out to 18 Irrigania users. A total of nine users responded. Responses are shown for each question, and comments when given.

Irrigania survey: use in the classroom and for outreach events	Responses	Comments
1. Have you used *Irrigania* in a classroom setting? (Yes/No)	Yes, 8 No, 1	
2. If yes, what educational level was it used for?	High school, 1 University, bachelor level, 4 University, graduate level (masters/PhD), 3	
3. What was the name of your course and what department/institute is it in?	Risk Analysis, School of Environmental Engineering, (Greece); Geography, Secondary 2 (high school; US); Geography, Oregon State University (US); Natural Resources Management and Integrated Water Resources Management (Italy); Engineering Systems Design (Singapore); Behavioral psychology, Dept. Psychology (US); Hydrology, Geography; Water resources, Environmental Engineering	Not all responded; country provided in brackets where given
4. If you have used *Irrigania* to teach about water concepts outside of a classroom setting, please let us know what kind of event it was, e.g., an outreach event or during a meeting.		No responses
If you have played *Irrigania* with students and/or other groups of players, please answer the questions below:		
5. How many students (or other players) played *Irrigania*?	Group size (number of replies) 1–10 (2) 11–20 (2) 21–50 (2) 50–80 (1) >80 (1)	
6a. How many times have you played *Irrigania* with the same group of students (or other players)?	0 (2) 1 (2) 3 (2) 3 games/same day (1) >10 (1)	
6b. How many times have you played *Irrigania* with different groups of students (or other players)?	0 (2) 1 (2) 2 (3) >5 (1)	
7. How long did the students (or other players) play *Irrigania*?	1 h (1) 2 h (2) 3 h (2) Over 1 week (2) Over 1 semester (1)	
8. Did the game hold their enthusiasm for this length of time, or could the session have been shorter/longer?	Longer (3) – Yes, the students were excited by *Irrigania* and wanted to play longer – yes, ideally it should be played for more than 2 h, e.g., 3–4 h. Shorter (2) – It is a wonderful game but the lack of visuals and graphics made it a little less engaging for the students, who are easily distracted and bored with things. – The session could have been a bit shorter as the students' enthusiasm decreased after they understood the mechanisms of the game.	

Table A1. Continued.

9. How interested/enthusiastic were the students (or other players) about the game?	Very interested (3) Very interested initially, but lost interest after ~ 1 h (2) Very interested in the game competition (2) Very interested in setting up different strategies and testing them, e.g., cooperative vs. non-cooperative (1)
10. How well in general did *Irrigania* teach about collaboration and conflicts with regard to shared water resources? (Very/Moderately/Not very successfully)	Very successful (5) Moderately successful (3) Not very successful (0)
11. Do you think there was improved understanding of shared resources like surface/river and groundwater?	Yes (8) Yes, but most did not get that far. Yes, but it is important to recall and consolidate these concepts in a debrief session.
12. Did you notice any interesting patterns that evolved when playing the game in a class?	– Cooperative behavior was improved among players – Yes. In the first rounds students were taking decisions a bit randomly. After this (testing phase), decisions started to be more rational and related to the objectives of the game.
13. Did you discuss game theoretical considerations related to water resource sharing? Before or after playing (each round)?	Before (3) After (2) Before and after (3)
14. Do you think *Irrigania* was successful in teaching students (or other players) about the tragedy of the commons?	Yes (6) Yes, more or less (2)
15. Have you used other educational games? If so, which ones? What differences did you find in teaching aspects compared to *Irrigania*?	No (4) Catchment Detox http://www.abc.net.au/science/catchmentdetox/files/home.htm
Please give any other information that might be useful in evaluating *Irrigania* as an innovative tool for learning about water resource sharing.	– Allow for more flexible groundwater levels – It has the potential to be a powerful educational tool, but it might need to be more engaging and more game-like. – It will be very useful for older children/young adults – Allow for more game settings, e.g., allow for different amounts of available water, rewards – Improve the user experience, include a nice interface with spatial representation of the villages. – Would be great if the results could be directly exported in some formats (e.g., Excel).

Acknowledgements. We thank the students in GEO401: Water Games, Marc Vic for helping to create and continue managing *Irrigania*, Sandra Pool for helping with pedagogical aspects of the Water Games course, and the H2K research group for playing games with us. Our water games can be found at http://www.geo.uzh.ch/en/units/h2k/services/water-games/.

Edited by: I. Stewart

References

Burton, M. A.: Experiences with the irrigation management game, Irrig. Drain. Syst., 3, 217–228, 1989.

Burton, M. A.: The irrigation management game: a role playing exercise for training in irrigation management, Irrig. Drain. Syst., 7, 305–318, 1994.

Corrigan, S., Zon, G. D. R., Maij, A., McDonald, N., and Martensson, L.: An approach to collaborative learning and the serious game development, Cogn. Tech. Work, 17, 269–278, doi:10.1007/s10111-014-0289-8, 2015.

Cuadrado, E., Tabernero, C., Luque, B., and Garcia, R.: Water use strategies under competition and cooperation conditions, in: Advances in Psychology and Psychological Trends Series: Psychology Applications and Developments, edited by: Pracana, C., in-Science Press, 289–296, 2014.

Deltares: World Water Game, http://world-water-game.de.softonic.com, last access: 18 November 2015.

Douven, W., Mul, M. L., Alvarez, B. F., Son, L. H., Bakker, N., Radosevich, G., and van der Zaag, P.: Enhancing capacities of riparian professionals to address and resolve transboundary issues in international river basins: experiences from the Lower Mekong River Basin, Hydrol. Earth Syst. Sci., 16, 3183–3197, doi:10.5194/hess-16-3183-2012, 2012.

FOEN: Der Umgang mit Wasser – mehr als ein Spiel, http://www.bafu.admin.ch/wassernutzung/07805/, last access: 18 November 2015.

Heins, B.: Nachhaltige Entwicklung – aus sozialer Sicht, Zeitschrift für angewandte Umweltforschung, 7, 19–25, 1994.

Hoekstra, A. Y.: Computer-supported games and role plays in teaching water management, Hydrol. Earth Syst. Sci., 16, 2985–2994, doi:10.5194/hess-16-2985-2012, 2012.

Hummel, H. G. K., van Houcke, J., Nadolski, R. J., van der Hiele, T., Kurvers, H., and Löhr, A.: Scripted collaboration in serious gaming for complex learning: Effects of multiple perspectives when acquiring water management skills, Brit. J. Educ. Technol., 42, 1029–1041, doi:10.1111/j.1467-8535.2010.01122.x, 2010.

Johnson, L., Adams, S., and Cummings, M.: The NMC Horizon Report: 2012 Higher education edition, Austin, Texas: The New Media Consortium, 36 pp., 2012.

Jones, N.: Video game: Playing with the planet, Nature Clim. Change, 1, 17–18, 2011.

Lecoutere, E., D'Exelle, B., and Van Campenhout, B.: Sharing Common Resources in Patriarchal and Status-Based Societies: Evidence from Tanzania, Fem. Econ., 21, 142–167, doi:10.1080/13545701.2015.1024274, 2015.

Madani, K.: Game theory and water resources, J. Hydrol., 381, 225–238, doi:10.1016/j.jhydrol.2009.11.045, 2010.

Magombeyi, M. S., Rollin, D., and Lankford, B.: The river basin game as a tool for collective water management at community level in South Africa, Phys. Chem. Earth, 33, 873–880, 2008.

Mansour, S. and El-Said, M.: Multi-player role playing educational serious games: a link between fun and learning, Int. J. Learn., 15, 229–240, 2008.

Medema, W., Furber, A., Adamowski, J., Zhou, Q., and Mayer, I.: Exploring the Potential Impact of Serious Games on Social Learning and Stakeholder Collaborations for Transboundary Watershed Management of the St. Lawrence River Basin, Water, 8, 175, doi:10.3390/w8050175, 2016.

Pierce, T. and Madani, K.: Online gaming for sustainable common pool resource management and tragedy of the commons prevention, in: Proc. 2013 IEEE International Conference on Systems, Man, and Cybernetics, 1765–1770, doi:10.1109/SMC.2013.304, 2013.

Prensky, M: Digital game-based learning, Comput. Entertain, 1, 1–21, 2003.

Rajabu, K. R. M.: Use and impacts of the river basin game in implementing integrated water resources management in Mkoji sub-catchment in Tanzania, Agr. Water Manage., 94, 63–72, doi:10.1016/j.agwat.2007.08.010, 2007.

Ruben, B. D.: Simulations, games, and experience-based learning: The quest for a new paradigm for teaching and learning, Simulat. Gaming, 30, 498–505, 1999.

Seibert, J. and Vis, M. J. P.: Irrigania – a web-based game about sharing water resources, Hydrol. Earth Syst. Sci., 16, 2523–2530, doi:10.5194/hess-16-2523-2012, 2012.

Wouters, P., van der Spek, E. D., and van Oostendorp, H.: Current Practices in Serious Game Research: A Review from a Learning Outcomes Perspective, in: Games-Based Learning Advancements for Multi-Sensory Human Computer Interfaces: Techniques and Effective Practices, IGI Global, 232–250, doi:10.4018/978-1-60566-360-9.ch014, 2009.

Wu, W., Chiou, W., Kao, H., Hu, C. A., and Huang, S.: Re-exploring game assisted learning research: The perspective of learning theoretical bases, Comp. Educ., 59, 1153–1161, 2012.

5

A gain–loss framework based on ensemble flow forecasts to switch the urban drainage–wastewater system management towards energy optimization during dry periods

Vianney Courdent[1,2], Morten Grum[1,a], Thomas Munk-Nielsen[1], and Peter S. Mikkelsen[2]

[1]Krüger Veolia, Søborg, 2860, Denmark
[2]Department of Environmental Engineering, Technical University of Denmark, Kgs. Lyngby, 2800, Denmark
[a]present address: WaterZerv, Environmental Services, Denmark

Correspondence to: Vianney Courdent (vatc@env.dtu.dk)

Abstract. Precipitation is the cause of major perturbation to the flow in urban drainage and wastewater systems. Flow forecasts, generated by coupling rainfall predictions with a hydrologic runoff model, can potentially be used to optimize the operation of integrated urban drainage–wastewater systems (IUDWSs) during both wet and dry weather periods. Numerical weather prediction (NWP) models have significantly improved in recent years, having increased their spatial and temporal resolution. Finer resolution NWP are suitable for urban-catchment-scale applications, providing longer lead time than radar extrapolation. However, forecasts are inevitably uncertain, and fine resolution is especially challenging for NWP. This uncertainty is commonly addressed in meteorology with ensemble prediction systems (EPSs). Handling uncertainty is challenging for decision makers and hence tools are necessary to provide insight on ensemble forecast usage and to support the rationality of decisions (i.e. forecasts are uncertain and therefore errors will be made; decision makers need tools to justify their choices, demonstrating that these choices are beneficial in the long run).

This study presents an economic framework to support the decision-making process by providing information on when acting on the forecast is beneficial and how to handle the EPS. The relative economic value (REV) approach associates economic values with the potential outcomes and determines the preferential use of the EPS forecast. The envelope curve of the REV diagram combines the results from each probability forecast to provide the highest relative economic value for a given gain–loss ratio. This approach is traditionally used at larger scales to assess mitigation measures for adverse events (i.e. the actions are taken when events are forecast). The specificity of this study is to optimize the energy consumption in IUDWS during low-flow periods by exploiting the electrical smart grid market (i.e. the actions are taken when no events are forecast). Furthermore, the results demonstrate the benefit of NWP neighbourhood post-processing methods to enhance the forecast skill and increase the range of beneficial uses.

1 Introduction

The primary objective of combined urban drainage systems (UDSs) and wastewater treatment plants (WWTPs) is to convey and treat waste water and to prevent flooding and combined sewer overflows (CSOs). In order to achieve these objectives, pipes and detention basins in combined UDSs are dimensioned to cope with relatively large rain events. Typically, surcharge of manholes and flooding is only allowed to occur on average every 10 years (as per the Danish regulations; Harremoës et al., 2005) whereas overflow occurs more frequently depending on the local environmental regulations, from 10 times per year to once in 10 years, for example. This means that during dry weather the flow is relatively low compared with the conveyance capacity of the UDS and that the storage capacity is left unused. Rainfall only occurs rarely, e.g. on the study case catchment (more details in Sect. 2.3.)

the raining period represents 7.2 % of the time. Hence, integrated urban drainage–wastewater systems (IUDWSs) are mostly under low-flow conditions. During these periods the IUDWS management objective can be switched from its priority operational focus on CSO and flood prevention towards other goals such as energy consumption and CO_2 emissions.

Denmark has the political ambitions to have a fossil fuel free energy system by 2050 which requires the development of renewable energy sources (Ministry of Foreign Affairs of Denmark, 2016). One of the main critiques towards renewable sources such as wind and solar energy is their intermittent nature. Therefore a key parameter for the transition to a green energy system is the implementation of an electric smart grid with flexible, proactive consumers to balance the fluctuating power production (Hadjsaïd and Sabonnodiere, 2012). The European Technology Platform for smart grids defines the concept of smart grids as an "electricity network that can intelligently integrate the actions of all users connected to it – generators, consumers and those that do both – in order to efficiently deliver sustainable, economic and secure electricity supplies" (www.smartgrids.eu/). Energy markets are developed, as part of the smart grid, to align electricity production and consumption through bids and offers. Hence the electricity price is based on supply and demand, creating an economic incentive to distribute the energy consumption in time (e.g. shifting non-essential energy consumption out of the consumption peaks). For further detailed history and description of electricity markets, see Weron (2006).

IUDWS can potentially be used actively to take advantage of the energy market variation. Wastewater, for example, contains organic matter which can be converted to biogas at the WWTP, and the biogas production process may provide some energy storage that is potentially useful in a smart grid context. Furthermore, during dry periods, the unused storage in the UDS can be used as a buffer to control the timing of the energy consumption associated with wastewater transportation and treatment. Figure 1 highlights that both wastewater production and energy consumption are driven by human activities and therefore have similar daily pattern. This means that the energy is generally more expensive when the need for wastewater transportation and treatment is peaking. The energy market is also influenced by other parameters (e.g. the solar and wind intensity) but on yearly average the impact of the daily consumption can be observed.

Aymerich et al. (2015) investigated the relation between the energy consumption and energy cost at a WWTP in regard to energy tariff structures (i.e. energy markets). The aeration process represents between 50 and 70 % of the WWTP process energy consumption (Rosso and Stenstrom, 2005). Leu et al. (2009) studied the impact of a varying wastewater load on the oxygen transfer efficiency and aeration costs, considering to the daily variation of the power rates, and showed that there is potential to reduce the average power costs, within the limitations of the WWTP storage capacity.

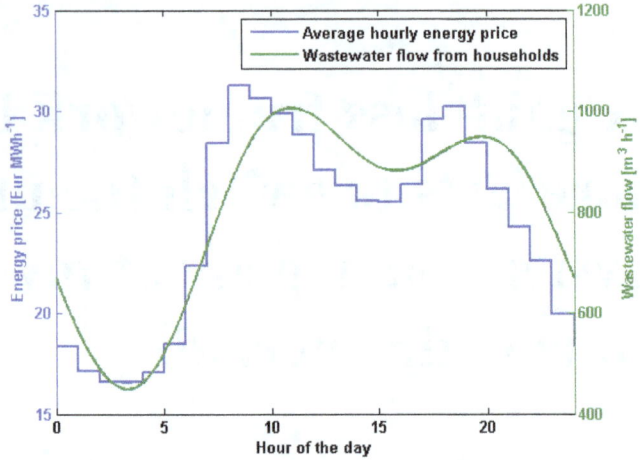

Figure 1. Yearly average (2015) of hourly energy price for the energy market DK2 covering the Copenhagen region (in blue, data from http://www.nordpoolspot.com/). Calibrated daily variation of the dry weather flow for the Damhuså catchment (green) used for demonstration in this paper; see further details in Sect. 2.3.

Bjerg et al. (2015) investigated the use of the storage volume in the pipe system upstream from a WWTP in Kolding (Denmark) to store wastewater and utilize the energy price fluctuations. However, such optimization requires information on the incoming loads (i.e. flow predictions), in order to know when it is safe to optimize the energy consumption (i.e. when the weather is dry) and when to prepare and operate the IUWDS to cope with large inflows during wet weather. Such flow predictions should ideally cover the forecast horizon of the smart grid market, i.e. 1 to 2 days (e.g. the day-ahead energy market; Zugno, 2013), which requires the use of numerical weather prediction (NWP) models.

NWPs are already in use in other fields such as wind and solar power production prediction (Bacher et al., 2009; Giebel et al., 2005), streamflow forecasting (Cuo et al., 2011; Shrestha et al., 2013), reservoir inflow prediction (Collischonn et al., 2007), flood forecasting (Damrath et al., 2000), and typhoon forecasting (Chang et al., 2015). Uncertainty is a challenge for NWP, especially for precipitation which is non-continuous and highly variable in both space and time. To tackle this problem, meteorologists commonly generate ensemble prediction systems (EPSs) by perturbing the initial conditions and the physics of the NWP models to generate a number of ensemble members (EMs) that represent an ensemble spread. The quality of an EPS can be quantitatively assessed based on various forecast characteristics. The relative operating characteristic (ROC; Mason, 1982) is used to measure the discrimination skill (i.e. the ability to discriminate between events and non-events) of an EPS, by plotting the empirical probability of detection (PoD) versus the probability of false detection (PoFD). Using an EPS increases the event discrimination skill by providing a larger range of predictions than an individual deterministic forecast.

The development of high-resolution limited-area NWP models has led to more realistic-appearing forecasts. Convective precipitations are described in an explicit and more detailed way using mesoscale atmospheric processes (Sun et al., 2014). These developments foster the opportunity of UDS applications which require fine temporal and spatial resolution. However, precipitation is one of the most difficult variables to forecast on an urban scale due to its large variability in space, time and intensity (Du, 2007). Precipitation forecast uncertainties increase rapidly with decreasing spatial grid size, as inevitable errors in the position and timing of rain cells are amplified with the increase in resolution. EPSs aim to describe this uncertainty, but are generally under-dispersive and unable to capture all sources of uncertainty. NWP post-processing methods (also called pre-processing from a hydrological modelling point of view) are thus necessary to obtain reliable probabilistic forecast as explained in WWRP/WGNE (2009). Courdent et al. (2017) described NWP post-processing methods for urban drainage flow forecasting and compared their event discrimination skills. The neighbourhood methods (Theis et al., 2005) can, for example, be used to enhance the forecast skill by accounting for potentially misplaced rain events. The "maximal threat" method NWP post-processing, used in this study, considered the highest rainfall prediction within a given area surrounding the catchment. The radius of the neighbouring area included is used as a parameter during the decision making, in addition to the fraction of EM f_{EM}.

This article presents a framework for objectively optimizing EPS forecast-based decision making in the management of IUDWSs by selecting the decision threshold f_{EM} and post-processing neighbourhood method, given the specific problem at hand. The relative economic value (REV) approach associates economic values to the outcomes of the decision system and assesses the forecast value relative to potential benefit resulting from a perfect forecast. The preferential management for a given EPS forecast is characterized by the highest REV. For example, we considered the decision-making problem of switching from normal operation focussing on flow management to dry weather operation focussing on energy optimization linking with the smart grid. To measure the usefulness of weather forecasts, the forecast skills have to be converted to potential economic benefits for the user decision making process. Richardson (2000) used the REV to assess the economic benefit of road gritting to prevent the formation of ice using weather models in comparison to using purely climatological information (i.e. the statistical behaviour of the weather, such as the return period of an event). Economic values were assigned to the different prediction outcomes described in a contingency table: (*a*) hits, (*b*) false alarms, (*c*) misses and (*d*) correct negatives. These economic values represent the benefit of taking actions (or non-action) when the forecast is revealed to be correct against the drawbacks of those actions (or non-action) in case of forecast error.

EPS provides a range of prediction skills characterized by the combined choice of post-processing method and decision threshold (f_{EM}) used to predict an event. The REV of each combination is quantified considering the occasions when the forecast proves to be beneficial, detrimental or neutral to the user, as well as the economic value associated with these situations. The higher the cost of inappropriate action relative to the potential gain, the more certainty the user requires about the forecast before he or she takes action.

Previous studies on REV analysis typically assessed the benefit of prevention measures mitigating severe weather events, such as frost (Richardson, 2000), intense precipitation (Atger, 2001), river floods (Roulin, 2007) and typhoons (Chang et al., 2015), expressed as a cost–loss ratio. This study develops a different perspective, assessing the potential benefit of optimizing IUDWS when the forecast predicts periods with low flow (i.e. dry weather when no events are forecast). Therefore, our decision model is not based on a cost–loss ratio but a gain–loss ratio. Furthermore, the studies mentioned consider a fixed ratio, whereas in our case (i) the gain depends on smart grid variations and (ii) the loss is related to the risk of CSO and the negative impact on the WWTP operation. Hence, the gain–loss ratio and the optimum combination of post-processing method and decision threshold need to be reassessed for each time step.

This paper is organized as follows: Sect. 2 introduces the DMI-HIRLAM-S05 weather model, which provides the rainfall forecast used in our study, the NWP post-processing method applied and the hydrological rainfall–runoff model. Section 3 describes the prediction performance evaluation methods used, including the ROC and the REV diagrams. Results and prediction examples are presented and discussed in Sect. 4. Finally, Sect. 5 provides the conclusions.

2 Material: NWP data, study case and hydrological model

As emphasized by Shrestha et al. (2013), the evaluation of NWP model precipitation forecasts for streamflow forecasting should be done with a hydrological perspective. Therefore, as recommended by Pappenberger et al. (2008), the evaluation of urban drainage flow forecasts is in this paper based on a coupled meteorological and hydrological model. Hence, the forecast skills are assessed based on discharge predictions and discharge observations rather than precipitation forecasts and precipitation observations. This methodology considers the importance of the dominant hydrological processes and the nonlinear error transformation by the hydrological model.

This section describes the NWP model and data used in the study. Then the post-processing neighbourhood methods are presented, the urban catchment study case is presented, the hydrological model is described, and finally the energy market data that was used is presented.

2.1 The EPS HIRLAM-DMI-S05 numerical weather prediction (NWP) model

The rainfall forecasts used in this study were generated by the DMI-HIRLAM-S05 model and were provided by the Danish Meteorological Institute (DMI). This NWP model has a horizontal resolution of 0.05° (approx. 5.6 km) and a forecast horizon of 54 h with hourly time-step predictions. New forecast are generated every 6 h, at 00:00, 06:00, 12:00 and 18:00 UTC. The DMI-HIRLAM-S05 ensemble is a 2-dimensional EPS comprising 25 members based on 5 different initial conditions and 5 different model structures. For further description of the processes and parameters mentioned above, see the HIRLAM technical documentation (Unden et al., 2002), the DMI technical report (Feddersen, 2009) and the HIRLAM website (http://www.hirlam.org/). This study uses 2 years of archived EPS NWP data (from June 2014 to May 2016).

2.2 Enhancing forecast by post-processing NWP EPS data

Two NWP post-processing methods developed in Courdent et al. (2017) were used in this study: (i) the realistic catchment "weighted areal overlap" method which only considers the grid cells overlapping the hydrologic catchment and weighs them based on the percentage of overlap and (ii) the maximal threat in the surroundings method, which considers cells within a defined radius around the catchment. The maximal threat method combines the worst-case-scenario approach and the neighbourhood method developed by Theis et al. (2005), and accounts for neighbourhood cells in the prediction as illustrated by Fig. 2. Hence, the maximal threat approach considers as input, for each EM, the highest rainfall intensity in the surroundings. This method keeps the same ensemble size as the weighted areal overlap method and reduces the number of missed events but increases the number of incorrectly predicted or over-predicted events.

2.3 Study case

The economic framework developed in this study was applied on the Damhuså urban drainage catchment (Copenhagen, Denmark). This 67 km^2 highly urbanized area composed of compact residential housing is equipped with a combined sewer system which conveys wastewater, rainfall runoff from paved surfaces and infiltration inflow, especially in the winter months. This catchment was chosen for the absence of major flow-control infrastructures affecting its hydraulic response in order to simplify the modelling approach needed for our demonstration. The Damhuså WWTP has a capacity of 350 000 PE (population equivalent). Its biological treatment has a maximal hydraulic capacity of 10 000 m^3 h^{-1}. In 2015, the WWTP treated 33 390 000 m^3 and consumed 8735 MWh of electricity, which correspond to

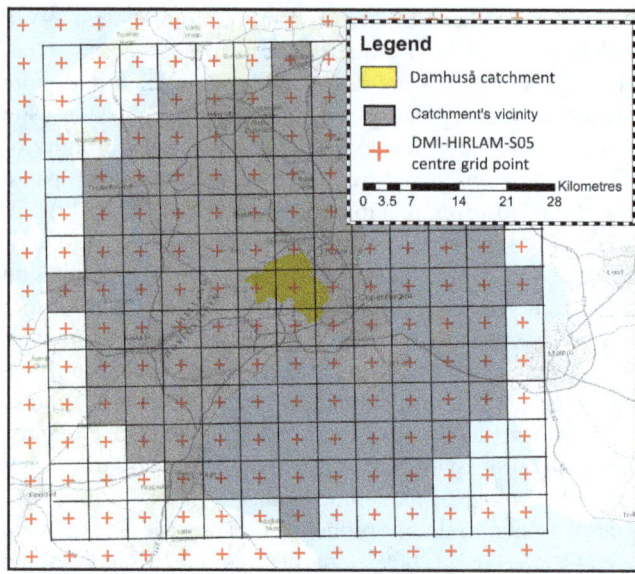

Figure 2. Illustration of the 6-grid-cell radius used by the maximal threat neighbourhood approach, for the Damhuså catchment used for demonstration in this paper (Courdent et al., 2017).

a ratio of 0.261 kWh m^{-3}. In parallel, the WWTP produced 8735 MWh of heat and 211 MWh of electricity from its biogas engine (BIOFOS, 2015).

Rainfall observation data were obtained from the national Danish SVK rain gauge network (blue circles in Fig. 3) which is operated by the Danish Meteorological Institute (DMI) and the Water Pollution Committee of the Danish Society of Engineers (SVK – Spildevandskomiteen, in Danish). The rainfall measurements were recorded with a 1 min temporal and a 0.2 mm volumetric resolution; for more information see Jørgensen et al. (1998). The catchment outlet (red hexagon in Fig. 3) is a combined sewer pipe interceptor with a maximum capacity of 10 000 m^3 h^{-1}. Once this threshold is reached, CSOs occur. The overflowing water is discharged, untreated, into a nearby small river (Damhuså) while the remaining flow is discharged through the interceptor pipe, which is monitored using an electromagnetic flow meter with a 2 min temporal resolution and operated by the utility company HOFOR.

This study is based on event prediction by characterizing the flow status in the IUDWS and distinguishing two domains: (i) periods with high flows during which the management objective is to maximize the hydraulic capacity of the WWTP to limit the impact of CSO, etc., and (ii) periods with low flows during which the management objectives can be switched to WWTP operational efficiency, minimizing energy consumption, etc. The event definition should be evaluated relatively to the specific IUDWS and low-flow optimiza-

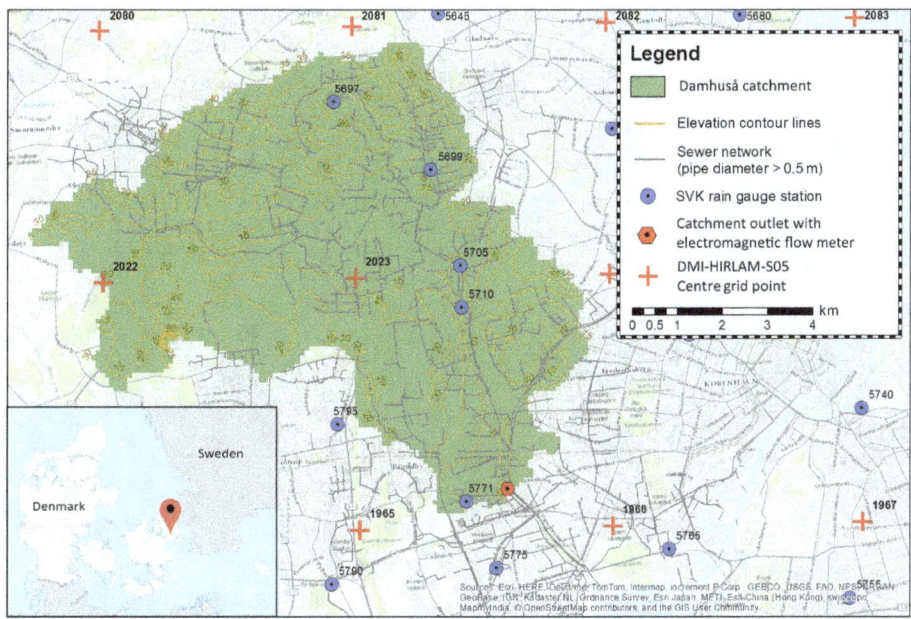

Figure 3. The Damhuså urban drainage catchment, Copenhagen, Denmark (contributing area: green area on the map).

tion scheme in focus. In this study the occurrence of an event is defined by a flow exceedance of $4000\,\mathrm{m}^3\,\mathrm{h}^{-1}$ over a 1 h period. For each NWP the occurrence (or non-occurrence) of a high-flow event is assessed for each hourly time step forecast.

2.4 Hydrological model description

The hydrological model is composed of three main conceptual parts: (i) the wastewater flow from households is modelled using second-order Fourier series (see, for example, Langergraber et al, 2008), (ii) the fast rainfall runoff from impervious areas is represented by a lumped conceptual model using the Nash linear reservoir cascade concept (Nash, 1957) and (iii) the slow runoff (caused e.g. by infiltration-inflow) is also modelled based on the Nash linear reservoir cascade concept using a wetness index characterized by the monthly potential evaporation and previous rainfall events. This hydrological model is further detailed in Courdent et al. (2017). The wastewater flow parameters were estimated first, using flow observations from summer periods without rainfall events to avoid influence from the two other processes. Then, using fixed wastewater parameters, the parameters of the fast rainfall runoff were estimated based on rain and flow data for rain events during summer months, to avoid influence from the slow runoff process, which was calibrated last for the full period (from November 2012 to November 2014). In all cases, the calibration was conducted using the differential evolution adaptive metropolis (DREAM) method (Laloy and Vrugt, 2012), considering the root mean square error as objective function.

2.5 Energy market data

This study used historical data from the day-ahead energy market provider Nord Pool. The day-ahead market has 24 h lead time. Buyers and suppliers submit bids and offers for each hour of the next day and each hourly market clearing price is set such that it balances supply and demand. The intra-day market, which only has 1 h lead time, is acting as a balancing market to support the day-ahead market. The hourly energy prices are defined over a geographical area. The geographical area corresponding to our case study is DK2 which covers the entire Zealand (http://www.nordpoolspot.com/).

3 Methodology

3.1 Contingency table

The probability that the flow will exceed a given threshold is estimated as the fraction of EMs predicting an event. The ensemble (probability) forecast can be converted to a single binary forecast by selecting a decision threshold (f_{EM}, threshold probability). If the fraction of EMs, predicting an event is higher or equal to the decision threshold (f_{EM}), then an event is forecast.

The empirical performance over a period of time of a binary forecast can be summarized in a 2×2 contingency table showing the number of correctly and incorrectly forecast events occurring or not occurring (Table 1). Hits (a) represent the correct positives, false alarms (b) represent the false positives, misses (c) represent the false negatives and the cor-

Table 1. Contingency table (with n the sample size).

Event forecast	Event observed		
	Yes	No	
Yes	hits (a)	false alarms (b)	$a+b$
No	misses (c)	correct negatives (d)	$c+d$
	$a+c$	$b+d$	$a+b+c+d=n$

Table 2. Verification measures based on the contingency table.

Score	Formula	Range	Perfect
Probability of detection, PoD	$a/(a+c)$	[0,1]	1
Probability of false detection, PoFD	$b/(b+d)$	[0,1]	0
Occurrence frequency of events, μ	$(a+c)/n$	[0,1]	n/a

n/a: not applicable.

rect negatives (d) represent the correct forecasts of no events occurring. Measures of performance of a sequence of binary forecasts can be formulated as a function of these four outcomes (a, b, c and d). Those four possible outcomes sum up to n, which corresponds to the total number of events assessed. Each event corresponds to the flow status of a given hourly time step forecast from a given NWP. The different lead times of the NWP are aggregated in the results.

Table 2 displays the verification measures used in this paper; a comprehensive review and further description of verification measures can be found in the meteorological literature, e.g. WWRP/WGNE (2009) and Wilks (2011).

The PoD is defined as the fraction of occurrences of events that were correctly forecast (i.e. hits), while the PoFD is the fraction of non-occurrences of events that were incorrectly forecast (i.e. false alarms). The empirical occurrence frequency (μ) expresses climatological information about the occurrence of events.

3.2 Brier skill score

The Brier score (Brier, 1950) assesses forecast quality of discrete probability forecasts predicting binary outcomes (i.e. "events" and "non-events") and is comparable to the mean square error. For a given tth hourly forecast time step, the forecast probability of an event ($0 \leq f_{\text{EM},t} \leq 1$) is compared to the observation (y_t). If the tth observation is an event (or non-event) then $y_t = 1$ (or $y_t = 0$).

$$\text{BS} = \frac{1}{n}\sum_{t=1}^{n}(f_{\text{EM},t} - y_t)^2 \qquad (1)$$

The Brier skill score (BSS) is formulated as a skill score related to a reference forecast, e.g. climatology in meteorology. In our case the reference forecast is based on the frequency of occurrence of events during the recorded forecast period (μ). A positive value of the BSS indicates that forecast is

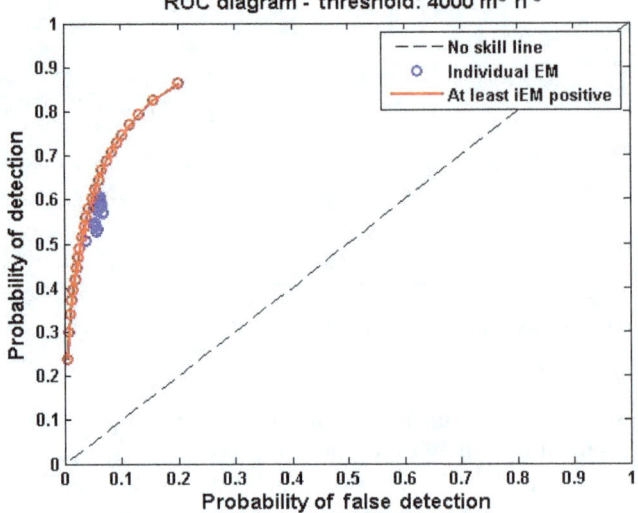

Figure 4. Example of a relative operating characteristic (ROC) diagram.

beneficial compared to the reference forecast.

$$\text{BSS} = 1 - \frac{\text{BS}}{\text{BS}_{\text{ref}}} \quad \text{with} \quad \text{BS}_{\text{ref}} = \frac{1}{n}\sum_{t=1}^{n}(\mu - y_t)^2 \qquad (2)$$

3.3 Relative operating characteristic (ROC)

The relative operating characteristic (ROC), which originates from signal detection theory (Mason, 1982), measures the discrimination ability (i.e. the ability to discriminate between events and non-events) of an EPS. The ROC plots the PoD versus the PoFD using a set of decreasing probability decision thresholds (Fig. 4). The selection of a lower decision threshold f_{EM} to convert the ensemble forecast to a single forecast is more conservative towards correctly predicting events. Therefore the PoD will be higher but the PoFD will increase as well.

The ROC diagram of the flow domain distinction using the weighted areal overlap NWP post-processing method is displayed in Fig. 4. The blue dots represent the discrimination skill of each individual EM. Figure 4 shows that all EMs have comparable discrimination skill. The red dots correspond to the discrimination skills from all decision thresholds, from $f_{\text{EM}} = 1$ at the bottom left (i.e. all EMs should agree on the event occurrence) to $f_{EM} = \frac{1}{N}$ on the top right (i.e. the prediction of an event from a single EM is enough to consider an occurrence). Figure 4 underlines that EPSs and decision thresholds provide a larger range of available prediction skills than an EM individually. The choice of a decision threshold represents a trade-off between predicting events correctly and generating false alarms.

The skill score of a ROC diagram is calculated based on the area under the curve (ROCA). The ROCA ranges from 0 to 1, with a score of 1 corresponding to a perfect forecast and a score of 0.5 corresponding to the skill of a random forecast based on the probability of occurrence (μ).

3.4 Relative economic value (REV)

A proper evaluation of the benefits of a forecast system should not only consider the forecasts skill, e.g. using PoD and PoFD, or BSS. A detailed knowledge of the decision-making process is needed to answer the question: "how does this skill translate to an economic value of a forecast?". Furthermore, when using ensemble forecasts, the following question should be answered as well: "which decision threshold and NWP post-processing method for the EPS is the most beneficial for my purpose?".

The economic benefit from a forecast depends on the alternative courses of action and their consequences. Each course of action is associated with a cost and leads to economic benefit or loss depending on the observed outcome. The task is thus to choose the appropriate actions that will maximize the expected gain or minimize the expected loss. The usefulness of the forecast can thus be quantified by considering the occasions when the forecast was beneficial, detrimental or neutral with respect to the process of decision making.

The relative economic value of our urban hydrological prediction system is here inspired by the relatively simple cost–loss ratio decision model introduced by Richardson (2000). Richardson developed this approach to assess the economic value of taking costly actions to mitigate the consequences of forecast adverse weather events in order to reduce the potential loss associated with them. The decision threshold that can empirically be shown to lead to the lowest expense in the long term should be adopted. Richardson illustrated his approach for the problem of road gritting to prevent the formation of ice. Subsequently Roulin (2007) used this approach to investigate the benefit of river-flow mitigation measures for two catchments in Belgium, and Chang et al. (2015) applied it to assess the relevance of typhoon mitigation measures in Taiwan.

All these studies consider adverse events which can be mitigated at a cost, reducing the loss associated with these events, and their decision models are therefore based on a cost–loss ratio. This study investigates a different perspective. Instead of taking mitigating measures when adverse events are predicted, the system is optimized when no events are predicted in order to achieve a positive gain, and left un-

Table 3. Economical value assigned to the different outcomes of the contingency table (L: loss; G: gain).

Event forecast	Event observed	
	Yes	No
Yes	0	0
No	L	G

der its traditional management when events are predicted. Therefore, our decision model is based on a gain–loss ratio. During low-flow periods, when no events are forecast, the management objective is switched to energy consumption by utilizing the smart grid energy market, leading to a gain (G). As a consequence, mis-predicted high-flow events will jeopardize the IUDWS, e.g. the detention basins may not be empty in time. These negative outcomes are represented by a loss (L). In the case of forecast events (hits and false alarms), the management objectives of the IUDWS remain unchanged. The economic outcome of these two situations remains the same and therefore a null value is assigned to them; see Table 3.

Furthermore Richardson (2000) used a static ratio, the cost of mitigation measures and reduction of loss associated were fixed. This study encompasses the possibility of a time-dependent gain–loss ratio. Indeed, the gain (G) from switching the management objectives to energy optimization depends on the state of the energy market at the given time. Similarly, the loss (L) resulting from mis-predicted events is related to the current status of the IUWDS, e.g. the volume of water stored.

Based on Tables 2 and 3 the expected economic value of using the forecast for decision making over one time step (n represents the total number of time steps) can be expressed empirically as follows:

$$E_{\text{forecast}} = \frac{d \cdot G - c \cdot L}{n}. \tag{3}$$

In case of a perfect forecast ($b = c = 0$) the economic value would be as follows:

$$E_{\text{perfect}} = d \cdot \frac{G}{n} = (1 - \mu) \cdot G. \tag{4}$$

If no forecasts are available, the optimal course of action can be determined based on the empirical frequency of occurrence of an event, μ (climatological information in case of weather event as for Richardson, 2000). The two possible courses of action are either to always optimize the system despite the losses or to never optimize the system. $E_{\text{statistic}}$ considers the highest economic value between these two courses of action (Eq. 5); never optimizing (i.e. the IUDWS management is unchanged) would lead to an null economic value whereas always optimizing would lead to a gain G associated to a loss L when events do occur.

$$E_{\text{statistic}} = \max(G - \mu \cdot L, 0) \tag{5}$$

The relative economic value (REV), as defined by Richardson (2000), compares the benefit of acting on a given forecast to the benefit which would be achieved by acting on a perfect forecast as a ratio (Eq. 6).

$$\text{REV} = \frac{E_{\text{forecast}} - E_{\text{statistic}}}{E_{\text{perfect}} - E_{\text{statisic}}} \tag{6}$$

The REV expressed by Eq. (6) can be reformulated using Eqs. (3), (4) and (5) and expressed as a function of the PoD, the PoFD, the frequency of occurrence (μ) and the gain–loss ratio ($\alpha = \frac{G}{L}$) as shown by Eq. (7) and displayed in Fig. 5.

$$\text{REV} = \tag{7}$$
$$\frac{\alpha \cdot (1 - \mu) \cdot (1 - \text{PoFD}) - (1 - \text{PoD}) \cdot \mu - \max(\alpha - \mu, 0)}{\alpha \cdot (1 - \mu) - \max(\alpha - \mu, 0)}$$

The possible value of the REV ranges from 1, corresponding to a perfect forecast, to minus infinity. In case of positive REV the use of the forecast is beneficial, whereas a negative REV indicates that using statistical information and either always or never optimizing the IUDWS yields a better economic value than using the weather forecast. Hence the REV can be divided in 3 domains: (i) the interval on the right of the curve in which it is preferable to always optimize (dotted domain on the right side of Fig. 5), (ii) the interval with positive REV covered by the curve in which using the forecast is beneficial (middle domain in Fig. 5) and (iii) the interval on the left in which it is preferable to never optimize (crosshatched domain on the left side of Fig. 5). Assuming that a perfect knowledge of the future yields a benefit β (compared to purely statistical information), then using the actual forecast provides a benefit to the user of $(100 \cdot \text{REV})$ % of β.

Figure 6 displays the ROC diagram and the REV-α relationship for flow forecast based on the catchment weighted areal overlap post-processing method. As explained in Sect. 3.3. the ROC diagram describes the EPS forecast discrimination skill for the different decision thresholds, f_{EM}. To support decision making the ROC diagram is converted to the REV-α relationship. Each point of the ROC diagram (Fig. 6a) represents a discrimination skill (PoD, PoFD) for a given decision threshold based on the fraction of EMs predicting an event (f_{EM}). For each of these points the REV can be determined as a function of the gain–loss ratio α (Eq. 7 and Fig. 5).

4   Results and discussion

4.1   ROC, REV and NWP post-processing methods.

The REV is closely related to the ROC diagram as indicated by Richardson (2000); Zhu et al. (2002) and illustrated in

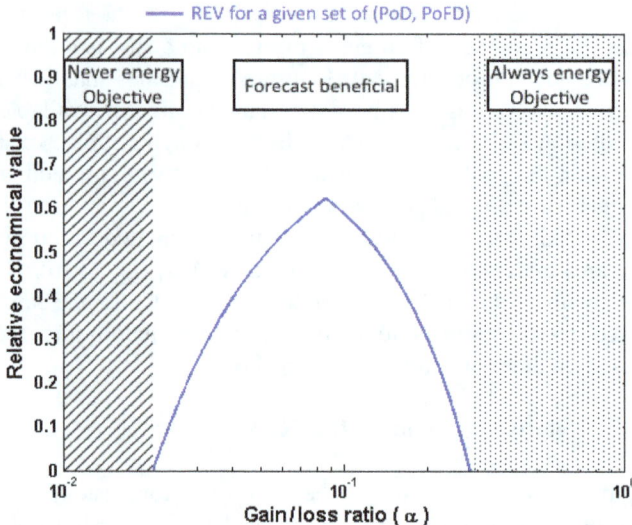

Figure 5. The 3 domains of operation of the REV curve as a function of the gain–loss ratio α.

Fig. 6. The curves in Fig. 6b show the REV-α relationship for the decision thresholds (f_{EM}) highlighted in Fig. 6a. The green dot (number 5) in Fig. 6a corresponds to a decision threshold $f_{\text{EM}} = 1/25$ and provides the highest PoD for this EPS; the REV associated with it, i.e. the green line (number 5) in Fig. 6b, leads to the highest REV for low α values (below 0.105) which corresponds to a high negative impact of missed events. Other decision thresholds yield better REV for higher α, e.g. the decision threshold $f_{\text{EM}} = 5/25$ corresponding to the red dot (legend 4) in Fig. 6a provides the highest REV (legend 4) for α within the range [0.16; 0.18]. Hence as demonstrated by Richardson (2000) the ensemble has better discrimination and can provide higher REV to a wider range of users (i.e. larger interval with positive REV) than any individual deterministic forecast (colour line) as illustrated by the envelope curve.

The implementation of the IUDWS energy consumption optimization scheme is challenged by potentially missed high-flow events. Indeed, these situations would lead to inappropriate management, jeopardizing the performance of the IUDWS. As explained in Sect. 2.2, post-processing methods can be applied to enhance the NWP, e.g. by accounting for potentially misplaced events which can have significant impact at an urban hydrology scale. Figure 7 displays the result considering the NWP maximal threat post-processing EPS method with a 6-grid-cell radius around the catchment. This approach is more conservative towards avoiding missed events and yields higher PoD at the cost of higher PoFD, which extends the ROC diagram. The ROC curves in Fig. 7a show that the two approaches are complementary; the areal overlap method provides better discrimination skill for low PoFD whereas the maximal threat EPS post-processing method provides better discrimination skill for higher PoFD.

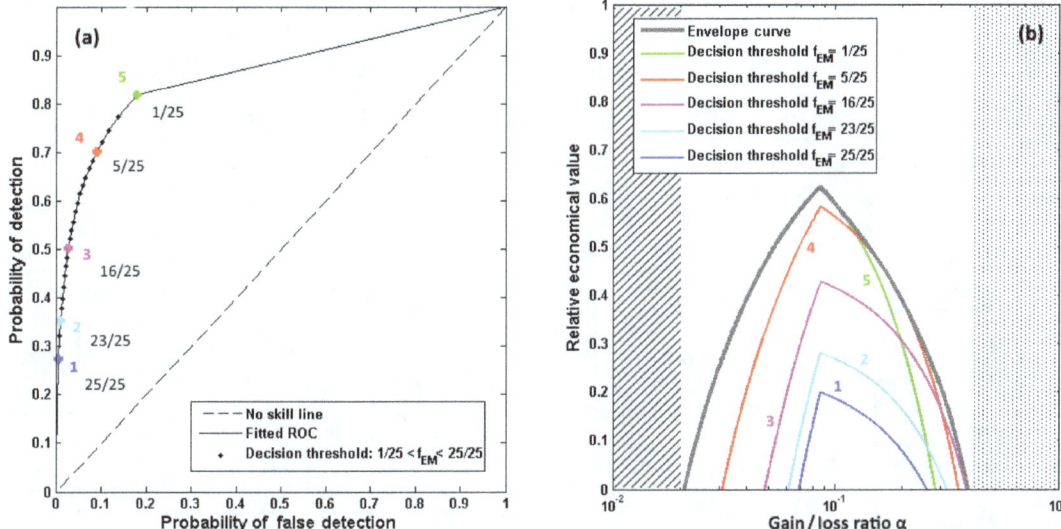

Figure 6. ROC and REV diagram for flow domain forecast based on catchment weighted areal overlap.

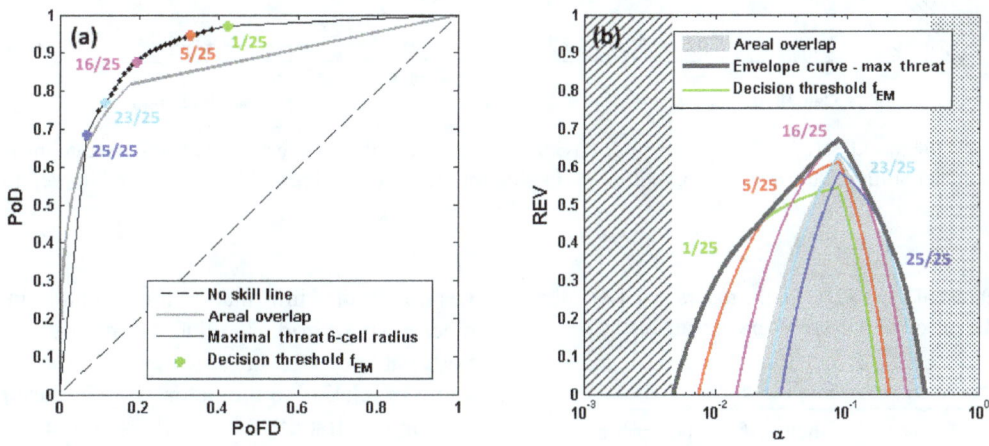

Figure 7. ROC and REV diagram for flow forecasts considering the two NWP post-processing methods: the maximal threat EPS method with a neighbourhood radius of 6 grid cells in colour and the catchment weighted areal overlap method in grey colour as background.

The ROCA of each approach is respectively 0.86 and 0.91 and the ROCA merging both approaches is 0.92.

This new ROC curve results in the extension of the α-interval with positive REV which characterizes the range of beneficial forecast use (Fig. 7b). To ease the comparison the area under the envelope curve of the areal overlap approach is displayed in grey colour as background in Fig. 7b, and Table 4 gives intervals of positive α for both approaches. The weighted areal overlap provides a slightly better upper bound whereas the maximal threat approach significantly expands the interval of positive REV for low α values. Therefore, using this NWP post-processing approach increases the range of beneficial forecast usages.

The comparison between these two NWP-post processing approaches using the Brier Skill Score (BSS) shows a deterioration of the forecast skill when using the maximal threat approach, which has a negative BSS indicating that the forecast performs worse than the reference forecast based on the frequency of occurrence of an event (μ). This decrease in performance can be explained by an increase in false alarms due to the precautions towards not missing a major rain event of this approach. This result underlines the need for an economical assessment rather than purely forecast skills to draw conclusions of the usefulness of a forecast for a given decision making situation.

4.2 Examples of EPS flow domain prediction

In order to illustrate the different situations of decision making taken as a starting point for this paper (i.e. when to switch from flow management to energy management and vice versa) a range of 4 theoretical α-values were consid-

Table 4. BSS and REV characteristics for the two different NWP post-processing methods.

	ROCA	α-interval		BBS
		Lower bound	Upper bound	
Weighted areal overlap	0.86	0.0208	0.3955	0.14
Maximal threat 6-cell radius	0.91	0.0049	0.3940	−1.52

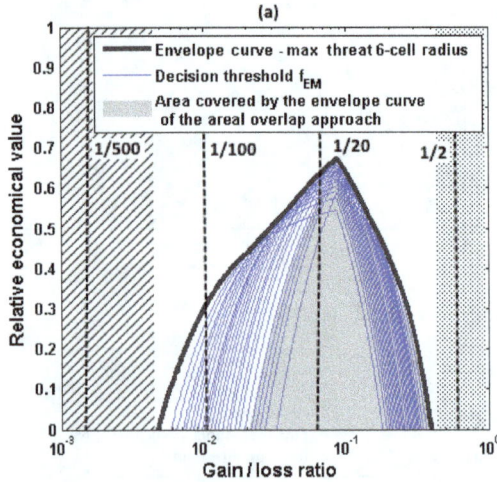 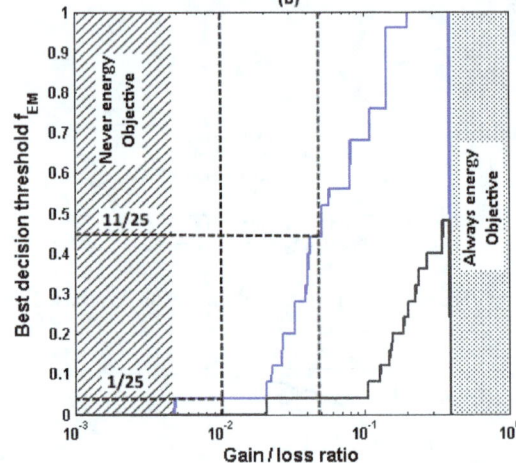

Figure 8. REV curves for the EPS NWP post-processing maximal threat in a radius of 6 grid cells from the catchment (**a**, left plot) and best decision threshold according to the α-value (**b**, right plot), in blue for the maximal threshold approach and in grey for the areal overlap approach.

Table 5. Decision threshold and REV for the theoretical 4 α-values considered, using the maximal threat post-processing method.

α	REV	Prediction criteria	
		Decision Threshold	NWP post-processing
1/2	Negative	Always energy objective	
1/20	0.59	$f_{EM} = 11/25$	Maximal Treat EPS
1/100	0.30	$f_{EM} = 1/25$	Maximal Treat EPS
1/500	Negative	Never energy objective	

ered, Table 5. The two outer α-values yield negative REV indicating that using the forecast data is not beneficial in these cases. The two other α-values yield positive REV indicating that using the forecast is beneficial in these cases. The decision threshold (f_{EM}) generating the highest relative benefit based on empirical data are displayed in Fig. 8 and in Table 5.

The coupled hydro-meteorological model provides an ensemble prediction of the flow at the catchment outlet for the incoming 2 days. Figure 9 provides an example of prediction. The first panel, Fig. 9a, displays the energy market during those two days, providing insight in the variation of the energy price and the CO_2 footprint through the proportion of wind energy. The shown data are based on historical val-

ues but similar information are available in real time on the electric smart grid. The fluctuation of the energy market for both parameters (Fig. 9a) illustrates the variation of the α-value in relation to the potential gain during a given period. During the first day (29 April 2015) the energy price ranges from 24 to 32 € MWh^{-1} and the proportion of wind energy varies from 15 to 49 %, whereas during the second day (30 April 2015) the energy price range from 23 to 41 € MWh^{-1} and the proportion of wind energy varies from 1 % to above 53 %. Hence the switch of consumption of 1 MWh can yield up to EUR 8 during the first day and up to EUR 18 during the second day. For comparison, the energy consumption per m^3 treated at Damhuså WWTP in 2015 was 0.261 kWh m^{-3} and in average 20 000 m^3 are treated during a dry day. Pumping and aeration of the biological treatment are the dominating energy users. The aeration of the bioreactor represents between 50 and 70 % of process energy consumption and largely depends on the inflow/load to the WWTP (Aymerich et al., 2015). The potential for energy switch highly dependents to the storage volume available upstream.

The North Pool Energy Market DK2, covering the Copenhagen area, has a Pearson correlation coefficient of −0.52 between energy price and proportion of wind energy in 2015, indicating a moderate negative linear relationship. Hence energy consumption optimizations based on economic objectives could also yield environmental benefits and vice versa.

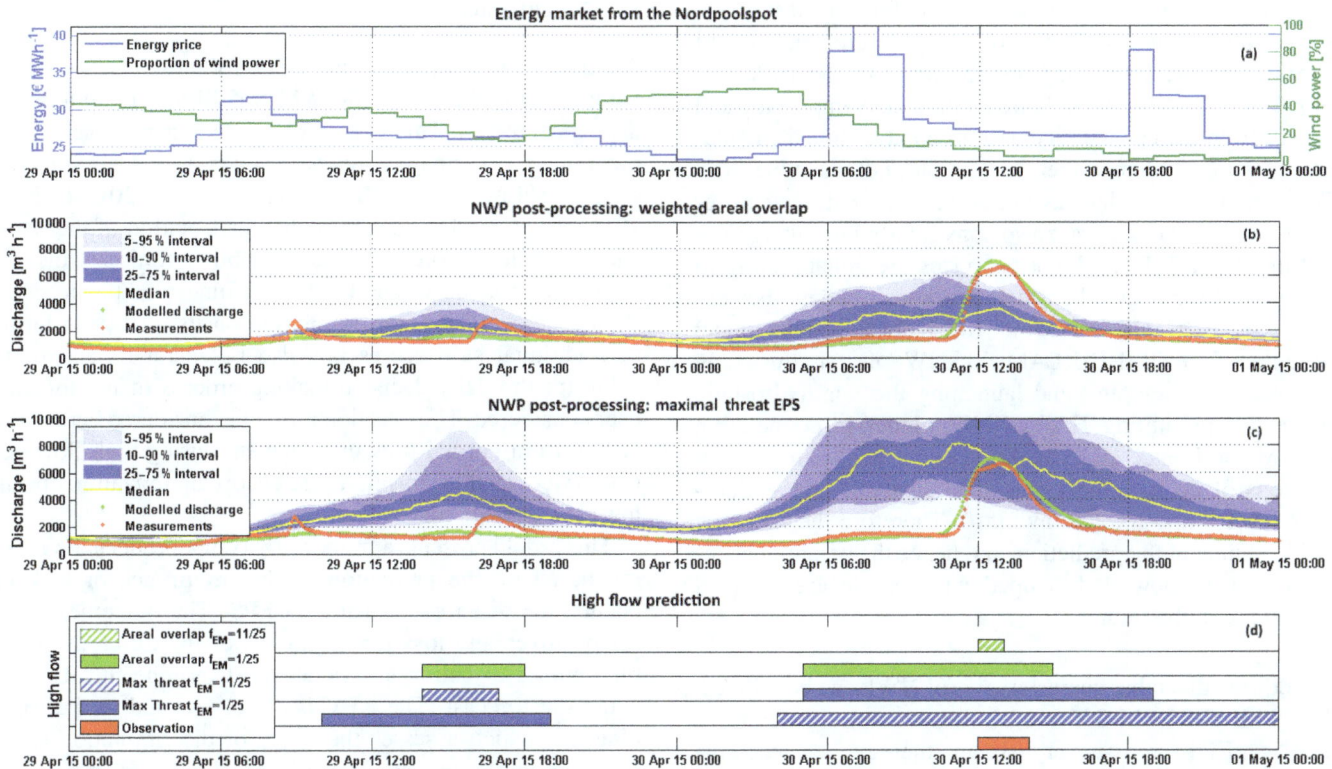

Figure 9. Example illustration of the EPS flow prediction system for 2 selected days, 29–30 April 2015. Energy market parameters, energy price and proportion of wind power (1, **a**), ensemble flow predictions using the areal average (**b**) and maximal threat (**c**) post-processing methods, and (**d**) flow domain predictions for the two post-processing methods and for each two decisions thresholds; cf. Table 5 (coloured areas imply that an event is predicted, otherwise not).

However it should be noticed that the control of the energy consumption based on the energy market can results in a decrease of the expenses together with an increase of the overall energy consumption as observed in Aymerich et al. (2015).

Figure 9b represents the flow forecast based on the catchment weighted areal overlap approach and Fig. 9c represents the flow forecast based on the maximal threat EPS approach with a 6-grid-cell radius. The measured flow during this period shows two minor rain events without significant flow impact the first day and a major rain event leading to high flows exceeding the $4000\,\mathrm{m}^3\,\mathrm{h}^{-1}$? in the IUDWS the second day. Figure 9b illustrates the difficulty of the prediction to have a correct timing, most EMs predict the high-flow event but often too early. It can be noticed that due to the conservativeness of this second approach the EPS plume of flow forecasts overestimates the observed flow (in red), which explains the worsening of the BSS when using this approach.

The best flow domain predictions, considering a given α, is provided by the decision threshold defined using the REV method presented in Sect. 3.4. As displayed in Table 5, the highest REV for $\alpha = 1/20$ (respectively $\alpha = 1/100$) is achieved using the NWP post-processing approach "Maximal Threat EPS" with $f_{\mathrm{EM}} = \frac{11}{25}$ (respectively $f_{\mathrm{EM}} = \frac{1}{25}$). The flow domain predictions based on these criteria and on

the EPS flow forecast displayed in Fig. 9b and c are shown by the blue hatched (respectively plain blue) colour in Fig. 9d.

4.3 Outlooks

As mentioned in Sect. 4.2, the potential benefit from the energy consumption optimization management is largely conditioned by the storage volume available upstream. A major project is currently under implementation to comply with new regulations on CSO. Two large pipes will be constructed just before the inlet of the WWTP with a volume equivalent to the daily dry weather flow to the WWTP. This large storage volume, soon available upstream from the WWTP, provides an opportunity for real world implementation of the concept developed in this paper. Halvgaard et al. (2017) present a model predictive control (MPC) to control the power consumption of pumps in a sewer system and the treatment power consumption according to electricity prices and effluent quality (nitrogen) based on a case study at Kolding. The controller is able to balance electricity costs and treatment quality during predicted dry weather flow periods.

The predictions and therefore the skills of the EPS are based on a coupled meteorological and hydraulic model. This study used a lumped conceptual hydraulic model; a more detailed hydrological model, including stochastic processes and on-line assimilation of flow measurements, might improve the prediction and thereby improve the REV further. Similarly, NWP models are continuously improving and benefit from the constant increase of computational calculation power to enhance their resolution and ensemble size. The techniques of data assimilation from radar measurement into NWP models are also consistently improving (Korsholm et al., 2015). Weather services are collaborating to continuously improve their meteorological models. For example, the HIRLAM consortium which developed the model structure of the DMI-HIRLAM-S05 NWP used in this study is currently developing and launching the non-hydrostatic convection-permitting HARMONIE model in cooperation with Météo-France and ALADIN, and EPSs with forecast horizons of up to 2 weeks are also available at the European level (http://www.ecmwf.int/). Therefore the accuracy and lead time of the prediction, and hence the potential benefit from the framework developed in this article, are expected to increase in the future.

Additionally, other characteristics of NWP can be utilized. The DMI-HIRLAM-S05 model, for example, generates a new 54 h EPS forecast every 6 h, and thereby the successive forecasts are overlapping each other. The forecast consistency, or in reverse the "forecast jump", provides valuable information on forecast uncertainty which could be utilized in the decision-making process. For example, the time-lagged method (Mittermaier, 2007) uses consecutive forecast overlapping to extend the EPS and enhance the predictions (i.e. the horizon of the forecast is reduced but its ensemble size is increased). This may increase the range of positive REV and allow use of the concept for decisions related to other problems than the energy optimization problem studied here.

Control systems can be decomposed into different layers in a hierarchy. Mollerup et al. (2016) presents a methodological approach to the design of optimized control strategies for sewer systems. The framework presented in this paper targets the upper layer of the hierarchy presented by Mollerup et al. (2016): the management of objectives where switching between different operational modes may take place. Completely different optimizing control strategies, including model predictive control techniques, may then run under different operation conditions – such as the "flow control" and "energy optimization" operational modes considered in this paper. Implementing such a switching system in practice requires that the gain–loss ratio expressing the economic consequences associated with the outcomes of the different courses of action used for the REV is quantified, which requires further research on monetization of non-market goods (e.g. CO_2 footprint or the environmental impact of CSOs) and may depend on local circumstances.

5 Conclusions

An ensemble flow prediction system for an IUDWS was developed using the DMI-HIRLAM-S05 EPS as input to a hydrological model. This system was tested on an urban catchment in the Copenhagen area based on recorded rainfall forecasts and flow data for the period from June 2014 to May 2016. Ensemble forecasting requires adaptation of the management rules in order to use probability forecasts instead of a deterministic forecast. The usefulness of the forecast should be evaluated not only based on its quality in terms of traditional skill scores but also based on its economic value for the daily decision-making process of the forecast user considered. The decision problem considered here is the switch from normal flow management during high-flow periods (wet weather) to smart grid energy optimization during low-flow periods (dry weather).

This article presents a framework to support decision making based on the prediction of the occurrence or lack of occurrence of an event using an EPS. The outcomes (gain for positives and loss for negatives) of the different possible courses of action are valued to determine the REV of using the forecast. The REV is closely related to the ROC diagram, which assesses the range of discrimination skills of an ensemble forecast. Hence, a REV curve, as a function of the gain / loss ratio α, can be generated for each probability threshold (f_{EM}) of the EPS. This method was developed in order to switch the IUDWS management objective from flow management to energy optimization, utilizing the electric smart grid when low-flow periods are predicted. This approach is based on daily optimization when non-events (dry weather) are forecast and differs from previous studies based on the REV concept, which investigated mitigation measures taking place when adverse events are forecast (e.g. flood, tornado) using a cost–loss ratio. In our approach for a given gain–loss ratio α, the probability threshold (f_{EM}) corresponding to the highest REV, symbolized by the envelope curve, should be applied to maximize the benefit of the optimization scheme. If the gain–loss ratio is outside the range of positive REV, then using the forecast is not beneficial. The gain–loss ratio α is a function of the potential gain from utilizing the variation of the smart grid energy market, which varies in time.

Two NWP post-processing methods were tested: (i) a realistic approach based on the weighted areal overlap between the NWP grid cells and the hydrological catchment, and (ii) a more conservative approach considering the maximal rainfall threat in the catchment vicinity. The second approach leads to a deterioration of classic forecast validation scores such as BSS due to a significant increase in the number of false alarms. However, this approach proves to be beneficial in regard to the decision-making process, especially when considering a low gain–loss ratio α for which missed forecast events are highly detrimental. Indeed, the maximal threat

NWP neighbourhood post-processing method improves the range of discrimination skill of the predictions shown on the ROC diagram and therefore provides a larger range of positive REV, increasing the range of beneficial forecast usage. This underlines the importance of assessing the forecast usefulness based on its potential economic value rather than solely on the usual forecast skills.

Competing interests. The authors declare that they have no conflict of interest.

Acknowledgements. This research was financially supported by the industrial PhD programme of the Innovation Fund Denmark. The catchment and flow data were kindly provided by Copenhagen Utility Company (HOFOR). We would like to thank the Danish Meteorological Institute (DMI), especially Henrik Feddersen, for providing EPS data from their NWP model DMI-HIRLAM-S05.

Edited by: P. Willems

References

Atger, F.: Verification of intense precipitation forecasts from single models and ensemble prediction systems, Nonlin. Processes Geophys., 8, 401–417, doi:10.5194/npg-8-401-2001, 2001.

Aymerich, I., Rieger, L., Sobhani, R., Rosso, D., and Corominas, L.: The difference between energy consumption and energy cost: Modelling energy tariff structures for water resource recovery facilities, Water Res., 81, 113–123, doi:10.1016/j.watres.2015.04.033, 2015.

Bacher, P., Madsen, H., and Nielsen, H. A.: Online short-term solar power forecasting, Sol. Energy, 83, 1772–1783, doi:10.1016/j.solener.2009.05.016, 2009.

BIOFOS: Miljøberetning, Copenhagen, available at: http://www.biofos.dk/wp-content/uploads/2014/11/Miljoeberetning-2015.pdf (last access: 1 April 2017), 2015.

Bjerg, J. E., Grum, M., Courdent, V., Halvgaard, R., Vezzaro, L., and Mikkelsen, P. S.: Coupling of Weather Forecasts and Smart Grid-Control of Wastewater inlet to Kolding WWTP (Denmark), in: 10th International Urban Drainage Modelling Conference, 47–59, Mont Sainte-Anne, Québec, Canada, 2015.

Brier, G. W.: Verification of forecasts expressed in terms of probability, Mon. Weather Rev., 78, 1–3, doi:10.1175/1520-0493(1950)078<0001:VOFEIT>2.0.CO;2, 1950.

Chang, H.-L., Yang, S.-C. and Yuan, H.: Analysis of the Relative Operating Characteristic and Economic Value Using the LAPS Ensemble Prediction System in Taiwan, Mon. Weather Rev., 143, 1833–1848, doi:10.1175/MWR-D-14-00189.1, 2015.

Collischonn, W., Morelli Tucci, C. E., Clarke, R. T., Chou, S. C., Guilhon, L. G., Cataldi, M., and Allasia, D.: Medium-range reservoir inflow predictions based on quantitative precipitation forecasts, J. Hydrol., 344, 112–122, doi:10.1016/j.jhydrol.2007.06.025, 2007.

Courdent, V., Grum, M., and Mikkelsen, P. S.: Distinguishing high and low flow domains in urban drainage systems 2 days ahead using numerical weather prediction ensembles, J. Hydrol., doi:10.1016/j.jhydrol.2016.08.015, in press, 2017.

Cuo, L., Pagano, T. C., and Wang, Q. J.: A Review of Quantitative Precipitation Forecasts and Their Use in Short- to Medium-Range Streamflow Forecasting, J. Hydrometeorol., 12, 713–728, doi:10.1175/2011JHM1347.1, 2011.

Damrath, U., Doms, G., Frühwald, D., Heise, E., Richter, B., and Steppeler, J.: Operational quantitative precipitation forecasting at the German Weather Service, J. Hydrol., 239, 260–285, doi:10.1016/S0022-1694(00)00353-X, 2000.

Du, J.: Uncertainty and Ensemble Forecast, National Weather Service, available at: http://www.nws.noaa.gov/ost/climate/STIP/STILecture1.pdf (last access: 1 April 2017), 2007.

Feddersen, H.: A Short-Range Limited Area Ensemble Prediction System, Danish Meteorological Institute, Copenhagen, available at: http://www.dmi.dk/fileadmin/Rapporter/TR/tr09-14.pdf (last access: 1 April 2017), 2009.

Giebel, G., Badger, J., Landberg, L., Nielsen, H. A., Nielsen, T. S., Madsen, H., Sattler, K., Feddersen, H., Vedel, H., Tøfting, J., Kruse, L. and Voulund, L.: Wind power prediction using ensembles, Roskilde, available at: http://orbit.dtu.dk/files/57134275/ris_r_1527.pdf (last access: 1 April 2017), 2005.

Hadjsaïd, N. and Sabonnodiere, J.-C.: Smart Grids, First, ISTE Ltd, London UK, 2012.

Halvgaard, R., Vezzaro, L., Mikkelsen, P. S., Grum, M., Munk-Nielsen, T., Tychsen, P., and Madsen, H.: Integrated Model Predictive Control of Wastewater Treatment Plants and Sewer Systems in a Smart Grid, 1–16, Control Eng. Pract., submitted, 2017.

Harremoës, P., Pedersen, C. M., Laustsen, A., Sørensen, S., Laden, B., Friis, K., Andersen, H. K., Linde, J. J., Mikkelsen, P. S., and Jakobsen, C.: Funktionspraksis for afløbssystemer under regn., IDA Spildevandskomiteen, 2005.

Jørgensen, H. K., Rosenorn, S., Madsen, H., and Mikkelsen, P. S.: Quality control of rain data used for urban runoff systems, Water Sci. Technol., 37, 113–120, 1998.

Korsholm, U. S., Petersen, C., Sass, B. H., Nielsen, N. W., Jensen, D. G., Olsen, B. T., Gill, R., and Vedel, H.: A new approach for assimilation of 2D radar precipitation in a high-resolution NWP model, Meteorol. Appl., 22, 48–59, doi:10.1002/met.1466, 2015.

Laloy, E. and Vrugt, J. A.: High-dimensional posterior exploration of hydrologic models using multiple-try DREAM (ZS) and high-performance computing, Water Resour. Res., 48, 1–18, doi:10.1029/2011WR010608, 2012.

Langergraber, G., Alex, J., Weissenbacher, N., Woerner, D., Ahnert, M., Frehmann, T., Halft, N., Hobus, L., Plattes, M., Spering, V., and Winkler, S.: Generation of diurnal variation for influent data for dynamic simulation, Water Sci. Technol., 57, 1483–1486, doi:10.2166/wst.2008.228, 2008.

Leu, S.-Y., Rosso, D., Larson, L. E., and Stenstrom, M. K.: Real-time aeration efficiency monitoring in the activated sludge process and methods to reduce energy consumption and operating costs, Water Environ. Res., 81, 2471–2481, doi:10.2175/106143009X425906, 2009.

Mason, I.: A model for assessment of weather forecasts, Aust. Met. Mag., 30, 291–303, 1982.

Ministry of Foreign Affairs of Denmark: Independent from fossil fuels by 2050, available at: http://denmark.dk/en/green-living/strategies-and-policies/independent-from-fossil-fuels-by-2050, last access: 1 April 2017.

Mittermaier, M. P.: Improving short-range high-resolution model precipitation forecast skill using time-lagged ensembles, Q. J. Roy. Meteor. Soc., 133, 1487–1500, doi:10.1002/qj.135 , 2007.

Mollerup, A. L., Mikkelsen, P. S., and Sin, G.: A methodological approach to the design of optimising control strategies for sewer systems, Environ. Model. Softw., 83, 103–115, doi:10.1016/j.envsoft.2016.05.004, 2016.

Nash, S. E.: The Form of the Instantaneous Unit Hydrograph, IASH Publ., 114–121, 1957.

Pappenberger, F., Scipal, K., and Buizza, R.: Hydrological aspect of meteorological verification, Atmos. Sci. Lett., 9, 43–52, doi:10.1002/asl.171, 2008.

Richardson, D. S.: Skill and relative economic value of the ECMWF ensemble prediction system, Q. J. Roy. Meteor. Soc., 126, 649–667, 2000.

Rosso, D. and Stenstrom, M. K.: Comparative economic analysis of the impacts of mean cell retention time and denitrification on aeration systems, Water Res., 39, 3773–3780, doi:10.1016/j.watres.2005.07.002, 2005.

Roulin, E.: Skill and relative economic value of medium-range hydrological ensemble predictions, Hydrol. Earth Syst. Sci., 11, 725–737, doi:10.5194/hess-11-725-2007, 2007.

Shrestha, D. L., Robertson, D. E., Wang, Q. J., Pagano, T. C., and Hapuarachchi, H. A. P.: Evaluation of numerical weather prediction model precipitation forecasts for short-term streamflow forecasting purpose, Hydrol. Earth Syst. Sci., 17, 1913–1931, doi:10.5194/hess-17-1913-2013, 2013.

Sun, J., Xue, M., Wilson, J. W., Zawadzki, I., Ballard, S. P., Onvlee-Hooimeyer, J., Joe, P., Barker, D. M., Li, P. W., Golding, B., Xu, M. and Pinto, J.: Use of nwp for nowcasting convective precipitation: Recent progress and challenges, B. Am. Meteorol. Soc., 95, 409–426, doi:10.1175/BAMS-D-11-00263.1, 2014.

Theis, S. E., Hense, A., and Damrath, U.: Probabilistic precipitation forecasts from a deterministic model: a pragmatic approach, Meteorol. Appl., 12, 257, doi:10.1017/S1350482705001763, 2005.

Unden, P., Rontu, L., Jarvinen, H., Lynch, P., Calvo, J., Cats, G., Cuxart, J., Eerola, K., Fortelius, C., Garcia-moya, J. A., Jones, C., Lenderlink, G., Mcdonald, A., Mcgrath, R., and Navascues, B.: HIRLAM-5 Scientific Documentation, Norrkoping, available at: http://hirlam.org/index.php/component/docman/doc_view/270-hirlam-scientific-documentation-december-2002?Itemid=70 (last access: 1 April 2017), 2002.

Weron, R.: Modeling and forecasting electricity loads and prices: A statistical approach, First Edn., John Wiley & Sons Ltd., 2006.

Wilks, D. S.: Statistical Methods in the Atmospheric Sciences, Elsevier, 2011.

WWRP/WGNE: Recommendations for the Verification and Intercomparison of QPFs and PQPFs from Operational NWP Models, Geneva, Switzerland, available at: http://www.wmo.int/pages/prog/arep/wwrp/new/documents/WWRP2009-1_web_CD.pdf (last access: 1 April 2017), 2009.

Zhu, Y., Toth, Z., Wobus, R., Richardson, D., and Mylne, K.: The economic value of ensemble–based weather forecasts, B. Am. Meteorol. Soc., 83, 73–83, 2002.

Zugno, M.: Optimization under uncertainty for management of renewables in electricity markets, University of Denmark, available at: http://orbit.dtu.dk/en/publications/optimization-under-uncertainty-for-management-of-renewables-in-electricity-markets(d314a0c4-185f-4983-95ff-21133defd41d).html (last access: 1 April 2017), 2013.

6

Robust global sensitivity analysis of a river management model to assess nonlinear and interaction effects

L. J. M. Peeters[1], G. M. Podger[2], T. Smith[2], T. Pickett[3], R. H. Bark[4], and S. M. Cuddy[2]

[1]CSIRO Land and Water, Water for a Healthy Country Flagship, Adelaide, Australia

[2]CSIRO Land and Water, Water for a Healthy Country Flagship, Canberra, Australia

[3]CSIRO Land and Water, Water for a Healthy Country Flagship, Brisbane, Australia

[4]CSIRO Ecosystem Sciences, Water for a Healthy Country Flagship, Brisbane, Australia

Correspondence to: L. J. M. Peeters (luk.peeters@csiro.au)

Abstract. The simulation of routing and distribution of water through a regulated river system with a river management model will quickly result in complex and nonlinear model behaviour. A robust sensitivity analysis increases the transparency of the model and provides both the modeller and the system manager with a better understanding and insight on how the model simulates reality and management operations.

In this study, a robust, density-based sensitivity analysis, developed by Plischke et al. (2013), is applied to an eWater Source river management model. This sensitivity analysis methodology is extended to not only account for main effects but also for interaction effects. The combination of sensitivity indices and scatter plots enables the identification of major linear effects as well as subtle minor and nonlinear effects.

The case study is an idealized river management model representing typical conditions of the southern Murray–Darling Basin in Australia for which the sensitivity of a variety of model outcomes to variations in the driving forces, inflow to the system, rainfall and potential evapotranspiration, is examined. The model outcomes are most sensitive to the inflow to the system, but the sensitivity analysis identified minor effects of potential evapotranspiration and nonlinear interaction effects between inflow and potential evapotranspiration.

1 Introduction

Water managers rely heavily on models to predict future water availability, optimize water use and evaluate water man-agement strategies in order to find a balance between environmental, social and economic demands on the system. It is therefore crucial to be aware of the ability of a model to capture the dynamics of the hydrological cycle relevant to the water management question. In recent decades, addressing this issue has been the focus of much research in hydrological model calibration and predictive uncertainty analysis (Gupta et al., 2012).

For a modeller, to arrive at a "well"-calibrated model or to produce sensible and robust prediction intervals, it is essential to have a thorough understanding of how the hydrological system works and how this system is represented in the model – how a variation in parameters, boundary conditions or driving forces will affect the prediction of interest. The knowledge gained from such sensitivity analysis is not only of relevance during model development, it also provides added value to the model as it can focus management and monitoring to those aspects of the system and model that are most important to the management of water resources (Saltelli et al., 2008). Additionally, discussing model sensitivities with stakeholders will remove the notion of the model being a "black box" and can provide stakeholders with a better appreciation of the accuracy of the model, which has proven to be a key aspect of adoption of model results by management (Patt, 2009; Bark et al., 2013).

River management models such as eWater Source (Welsh et al., 2013) are increasingly used, especially in Australia, in the development of basin-wide water allocation plans. As these plans directly affect the livelihood of people and the health of ecosystems, it is essential that the models under-pinning these plans have wide support and are robust. It is therefore essential that practitioners have a set of tools for sensitivity analysis available, tailored to the needs of water allocation modelling. The most straightforward sensitivity analysis technique is One-At-a-Time (OAT) sensitivity analysis in which one model aspect is changed while the others are fixed. The sensitivity of the model output to variation of the tested parameter is proportional to the gradient of the response surface. This is formalized in gradient-based calibration routines, such as Levenberg–Marquardt optimization. Examples of such OAT sensitivity analysis are Doherty and Hunt (2009), Foglia et al. (2009), Castaings et al. (2009) and Peeters et al. (2011). This methodology is attractive as it requires a very limited number of model runs, about two or three model runs per parameter evaluated, and, as long as the model behaves linearly, parameter interaction effects can be explored (Hill and Tiedeman, 2007). Saltelli and Annoni (2010) highlight that OAT sensitivity analysis only provides reliable and robust results if it can be shown that the model behaviour is linear. This condition is seldom satisfied for hydrological models or even known before a sensitivity analysis. The Elementary Effects method (Campolongo et al., 2007) is more robust against nonlinearity in the model behaviour, whilst still being frugal in the number of model runs.

Global sensitivity analysis techniques however do not require the model behaviour to be linear (Saltelli et al., 2008). The most straightforward global sensitivity analysis is either random or density-based sampling of parameter space and visualizing scatter plots of the parameter value against the prediction of interest (Wagener and Kollat, 2007; Peeters et al., 2013). Variance-based methods, such as Sobol' sensitivity analysis (Saltelli and Annoni, 2010; Nossent et al., 2011), use a scheme of structured resampling of a random base sampling to decompose the variance of the metric of interest into the main effects of a parameter and interaction effects of other parameters.

The main drawback of variance-based methods is that it assumes that the entire effect of a parameter can be summarized by the variance (Borgonovo, 2007; Borgonovo et al., 2011). Variance-based sensitivity indices will therefore be less reliable if the response to a parameter has a skewed or multi-modal distribution. Density-based sensitivity analysis techniques attempt to account for this by incorporating the entire distribution of the response of a prediction of interest in the metric in a way that does not require any assumptions on the shape of the distribution. The methodology suggested by Plischke et al. (2013) implements such a density-based sensitivity analysis technique which is independent of the parameter sampling scheme. This has the added benefit that as no model runs need to be devoted to the resampling of a base sampling, more computing resources can be directed to exploration of parameter space.

The goal of this study is to apply a density-based sensitivity analysis in a river management modelling context to assess its capability to identify and quantify nonlinear effects and to extend the methodology to account for interaction effects. An idealized, hypothetical river management model implemented in the eWater Source platform (Welsh et al., 2013) serves as testing platform to assess the ability of the sensitivity analysis methodology to quantify the influence of a small number of forcing variables upon a variety of model outcomes.

The next section presents the theoretical background and numerical implementation of the Plischke et al. (2013) global sensitivity analysis method. The river management model is briefly introduced before presenting the results of the sensitivity analysis and summarizing the findings in the discussion and conclusion sections.

2   Methods

The sensitivity analysis introduced in Plischke et al. (2013) provides a robust, global density-based sensitivity analysis, independent of sampling strategy. This section provides a short summary of this methodology. For a detailed overview the interested reader is referred to Plischke et al. (2013).

Consider X and Y the set of variables that comprise the input and output respectively of a river system model. Fixing X to a single realization, the parameter combination x, results in a conditional cumulative distribution of Y equal to $F_{Y|X=x}(y)$ and an equivalent density function $f_{Y|X=x}(y)$. The importance of fixing X to x can be quantified by the separation between the unconditional $F_Y(y)$ and the conditional $F_{Y|X=x}(y)$ or, similarly, the separation between $f_Y(y)$ and $f_{Y|X=x}(y)$. Using the L1-norm, the separation between the two density functions can be written as

$$s(x) = \int_Y |f_Y(y) - f_{Y|X=x}(y)| \mathrm{d}y. \tag{1}$$

The importance of factor X on outcome Y can then be defined as

$$\delta(Y, X) = \frac{1}{2}\mathrm{E}[s(X)]$$
$$= \frac{1}{2}\int_X f_X(x) \int_Y |f_Y(y) - f_{Y|X=x}(y)| \mathrm{d}y \mathrm{d}x. \tag{2}$$

The sensitivity index $\delta(X, Y)$ varies between 0 and 1 and it can be shown that this index is zero when X and Y are completely independent (Plischke et al., 2013).

To compute $\delta(X, Y)$ the integrals in Eq. (2) need to be approximated numerically. This can be achieved by taking n samples of the parameter space X and computing the corresponding values for Y. The method does not impose any restrictions on the sampling strategy of the parameter space.

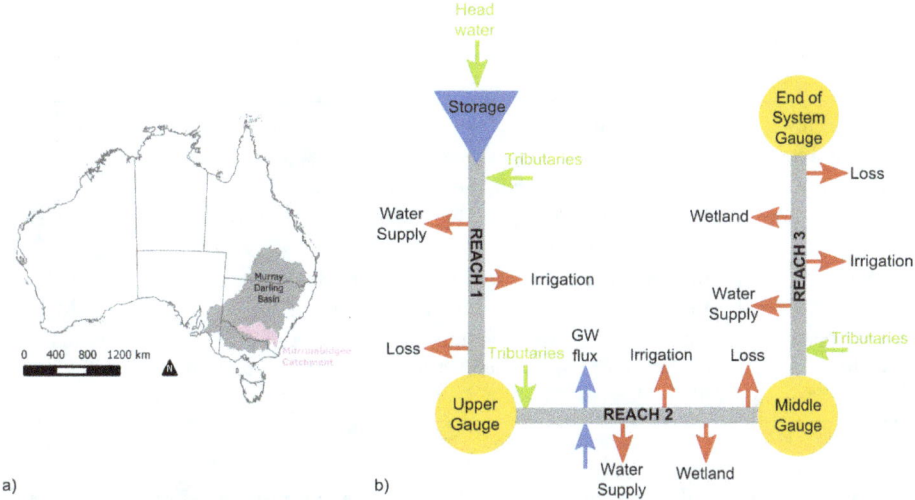

Figure 1. (a) Map showing the extent (indicated by pink shading) of the idealized river system model within the Murray–Darling Basin and **(b)** schematic structure of the river management model.

This implies that the methodology can be applied with random sampling, quasi-random sampling (e.g. Latin Hypercube Sampling or Sobol' sequences) or Markov chain Monte Carlo simulation.

The resulting data set is partitioned into M classes C_m with $m = 1, \ldots, M$. For each class C_m, the density function can be approximated with a kernel smoothing function with kernel $K(.)$ and bandwidth α (Devroye and Gyorfi, 1985):

$$\hat{f}_Y(y) = \frac{1}{n} \sum_{i=1}^{n} \frac{1}{\alpha} K\left(\frac{y - y_i}{\alpha}\right)$$

$$\hat{f}_{Y|C_m}(y) = \frac{1}{n_m} \sum_{i:x_i \in C_m}^{n_m} \frac{1}{\alpha_m} K\left(\frac{y - y_i}{\alpha_m}\right), \tag{3}$$

where n_m is the number of samples in class C_m and α_m the corresponding bandwidth for the kernel smoothing function.

The next step is to approximate the L1-norm between the two distributions for each class. Using a predefined number of quadrature points $\{\tilde{y}_j, \ j = 1, \ldots, l\}$, the separation can be computed as

$$s_{m,j} = \hat{f}_Y(\tilde{y}_j) - \hat{f}_{Y|C_m}(\tilde{y}_j)$$

$$\hat{S}_m = \frac{1}{2} \sum_{j=1}^{l-1} \left(|s_{m,j+1}| + |s_{m,j}|\right)\left(\tilde{y}_{j+1} - \tilde{y}_j\right). \tag{4}$$

The sensitivity index δ can then be approximated by

$$\hat{\delta} = \frac{1}{2n} \sum_{m=1}^{M} n_m \hat{S}_m. \tag{5}$$

To avoid bias in the sensitivity index and to assess the robustness of the sensitivity index estimate, it is recommended to perform a bootstrap of the sensitivity index (Efron, 1977)

and to adjust $\hat{\delta}$ with the mean of the bootstrap $\bar{\delta}^*$:

$$\hat{\hat{\delta}} = 2\hat{\delta} - \bar{\delta}^*. \tag{6}$$

$\hat{\hat{\delta}}$ provides the sensitivity index of the main effect of a variable. Plischke et al. (2013) however does not provide a method to explore second-order effects, i.e. the interaction between two variables. To estimate second-order effects between variables X_1 and X_2, the samples are subdivided into n groups of equal intervals for X_1. The sensitivity index $\hat{\delta}$ for X_2, $\hat{\delta}_{X_2}$, is computed for each interval. If there is no interaction effect between X_1 and X_2, then $\hat{\delta}_{X_2}$ will not vary with the level of X_1. To quantify this, the variance of $\hat{\delta}_{X_2}$ is computed over all n levels of X_1. Small variances indicate small interaction effects and vice versa.

3   Model description and setup

The case study is a hypothetical river system model (Fig. 1), based on a simplified version of the Murrumbidgee River model in New South Wales, Australia (Dutta et al., 2012; Podger et al., 2014). Using the full version of the Murrumbidgee River model was not warranted, not only because of the complexity of the system and the management rules, but, more importantly, because of legal issues with regard to model licensing and confidentiality. The idealized, hypothetical model retains most of the relevant complexity practitioners encounter when creating water allocation models, which is more than sufficient to illustrate the sensitivity analysis methodology.

In the model, water is routed from a storage reservoir through three river reaches. Routing starts in reach 1 at the storage reservoir with hydropower generators that receive water from a single tributary inflow. In reach 1, water is taken

Table 1. Output variables of the Source river system model.

Name	Description	Units
UpperFlow	Flow rate at the gauge at the end of the first reach	$m^3\,s^{-1}$
MiddleFlow	Flow rate at the gauge at the end of the middle reach	$m^3\,s^{-1}$
EndFlow	Flow rate at the gauge at the end of the final reach	$m^3\,s^{-1}$
$AlgalBloom	Monetary value generated by recreation as function of the risk of algal blooms	10^6 AUD
$Stor	Monetary value generated by recreation on storages	10^6 AUD
$TotalAg	Monetary value generated by irrigated agriculture	10^6 AUD
Hydropower	Electricity generated from the storage reservoir	kWh
GenSec	Percentage of time general security licences receive their full entitlement	%

from the system for town water supply and irrigation and water is received from unregulated rain-fed tributaries. From the Upper Gauge at the end of reach 1, water is routed through reach 2. In this reach, interaction with groundwater is taken into account by an exchange flux. As in reach 1, water is received from unregulated, rain-fed tributaries and water is taken out for irrigation and town water supply. In addition to these offtake, water is diverted into an off-river wetland system. Reach 3 starts at the middle gauge and is similar to reach 2. It also has offtake for town water supply, irrigation and off-river wetlands and receives inflow from rainfed tributaries. Groundwater–surface water interaction is not taken into account in this reach. Each reach has a term representing unaccounted losses. The loss relationships are taken from the more complex model. The total travel time from headwater to end-of-system is 18 days (3 days for reach 1, 6 days for reach 2 and 9 days for reach 3). These values, together with the other parameters influencing routing of water are also taken and aggregated from the more complex model.

Daily time series of rainfall and evaporation from 1895 to 2006 are obtained from SILO (http://www.longpaddock.qld.gov.au/silo/) for sites representative of each of the three reaches. These time series are used to simulate inflow from tributaries and compute irrigation demand. Inflow into the main storage in the model is taken from daily gauged data from 1895 to 2006.

The town water demands are based on a fixed annual pattern (8.8, 3.0 and $1.2 \times 10^6\,m^3\,year^{-1}$ for reaches 1, 2 and 3 respectively). Irrigation demands are based on a reach-based aggregation of irrigation use as well as rationalizing of crop types. There are environmental demands for the wetlands in reach 2 and 3, which are designed to establish and maintain favourable habitat conditions for indigenous fauna and flora (Janssen, 2012).

Two aspects of water management are considered: $347\,m^3\,s^{-1}$ order constraint on storage releases, i.e. the maximum flow that can be requested by water users in the system of the storage, and an annual allocation system. The allocation system comprises high and general security order debit

annual accounting schemes. Water is first allocated from the storage to high security entitlement holders and only once these are fulfilled is water allocated to general security entitlement holders. The start of the water year is 1 July with allocations updated continuously throughout the year, where these include allowances for minimum tributary inflows and delivery losses. At the end of the water year accounts are reset to zero. Licence entitlements were aggregated on a reach basis. Two socio-economic indicators have been included to indicate the impacts of storage volumes on recreational usage and mid-river flows on algal blooms and the associated impact on recreational usage. There are three storage volume categories (< 10, < 50 and $> 50\%$) for recreational usage based on visitor numbers. Recreational benefits are calculated for periods of time the model is at each threshold, using the Crase and Gillespie (2008) 100 000 visitor estimate to Lake Hume. Estimates of visitor numbers at high and low storage volumes are based on this estimate and the actual Tourism Research Australia (TRA) average, low and high visitor numbers in the Murrumbidgee catchment in the period 2003–2010 (DRET, 2010). Benefit transfer recreation values are taken from the same study (updated to 2012 Australian dollars (AUD) using the Australian Consumer Price Index, CPI). There are three risk of algal bloom categories (no bloom, alert and bloom) – no bloom occurs if there is a flow of at least $11.6\,m^3\,s^{-1}$ in the previous 7 days and alert if this flow occurs within the previous 14 days; if flow does not exceed $11.6\,m^3\,s^{-1}$ in the previous 14 days, algal bloom is simulated to occur. Australian dollars have been associated with loss of amenity in the weeks when there is an alert or bloom using the thresholds, estimated visitor numbers using TRA data and high and low estimates of river recreation based on survey data (DRET, 2010), and benefit transfer of general recreation benefits from Morrison and Hatton MacDonald (2010) (2010 AUD values are updated to 2012 AUD using the CPI and where the full value is used for no bloom, a proportion based on Crase and Gillespie (2008) for an alert and 0 AUD for a bloom).

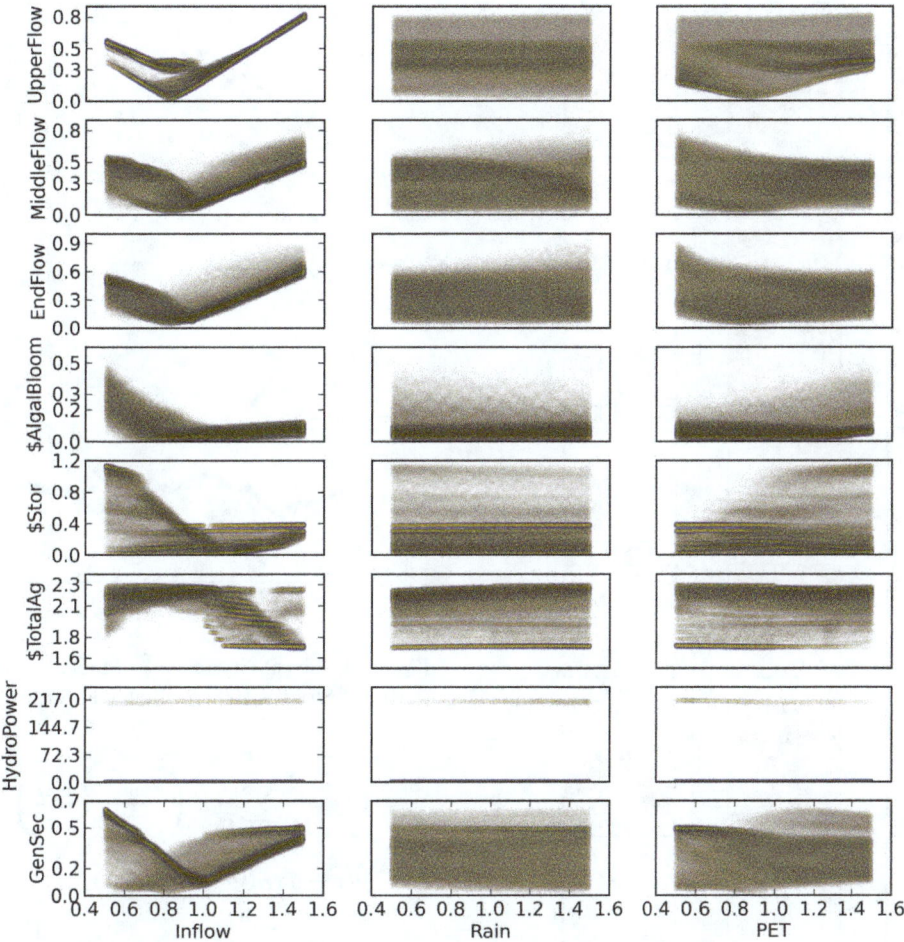

Figure 2. Scatter plots of $\hat{M}$, the difference between kernel density estimates for each simulation and the kernel density estimate of the reference simulation for all forcing data and model output variables for the eWater Source hypothetical river management model.

4 Results

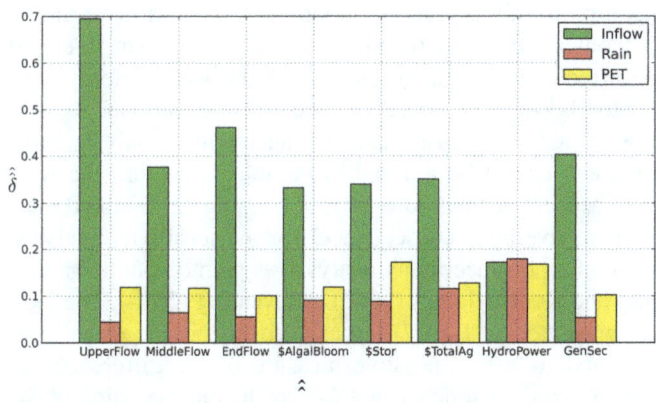

Figure 3. Sensitivity indices, δ, for all forcing data and model output variables for the eWater Source hypothetical river management model.

In the sensitivity analysis, the three main forcing variables are considered: the system inflow (Inflow), the precipitation (Rain) and the potential evapotranspiration (PET). The latter two affect the inflow into the reaches and the irrigation demand. Inspired by the work of Leblanc et al. (2012), the forcing variables are changed through a multiplier to the corresponding input time series with the range of the multiplier for each variable between 0.5 and 1.5. This range encompasses both historical variation in hydrological input and output, as well as the expected change under various climate change models and scenarios. While elaborate schemes are available to perturb hydrological time series, this is not warranted in this study as the focus is on metrics that integrate the entire flow time series. As such, the emphasis of this research is on changes in total flow in or out of the model, rather than in changes of the timing of flow.

Using Sobol' sequences (Sobol, 1976), 100 000 quasi-random samples of the three input variables are generated. For each of these samples a range of output time series is

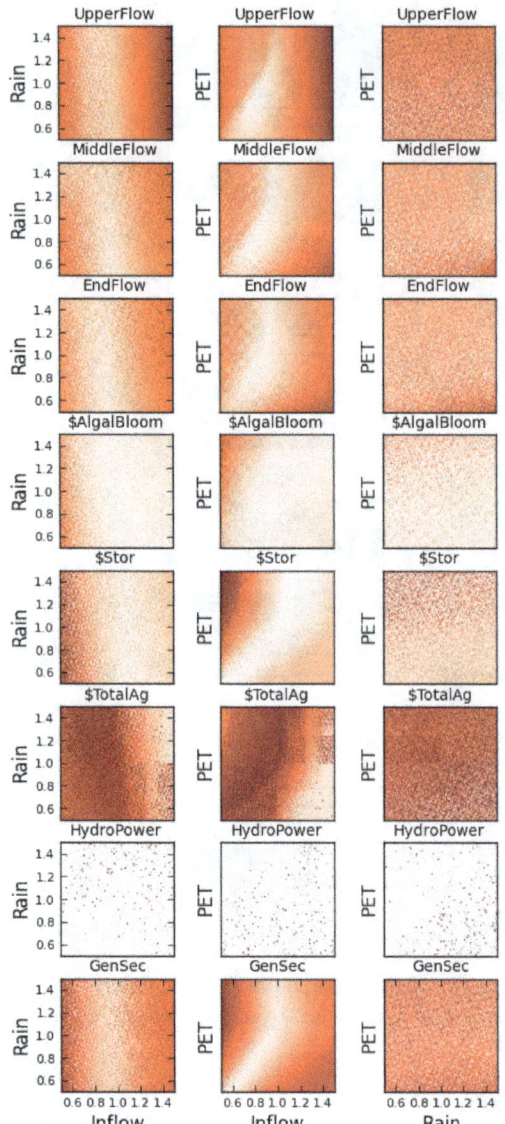

Figure 4. Scatter plots of interaction of the driving forces. The intensity of the colour scale is proportional to the model outcome value, where dark red indicates high values and light red indicates low values.

calculated (Pickett et al., 2013). Table 1 lists the names of the output series and a short description.

Each of the output variables in Table 1 is a daily time series. The metric for the sensitivity for different forcing data ($\hat{M}$) is the difference between the kernel density estimate of the daily times series of a randomly selected reference simulation ($\hat{f}_{Y\text{ref}}(y)$) and the kernel density estimate of the daily time series for the changed forcing data ($\hat{f}_{Y\text{sim}}(y)$):

$$\hat{f}_{Y\text{ref}}(y) = \frac{1}{n} \sum_{j=1}^{n} \frac{1}{\alpha} K \left(\frac{y_{\text{ref}} - y_{\text{ref},i}}{\alpha} \right)$$

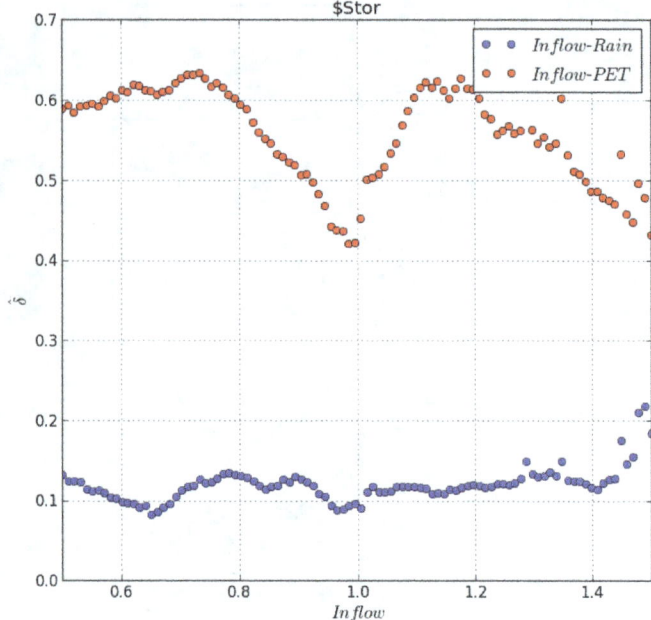

Figure 5. Sensitivity index $\hat{\delta}$ of the effect of Rain (blue) and PET (red) on \$Stor for 100 equal intervals of Inflow.

$$\hat{f}_{Y\text{sim}}(y) = \frac{1}{n} \sum_{j=1}^{n} \frac{1}{\alpha} K \left(\frac{y_{\text{sim}} - y_{\text{sim},i}}{\alpha} \right)$$

$$d_j = \hat{f}_{Y\text{ref}}(\tilde{y}_j) - \hat{f}_{Y\text{sim}}(\tilde{y}_j)$$

$$\hat{M} = \frac{1}{2} \sum_{j=1}^{l-1} \left(d_{j+1} + d_j \right) \left(|\tilde{y}_{j+1} - \tilde{y}_j| \right) \quad (7)$$

The choice of this metric is motivated by the fact that, since the case study is an idealized, hypothetical model, it is not possible to directly compare the results with observations. In addition to this, and more importantly, the variety of model outcomes examined in this study are more than likely to be affected by different aspects of the hydrograph. Similar to choosing an objective function in traditional calibration or a likelihood function in uncertainty analysis, such metric needs to be tailored to be able to capture the relevant aspects of the hydrograph. Choosing an ill-suited metric can have huge consequences for the sensitivity analysis, calibration or uncertainty analysis, as pointed out in Montanari and Koutsoyiannis (2012) and Nearing (2014). The metric presented in Eq. (7) is designed to provide an as general and robust as possible measure of the difference between two time series as not to bias the interpretation of the sensitivity analysis.

4.1 Main effects

Figure 2 shows the scatter plots of sensitivity metric $\hat{M}$ for all combinations of forcing data and output variables. It is clear that the dominant influencing driving variable is Inflow, as

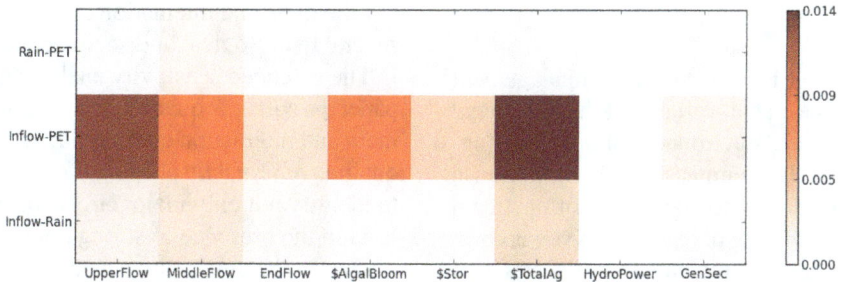

Figure 6. $\text{Var}(\hat{\delta}_{X1-X2})$ for all combinations of driving forces for all model outcomes. High values indicate potential interaction between $X1$ and $X2$. The values for Hydropower are omitted in order not to distort the visualization.

a strong response is noticeable for variations in this driving variable for all output variables with the exception of HydroPower. The effects of "Rain" and "PET" are less pronounced. A very striking feature are the many nonlinearities in the response surface of the hypothetical model. This is mostly due to a number of threshold values used in the management rules of the river management system. For instance, generation of hydro-power is only possible when the storage level in the dam exceeds a predefined threshold related to the height of the water intake point for the turbines.

Figure 3 shows a barplot of the sensitivity indices $\hat{\delta}$ for all main effects. These indices confirm the dominant influence of Inflow on most output variables. They provide a relative ranking of the influence of the input variable Inflow on the various output variables. MiddleFlow, EndFlow and GenSec respond to a similar degree to changes in Inflow and the same is true for the output variables related to monetary value ($AlgalBloom, $Stor and $TotalAg). HydroPower is least influenced by Inflow, which, from Fig. 2, is clearly related to the threshold-induced nonlinear behaviour.

The methodology is also able to quantify the often small and nonlinear effects of the other forcing variables. This is especially noticeable for PET. There is a clear but highly nonlinear effect of PET on $Stor, which is reflected in a higher $\hat{\delta}$. The output variable HydroPower has a bimodal distribution where the majority of simulations have an $\hat{M}$ close to zero. Nevertheless, the global sensitivity method is able to distinguish and quantify the subtle trends in the non-zero values for the different input variables.

4.2 Interaction effects

The previous section established the importance of Inflow as the main driving variable. It is however from both a management and modelling perspective interesting to have an understanding of how the interaction between variables affects the model outcome.

Figure 4 shows plots with the factor values on the x- and y-axis, with a colour scale to visualize $\hat{M}$ for the three combinations of interaction of the driving forces (Inflow-Rain, Inflow-PET and Rain-PET) for all eight model outcomes.

The first column shows that the effect of Inflow on most of the model outputs does not vary with the value of Rain. There is however a clear interaction between Inflow and PET for most of the model outputs; while the Inflow response is the dominant feature in the plots, the shape of this response depends on the value of PET. HydroPower is a noted exception as it displays very little structure in the scatter plots. This is because hydropower is generated by release of water from the reservoir in function of the demand and the water level in the reservoir. These management rules create a buffer to immediate impact from rainfall and inflow and also result in nonlinear, threshold related behaviour.

Very little structure is noticeable in the third column of Fig. 4, which shows the interaction between Rain and PET, reflecting the limited influence both driving forces have as a main effect.

To quantify the interaction effect for each interaction combination in Fig. 4, the variance of the $\hat{\delta}$ of the variable on the y-axis is computed for 100 equal intervals of the variable on the x-axis. By using Sobol' sequences to generate the 100 000 samples of the parameter space, each equal interval of the x-axis variable has approximately 1000 samples to compute the $\hat{\delta}$.

Figure 5 illustrates this for the interaction effects of Inflow, Rain and PET on $Stor. The sensitivity index values for Rain are low and hardly vary for different levels of Inflow, which is an indication of very limited interaction between Rain and Inflow, as confirmed by the scatter plot (Fig. 4). The $\hat{\delta}$ values for PET vary markedly with the level of Inflow. This sensitivity index reaches a minimum for Inflow values close to 1, while reaching peaks close to values of 0.75 and 1.1. This is reflected in the variance of the $\hat{\delta}$ values which is 4.5×10^{-4} for the Inflow–Rain couple and 3.5×10^{-3} for Inflow–PET. Figure 6 shows the variance of the sensitivity indices for all interaction pairs for all model outcomes. The values for Hydropower are much higher than for the other model outcomes due to the nonlinear behaviour. They were omitted from Fig. 6 as they distorted the visualization.

The most dominant interaction effects are between Inflow and PET for $TotalAg and UpperFlow, followed by $AlgalBloom, $Stor and MiddleFlow.

5   Discussion

The sensitivity analysis of the hypothetical river management model highlights inflow as a crucial variable of the model and how this affects the economic, environmental and sociological functions of the river. This emphasizes the importance of an accurate characterization of the flow rates of upstream areas when modelling flow routing in regulated systems comparable to the case study, i.e. the regulated river systems of the Murray–Darling Basin in Australia. An accurate characterization of flow rates not only entails maintaining a dense river gauge network, it also means adequately describing the measurement uncertainty in the flow rates, not in the least the uncertainty introduced by the rating curve that describes the stage–discharge relationship (Tomkins, 2012). The work of Hughes et al. (2014) illustrates this as they identify the inflow from ungauged catchment as crucial in the calibration of river management models.

Direct precipitation in the storage, wetlands and irrigation areas has a very minor influence on the model outcomes. This is mostly due to the small volume of rainfall ($0.633\,\mathrm{km}^3\,\mathrm{yr}^{-1}$) compared to the inflow volume ($4.4\,\mathrm{km}^3\,\mathrm{yr}^{-1}$) and the correlation between the inflow volume and rainfall. Any effect of rainfall will therefore be dwarfed by the effect of inflow to the system. The interaction effect of Inflow and PET is mostly due to the feedback mechanism as irrigation requirements increase with increasing potential evapotranspiration.

Such parameter interaction is well known in other areas of hydrological modelling, such as in rainfall–runoff modelling (Gallagher and Doherty, 2007; Zhang et al., 2013; Peeters et al., 2013) and in groundwater modelling (Doherty and Hunt, 2009), although it has not received much attention in river system modelling. Letcher et al. (2007) discuss the importance of interacting effects in water allocation models, without however providing a rigorous quantitative framework to evaluate the effects.

The sensitivity analysis in this study was limited to multiplying factors on three driving forces. It would be very insightful to include other model parameters in the sensitivity analysis, especially those controlling storage volumes and irrigation requirements. Along the same lines, including the parameters of the management rules, e.g. rules on allocations, in the sensitivity analysis can yield additional understanding of the operational management of the river system, as shown by Micevski et al. (2011).

6   Conclusions

The density-based sensitivity analysis of Plischke et al. (2013) has been applied to a river management model representing an idealized regulated river system representative of the southern Murray–Darling Basin in Australia to iden-

tify the main and interaction effects of three driving forces on several hydrological and socio-economic model outcomes.

The extended sensitivity analysis method presented in this paper provides a quantitative measure of sensitivity of the main and interaction effects and, through a combination with qualitative visual inspection of scatter plots, proved to be able to identify not only major effects but also subtle interactions, even in the presence of strong nonlinearities.

Due to the small dimensionality of the case study, it was possible to visualize all main effects and their interactions through scatter plots for all model outcomes. Although this will be challenging for higher-dimensional problems, the visual inspection of scatter plots is an invaluable complement to the sensitivity indices.

Understanding the dynamics of river system models is often not intuitive, especially in larger or basin-scale models (Johnston and Smakhtin, 2014). A robust and comprehensive sensitivity analysis is an invaluable step in model development to elucidate the often intricate interactions between driving forces, management rules and parameters. Increased understanding of the model will not only lead to improvements in calibration and prediction, it also has enormous potential in establishing the credibility and understanding of models.

Acknowledgements. The authors thank Russell Crosbie and Dave Penton for their constructive comments.

Edited by: H. Cloke

References

Bark, R., Peeters, L., Lester, R., Pollino, C., Crossman, N., and Kandulu, J.: Understanding the sources of uncertainty to reduce the risks of undesirable outcomes in large-scale freshwater ecosystem restoration projects: An example from the Murray-Darling Basin, Australia, Environ. Sci. Pol., 33, 97–108, 2013.

Borgonovo, E.: A new uncertainty importance measure, Reliability Eng. Syst. Safety, 92, 771–784, 2007.

Borgonovo, E., Castaings, W., and Tarantola, S.: Moment Independent Importance Measures: New Results and Analytical Test Cases, Risk Anal., 31, 404–428, 2011.

Campolongo, F., Cariboni, J., and Saltelli, A.: An effective screening design for sensitivity analysis of large models, Environ. Modell. Softw., 22, 1509–1518, 2007.

Castaings, W., Dartus, D., Le Dimet, F.-X., and Saulnier, G.-M.: Sensitivity analysis and parameter estimation for distributed hydrological modeling: potential of variational methods, Hydrol. Earth Syst. Sci., 13, 503–517, doi:10.5194/hess-13-503-2009, 2009.

Crase, L. and Gillespie, R.: The impact of water quality and water level on the recreation values of Lake Hume, Aust. J. Environ. Manage., 15, 21–29, 2008.

Devroye, L. and Gyorfi, L.: Nonparametric Density Estimation: The L1 View, John Wiley & Sons, Ltd New York, NY., 1985.

Doherty, J. and Hunt, R. J.: Two statistics for evaluating parameter identifiability and error reduction, J. Hydrol., 366, 119–127, 2009.

DRET: Destination visitor survey: strategic regional research report – New South Wales, Victoria and South Australia. Impact of the drought on tourism in the Murray River Region., Tech. rep., Department of Resources, Energy and Tourism, Tourism Research Australia, 57 pp., 2010.

Dutta, D., Hughes, J., Vaze, J., Kim, S., Yang, A., and Podger, G.: A daily river system model for the Murray-Darling Basin: development, testing and implementation, in: 34th Hydrology and Water Resources Symposium 2012, Sydney, 19–22 November 2012, 1057–1066, available at: http://www.hwrs2012.org.au/, 2012.

Efron, B.: Bootstrap methods: another look at the jackknife, The Annal. Stat., 7, 1–26, 1977.

Foglia, L., Hill, M. C., Mehl, S. W., and Burlando, P.: Sensitivity analysis, calibration, and testing of a distributed hydrological model using error-based weighting and one objective function, Water Resour. Res., 45, W06427, doi:10.1029/2008WR007255, 2009.

Gallagher, M. R. and Doherty, J.: Parameter interdependence and uncertainty induced by lumping in a hydrologic model, Water Resour. Res., 43, W05421, doi:10.1029/2006WR005347, 2007.

Gupta, H. V., Clark, M. P., Vrugt, J. A., Abramowitz, G., and Ye, M.: Towards a comprehensive assessment of model structural adequacy, Water Resour. Res., 48, W08301, doi:10.1029/2011WR011044, 2012.

Hill, M. C. and Tiedeman, C. R.: Effective groundwater model calibration, Wiley, 2007.

Hughes, J., Dutta, D., Vaze, J., Kim, S., and Podger, G.: An automated multi-step calibration procedure for a river system model, Environ. Modell. Softw., 51, 173–183, 2014.

Janssen, V.: Indirect Tracking of Drop Bears Using GNSS Technology, Aust. Geogr., 43, 445–452, doi:10.1080/00049182.2012.731307, 2012.

Johnston, R. and Smakhtin, V.: Hydrological Modeling of Large river Basins: How Much is Enough?, Water Resour. Manage., 28, 1–36, doi:10.1007/s11269-014-0637-8, 2014.

Leblanc, M., Tweed, S., Van Dijk, A., and Timbal, B.: A review of historic and future hydrological changes in the Murray-Darling Basin, Global Planet. Change, 80–81, 226–246, 2012.

Letcher, R. A., Croke, B. F. W., and Jakeman, A. J.: Integrated assessment modelling for water resource allocation and management: A generalised conceptual framework, Environ. Modell. Softw., 22, 733–742, 2007.

Micevski, T., Lerat, J., Kavetski, D., Thyer, M., and Kuczera, G.: Exploring the utility of multi-response calibration in river system modelling, 19th International Congress On Modelling and Simulation (modsim2011), 12–16 December 2011, Perth, Australia, 3889–3895, 2011.

Montanari, A. and Koutsoyiannis, D.: A blueprint for process-based modeling of uncertain hydrological systems, Water Resour. Res., 48, W09555, doi:10.1029/2011WR011412, 2012.

Morrison, M. and Hatton MacDonald, D.: Economic valuation of environmental benefits in the Murray-Darling Basin, Tech. rep., A report to the Murray Darling Basin Authority, 2010.

Nearing, G.: Comment on "A blueprint for process-based modeling of uncertain hydrological systems" by Monta-

nari and Koutsoyiannis, Water Resour. Res., 50, 6260–6263, doi:10.1002/2013WR014812, 2014.

Nossent, J., Elsen, P., and Bauwens, W.: Sobol' sensitivity analysis of a complex environmental model, Environ. Modell. Softw., 26, 1515–1525, 2011.

Patt, A.: Uncertainties in environmental modelling and consequences for policy making, chap. Communicating uncertainty to policy makers, pp. 231–251, NATO science for peace and security series – C: Environmental Security, Springer, available at: http://www.pik-potsdam.de/news/public-events/archiv/alter-net/former-ss/2009/14.09.2009/patt/literature/bavaye-book.pdf, 2009.

Peeters, L., Lerat, J., and Rassam, D.: Improving parameter estimation in transient groundwater models through temporal differencing, in: MODSIM 2011: Sustaining Our Future, 12–16 December 2011, Perth, Australia, p. 7, 2011.

Peeters, L., Crosbie, R., Doble, R., and Van Dijk, A.: Conceptual evaluation of continental land-surface model behaviour, Environ. Modell. Softw., 43, 49–59, 2013.

Pickett, T., Smith, T., Bulluss, B., Penton, D., Peeters, L., Podger, G., and Cuddy, S.: Approaches to distributed execution of hydrologic models: methods for ensemble Monte Carlo risk modelling with and without workflows, in: MODSIM 2013 20th International Congress on Modelling and Simulation, 1–6 December 2013, Adelaide, Australia, p. 7, 2013.

Plischke, E., Borgonovo, E., and Smith, C. L.: Global sensitivity measures from given data, Eur. J. Operational Res., 226, 536–550, 2013.

Podger, G., Cuddy, S., Peeters, L., Smith, T., Bark, R., and Black, D.: Risk management frameworks: supporting the next generation of Murray-Darling Basin water sharing plans, in: Evolving Water Resources Systems: Understanding, Predicting and Managing Water-Society Interactions, Proceedings of ICWRS2014, Bologna, Italy, June 2014 (IAHS Publ. 364, 2014, 452–457), 364, 452–457, 2014.

Saltelli, A. and Annoni, P.: How to avoid a perfunctory sensitivity analysis, Environ. Modell. Softw., 25, 1508–1517, 2010.

Saltelli, A., Ratto, M., Andres, T., Campolongo, F., Cariboni, J., Gatelli, D., Saisana, M., and Tarantola, S.: Global Sensitivity Analysis. The Primer, John Wiley & Sons, Ltd, doi:10.1002/9780470725184, 2008.

Sobol, I.: Uniformly distributed sequences with an additional uniform property, USSR Comput. Mathem. Mathem. Phys., 16, 236–242, 1976.

Tomkins, K. M.: Uncertainty in streamflow rating curves: methods, controls and consequences, Hydrol. Process., 28, 464–481, doi:10.1002/hyp.9567, 2012.

Wagener, T. and Kollat, J.: Numerical and visual evaluation of hydrological and environmental models using the Monte Carlo analysis toolbox, Environ. Modell. Softw., 22, 1021–1033, 2007.

Welsh, W. D., Vaze, J., Dutta, D., Rassam, D., Rahman, J. M., Jolly, I. D., Wallbrink, P., Podger, G. M., Bethune, M., Hardy, M. J., Teng, J., and Lerat, J.: An integrated modelling framework for regulated river systems, Environ. Modell. Softw., 39, 81–102, 2013.

Zhang, C., Chu, J., and Fu, G.: Soboĺ sensitivity analysis for a distributed hydrological model of Yichun River Basin, China, J. Hydrol., 480, 58–68, 2013.

7

Scaling properties reveal regulation of river flows in the Amazon through a "forest reservoir"

Juan Fernando Salazar[1], **Juan Camilo Villegas**[1,2], **Angela María Rendón**[1], **Estiven Rodríguez**[1], **Isabel Hoyos**[3,4], **Daniel Mercado-Bettín**[1], **and Germán Poveda**[5]

[1]GIGA, Escuela Ambiental, Facultad de Ingeniería, Universidad de Antioquia, Medellín, Colombia
[2]School of Natural Resources and the Environment, University of Arizona, Tucson, USA
[3]GAIA, Escuela Ambiental, Facultad de Ingeniería, Universidad de Antioquia, Medellín, Colombia
[4]Instituto de Física, Universidad de Antioquia, Medellín, Colombia
[5]Universidad Nacional de Colombia, Sede Medellín, Departamento de Geociencias y Medio Ambiente, Facultad de Minas, Medellín, Colombia

Correspondence: Juan Fernando Salazar (juan.salazar@udea.edu.co)

Abstract. Many natural and social phenomena depend on river flow regimes that are being altered by global change. Understanding the mechanisms behind such alterations is crucial for predicting river flow regimes in a changing environment. Here we introduce a novel physical interpretation of the scaling properties of river flows and show that it leads to a parsimonious characterization of the flow regime of any river basin. This allows river basins to be classified as regulated or unregulated, and to identify a critical threshold between these states. We applied this framework to the Amazon river basin and found both states among its main tributaries. Then we introduce the "forest reservoir" hypothesis to describe the natural capacity of river basins to regulate river flows through land–atmosphere interactions (mainly precipitation recycling) that depend strongly on the presence of forests. A critical implication is that forest loss can force the Amazonian river basins from regulated to unregulated states. Our results provide theoretical and applied foundations for predicting hydrological impacts of global change, including the detection of early-warning signals for critical transitions in river basins.

1 Introduction

Mean and extreme river flows are global-change-sensitive components of river flow regimes that are determinant for many ecological and societal processes (Zhang et al., 2016; Lima et al., 2014; Sterling et al., 2013; Coe et al., 2009; Piao et al., 2007; Mahe et al., 2005). Landscape and climate alterations foreshadow shifts in precipitation and river flow regimes (Boers et al., 2017; Khanna et al., 2017; Zemp et al., 2017; Lawrence and Vandecar, 2015; Botter et al., 2013; Davidson et al., 2012; Hirota et al., 2011; Sampaio et al., 2007). The conversion of precipitation into river flow through the accumulation of runoff depends on a suite of complex and heterogeneous biophysical processes and attributes of river basins, on different scales (Blöschl et al., 2007; McDonnell et al., 2007). This conversion results in spatial scaling properties – properties that do not vary within a wide range of scales – observable through river flow records (Gupta et al., 2007; Gupta and Waymire, 1990). The existence of scaling properties in river basins implies a power law correlation between the system response (river flows) and a scale parameter (typically the drainage area) (Gupta et al., 2007). Power laws go beyond statistical fitting; they indicate scale invariance as a fundamental emergent property arising from the self-organization of many complex systems in nature (Kéfi et al., 2007; Sivapalan, 2005; Brown et al., 2002). Scaling properties are common to river basins with very different environmental conditions (Gupta et al., 2010; Poveda et al., 2007). This suggests that the spatial scaling properties of river flows have a common, mechanistic origin, which has been related to conservation principles and the fractal nature of river networks (Gupta et al., 2007; Sivapalan, 2005).

The values of the scaling parameters – the scaling exponent and coefficient of a given power law – are neither universal nor static features of river basins, because they depend on runoff production processes that are spatially heterogeneous (Blöschl et al., 2007; McDonnell et al., 2007) and sensitive to both climate and land cover change (Sterling et al., 2013; Coe et al., 2009; Piao et al., 2007; Mahe et al., 2005). Understanding the mechanisms behind the scaling parameters in river basins, as well as their sensitivity to global change, is a crucial step for enabling the use of the scaling theory in hydrological *prediction in ungauged basins* (the "PUB problem"; Hrachowitz et al., 2013) and, more generally, in a changing environment where the processes governing the hydrological cycle are not static (the "Panta Rhei–Everything Flows" debate; Montanari et al., 2013). We address this problem by linking the scaling properties of river flows to the capacity of river basins for regulating their hydrological response.

2   Scaling properties reveal river flow regulation

The scaling properties of river flows are evidenced through power laws of the form (Gupta and Waymire, 1990)

$$E\left[Q_i^k\right] = \alpha_i S^{\beta_i}, \tag{1}$$

where $E[Q_i^k]$ is the kth-order statistical moment of the probability distribution function of river flows, S is a scale parameter, and α_i and β_i are the scaling coefficient and exponent, respectively. Q_i can be floods ($i = F$), mean flows ($i = M$) or low flows ($i = L$). The scaling parameters (α_i and β_i) vary among river basins and flow types and are always positive because river flows cannot be negative and increase downstream as a consequence of mass continuity.

The state of a river basin can be classified as *regulated* or *unregulated* depending on its river flow regime, which determines how the scaling exponents for floods (β_F), mean flows (β_M) and low flows (β_L) are organized. Regulation is defined here as the capacity of river basins to attenuate the amplitude of the river flow regime, that is, to reduce the difference between floods and low flows. A river basin is regulated if $\beta_L > \beta_M > \beta_F$ or unregulated if $\beta_L < \beta_M < \beta_F$. A metric of the amplitude of the extremes is the difference (Δ_Q) between long-term average floods ($E[Q_F]$) and low flows ($E[Q_L]$), relative to mean flows ($E[Q_M]$):

$$\Delta_Q = \frac{E[Q_F] - E[Q_L]}{E[Q_M]} = \frac{\alpha_F S^{\beta_F} - \alpha_L S^{\beta_L}}{\alpha_M S^{\beta_M}}. \tag{2}$$

Our distinction between regulated and unregulated states is consistent with the definition of regulation in artificial reservoirs, whereby a reservoir regulates river flows by either mitigating floods through water retention or enhancing low flows through water release (Magilligan and Nislow, 2005). The amplitude of the extremes is dampened in the regulated state (Δ_Q is reduced as S increases) or amplified in the unregulated state (Δ_Q is increased as S increases), as a consequence of how river flows grow downstream in a river basin. These

contrasting behaviours are reflected by the scaling exponents through the spatial rate of change

$$\frac{\partial \Delta_Q}{\partial S} = \frac{\alpha_F S^{\beta_F} (\beta_F - \beta_M) + \alpha_L S^{\beta_L} (\beta_M - \beta_L)}{\alpha_M S^{\beta_M + 1}}$$

$$\begin{cases} < 0, & \text{if } \beta_L > \beta_M > \beta_F \text{ (regulated state)} \\ = 0, & \text{if } \beta_L = \beta_M = \beta_F \text{ (critical threshold)} \\ > 0, & \text{if } \beta_L < \beta_M < \beta_F \text{ (unregulated state)} \end{cases} \tag{3}$$

The difference between the regulated and unregulated states is evidenced by the theoretical limit

$$\lim_{S \to \infty} \Delta_Q =$$

$$\begin{cases} 0, & \text{if } \beta_L > \beta_M > \beta_F \text{ (regulated state)} \\ (\alpha_F - \alpha_L)/\alpha_M \text{ (a positive constant)}, & \text{if } \beta_L = \beta_M = \beta_F \text{ (critical threshold)} \\ \infty, & \text{if } \beta_L < \beta_M < \beta_F \text{ (unregulated state)} \end{cases} \tag{4}$$

In the regulated state, the flow regime tends to the limit of complete regulation (constant flow: $E[Q_F] = E[Q_M] = E[Q_L]$), owing to the capacity of the river basin to dampen extremes ($\Delta_Q \to 0$). The opposite occurs in the unregulated state: the extremes are amplified ($\Delta_Q \to \infty$) and, hence, $E[Q_F] \gg E[Q_M] \gg E[Q_L]$. Therefore, in a given river basin, reversing the direction of the inequality from $\beta_L > \beta_M > \beta_F$ to $\beta_L < \beta_M < \beta_F$ indicates a shift between the regulated and unregulated states, with $\beta_L = \beta_M = \beta_F$ being a critical threshold. This agrees with the definition of a tipping point as "the corresponding critical point – in forcing and a feature of the system – at which the future state of the system is qualitatively altered" (Lenton, 2011). The difference ($\beta_L - \beta_F$) denotes a metric of the regulation level that indicates the proximity to the critical threshold in a river basin. Everything else being equal, a reduction in β_L indicates an increased severity of low flows, whereas an increase in β_F indicates an increase in flood severity.

The occurrence of regulated or unregulated states depends on the combined effect of *dampening* and *amplification* processes operating within a river basin. Both processes can coexist in a regulated river basin because higher regulation implies both reducing floods through a dampening effect produced by water retention within the basin, and increasing low flows through an amplification effect resulting from the release of water stored within the basin. The occurrence of either of these effects is described by how the rate of change

$$\frac{\partial E\left[Q_i^k\right]}{\partial S} = \alpha_i \beta_i S^{\beta_i - 1} \tag{5}$$

grows with increasing scale. If $\partial E[Q_i^k]/dS$ decreases with S – i.e. the power law (Eq. 1) is convex in S, – then the flows are dampened within the river basin, meaning that the production of runoff per unit area decreases downstream along the river network. The opposite occurs if $\partial E[Q_i^k]/dS$ increases with S – i.e. the power law (Eq. 1) is concave in S. Whether $\partial E[Q_i^k]/dS$ increases or decreases with increasing S is determined by the value of the scaling exponent β_i relative to 1, as given by

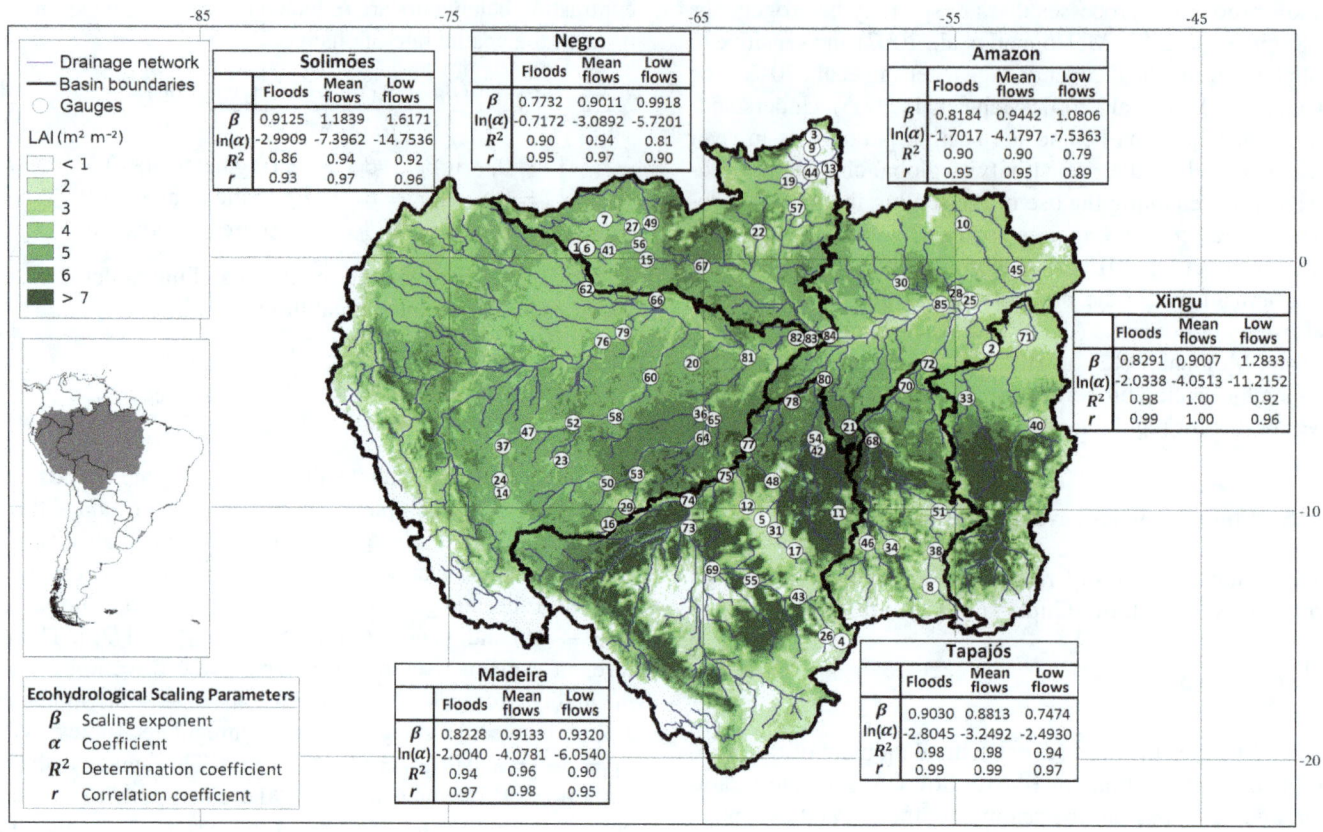

Solimões

	Floods	Mean flows	Low flows
β	0.9125	1.1839	1.6171
$\ln(\alpha)$	-2.9909	-7.3962	-14.7536
R^2	0.86	0.94	0.92
r	0.93	0.97	0.96

Negro

	Floods	Mean flows	Low flows
β	0.7732	0.9011	0.9918
$\ln(\alpha)$	-0.7172	-3.0892	-5.7201
R^2	0.90	0.94	0.81
r	0.95	0.97	0.90

Amazon

	Floods	Mean flows	Low flows
β	0.8184	0.9442	1.0806
$\ln(\alpha)$	-1.7017	-4.1797	-7.5363
R^2	0.90	0.90	0.79
r	0.95	0.95	0.89

Xingu

	Floods	Mean flows	Low flows
β	0.8291	0.9007	1.2833
$\ln(\alpha)$	-2.0338	-4.0513	-11.2152
R^2	0.98	1.00	0.92
r	0.99	1.00	0.96

Madeira

	Floods	Mean flows	Low flows
β	0.8228	0.9133	0.9320
$\ln(\alpha)$	-2.0040	-4.0781	-6.0540
R^2	0.94	0.96	0.90
r	0.97	0.98	0.95

Tapajós

	Floods	Mean flows	Low flows
β	0.9030	0.8813	0.7474
$\ln(\alpha)$	-2.8045	-3.2492	-2.4930
R^2	0.98	0.98	0.94
r	0.99	0.99	0.97

Legend:
- Drainage network
- Basin boundaries
- Gauges

LAI (m² m⁻²)
- < 1
- 2
- 3
- 4
- 5
- 6
- > 7

Ecohydrological Scaling Parameters
- β Scaling exponent
- α Coefficient
- R^2 Determination coefficient
- r Correlation coefficient

Figure 1. The Amazon basin and its major sub-basins. The map shows the long-term leaf area index averaged over the period 1981–2012, boundaries and drainage network of the sub-basins, and river flow gauges provided by the SO-HYBAM project (http://www.ore-hybam.org). Detailed information about the gauges is in Table S1. Tables show the parameters of power laws for mean and extreme river flows in each basin.

$$\frac{\partial^2 E\left[Q_i^k\right]}{\partial S^2} = \alpha_i \beta_i (\beta_i - 1) S^{\beta_i - 2}$$

$$\begin{cases} < 0, & \text{if } 0 < \beta_i < 1 \text{ (dampening process)} \\ = 0, & \text{if } \beta_i = 1 \text{ (critical point)} \\ > 0, & \text{if } \beta_i > 1 \text{ (amplification process)}, \end{cases} \qquad (6)$$

whereby $0 < \beta_i < 1$ and $\beta_i > 1$ represent, respectively, the dampening and amplification processes, and $\beta_i = 1$ is a critical value around which the curvature of the power law (Eq. 1) – and therefore the sign of its second derivative – changes. Higher regulation leads to dampened floods ($0 < \beta_F < 1$) and enhanced low flows ($\beta_L > 1$).

3   Regulated and unregulated basins in the Amazon

We tested our physical interpretation of the scaling properties in the Amazon river basin as a whole, and in its major sub-basins treated as independent systems (Fig. 1). Large-scale forest degradation or loss is a major driver of environmental change in these river basins (Boers et al., 2017; Khanna et al., 2017; Zemp et al., 2017; Lawrence and Vandecar, 2015; Lima et al., 2014; Davidson et al., 2012; Hi-

rota et al., 2011; Coe et al., 2009). The capacity to maintain high evapotranspiration rates is a key attribute of Amazonian forests associated with their large cumulative area of leaves (Caldararu et al., 2012; von Randow et al., 2012; Da Rocha et al., 2009). We take this into account by setting the scaling parameter as $S = \mathrm{LA} = A \times \overline{\mathrm{LAI}}$, where $\overline{\mathrm{LAI}}$ is the leaf area index averaged over the drainage area A of each basin, so the power law (Eq. 1) becomes

$$E\left[Q_i^k\right] = \alpha_i \mathrm{LA}^{\beta_i}. \qquad (7)$$

We tested the consistency of our results when using $E[Q_i^k] = \gamma_i A^{\delta_i}$ instead of Eq. (7), i.e. by setting A as the scale parameter (results are included in the Supplement). Using basin topographic data and daily river flow records from 85 gauges from the SO-HYBAM project (Cochonneau et al., 2006, Fig. S1 and Table S1 in the Supplement), and LAI data (Liu et al., 2012) averaged for 1981–2012 (Fig. 1), we found that annual mean and extreme river flows ($E[Q_i^k]$ with $k = 1$) in the Amazonian basins exhibit significant ($p < 0.05$, t-test results are in Table S2) scaling properties through power laws of the form of Eq. (7) (Fig. 2). Likewise, the scaling

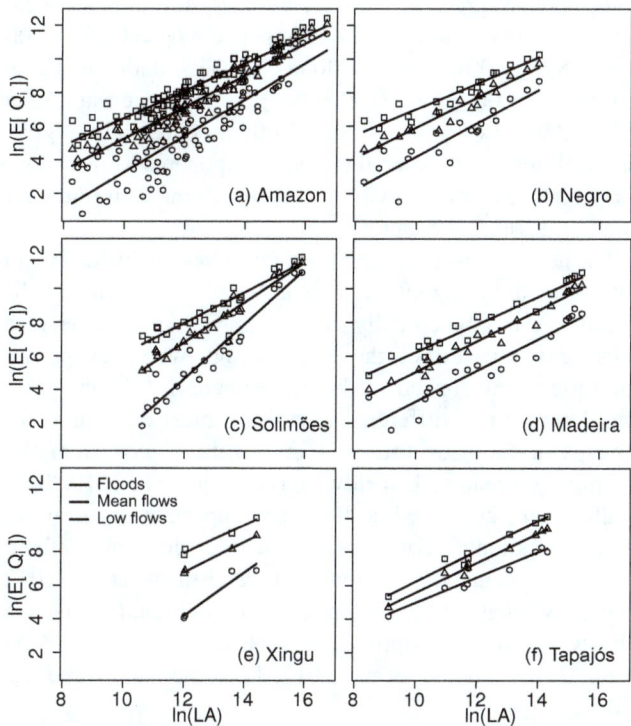

Figure 2. Power laws of the form $E[Q_i] = \alpha_i LA^{\beta_i}$ (Eq. 7 with $k = 1$) for low flows ($i = L$), mean flows ($i = M$) and floods ($i = F$). Points are observed river flows and lines are the scaling relations (in all cases $r > 0.88$ and $p < 0.05$). **(a)** Amazon: $E[Q_L] = \exp(-7.53)LA^{1.08}$; $E[Q_M] = \exp(-4.18) \; LA^{0.94}$; $E[Q_F] = \exp(-1.70)LA^{0.82}$. **(b)** Negro: $E[Q_L] = \exp(-5.72)LA^{0.99}$; $E[Q_M] = \exp(-3.09)$ $LA^{0.90}$; $E[Q_F] = \exp(-0.71)LA^{0.77}$. **(c)** Solimões: $E[Q_L]$ $= \exp(-14.75)LA^{1.62}$; $E[Q_M] = \exp(-7.40)LA^{1.18}$; $E[Q_F] = \exp(-2.99)LA^{0.91}$. **(d)** Madeira: $E[Q_L] = \exp(-6.05)LA^{0.93}$; $E[Q_M] = \exp(-4.08)LA^{0.91}$; $E[Q_F] = \exp(-2.00)LA^{0.82}$. **(e)** Xingu: $E[Q_L] = \exp(-11.22)LA^{1.28}$; $E[Q_M] = \exp(-4.05)$ $LA^{0.90}$; $E[Q_F] = \exp(-2.03)LA^{0.83}$. **(f)** Tapajós: $E[Q_L]$ $= \exp(-2.49)LA^{0.75}$; $E[Q_M] = \exp(-3.25)LA^{0.88}$; $E[Q_F] = \exp(-2.80)LA^{0.90}$. For convenience, α_i is expressed as $\exp(\ln(\alpha_i))$.

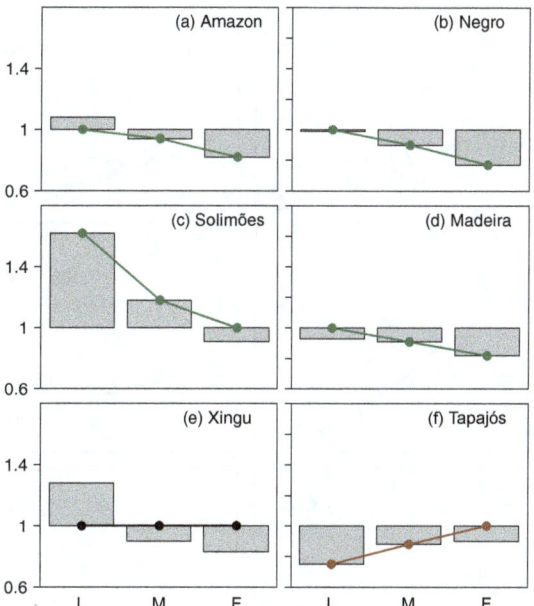

Figure 3. Observed patterns of the values of the scaling exponents (β_i) for low flows (L), mean flows (M) and floods (F) in the Amazon basin and its six major sub-basins. Dots over the bars indicate whether the scaling exponent is significantly different to 1 ($p < 0.05$, the dot is not over 1) or not (the dot is over 1). Details about the t tests are in Tables S3 to S8. In regulated states (green, **a–d**), the exponents decrease from low flows to floods, whereas in unregulated states (brown, **f**), the exponents increase from low flows to floods. In the Xingu river basin **(e)**, the hypothesis that all exponents are equal to 1 cannot be rejected ($p > 0.05$) because of the small number of degrees of freedom (gauges).

properties are evident when using A as the scale parameter (Figs. S1–S7 and Table S9).

Estimated values of the scaling exponents reveal the existence of both regulated and unregulated basins within the Amazon (Figs. 3 and S8). The Amazon, Negro, Solimões and Madeira river basins are regulated as indicated by their scaling exponents: $\beta_L > \beta_M > \beta_F$. The statistical significance of the comparisons between the scaling exponents in the Xingu river is limited because of the few degrees of freedom determined by the number of gauges, so we excluded this basin from this analysis. In these regulated basins, Δ_Q decreases with the spatial scale, as given by Eqs. (3) and (4) with $\beta_L > \beta_M > \beta_F$ (Figs. 4 and S9). In contrast, the scaling exponents ($\beta_L < \beta_M < \beta_F$) indicate that the Tapajós river basin

has already transitioned into the unregulated state, whereby Δ_Q is not reduced with the spatial scale (Fig. 4f).

River basins can be classified by their regulation level: $\beta_L - \beta_F$ (Table 1). The Solimões is the more regulated basin ($\beta_L - \beta_F = 0.70 > 0$), while the Madeira is still regulated but close to the critical threshold ($\beta_L - \beta_F = 0.11 > 0$) and the Tapajós basin has already transitioned into the unregulated state ($\beta_L - \beta_F = -0.16 < 0$). The Amazon as a whole is in the regulated state, but it is less regulated than the Solimões ($\beta_L - \beta_F = 0.26$), consistent with the presence of the less regulated basins within the whole Amazon. In the following section, we explore the physical mechanisms behind the occurrence of different regulation states.

4 Discussion

4.1 The use of LA as scale parameter

Our general idea about the classification of river basins is independent of using LA as the scale parameter. The interpretation of the scaling properties presented in Section 2 is based only on the assumption that river flows in a given river

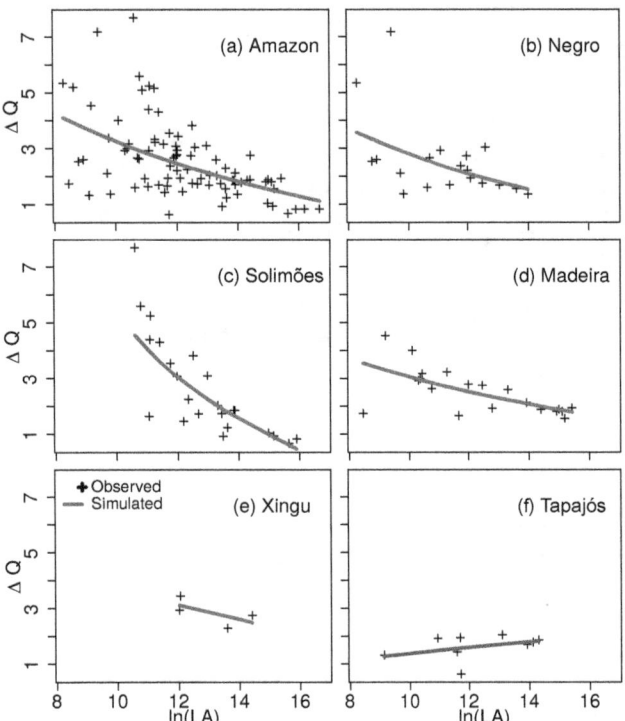

Figure 4. Amplitude of the extremes, $\Delta_Q = (E[Q_F] - E[Q_L])/E[Q_M]$, as observed (crosses) and simulated (lines) by $(\alpha_F LA^{\beta_F} - \alpha_L LA^{\beta_L})/\alpha_M LA^{\beta_M}$ (from Eq. 2 with $S = LA$), using the scaling parameters of each basin. Δ_Q either decreases or increases with spatial scale (LA) depending on whether the river basin is regulated ($\beta_L > \beta_M > \beta_F$, e.g. Solimões) or unregulated ($\beta_L < \beta_M < \beta_F$, e.g. Tapajós).

basin exhibit scaling properties through power laws of the form of Eq. (1). This does not require the use of LA as the scale parameter. Instead, it allows the investigation of the use of different scale parameters (e.g. Poveda et al., 2007): all of the equations in Sect. 2 use S as a general scale parameter that could be replaced by different factors depending on the case study. LA was introduced as the scale parameter for the application of our general framework (Sect. 2) to the particular case of the Amazon (Sect. 3). The idea is not that LA must be used as the scale parameter in any river basin, but to show that it can be successfully used as a scale parameter in the Amazon.

Although using LA as the scale parameter does not always improve R^2 in the scaling power laws (Figs. S1–S6), the main results of our study are statistically significant and consistent among the two scaling models: $E[Q_i^k] = \alpha_i LA^{\beta_i}$ (using LA) and $E[Q_i] = \gamma_i A^{\delta_i}$ (using A). Both models agree in the ordering of basins by their regulation level, and that the Tapajós basin is unregulated (Table 1). The most conspicuous difference between the models is that they do not fully agree in the description of amplifying and dampening processes in the Tapajós basin (Table 1). However,

both models agree that, in this basin (i) low flows are not amplified and can even be dampened ($\beta_L = 0.75 < 1.00$; $\delta_L = 0.89 \leq 1.00$), and (ii) floods are less dampened than low flows ($1.00 \geq \beta_F = 0.90 > \beta_L = 0.75$) or even amplified ($\delta_F = 1.09 > 1.00 \geq \delta_L = 0.89$). Both models show significant differences between the scaling exponents for low flows and floods ($\beta_L < \beta_F$ and $\delta_L < \delta_F$), consistent with unregulation in the Tapajós basin.

The use of A as the scale parameter relies on the idea that it represents the horizontal area over which precipitation falls. Using LA is conceptually consistent with this same idea, because LA describes the area through which evapotranspiration is transferred to the atmosphere. LA is an important descriptor of differences between forest and non-forest cover. Our focus on forests is due to these ecosystems being highly threatened worldwide (Hansen et al., 2010, 2013; Malhi et al., 2014), while there are important uncertainties about the potential consequences of forest loss on continental water balances (e.g. Bonan, 2008; Ellison et al., 2012; Makarieva et al., 2013; Zhang et al., 2016), including the possibility of forest loss tipping points (Boers et al., 2017; Zemp et al., 2017; Khanna et al., 2017; Lawrence and Vandecar, 2015).

Using LA instead of A as the scale parameter has practical implications for future studies. Using LA allows the influence of a changing scale parameter to be explored. LA is much more sensitive to global change than A, on timescales that are relevant for decision-making processes. Although studying this sensitivity is out of the scope of our present study, present results provide a basis for future studies.

4.2 The "forest reservoir" hypothesis

The less regulated river basins, Tapajós and Madeira, are also the ones with the less forest cover (Fig. 5a). Forest cover is not a static characteristic of river basins, so different values of the forest cover fraction can be assigned to each basin depending on the selected data source and time: we use 2003 data from Soares-Filho et al. (2006) – 2003 is within the range of all of the studied river flow records. However, what is important to our argument is not the precise value of the forest cover fraction in each basin, but the observation that, among the Amazon tributaries, the Tapajós and Madeira river basins have experienced large forest cover reductions mainly as a result of forest loss and/or degradation along the so-called arc of deforestation in south-southeastern Amazonia (Coe et al., 2013; Asner et al., 2010; Costa and Pires, 2010; Soares-Filho et al., 2006). Using 2002–2014 land water data (GRACE, CSR-v 5.0; Tapley et al., 2004), and 2002–2014 atmospheric water data (ERA-Interim reanalysis; Balsamo et al., 2015), we also observed that Tapajós and Madeira are the river basins with the higher long-term average variability of the terrestrial water storages (amplitude of the liquid water equivalent thickness, LWET, Fig. 5b), and the lower long-term average amount of water stored in the

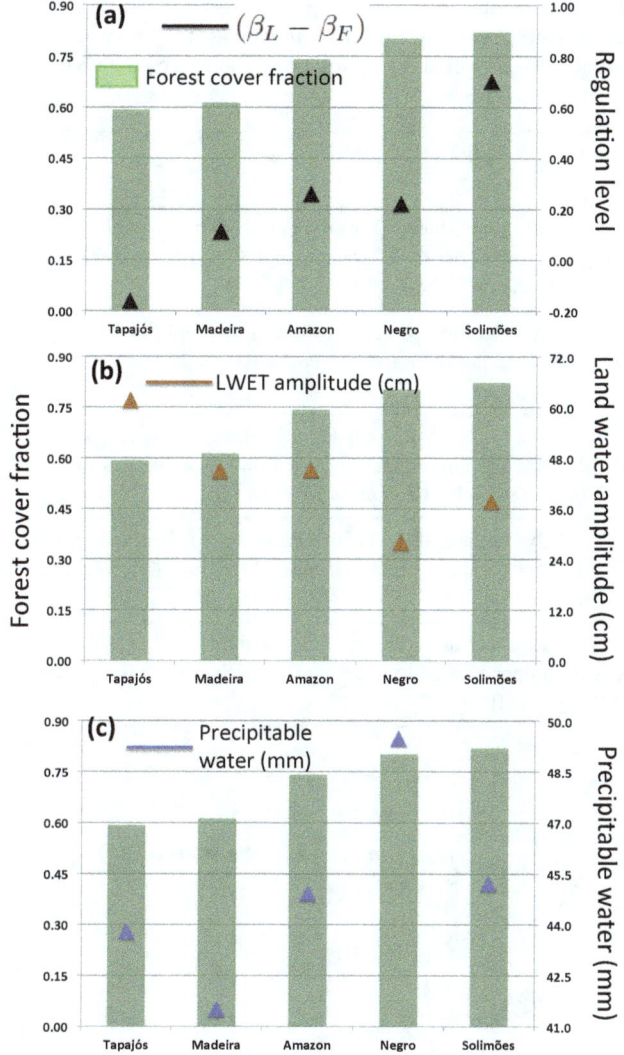

Figure 5. Forest cover fraction (2003 data from Soares-Filho et al., 2006) and **(a)** the regulation level ($\beta_L - \beta_F$, this study); **(b)** the long-term (2002–2014) average variability of the land water storages as indicated by the amplitude of the liquid water equivalent thickness, LWET (data from GRACE, CSR-v 5.0), and **(c)** the long-term (2002–2014) average amount of atmospheric water as indicated by the column-integrated precipitable water (data from ERA-Interim reanalysis). The Xingu was excluded because the scaling exponents are not significantly different from 1 (Fig. 3e).

atmosphere (column-integrated precipitable water, Fig. 5c). Taken together, these characteristics are consistent with a river basin with lower capacity to store water within the coupled land–atmosphere system. These observations led us to propose the "forest reservoir" hypothesis that relates the regulation level of the Amazonian river basins with their forest cover.

The physical causes for a river basin to be regulated or unregulated are summarized by its capacity for storing wa-

ter and controlling its release. Analogously, the capacity of artificial reservoirs to regulate river flows depends on its capacity for storing water and operation rules about how to release it (Magilligan and Nislow, 2005). River basins have natural mechanisms to implement these processes of water handling. These mechanisms depend not only on relatively invariant physical attributes (e.g. geomorphological and geological properties), but also on biophysical processes and characteristics of river basins that can be highly sensitive to global change on policy-relevant timescales, such as forest cover in the Amazon (Malhi et al., 2008; Soares-Filho et al., 2006; Guimberteau et al., 2017). Identifying those factors that are both highly sensitive to global change and strongly influential on runoff production is crucial for predicting the potential effects of global change on river flow regimes. Vegetation cover and vegetation-related processes meet these two conditions in many river basins of the world (Sterling et al., 2013; Coe et al., 2009; Piao et al., 2007), and particularly in the Amazon where the role of forests is so relevant that forest loss could force the system beyond a tipping point (Boers et al., 2017; Khanna et al., 2017; Zemp et al., 2017; Lawrence and Vandecar, 2015; Davidson et al., 2012; Hirota et al., 2011; Sampaio et al., 2007).

Forests can exert strong effects on the store and release of water through a variety of mechanisms. These mechanisms include large evapotranspiration fluxes (Caldararu et al., 2012; von Randow et al., 2012; Da Rocha et al., 2009; Carmona et al., 2016) linked to large precipitation recycling ratios (Van der Ent et al., 2010; Eltahir and Bras, 1994), accumulation and redistribution of soil moisture by root systems (Nadezhdina et al., 2010; Lee et al., 2005; Nepstad et al., 1994), strong capacity for stomatal regulation due to the large cumulative surface area of leaves (Berry et al., 2010; Costa and Foley, 1997), production of biogenic cloud condensation nuclei (Pöschl et al., 2010), below-canopy shading and temperature inversions that restrict direct soil evaporation (Henao et al., 2018), and the surface drag that is caused by the large height of trees and affects the flow of air over the forests (Khanna et al., 2017).

Collectively, these mechanisms imply that forests have a strong potential to enhance the capacity of river basins for storing water and controlling its release, as well as for producing contrasting and time-variable (e.g. seasonally different) effects on the water balance components. These dual and dynamic effects are key for regulation because it requires opposite effects on low flows (amplification) and floods (dampening). The forest reservoir describes *the natural capacity of river basins (in the Amazon or similar basins) to store water and control its release through land–atmosphere interactions (mainly precipitation recycling) that depend strongly on the presence of forests.* This hypothesis considers a river basin as the coupled land–atmosphere system comprising not only the terrestrial fluxes and storages of water but also the atmospheric ones (Fig. 6). Although the capacity of the atmosphere to store water is relatively small, its capacity to

Table 1. River flow regulation state and level in each basin as revealed by the scaling exponents of power laws $E[Q_i] = \alpha_i \mathrm{LA}^{\beta_i}$ (or $E[Q_i] = \gamma_i A^{\delta_i}$). Difference $\beta_L - \beta_F$ (or $\delta_L - \delta_F$) indicates both the regulation state (regulated if positive, unregulated if negative) and the proximity to the critical threshold or regulation level (magnitude of the difference). Basins are ordered from top to bottom by their regulation level.

River basin	$\beta_L - \beta_F$ ($\delta_L - \delta_F$)	State	Behaviour of the extremes with increasing spatial scale (LA or A)
Solimões	0.70 (0.67)	Regulated	The amplitude of the extremes (Δ_Q) is greatly reduced (Figs. 4c and S9c) because of a strong capacity of the basin for amplifying low flows ($\beta_L = 1.62 \gg 1.00$ and $\delta_L = 1.55 \gg 1.00$) while not amplifying floods ($\beta_F = 0.91 \le 1.00$ and $\delta_F = 0.88 \le 1.00$).
Amazon	0.26 (0.31)	Regulated	Δ_Q is reduced (Figs. 4a and S9a) due to the combined effect of low-flow amplification ($\beta_L = 1.08 \ge 1.00$ and $\delta_L = 1.17 > 1.00$) and flood dampening ($\beta_F = 0.82 < 1.00$ and $\delta_F = 0.86 < 1.00$).
Negro	0.22 (0.17)	Regulated	Δ_Q is reduced (Figs. 4b and S9b) because of the basin's capacity for dampening floods ($\beta_F = 0.77 < 1.00$ and $\delta_F = 0.90 < 1.00$) while not dampening low flows. Low flows grow approximately linearly with scale ($\beta_L = 0.99 \approx 1.00$ and $\delta_L = 1.07 \ge 1.00$).
Madeira	0.11 (0.14)	Regulated	Δ_Q is reduced (Figs. 4d and S9d) mainly because of the basin's capacity for dampening floods ($\beta_F = 0.82 < 1.00$ and $\delta_F = 0.86 < 1.00$). Low flows are not amplified ($\beta_L = 0.93 \le 1.00$ and $\delta_L \approx 1.00$).
Tapajós	−0.16 (−0.20)	Unregulated	Δ_Q is increased (Figs. 4f and S9f) because low flows are not amplified ($\beta_L = 0.75 < 1.00$, $\delta_L = 0.89 \le 1.00$) and floods are less dampened than low flows ($1.00 \ge \beta_F = 0.90 > \beta_L = 0.75$) or even amplified ($\delta_F = 1.09 > 1.00 \ge \delta_L = 0.89$).

transport water within or outside a system is huge (Trenberth et al., 2007). Indeed, in the long term, all continental water comes from the ocean through the atmosphere because the atmospheric fluxes of water are the only ones that flow upstream in river networks, while terrestrial fluxes are directed into the ocean by gravitational forces.

The water balance equation for the forest reservoir control volume (Fig. 6),

$$\frac{\mathrm{d}(S_l + S_a)}{\mathrm{d}t} = \nabla Q - R, \qquad (8)$$

establishes that changes in water storage – including both land (S_l) and atmospheric (S_a) components – are governed by differences between the net atmospheric moisture convergence (∇Q, the only input flux) and runoff (R, including both surface and sub-surface fluxes, the only output flux). P (precipitation), ET (evapotranspiration) and I (infiltration) are not external fluxes but components of complex land–atmosphere interactions (e.g. precipitation recycling) that occur within the system and, therefore, are fundamental to the mechanisms that can explain the capacity of a basin system for regulating river flows. Although external forcings (e.g. climate change or variability effects) do affect the response of the system (R is not independent of ∇Q), the capacity for regulating river flows can only be a consequence of the system's internal dynamics. Otherwise, if the response of a system simply follows external forcings (if R were entirely governed by ∇Q), then there would be no capacity for regulation. Variations in the internal dynamics of water stor-

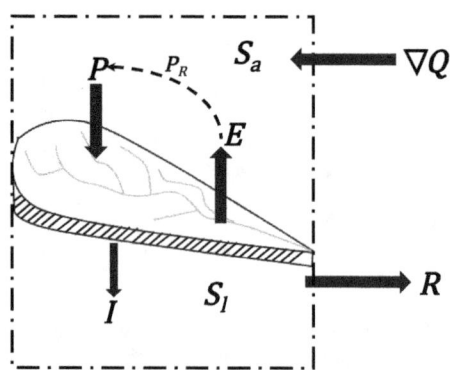

Figure 6. Forest reservoir control volume including the coupled land–atmosphere basin system. The system exchanges water with its exterior through the net atmospheric moisture convergence (∇Q) and runoff (R which includes surface and sub-surface fluxes). P (precipitation), ET (evapotranspiration) and I (infiltration) are internal fluxes that determine the distribution of water storage between land (S_l) and atmospheric components (S_a). Precipitation recycling (P_R) can occur within the system.

age allow for the occurrence of different river flow regimes under the same external forcings.

The occurrence of floods or low flows is related, respectively, to the abundance or scarcity of water, which depend on external forcings that determine whether ∇Q is large or small during any given time period (e.g. wet and dry seasons). Flood dampening depends on the capacity of the basin to retain water when ∇Q is large (wet season), which implies increasing water storage, consistent with

$$\frac{\mathrm{d}\,(S_l + S_a)}{\mathrm{d}t}$$
$$\begin{cases} > 0, & \text{if } \nabla Q > R \text{ (floods dampening via water storage)} \\ \leq 0, & \text{if } \nabla Q \leq R \text{ (no dampening or even amplification of floods)} \end{cases} \quad (9)$$

Analogously, low-flow amplification depends on the basin's capacity for releasing previously stored water when ∇Q is small (dry season), therefore reducing water storage as described by

$$\frac{\mathrm{d}\,(S_l + S_a)}{\mathrm{d}t}$$
$$\begin{cases} \geq 0, & \text{if } \nabla Q \geq R \text{ (no amplification or even dampening of low flows),} \\ < 0, & \text{if } \nabla Q < R \text{ (low flows amplification via water release).} \end{cases} \quad (10)$$

The importance of forests for the system's internal dynamics of water storage is highlighted by their relation with precipitation. Precipitation is not entirely determined by external forcings nor independent of the presence of forests. If precipitation regimes were independent of forest-related processes, then those regimes should not significantly change in response to forest cover change. This is contradicted by an increasing body of scientific evidence indicating that forest cover change can significantly alter precipitation regimes in the Amazon (Zemp et al., 2017; Lawrence and Vandecar, 2015; Spracklen and Garcia-Carreras, 2015; Lima et al., 2014; Makarieva et al., 2013; Stickler et al., 2013; Costa and Pires, 2010; Coe et al., 2009; Makarieva and Gorshkov, 2007). Through its impact on precipitation, forest cover change can affect all other water balance fluxes (e.g. river flows; Lima et al., 2014; Stickler et al., 2013; Coe et al., 2009), as well as terrestrial and atmospheric storages. Notably, the simulated impacts of deforestation on river flows can be opposite depending on whether the precipitation response to deforestation is included or not (Lima et al., 2014; Coe et al., 2009).

Recycled precipitation (P_R) is a key factor for regulation because it represents a potentially large amount of water that can be retained within the system through land–atmosphere circulation (Fig. 6). Therefore, in largely forested basins, the precipitation recycling ratio is indicative of the importance for regulation of the forest-mediated land–atmosphere interactions. Global estimates indicate that land evaporation accounts for about half of continental precipitation (Gimeno et al., 2012; Van der Ent et al., 2010), of which forests are major contributors (Schlesinger and Jasechko, 2014; Bonan, 2008). In the Amazon river basin, recycled precipitation also accounts for about half of the total precipitation (Eltahir and Bras, 1994). With this amount of forest-related precipitation, a disruption of the recycling mechanism has a strong potential to modify the internal dynamics of water transport and storage, which control river flow regulation (e.g. Zemp et al., 2017).

Precipitation recycling is not a dominant process on all spatial and temporal scales in every basin of the world. It is difficult to quantify the degree to which terrestrial evapotranspiration supports the occurrence of precipitation within a certain region, partly because this mechanism has characteristic time and length scales and depends on the size, shape and location of basins, as well as on the atmospheric pathways of moisture transport (van der Ent and Savenije, 2011). However, it is widely recognized that precipitation recycling is a crucial process in the hydrological cycle of the Amazon and neighbouring basins (Martinez and Dominguez, 2014; Zemp et al., 2014; Eltahir and Bras, 1994). All of the studied large basins are sinks (receive recycled precipitation) and sources (feed recycled precipitation through evapotranspiration) of significant amounts of continental moisture, with impacts that can be spread throughout the continent by complex cascading effects that are sensitive to forest cover change (Zemp et al., 2017, 2014). Global estimates indicate the length scale of precipitation recycling can be as low as 500 km in tropical regions (van der Ent and Savenije, 2011), which is not excessively large compared with the size of the basins. The observed seasonal variability of atmospheric moisture pathways over South America allows for the occurrence of significant precipitation recycling all over the Amazon basin (Zemp et al., 2014; Arraut et al., 2012).

Our conclusion that the Madeira and Tapajós are the less regulated basins, with Tapajós being unregulated (Table 1), relies only on the observed values of the scaling exponents, following the theoretical framework developed in Sect. 2. Therefore, this conclusion does not ignore the important role of geological and geomorphological processes (Miguez-Macho and Fan, 2012; Bruijnzeel, 2004). Depending on the case study, different levels of regulation or transitions between states could be attributed to different causes. The forest reservoir hypothesis provides a potential explanation linking forest cover and river flow regulation. The idea is not that the effect of land cover (particularly forest cover in the Amazon) on river flow regulation is stronger than any other effect (e.g. geological and geomorphological effects), but that the role of land cover is not negligible and critically important because of its sensitivity to global change, especially in a region such as the Amazon where forest ecosystems are highly threatened and forest-related precipitation recycling plays a major role (Davidson et al., 2012). We foresee a potential danger in the assumption that the regulation capacity of river basins depends on geomorphological and geological processes with land cover playing a negligible role. Under this assumption, land cover change (e.g. forest loss) would not change the capacity of river basins to regulate river flows.

The forest reservoir mechanisms may have been previously overlooked because the size of the atmospheric storage is much smaller than that of the terrestrial storage ($S_a \ll S_l$; Trenberth et al., 2007), and also because the size of the terrestrial storage (e.g. aquifer systems) is mainly determined by geological and geomorphological properties. However, the key factor for regulation is not the size of the atmospheric storage but the possibility of retaining large amounts of water within the system through land–atmosphere interactions.

4.3   Forest loss effects on regulation: a potential critical threshold

Forest loss does not reduce or increase river flows in every basin on every temporal and spatial scale (Zhang et al., 2016; Ellison et al., 2012; Zhou et al., 2015). Fundamental reasons for this are that forests have an inherent capacity to either increase or decrease the water balance components, and that these effects have a complex and dynamic nature. For instance, forests can increase or decrease ET via opening or closing stomata, respectively, which is related to water availability: stomatal aperture tends to be increased during drought stress and decreased during excessive water stress (Cornic, 2000; Lambers et al., 2008). Further, forest loss can significantly alter the hydraulic properties of soils, especially by reducing infiltrability (Zimmermann et al., 2006). Through these impacts, forest loss can alter all the water balance components in complex ways. If the effect of forest loss were always to reduce ET (due to reduction in the cumulative leaf area) with no impact on P (as implicitly assumed in hydrological models that use P as a fixed input) nor on the hydraulic properties of soils and regulation capacity of the basin, then forest loss should be always associated with increased R and, therefore, increased floods and low flows. Likewise, if the effect of forest loss were always to increase ET (related to weaker stomatal regulation, disruption of below canopy shading and stability, and increased wind speed over the surface, for example) with no other effects, then forest loss should always lead to reduced R and, therefore, reduced floods and low flows. In both cases, the effect of forest loss on extreme river flows would always be in the same direction. In contrast, the forest reservoir hypothesis considers that forest loss can have contrasting effects on low flows and floods, mainly because the production of these extreme flows is governed by different processes occurring during different seasons.

The forest reservoir hypothesis implies that the regulation capacity of a river basin can be especially sensitive to forest cover change. The size of artificial reservoirs determines their regulatory capacity. Likewise, the regulatory capacity of the forest reservoir depends on its size, which is related to the extent of forest cover. This implies that forest loss weakens regulation. The lower levels of regulation in the Madeira and Tapajós river basins (Table 1) are consistent with a weaker forest reservoir (these two basins are the less forested ones, Fig. 5a), likely related to extensive forest loss that has occurred along the arc of deforestation (Coe et al., 2013; Asner et al., 2010; Costa and Pires, 2010; Soares-Filho et al., 2006). Notably, these less regulated basins are also the ones with more large artificial reservoirs in operation (http://dams-info.org/). The introduction of artificial reservoirs can cause contrasting effects on regulation. As-

suming that an artificial reservoir is operated so as to reduce floods and increase low flows, its introduction in a river basin should enhance river flow regulation. However, the construction of reservoirs is usually linked to other human activities – e.g. road construction, and associated agricultural expansion and deforestation (Soares-Filho et al., 2006; Mahe et al., 2005) – that can reduce the natural capacity of river basins to regulate river flows. Our results suggest that this is the case in the Madeira and Tapajós basins.

Forest loss does not weaken regulation because it changes the capacity of the atmospheric and terrestrial water storages, but mainly because it reduces the capacity of the basin system (Fig. 6) to retain water through its complex internal dynamics of land–atmosphere interactions. Figure 7 shows a conceptual example of how forest loss can disrupt river flow regulation (increase the amplitudes of extremes) via weakening the forest reservoir. Forest loss can exacerbate floods by increasing R through reduction in ET and I during the wet season when P is large due to large ∇Q (Figs. 6 and 7a). ET and I reduction can be associated, respectively, with reduced leaf area and infiltrability. ET reduction can weaken P recycling as a mechanism for dampening floods by recirculating water within the system. These effects are consistent with an enhanced conversion of P into R during the wet season and, therefore, enhanced floods and reduced water storage. This is described by Eq. (9) where floods are not dampened if water storage $(S_1 + S_a)$ is not increased. Water storage reduction during the wet season results in a decreased capacity of the system to amplify low flows via base flow during the dry season (Fig. 7b). Amplifying low flows when ∇Q is relatively small (the dry season) requires the release of water that has been previously stored, consistent with $d(S_1 + S_a)/dt < 0$ in Eq. (10). Deforestation-induced reduction in P (Spracklen and Garcia-Carreras, 2015) or lengthening of the dry season (Lima et al., 2014; Costa and Pires, 2010), consistent with a disruption of the wet season onset (Wright et al., 2017), can further reduce low flows.

The forest reservoir hypothesis implies that forest loss can increase floods while reducing low flows (Fig. 7). This is not inconsistent with increasing scientific evidence that large-scale forest loss will reduce P over the Amazon (Spracklen and Garcia-Carreras, 2015). Reduced P can explain a decrease in low flows but does not necessarily imply a decrease in floods too. Floods strongly depend not only on the total amount of P but also on its temporal distribution (rainfall intensity and duration) and the hydraulic properties of the surface (Reed, 2002). Variations in the capacity of the basin system for retaining and releasing water during wet and dry seasons allow for the occurrence of larger floods with smaller P. A comparable situation has been observed in the Nakambé River in Africa where reduced precipitation has lead to the counter-intuitive effect of increased floods, even despite an increase in the number of dams in the river basin (Mahe et al., 2005).

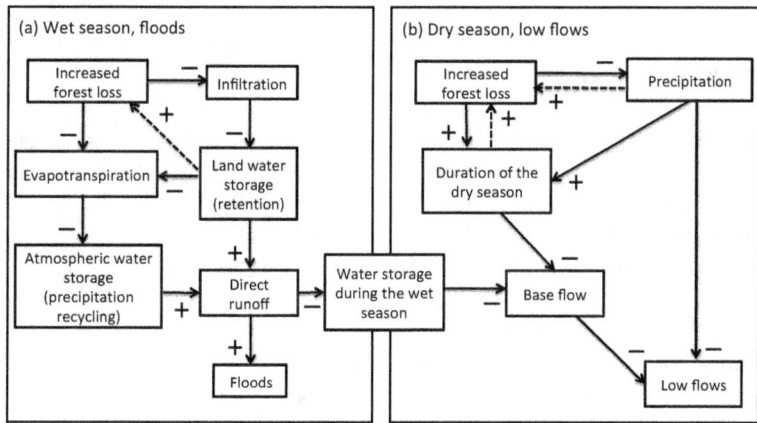

Figure 7. Potential weakening of the forest reservoir due to forest loss. **(a)** The loss of forests can exacerbate floods through increases in the direct runoff associated with reductions in the evapotranspiration and infiltration fluxes. These effects are associated with a reduction in the capacity of the coupled land–atmosphere system for retaining water during the wet season. **(b)** Less water retention during the wet season can reduce the base flow during the dry season. The loss of forests can reduce low flows through reductions in base flow and precipitation, both of them associated with a reduction in the capacity of the coupled land–atmosphere system for storing and releasing water during different periods of time. Dashed arrows indicate potential positive feedbacks to forest loss.

The identification of alternative regulation states from scaling properties in river basins (Sect. 2), together with the hypothesis that forest loss weakens the regulatory capacity, implies that forest loss can cause a transition from the regulated state to the unregulated state. This also implies that there is a forest cover critical threshold where the transition occurs. In our results, the forest cover fraction in the less regulated basins is ~ 0.60, while in the more regulated basins it is > 0.70 (Fig. 5a), which suggests a possible range for the critical threshold. Although more-detailed studies are essential to understand regulation dynamics in different regions, as well as to identify potential critical thresholds, our analysis shows that scaling patterns may be used to characterize regulation states and infer transitions in river basins. Such empirical approaches are essential (e.g. Hirota et al., 2011) because it is becoming clear that accurate mechanistic models to predict critical thresholds (or tipping points) are currently beyond our reach (Scheffer et al., 2009), and the detection of early-warning signals for critical transitions in complex systems (e.g. river basins) remains a fundamental challenge in environmental science today (Scheffer et al., 2009; Lenton, 2011).

5 Conclusions

We have shown how the scaling properties of mean and extreme river flows are a signature of the river flow regime in any river basin. Through the values of the scaling exponents, a river basin can be classified as regulated or unregulated, depending on whether it dampens or amplifies extreme river flows, respectively. These scaling exponents are sensitive to global change, so a river basin can shift from the regulated to the unregulated state. The scaling exponents provide a metric for the proximity to the critical threshold. Our results indicate that environmental perturbations that reduce the natural capacity of river basins to regulate river flows tend to increase the scaling exponent for floods and to decrease that for low flows. This provides a prediction of the direction of change in the scaling exponents of river basins as a result of global change, which can be used to design and simulate scenarios of future river flow regimes. The theoretical basis of our physical interpretation of the scaling properties of river flows is generally applicable to any river basin.

We have applied the proposed interpretation of river flow scaling properties to the Amazon river basin and found both the regulated (all except the Tapajós) and unregulated (the Tapajós) states among its main tributaries. Then we proposed the forest reservoir hypothesis to describe the natural capacity of river basins to regulate river flows through land–atmosphere interactions (mainly precipitation recycling) that depend strongly on the presence of forests, especially in the Amazon. A critical implication of this hypothesis is that forest loss can force the Amazonian river basins from regulated to unregulated states. This provides further evidence about the possible outcome of widespread forest loss in the Amazon, potentially involving forest loss critical thresholds, a matter of great uncertainty and concern (Boers et al., 2017; Khanna et al., 2017; Zemp et al., 2017; Lawrence and Vandecar, 2015; Davidson et al., 2012; Hirota et al., 2011).

These results provide foundations and a quantitative basis for using the scaling theory in solving four fundamental challenges in river basin science: the "PUB problem" that extends to every river basin in a changing environment (Hrachowitz et al., 2013; Gupta et al., 2007); the detection of early-warning signals of critical thresholds in river basins (Lenton, 2011; Scheffer et al., 2009); the production of parsimonious river basin classifications based on dimensionless similarity indices (the scaling exponents) or dominant processes (amplification or dampening of extreme river flows) (McDonnell et al., 2007); and the exploration of the organizing principles that underlie the heterogeneity and complexity of river flow production processes in river basins with different hydroclimatic regimes and on different scales (Blöschl et al., 2007; McDonnell et al., 2007). We addressed this by advancing from observed patterns (Figs. 2–5) to processes: the forest reservoir hypothesis (Figs. 6 and 7), as recommended by Sivapalan (2005).

Author contributions. JFS, JCV, AMR and GP designed the research. JFS and AMR developed the mathematical model. JFS, ER, IH and DM performed data analysis. JFS developed the forest reservoir hypothesis and wrote the paper with input from other authors. All authors discussed the results and conclusions.

Competing interests. The authors declare that they have no conflict of interest.

Acknowledgements. We gratefully acknowledge constructive comments from editor Patricia Saco and two anonymous referees. Funding was provided by "Programa de investigación en la gestión de riesgo asociado con cambio climático y ambiental en cuencas hidrográficas" (UT-GRA), Convocatoria 543-2011 Colciencias. Juan Fernando Salazar was partially supported by the IAI-INPE Internship program "Understanding Climate Change and Variability in the Americas". Angela María Rendón was partially supported by Colciencias grant 115-660-44588. Juan Camilo Villegas was partially supported by NSF-EF-1340624 through the University of Arizona.

Edited by: Patricia Saco

References

Arraut, J. M., Nobre, C., Barbosa, H. M., Obregon, G., and Marengo, J.: Aerial rivers and lakes: looking at large-scale moisture transport and its relation to Amazonia and to subtropical rainfall in South America, J. Climate, 25, 543–556, 2012.

Asner, G. P., Powell, G. V., Mascaro, J., Knapp, D. E., Clark, J. K., Jacobson, J., Kennedy-Bowdoin, T., Balaji, A., Paez-Acosta, G., Victoria, E., Secada, L., Valqui, M., and Hughes, R. F.: High-

resolution forest carbon stocks and emissions in the Amazon, P. Natl. Acad. Sci. USA, 107, 16738–16742, 2010.

Balsamo, G., Albergel, C., Beljaars, A., Boussetta, S., Brun, E., Cloke, H., Dee, D., Dutra, E., Muñoz-Sabater, J., Pappenberger, F., de Rosnay, P., Stockdale, T., and Vitart, F.: ERA-Interim/Land: a global land surface reanalysis data set, Hydrol. Earth Syst. Sci., 19, 389–407, https://doi.org/10.5194/hess-19-389-2015, 2015.

Berry, J. A., Beerling, D. J., and Franks, P. J.: Stomata: key players in the earth system, past and present, Curr. Opin. Plant Biol., 13, 232–239, 2010.

Blöschl, G., Ardoin-Bardin, S., Bonell, M., Dorninger, M., Goodrich, D., Gutknecht, D., Matamoros, D., Merz, B., Shand, P., and Szolgay, J.: At what scales do climate variability and land cover change impact on flooding and low flows?, Hydrol. Process., 21, 1241–1247, 2007.

Boers, N., Marwan, N., Barbosa, H. M., and Kurths, J.: A deforestation-induced tipping point for the South American monsoon system, Scient. Rep., 7, 41489, https://doi.org/10.1038/srep41489, 2017.

Bonan, G. B.: Forests and climate change: forcings, feedbacks, and the climate benefits of forests, Science, 320, 1444–1449, 2008.

Botter, G., Basso, S., Rodriguez-Iturbe, I., and Rinaldo, A.: Resilience of river flow regimes, P. Natl. Acad. Sci. USA, 110, 12925–12930, 2013.

Brown, J. H., Gupta, V. K., Li, B.-L., Milne, B. T., Restrepo, C., and West, G. B.: The fractal nature of nature: power laws, ecological complexity and biodiversity, Philos. T. Roy. Soc. Lond. B, 357, 619–626, 2002.

Bruijnzeel, L. A.: Hydrological functions of tropical forests: not seeing the soil for the trees?, Agr. Ecosyst. Environ., 104, 185–228, 2004.

Caldararu, S., Palmer, P. I., and Purves, D. W.: Inferring Amazon leaf demography from satellite observations of leaf area index, Biogeosciences, 9, 1389–1404, https://doi.org/10.5194/bg-9-1389-2012, 2012.

Carmona, A. M., Poveda, G., Sivapalan, M., Vallejo-Bernal, S. M., and Bustamante, E.: A scaling approach to Budyko's framework and the complementary relationship of evapotranspiration in humid environments: case study of the Amazon River basin, Hydrol. Earth Syst. Sci., 20, 589–603, https://doi.org/10.5194/hess-20-589-2016, 2016.

Cochonneau, G., Sondag, F., Guyot, J.-L., Geraldo, B., Filizola, N., Fraizy, P., Laraque, A., Magat, P., Martinez, J.-M., Noriega, L., Oliveira, E., Ordonez, J., Pombosa, R., Seyler, F., Sidgwick, J., and Vauchel, P.: The Environmental Observation and Research project, ORE HYBAM, and the rivers of the Amazon basin, in: Climate variability and change: hydrological impacts, IAHS Publications 308, IAHS, the Netherlands,, 44–50, 2006.

Coe, M. T., Costa, M. H., and Soares-Filho, B. S.: The influence of historical and potential future deforestation on the stream flow of the Amazon River–Land surface processes and atmospheric feedbacks, J. Hydrol., 369, 165–174, 2009.

Coe, M. T., Marthews, T. R., Costa, M. H., Galbraith, D. R., Greenglass, N. L., Imbuzeiro, H. M., Levine, N. M., Malhi, Y., Moorcroft, P. R., Muza, M. N., Powell, T. L., Saleska, S. R., Solorzano, L. A., and Wang, J.: Deforestation and climate feedbacks threaten the ecological integrity of south–southeastern Amazonia, Philos. T. Roy. Soc. B, 368, 20120155, https://doi.org/10.1098/rstb.2012.0155, 2013.

Cornic, G.: Drought stress inhibits photosynthesis by decreasing stomatal aperture – not by affecting ATP synthesis, Trends Plant Sci., 5, 187–188, 2000.

Costa, M. H. and Foley, J. A.: Water balance of the Amazon Basin: Dependence on vegetation cover and canopy conductance, J. Geophys. Res.-Atmos., 102, 23973–23989, 1997.

Costa, M. H. and Pires, G. F.: Effects of Amazon and Central Brazil deforestation scenarios on the duration of the dry season in the arc of deforestation, Int. J. Climatol., 30, 1970–1979, 2010.

Da Rocha, H. R., Manzi, A. O., Cabral, O. M., Miller, S. D., Goulden, M. L., Saleska, S. R., R-Coupe, N., Wofsy, S. C., Borma, L. S., Artaxo, P., Vourlitis, G., Nogueira, J. S., Cardoso, F. L., Nobre, A. D., Kruijt, B., Freitas, H. C., von Randow, C., Aguiar, R. G., and Maia, J. F.: Patterns of water and heat flux across a biome gradient from tropical forest to savanna in Brazil, J. Geophys. Res.-Biogeo., 114, G00B12, https://doi.org/10.1029/2007JG000640, 2009.

Davidson, E. A., de Araújo, A. C., Artaxo, P., Balch, J. K., Brown, I. F., Bustamante, M. M., Coe, M. T., DeFries, R. S., Keller, M., Longo, M., Munger, J. W., Schroeder, W., Soares-Filho, B. S., Souza, C. M., and Wofsy, S. C.: The Amazon basin in transition, Nature, 481, 321–328, 2012.

Ellison, D., N Futter, M., and Bishop, K.: On the forest cover–water yield debate: from demand-to supply-side thinking, Global Change Biol., 18, 806–820, 2012.

Eltahir, E. A. and Bras, R. L.: Precipitation recycling in the Amazon basin, Q. J. Roy. Meteorol. Soc., 120, 861–880, 1994.

Gimeno, L., Stohl, A., Trigo, R. M., Dominguez, F., Yoshimura, K., Yu, L., Drumond, A., Durán-Quesada, A. M., and Nieto, R.: Oceanic and terrestrial sources of continental precipitation, Rev. Geophys., 50, RG4003, https://doi.org/10.1029/2012RG000389, 2012.

Guimberteau, M., Ciais, P., Ducharne, A., Boisier, J. P., Dutra Aguiar, A. P., Biemans, H., De Deurwaerder, H., Galbraith, D., Kruijt, B., Langerwisch, F., Poveda, G., Rammig, A., Rodriguez, D. A., Tejada, G., Thonicke, K., Von Randow, C., Von Randow, R. C. S., Zhang, K., and Verbeeck, H.: Impacts of future deforestation and climate change on the hydrology of the Amazon Basin: a multi-model analysis with a new set of land-cover change scenarios, Hydrol. Earth Syst. Sci., 21, 1455–1475, https://doi.org/10.5194/hess-21-1455-2017, 2017.

Gupta, V. K. and Waymire, E.: Multiscaling properties of spatial rainfall and river flow distributions, J. Geophys. Res.-Atmos., 95, 1999–2009, 1990.

Gupta, V. K., Troutman, B. M., and Dawdy, D. R.: Towards a nonlinear geophysical theory of floods in river networks: an overview of 20 years of progress, in: Nonlinear dynamics in geosciences, Springer, New York, USA, 121–151, 2007.

Gupta, V. K., Mantilla, R., Troutman, B. M., Dawdy, D., and Krajewski, W. F.: Generalizing a nonlinear geophysical flood theory to medium-sized river networks, Geophys. Res. Lett., 37, L11402, https://doi.org/10.1029/2009GL041540, 2010.

Hansen, M. C., Stehman, S. V., and Potapov, P. V.: Quantification of global gross forest cover loss, P. Natl. Acad. Sci. USA, 107, 8650–8655, 2010.

Hansen, M. C., Potapov, P. V., Moore, R., Hancher, M., Turubanova, S., Tyukavina, A., Thau, D., Stehman, S., Goetz, S., Loveland, T., Kommareddy, A., Egorov, A., Chini, L., Justice, C. O., and Townshend, J. R. G.: High-resolution global maps of 21st-century forest cover change, Science, 342, 850–853, 2013.

Henao, J. J., Salazar, J. F., Villegas, J. C., and Rendón, A. M.: Amazon forest controls surface moisture via below-canopy temperature inversion: Potential forest loss-induced ecohydrological impacts, Agr. Forest Meteorol., submitted, 2018.

Hirota, M., Holmgren, M., Van Nes, E. H., and Scheffer, M.: Global resilience of tropical forest and savanna to critical transitions, Science, 334, 232–235, 2011.

Hrachowitz, M., Savenije, H. H. G., Blöschl, G., McDonnell, J. J., Sivapalan, M., Pomeroy, J. W., Arheimer, B., Blume, T., Clark, M. P., Ehret, U., Fenicia, F., Freer, J. E., Gelfan, A., Gupta, H. V., Hughes, D. A., Hut, R. W., Montanar, A., Pande, S., Tetzlaff, D., Troch, P. A., Uhlenbrook, S., Wagener, T., Winsemius, H. C., Woods, R. A., Zehe, E., and Cudennec, C.: A decade of Predictions in Ungauged Basins (PUB) – a review, Hydrolog. Sci. J. 58, 1198–1255, 2013.

Kéfi, S., Rietkerk, M., Alados, C. L., Pueyo, Y., Papanastasis, V. P., ElAich, A., and De Ruiter, P. C.: Spatial vegetation patterns and imminent desertification in Mediterranean arid ecosystems, Nature, 449, 213–217, 2007.

Khanna, J., Medvigy, D., Fueglistaler, S., and Walko, R.: Regional dry-season climate changes due to three decades of Amazonian deforestation, Nat. Clim. Change, 7, 200–204, 2017.

Lambers, H., Pons, T., and Chapin III, F.: Plant physiological ecology, Springer, New York, USA, 2008.

Lawrence, D. and Vandecar, K.: Effects of tropical deforestation on climate and agriculture, Nat. Clim. Change, 5, 27–36, 2015.

Lee, J.-E., Oliveira, R. S., Dawson, T. E., and Fung, I.: Root functioning modifies seasonal climate, P. Natl. Acad. Sci. USA, 102, 17576–17581, 2005.

Lenton, T. M.: Early warning of climate tipping points, Nat. Clim. Change, 1, 201–209, 2011.

Lima, L. S., Coe, M. T., Soares Filho, B. S., Cuadra, S. V., Dias, L. C., Costa, M. H., Lima, L. S., and Rodrigues, H. O.: Feedbacks between deforestation, climate, and hydrology in the Southwestern Amazon: implications for the provision of ecosystem services, Landscape Ecol., 29, 261–274, 2014.

Liu, Y., Liu, R., and Chen, J. M.: Retrospective retrieval of long-term consistent global leaf area index (1981–2011) from combined AVHRR and MODIS data, J. Geophys. Res.-Biogeo., 117, G04003, https://doi.org/10.1029/2012JG002084, 2012.

Magilligan, F. J. and Nislow, K. H.: Changes in hydrologic regime by dams, Geomorphology, 71, 61–78, 2005.

Mahe, G., Paturel, J.-E., Servat, E., Conway, D., and Dezetter, A.: The impact of land use change on soil water holding capacity and river flow modelling in the Nakambe River, Burkina-Faso, J. Hydrol., 300, 33–43, 2005.

Makarieva, A. M. and Gorshkov, V. G.: Biotic pump of atmospheric moisture as driver of the hydrological cycle on land, Hydrol. Earth Syst. Sci., 11, 1013–1033, https://doi.org/10.5194/hess-11-1013-2007, 2007.

Makarieva, A. M., Gorshkov, V. G., and Li, B. L.: Revisiting forest impact on atmospheric water vapor transport and precipitation, Theor. Appl. Climatol., 111, 79–96, 2013.

Malhi, Y., Roberts, J. T., Betts, R. A., Killeen, T. J., Li, W., and Nobre, C. A.: Climate change, deforestation, and the fate of the Amazon, Science, 319, 169–172, 2008.

Malhi, Y., Gardner, T. A., Goldsmith, G. R., Silman, M. R., and Zelazowski, P.: Tropical forests in the Anthropocene, Annu. Rev. Environ. Resour., 39, 125–159, 2014.

Martinez, J. A. and Dominguez, F.: Sources of atmospheric moisture for the La Plata River basin, J. Climate, 27, 6737–6753, 2014.

McDonnell, J. J., Sivapalan, M., Vaché, K., Dunn, S., Grant, G., Haggerty, R., Hinz, C., Hooper, R., Kirchner, J., Roderick, M. L., Selker, J., and Weiler, M.: Moving beyond heterogeneity and process complexity: A new vision for watershed hydrology, Water Resour. Res., 43, W07301, https://doi.org/10.1029/2006WR005467, 2007.

Miguez-Macho, G. and Fan, Y.: The role of groundwater in the Amazon water cycle: 1. Influence on seasonal streamflow, flooding and wetlands, J. Geophys. Res.-Atmos., 117, D15113, https://doi.org/10.1029/2012JD017539, 2012.

Montanari, A., Young, G., Savenije, H. H. G., Hughes, D., Wagener, T., Ren, L. L., Koutsoyiannis, D., Cudennec, C., Toth, E., Grimaldi, S., Blöschl, G., Sivapalan, M., Beven, K., Gupta, H., Hipsey, M., Schaefli, B., Arheimer, B., Boegh, E., Schymanski, S. J., Di Baldassarre, G., Yu, B., Hubert, P., Huang, Y., Schumann, A., Post, D. A., Srinivasan, V., Harman, C., Thompson, S., Rogger, M., Viglione, A., McMillan, H., Characklis, G., Pang, Z., and Belyaev, V.: "Panta Rhei – everything flows": change in hydrology and society – the IAHS scientific decade 2013–2022, Hydrolog. Sci. J., 58, 1256–1275, 2013.

Nadezhdina, N., David, T. S., David, J. S., Ferreira, M. I., Dohnal, M., Tesař, M., Gartner, K., Leitgeb, E., Nadezhdin, V., Cermak, J., Jimenez, M. S., and Morales, D.: Trees never rest: the multiple facets of hydraulic redistribution, Ecohydrology, 3, 431–444, 2010.

Nepstad, D. C., de Carvalho, C. R., Davidson, E. A., Jipp, P. H., Lefebvre, P. A., Negreiros, G. H., da Silva, E. D., Stone, T. A., Trumbore, S. E., and Vieira, S.: The role of deep roots in the hydrological and carbon cycles of Amazonian forests and pastures, Nature, 372, 666–669, 1994.

Piao, S., Friedlingstein, P., Ciais, P., de Noblet-Ducoudré, N., Labat, D., and Zaehle, S.: Changes in climate and land use have a larger direct impact than rising CO_2 on global river runoff trends, P. Natl. Acad. Sci. USA, 104, 15242–15247, 2007.

Pöschl, U., Martin, S. T., Sinha, B., Chen, Q., Gunthe, S. S., Huffman, J. A., Borrmann, S., Farmer, D. K., Garland, R. M., Helas, G., Jimenez, J. L., King, S. M., Manzi, A., Mikhailov, E., Pauliquevis, T., Petters, M. D., Prenni, A. J., Roldin, P., Rose, D., Schneider, J., Su, H., Zorn, S. R., Artaxo, P., and Andreae, M. O.: Rainforest aerosols as biogenic nuclei of clouds and precipitation in the Amazon, Science, 329, 1513–1516, 2010.

Poveda, G., Vélez, J. I., Mesa, O. J., Cuartas, A., Barco, J., Mantilla, R. I., Mejía, J. F., Hoyos, C. D., Ramírez, J. M., Ceballos, L. I., Zuluaga, M. D., Arias, P. A., Botero, B. A., Montoya, M. I., Giraldo, J. D., and Quevedo, D. I.: Linking long-term water balances and statistical scaling to estimate river flows along the drainage network of Colombia, J. Hydrol. Eng., 12, 4–13, 2007.

Reed, D. W.: Reinforcing flood–risk estimation, Philos. T. Roy. Soc. Lond. A, 360, 1373–1387, 2002.

Sampaio, G., Nobre, C., Costa, M. H., Satyamurty, P., Soares-Filho, B. S., and Cardoso, M.: Regional climate change over eastern Amazonia caused by pasture and soybean cropland expansion, Geophys. Res. Lett., 34, L17709, https://doi.org/10.1029/2007GL030612, 2007.

Scheffer, M., Bascompte, J., Brock, W. A., Brovkin, V., Carpenter, S. R., Dakos, V., Held, H., Van Nes, E. H., Rietkerk, M., and Sugihara, G.: Early-warning signals for critical transitions, Nature, 461, 53–59, 2009.

Schlesinger, W. H. and Jasechko, S.: Transpiration in the global water cycle, Agr. Forest Meteorol., 189, 115–117, 2014.

Sivapalan, M.: Pattern, Process and Function: Elements of a Unified Theory of Hydrology at the Catchment Scale, in: Encyclopedia of Hydrological Sciences, edited by: Anderson, M. G., John Wiley & Sons, Ltd, New York, USA, 193–219, https://doi.org/10.1002/0470848944.hsa012, 2005.

Soares-Filho, B. S., Nepstad, D. C., Curran, L. M., Cerqueira, G. C., Garcia, R. A., Ramos, C. A., Voll, E., McDonald, A., Lefebvre, P., and Schlesinger, P.: Modelling conservation in the Amazon basin, Nature, 440, 520–523, 2006.

Spracklen, D. and Garcia-Carreras, L.: The impact of Amazonian deforestation on Amazon basin rainfall, Geophys. Res. Lett., 42, 9546–9552, 2015.

Sterling, S. M., Ducharne, A., and Polcher, J.: The impact of global land-cover change on the terrestrial water cycle, Nat. Clim. Change, 3, 385–390, 2013.

Stickler, C. M., Coe, M. T., Costa, M. H., Nepstad, D. C., McGrath, D. G., Dias, L. C., Rodrigues, H. O., and Soares-Filho, B. S.: Dependence of hydropower energy generation on forests in the Amazon Basin at local and regional scales, P. Natl. Acad. Sci. USA, 110, 9601–9606, 2013.

Tapley, B. D., Bettadpur, S., Watkins, M., and Reigber, C.: The gravity recovery and climate experiment: Mission overview and early results, Geophys. Res. Lett., 31, L09607, https://doi.org/10.1029/2004GL019920, 2004.

Trenberth, K. E., Smith, L., Qian, T., Dai, A., and Fasullo, J.: Estimates of the global water budget and its annual cycle using observational and model data, J. Hydrometeorol., 8, 758–769, 2007.

van der Ent, R. J. and Savenije, H. H. G.: Length and time scales of atmospheric moisture recycling, Atmos. Chem. Phys., 11, 1853–1863, https://doi.org/10.5194/acp-11-1853-2011, 2011.

Van der Ent, R. J., Savenije, H. H., Schaefli, B., and Steele-Dunne, S. C.: Origin and fate of atmospheric moisture over continents, Water Resour. Res., 46, W09525, https://doi.org/10.1029/2010WR009127, 2010.

von Randow, R. C., von Randow, C., Hutjes, R. W., Tomasella, J., and Kruijt, B.: Evapotranspiration of deforested areas in central and southwestern Amazonia, Theor. Appl. Climatol., 109, 205–220, 2012.

Wright, J. S., Fu, R., Worden, J. R., Chakraborty, S., Clinton, N. E., Risi, C., Sun, Y., and Yin, L.: Rainforest-initiated wet season onset over the southern Amazon, P. Natl. Acad. Sci. USA, 114, 8481–8486, https://doi.org/10.1073/pnas.1621516114, 2017.

Zemp, D. C., Schleussner, C.-F., Barbosa, H. M. J., van der Ent, R. J., Donges, J. F., Heinke, J., Sampaio, G., and Rammig, A.: On the importance of cascading moisture recycling in South America, Atmos. Chem. Phys., 14, 13337–13359, https://doi.org/10.5194/acp-14-13337-2014, 2014.

Zemp, D. C., Schleussner, C.-F., Barbosa, H. M., Hirota, M., Montade, V., Sampaio, G., Staal, A., Wang-Erlandsson, L., and Rammig, A.: Self-amplified Amazon forest loss due to vegetation-atmosphere feedbacks, Nat. Commun., 8, 14681, https://doi.org/10.1038/ncomms14681, 2017.

Zhang, M., Liu, N., Harper, R., Li, Q., Liu, K., Wei, X., Ning, D., Hou, Y., and Liu, S.: A global review on hydrological responses to forest change across multiple spatial scales: importance of scale, climate, forest type and hydrological regime, J. Hydrol., 546, 44–59, 2016.

Zhou, G., Wei, X., Chen, X., Zhou, P., Liu, X., Xiao, Y., Sun, G., Scott, D. F., Zhou, S., Han, L., and Su, Y.: Global pattern for the effect of climate and land cover on water yield, Nat. Commun., 6, 5918, https://doi.org/10.1038/ncomms6918, 2015.

Zimmermann, B., Elsenbeer, H., and De Moraes, J. M.: The influence of land-use changes on soil hydraulic properties: implications for runoff generation, Forest Ecol. Manage., 222, 29–38, 2006.

8

Practitioners' viewpoints on citizen science in water management: a case study in Dutch regional water resource management

Ellen Minkman[1,2,3], **Maarten van der Sanden**[2], **and Martine Rutten**[1]

[1]Department of Water Resource Management, Delft University of Technology, Delft, 2628 CN, the Netherlands

[2]Department of Science Education and Communication, Delft University of Technology, Delft, 2628 CJ, the Netherlands

[3]Presently at Department of Public Administration and Sociology, Erasmus University Rotterdam, Burgemeester Oudlaan 50, 3062 PA Rotterdam, the Netherlands

Correspondence to: Martine Rutten (m.m.rutten@tudelft.nl)

Abstract. In recent years, governmental institutes have started to use citizen science as a form of public participation. The Dutch water authorities are among them. They face pressure on the water governance system and a water awareness gap among the general public, and consider citizen science a possible solution. The reasons for practitioners to engage in citizen science, and in particular those of government practitioners, have seldom been studied. This article aims to pinpoint the various viewpoints of practitioners at Dutch regional water authorities on citizen science. A Q-methodological approach was used because it allows for exploration of viewpoints and statistical analysis using a small sample size. Practitioners (33) at eight different water authorities ranked 46 statements from agree to disagree. Three viewpoints were identified with a total explained variance of 67 %. Viewpoint A considers citizen science a potential solution that can serve several purposes, thereby encouraging citizen participation in data collection and analysis. Viewpoint B considers citizen science a method for additional, illustrative data. Viewpoint C views citizen science primarily as a means of education. These viewpoints show water practitioners in the Netherlands are willing to embrace citizen science at water authorities, although there is no support for higher levels of citizen engagement.

1 Introduction

The OECD (Organization of Economic Cooperation and Development) named the Netherlands "an international example" of water resource management in their 2014 report, but warns against "a striking awareness gap among Dutch citizens related to key water management functions, how they are performed and by whom" (OECD, 2014, p. 21). The main causes of this awareness gap are the absence of major water calamities in the past 60 years and the improvement of water quality over the past decades. Dutch citizens take the excellent water resource management for granted (OECD, 2014), which poses social challenges for Dutch water resource management. Citizens' behaviour counteracts efforts of water authorities; flood defences are violated by property development and civic pollution is common (OECD, 2014). Citizens and interest groups do not recognize water threats (Tielrooij, 2000), which causes a decreasing support for investments in flood defence and water quality management (OECD, 2014; Tielrooij, 2000; UvW, 2015a). Other countries also experience a "lack of citizen concern about water policy and low involvement of water users' associations" (OECD, 2011, p. 60). Half of the reviewed OECD countries across the globe face such challenges, including Chile, Italy, Korea and Mexico (OECD, 2011, p. 61). The Dutch water authorities (Unie van Waterschappen, UvW) governing body concluded that collaboration with other government layers, industry, interest groups and citizens is needed (UvW, 2015a). The UvW envisions increased public participation, with citizen science as a form of such participation (UvW, 2015b). In addition to raising awareness, citizen science could contribute to data collection and help water authorities to enhance their monitoring programmes, particularly with respect to the Water Framework Directive.

Definitions of citizen science can be narrow and focussed solely on data collection for academic purposes. Silvertown (2009) describes citizen science in a broader perspective, applicable to citizen science in practitioners' activities considered in this article. "Today, most citizen scientists work with professional counterparts on projects that have been specifically designed or adapted to give amateurs a role, either for the educational benefit of the volunteers themselves or for the benefit of the project. The best examples benefit both" (Silvertown, 2009, p. 467). To prevent confusion, a distinction is made within this category of professional counterparts (Silvertown, 2009). The professional counterparts in Silvertown's definition include scientists, conservation professionals and government practitioners. We define scientists as those involved in academia. Conservation professionals are those working for nature managers or conservation organizations. Government practitioners are defined as those working at a government agency or at the local government level.

Citizen science in water resource management is upcoming but lingering on the verge of a breakthrough (Buytaert et al., 2014; Cohn, 2008; Fraternali et al., 2012). The rise of robust, cheap and low-maintenance sensors enhances opportunities for citizen science in the complex arena of water resource management (Buytaert et al., 2014). New hydrological modelling frameworks (e.g. Clark et al., 2015) and specifically uncertainty quantification (e.g. Shoaib et al., 2016) can allow for more effective design of citizen science campaigns to systematically reduce model uncertainty, as well as more effective use of citizen-collected data. These data are often considered highly uncertain by practitioners, as they indicated in the exploratory interviews held at the start of this research. Fraternali et al. (2012) give an overview of the potential of amateurs taking part in data collection, data analysis and the process of decision-making in water resource management.

This potential is also recognized in the Netherlands. In November 2014 water authority Delfland organized a workshop[1] on big data and citizen science in Delft, the Netherlands. Dutch regional water managers expressed their interest in citizen science during this workshop, although they also indicated their doubts regarding this approach. Their main questions were (a) "what motivates citizens to participate?", (b) "what should be the role of citizens?", and (c) "why should a water authority engage in citizen science?"

Citizens' motivations have been studied extensively in a diverse set of citizen science projects, such as online crowd-sourcing (e.g. Chandler and Kapelner, 2013; Raddick et al., 2010; Rogstadius et al., 2011), environmental monitoring (e.g. Hobbs and White, 2012; Roy et al., 2012) and meteorology (e.g. Gharesifard and Wehn, 2016). It has been acknowledged that the idea of "the public" does not exist (e.g. Varner, 2014), since "the public" consists of a wide variety of people with different backgrounds, interests, traits, values and beliefs. Nevertheless, existing studies of citizen science in field and online projects, despite this diversity, reveal the same dominant motivations over a wide range of projects and participants. The most mentioned reasons for citizens to engage in citizen science are because they think it is fun, because the topic interests them, and because the topic matters to them, e.g. they want to contribute to science or nature conservation (e.g. Chandler and Kapelner, 2013; Hobbs and White, 2012; Raddick et al., 2010; Rogstadius et al., 2011; Roy et al., 2012). Citizens are motivated to continue to contribute by (increasing) the extent of their involvement (Rotman et al., 2012; Roy et al., 2012), offering feedback concerning the work at three levels (individual contribution, group contribution and the use of data) and building a relationship based on trust between scientists and citizens (Rotman et al., 2012). The importance of trust is stressed by authors in the field of water management as well (Buytaert et al., 2014; Gharesifard and Wehn, 2016).

The role of citizens varies depending on the purpose of the citizen science project. Tulloch et al. (2013) studied the purpose of citizen science projects in bird watching, a research field with a century-long history of citizen science. The most common purpose of citizen science is knowledge generation (knowledge also new to the involved professionals), followed by improving monitoring methods and raising awareness among citizens. In water resource management, citizen science can enhance knowledge about the water system, for example by generating knowledge on different spatial or temporal scales. It may also be used to add or test new monitoring methods for the existing monitoring network (mentioned in exploratory interviews held at the start of this study). Citizen science can also be used to increase the water awareness that was dubbed absent in the OECD report (OECD, 2014). According to Tulloch et al. (2013), citizen science can be used to improve management practice as well. In water resource management such improved management can be the result of more frequent monitoring. The literature mentions public education as an important purpose of citizen science (e.g. Cohn, 2008), but Tulloch et al. (2013) found that public education was rarely the main purpose. Other identified, yet also more rare, purposes include doing social research (e.g. on human behaviour), offering recreation and serendipity (i.e. unexpected discoveries). A more recent study in the field of ecology specified the potential of citizen science for the purpose of policy development (Hollow et al., 2015). In early stages of a policy development process, citizen science can be used to discover alternative management actions. In case there is a range of alternatives, citizen science

---

[1]Part of the symposium "De fysieke Digitiale Delta" (the Physical Digital Delta; see also http://www.digitaldelta.nu/en/events/item14).

can be used to gauge public opinion. In later stages it can be used to persuade public opinion towards a desired alternative or to provide a legal justification for the chosen policy. Citizen science-based data can be used for decision-making in water resource management as well (Macknick and Enders, 2012). Projects can have one or multiple purposes, as the iSPEX project demonstrates. This project served knowledge generation, public education and method improvement (Land-Zandstraet al., 2015; Snik et al., 2014).

For the purposes of knowledge generation, improving monitoring methods improves management and policy development. Bonney et al. (2009) provide a useful classification of citizens' roles. They suggest there are basically three levels of citizen involvement possible: contribution, collaboration and co-creation. In a contributory project, citizens are mainly involved in data collection; the research question and design are done by scientists or experts. In collaborative projects citizens are involved in the analysis and can be involved in the design and dissemination of results as well. In co-created projects citizens are involved in all steps of the research process and may even initiate the project. The vast majority of studies in the overview presented by Bonney et al. (2009) considered contributory projects. Even the occasionally co-created projects were part of multi-case studies in which contributory projects dominated the results. Citizens' involvement in activities other than data collection may serve different purposes; for example, citizen-based goal-setting could enhance adaptive management practices (Cooper et al., 2007). Bonney et al. (2009) is frequently cited (e.g. Rotman et al., 2012; Roy et al. 2012) and can be considered a typical classification. The levels of involvement align to a large extent with the governance structures defined by Conrad and Hilchey (2011): consultative, collaborative and transformative governance. In transformative governance citizens initiate a project. An example can be found in the global community monitor that measures air and water quality on a global scale (Conrad and Hilchey, 2011).

For the purposes of public education, raising awareness, improving management and policy development, science communication literature (e.g. Varner, 2014) provides a more useful classification of the interaction between citizens and professional counterparts. We use the term science communication here in a broad sense and include communication between public and all professional counterparts referred to by Silvertown (2009), including practitioners. Citizen science is often viewed as a form of informal science education, contributing to public awareness of science (PAS) and public understanding of science (PUS). We think this view is too limited, as it only encompasses the deficit model of science communication. Higher levels of involvement public engagement of science (PES) and public participation in science (PPS) are possible (Van der Auweraert, 2005), particularly in the collaborative and co-created projects in Bonney's definition. An example of such participation is the water quality pro-

gramme in Rhode Island mentioned in the review by Conrad and Hilchey (2011).

The motivation of professional counterparts to engage in citizen science has been less frequently studied than the motivation of citizens, and to our best knowledge research on professionals' motivation is limited to scientists. Scientists' motivations are primarily to advance science as well as develop their careers (Rotman et al., 2012). This is aligned with citizens' motivational desire to contribute to science and conservation or to engage in exploring a topic of their interest further (e.g. Rotman et al., 2012). Weng (2015) identifies three areas of friction between the vision of scientists and volunteers with regard to citizen science. The first area of friction is often short-term participation of volunteers that conflicts with scientists' interest in long-term processes. The second area concerns the limits of what volunteers can do and their dissatisfaction with the research processes. The third area regards a power hierarchy between citizens and scientists. Rotman et al. (2012) found that while the motivations of citizens and scientists are complementary, they can also change over time. Therefore, continued attention by those who are managing citizen science projects with regard to matching these motivations is crucial.

The motivation of scientists cannot be translated one-to-one to practitioners or (local) government representatives for two reasons. First, scientists are concerned with scientific data collection (Rotman et al., 2012), while practitioners are often interested in improving management practices (Weng, 2015) and government agencies are concerned with policy-making (Hollow et al., 2015). Second, the different role of authorities leads to different expectations. Water authorities believe that citizens see water resource management as a task for authorities only, which implies that citizens do not want to be involved. Nevertheless, most water authorities agree that they need the observations of citizens for their work, as expressed in several studies related to flood risk management (Wehn and Evers, 2014; Wehn et al., 2015).

This study aims to explore perspectives of government practitioners regarding citizen science. We explore Dutch water practitioners' perceptions of citizen motivation, acceptance in their organization, (potential) purposes and level of citizen engagement. We address apparent knowledge gaps regarding the motivation of professionals to embrace citizen science and specifically the link with citizen participation. Perceptions were explored using Q methodology. With Q methodology we identified a set of viewpoints that describe the variation in opinions. This variation we consider meaningful for the design and implementation of citizen science at water boards in the Netherlands. In addition to providing insight into the viewpoints of Dutch practitioners, this study aims to develop a methodology to study perspectives regarding citizen science in a wider range of countries and professionals.

2 Method

The study uses Q methodology (Van Exel and De Graaf, 2005; Watts and Stenner, 2012) to find viewpoints on citizen science among employees of the Dutch water authorities. Q methodology is a relatively uncommon method in water resource management (e.g. Raadgever et al., 2008), but it is a popular method in social sciences fields, such as political science and psychology (Cools et al., 2009). The strengths of the Q methodology are that it combines qualitative and quantitative aspects and that it is statistically robust, with small samples of 30–40 people (Van Exel and De Graaf, 2005; Watts and Stenner, 2012). For this research Q methodology has the advantage over fully quantitative approaches in that we could explore a wide spectrum of viewpoints, whereas quantitative methods would arrive easily at averaged values. Quantitative methods would likely not reveal the distinguishing elements that are in our opinion crucial to be aware of for effective implementation of citizen science. Q methodology has the advantage over fully qualitative methods that it reduces the variation in opinions to a representative small set of viewpoints in a trackable, relatively little biased way (further discussed below).

This section provides a short description of the methodology and research-specific details. We kindly refer readers interested in more methodological details to Van Exel and De Graaf (2005) for a quick introduction, and Watts and Stenner (2012) or Brown (1980) for an elaboration on the philosophy and statistical basis. In conducting this research we closely followed the guidelines of Watts and Stenner (2012). Our Q methodological research consisted of seven stages summarized in Fig. 1 and elaborated below.

The first stage aimed at collecting as wide a range as possible of opinions on the topic of citizen science for water quality monitoring (discourse) and documenting them in statements (concourse). We held 10 semi-structured interviews with employees of water authorities, nature managers and citizen organizations, from which 181 statements were derived. Additionally, we organized a structured group discussion about citizen science with water professionals at the afore-mentioned workshop, which resulted in an additional 21 statements. To collect a wider range of opinions, we organized a focus group meeting with five middle-aged woman of an informal walking club with high potential to participate in citizen science, which resulted in an additional 20 statements. Finally, benefits and downsides of citizen science were extracted from the literature, which resulted in an additional seven statements. To reduce researcher bias, we based statements as much and as closely as possible on quotes. The final concourse (all possible statements on a topic) consisted of 229 statements. Four themes could be identified in the concourse: (I) citizen motivation, (II) acceptance of citizen science at the water authority, (III) purposes of citizen science for professionals and (IV) level of citizen engagement.

In Stage 2 the concourse was reduced to the so-called Q-set set of 48 statements that still reflected the full discourse. We first reduced the concourse to a preliminary Q-set of 65 statements by excluding or reframing statements that were similar to others, out of scope or ambiguous. Six masters students between the ages of 22 and 25 tested this preliminary Q-set. Two female students had a major in water resource management; the other two female and two male students had a different major (medicine, mathematics, mechanical engineering and management studies). We instructed them to sort the statements and list statements that they did not understand, found similar in meaning or considered irrelevant, and improved the preliminary Q-set based on their feedback. Table 1 contains the final Q-set of 46 statements; the Roman numerals indicate to what theme the statement contributes.

In Stage 3, the P-set (i.e. the group of participants) was sampled using both a structured approach with three criteria and snowball sampling. Flood risk was a first criterion, as water authorities with a high flood risk face different challenges than water authorities with a lower flood risk. Age (expressed in years since the last reform or merger with another water authority) was a second criterion, as a recent reform suggests the organization may be more susceptible to innovation, such as citizen science. Location (within or outside the urban conglomerate Randstad) was a third criterion, because interviewees in the semi-structured interviews held to define the discourse suggested a different relationship between water authority and population in rural and urban areas. The structured approach resulted in eight water authorities representing a combination of these criteria. In each water authority we approached the ecologists because we expected them to be more familiar with the concept of citizen science and we had access to a list of ecologists per water authority. Additionally we used snowball sampling. We asked the participating ecologists to recruit colleagues with a similar opinion, to enhance overlap between individual opinions, and, with different opinions, to increase the diversity of opinions. Also, we asked participants to recommend someone with an opposing opinion, in order to discover as many viewpoints as possible. Participants 20, 24, 25, 30 and 31 out of 33 were recruited with this strategy, with the aim that they would belong to different (new) viewpoints. Two to six people with different positions were interviewed per water authority, which resulted in interviews with 1 politician, 20 policy advisors, 10 ecologists/hydrologists and 2 field staff members.

Next, in Stage 4, the Q-sorts (the actual arranging sorting process) took place in four sub-steps, taking a total time of 60 to 75 min. First, the first author gave three examples of consultative citizen science to all participants, to ensure everyone had a basic level of understanding of citizen science. These examples were

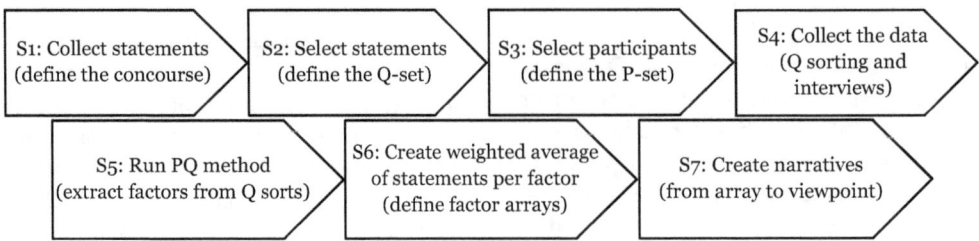

Figure 1. Flowchart of the steps of the Q methodological research approach. Based on Watts and Stenner (2012). For clarity reasons, we choose to describe Stage 5 in Watts and Stenner (2012) as three separate stages.

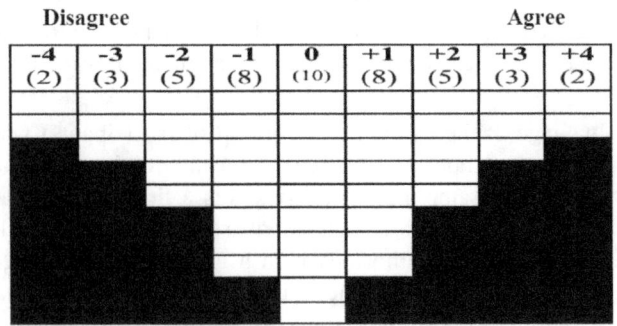

Figure 2. The fixed distribution used in this study. The participant places the two statements that he agrees most with in the +4 column, the next three statements in the +3 column, etc. The process is repeated for the disagree statements.

– the Dutch garden bird count (www.tuintelling.nl);

– iSPEX: citizens measured particulate matter with a smartphone device called iSPEX (Snik et al., 2014, p. 7351); and

– water level monitoring by citizens in a Dutch water authority (UvW, 2015b, p. 15).

Second, participants pre-sorted the statements into three piles: agree, disagree and neutral. Third, they made a final sorting of the statements in a fixed distribution (see Fig. 2). Finally, the first author held a structured post-sorting interview. Post-sorting interviews were included in this study, because they can provide in-depth insight into the beliefs and values underlying the sorts and allow for an analysis based on the participants' rationale rather than on the available literature or the researcher's bias (Gallagher and Porock, 2010). Discussing all statements with participants would have been preferable, but was not feasible given the available time for this study and the geographic spreading of participants. In a structured interview, participants explained their reasoning for the statements in categories +4 and −4 and (if time allowed) any statement of their choice. Participant's afterthoughts were recorded, transcribed and categorized per statement and per factor.

Next, in Stage 5, a factor analysis was performed with software package PQMethod, version 3.2.1. PQMethod is used to perform a factor analysis and is frequently used in Q methodological research (Van Exel and De Graaf, 2005; Cools et al., 2009; Raadgever et al., 2008; Watts and Stenner, 2012). A factor analysis is a statistical method to describe variability in a set of correlated variables, in this case the ranking of statements by individuals, by a smaller number of factors. We included three factors with an eigenvalue above 1 (recommended by Watts and Stenner, 2012) for further analysis. The factor analysis also showed which people load on which factor, i.e. which people have a perspective resembling that factor. People can load on none, one or multiple factors. For this a threshold was used, the significant factor loading (SFL). In this study we followed Watts and Stenner (2012) and used a SFL of 0.38. Factors were optimized using factor rotation, which aims to have as many people as possible load on a single factor rather than loading on two factors simultaneously. The rotating process does not alter the results themselves, but changes the researcher's observation position in order to optimize the loading of each Q-sort on a single factor (Watts and Stenner, 2012, p. 118). A manual rotation was preferred above the built-in Varimax rotation of the PQMethod, because it has a lower inter-factor correlation and thus results in more distinct viewpoints.

Next, in Stage 6, a weighted average was used to create an illusory person with a factor loading of 1.0 per factor, i.e. a hypothetical person who has fully adopted this factor. Following Watts and Steiner (2012) Q-sorts with loadings that exceed the SFL of 0.38 for a factor were incorporated to compute the weighted average for that factor. See Table 2 for the factor arrays and Table 3 for the final factor loadings. It must be noted that the number of people loading on a factor cannot be used to determine the distribution of viewpoints in the total population without additional (quantitative) research.

The final stage, Stage 7, is data analysis, where the factor arrays were translated into a viewpoint narrative. We followed the guidelines of Watts and Stenner (2012), Gallagher and Porock (2010) and Cools et al. (2009), and used distinguishing statements. We created a narrative of the +4 and −4 ranked statements and the statements ranked highest in a single factor array, meaning this statement is ranked lower in all

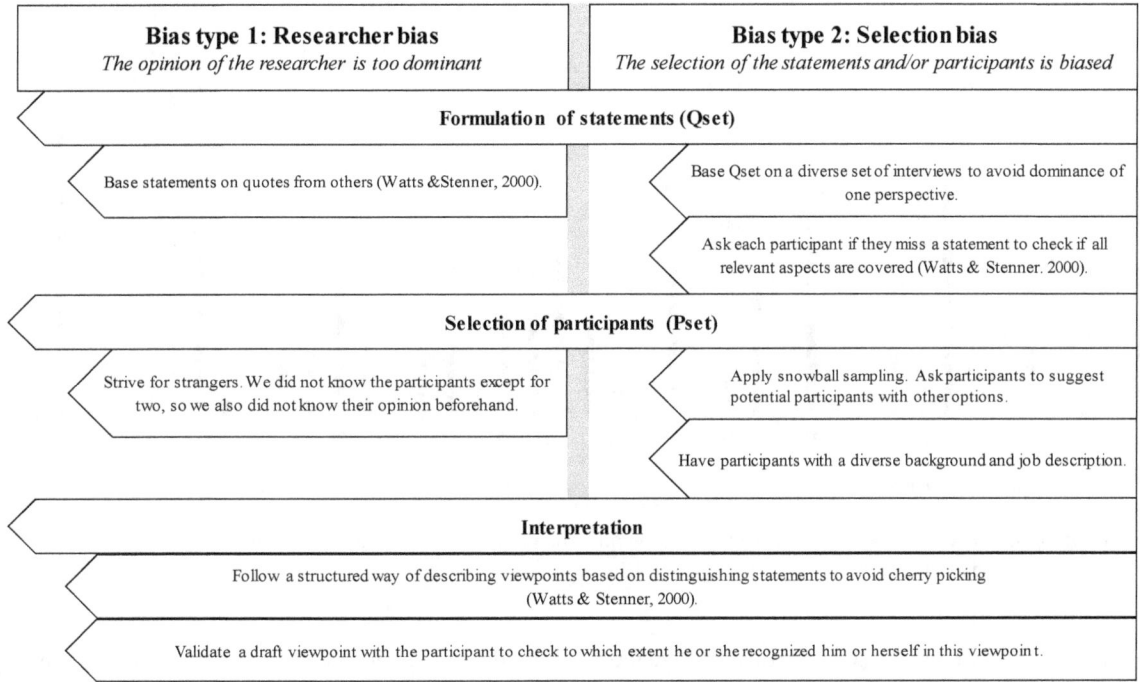

Figure 3. A summary of the mechanisms used to reduce bias. In two columns actions are listed to reduce researcher bias (left) and selection bias (right).

other factors, and vice versa. For example, in Factor A, Statements 2 and 9 ranked with +4 and Statements 10 and 41 with −4, and Statement 34 is an example of a Statement ranked highest in factor A. Factor A ranks it +2, compared to −1 and +1 in Factors B and C. Hence, Statements 2, 9, 10, 41 and 34 were included in the viewpoint narrative of Factor A. In addition to this mechanism of distinguishing statements, two other mechanisms were introduced to reduce researcher bias. First, we conducted post-sorting interviews in order to be able to incorporate participant's underlying values and assumptions in the process of interpretation. During these interviews we also checked whether in their opinion all relevant aspects were covered by the Q-set. Second, we showed all participants an initial version of the narratives and asked whether they recognized themselves in their assigned narrative viewpoint and why (or why not). In addition we presented the results to employees of the Dutch water authorities on two occasions (Delfland Scriptieprijs uitreiking 21 January 2016 and STOWA Monitoringcongres 19 April 2016) to collect feedback. An overview of mechanisms used to reduce bias in all steps of the research is given in Fig. 3.

3 Results

The 33 Q-sorts resulted in the identification of three factors from which three viewpoints were derived: A "Citizen participation for data application", B "Water authority in control", and C "Education and sharing local knowledge". The choice

for three factors was based on the explained variance of the first four factors before rotation that displayed with 53, 8, 6 and 1 % a cut-off after the third factor that was supported by qualitative information from the interviews. Table 1 and the radar charts presented in Fig. 5 show how an individual would rank the items if that person were representing that factor 100 %. For example, Statement 9 ("Citizen Science enables the collection of large amounts of measurements", Theme III Purposes of Citizen Science) would be placed under most agree (column +4) by a person with Factor A, under agree (column +2) for Factor B and under neutral (column 0) for Factor C. None of the viewpoints disagrees with Statement 9, but the difference between Viewpoints A and C is evident. Except for Statement 2 ("Citizen Science is important, since it contributes to increasing water awareness"), with which all viewpoints fully agree, and Statement 35 ("If citizens are structurally contributing, they should be compensated for that"), differences were found between the viewpoints that will be further discussed below. The factors provided a quite clear separation of the participating practitioners in groups given Table 2. Out of 33 participants, 21 loaded significantly and uniquely on Factor A, 4 on Factor B and 2 on Factor C. Three participants loaded significantly on Factors A and C, one on Factors A and B and one on Factors B and C. One participant did not load significantly on any of the factors.

The remainder of this section contains the three viewpoint narratives. The term viewpoint is used to refer to the factor's

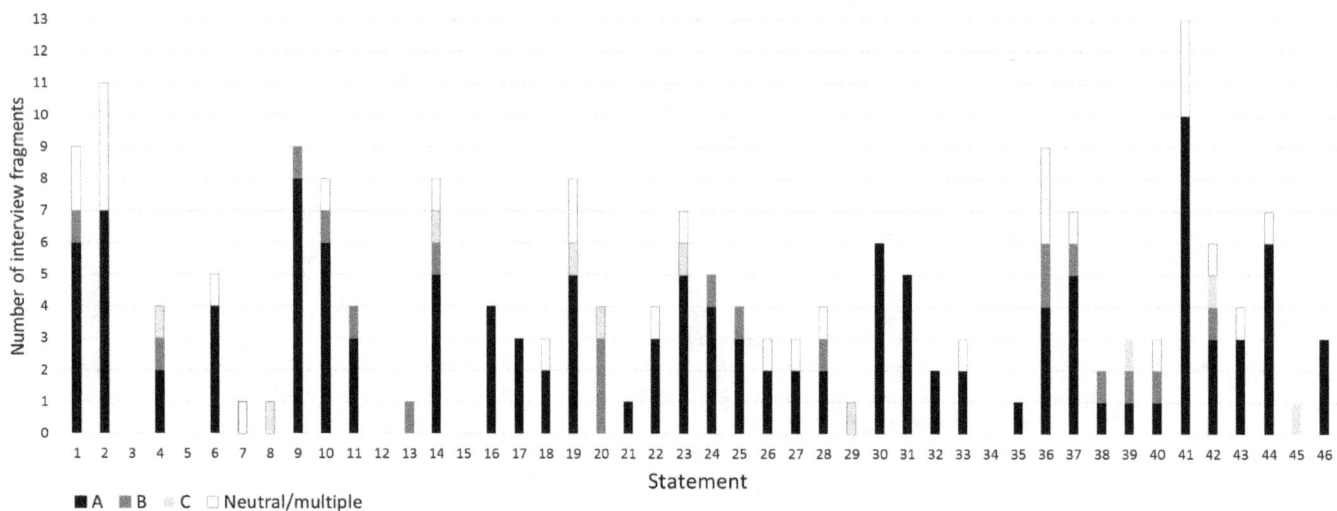

Figure 4. Distribution of interview fragments available per viewpoint and per statement. Interview fragments of people that load to no or two viewpoints were categorized as "neutral".

interpretation, for which we made use of the quantitative factor arrays in Table 1 and qualitative quotes from participants loading significantly on that factor (see Table 2). The narratives are based on absolute results (agree or disagree), the relative results (an item is ranked higher or lower in Viewpoint A than Viewpoints B and C) and characteristic interview items. Item rankings are presented in the following format: (item number: item ranking) such that (2: +4) means item 02 is ranked +4 in this viewpoint. Interview fragments are integrated in the narratives as a quote followed by the letter Q and a number indicating the source. For example ("quote" – Q1) means the quote comes from the Q-sort and thus participant 1. Figure 4 shows the availability of interview fragments per factor and per statement. Most fragments (125) were available to interpret Factor A as most people loaded significantly to that Factor A and less for Factors B (20 fragments) and C (10 fragments). As an intended result of the interview technique (see Stage 4 in the Method Section), most fragments were available for statements with particularly low or high rankings, and there was a positive relation between the level of disagreement among factors and the number of interview fragments. The higher the total absolute difference between the Factors A, B and C, the higher the number of interview fragments.

3.1 Viewpoint A: "Citizen participation for data application"

The people loading on Factor A (see Table 2) are a mix of hydrologists, advisors, policy advisors, field staff and a politician. In this group are 14 men and 11 women. Eleven people are middle aged (between 45 and 65). They represent all eight incorporated water authorities, which are located within and outside the Randstad and have a mixture of higher and lower

flood risk. Nineteen people work at a water authority that has recently gone through an organizational reform.

People with Viewpoint A think that citizen science is important for water authorities to increase water awareness (2: +4), because citizens are unacquainted with the work of the water authority: "People often do not know what the water authority is doing exactly and we do not really stand out. Citizens sometimes really wonder what they pay tax for [...]" (Q5) and "what they can do themselves to improve water quality" (Q27). They believe that their water authority should actively incorporate citizen science in its policy (34: +2) and that the water authority should not wait to invest in citizen science until it is included in top-level policies (28: −3).

Additionally, people with this viewpoint value citizen science for the collection of large amounts of data (9: +4) and for conducting measurements more frequently (8: +3). "This data, they are an opportunity to have an area covering insight in dynamics of water quality and ecology" – Q26. They think that citizens can be trusted to conduct these measurements (21: −1). Although citizen science is a social innovation and the acquired data are less accurate, it should be accepted by the water authority (19: −3) ("I mainly disagree strongly with the latter part of this statement" – Q13 [(...) and should not be accepted by the water authority]) and will be a valuable addition to the official monitoring network (10: −4). These people do not prefer the smart use of existing data to citizen science data (12: −1). The organization is expected to have sufficient capacity to analyse all the data (33: −2) at the moment, but the water authority has to learn how to handle the uncertainty of these alternative (often more economical) measurements (18: +2). They do not believe that the water authority needs to maintain control of monitoring, even though water authorities are in the end responsible for monitoring (42: −2). "This is nonsense, because a lot is al-

Table 1. Final factor arrays; the numbers in columns A, B and C are the theoretical item score for a person whose viewpoint is 100 % that factor. Roman numerals indicate the theme category (cat) of the item. I is citizen motivation, II is water authority acceptance of citizen science, III is purposes, and IV is level of engagement.

	Statements (QSet)	Cat	A	B	C
1	Providing citizens with insight in water quality will only lead to unnecessary panic and questions.	II	−3	−4	−4
2	Citizen Science is important, since it contributes to increasing water awareness.	III	+4	+4	+4
3	Citizen Science is a solution to explain why you take certain measures as a water authority.	III	+1	−1	−1
4	Water quality is an abstract concept, citizens will not understand what they measure.	IV	−1	−2	−3
5	It is important to have proper communications to citizens about why values deviate from the norm and what the uncertainty in the measured value is.	III	+1	0	+1
6	I would not know why citizens would not be interested in monitoring water quality.	I	−1	−2	0
7	Citizen Science is an economical way to collect (extra) measurements.	II	+1	+1	−1
8	Citizen Science enables the collection of more measurements by conducting them more frequently.	III	+3	+3	+1
9	Citizen Science enables the collection of large amounts of measurements.	III	+4	+2	0
10	Measurements and observations by citizens are no valuable addition to the official monitoring network.	III	−4	−2	−1
11	The most important goal is that the measurement data provide value to the water authority because the organization has invested its time and energy.	III	0	+1	+1
12	I would rather make (smart) use of existing measurements than let citizens conduct more measurements.	II	−1	0	0
13	The greatest challenge is how to teach people something, if they can or want to spend little time on it.	III	0	0	−1
14	Schools are especially suitable target groups to conduct these measurements, for example during a "water lesson".	III	0	0	2
15	The most important goal of Citizen Science is to teach people something about the environment they live in.	III	+1	+2	+3
16	Citizen Science is an interesting social innovation, but not suitable for actually collecting useful data.	II	−2	−2	0
17	Citizens' abilities are often underestimated; they are better educated and smarter than we think.	II	+1	+1	0
18	As a water authority we need to learn how to handle the uncertainty of alternative (cheap) measurements that originate from Citizen Science.	II	+2	+1	+1
19	Data collection by citizens is unreliable and should not be accepted by the water authority.	III	−3	−2	−1
20	Citizens will only participate in Citizen Science if participation is in their own interest.	I	0	+2	−2
21	Not all citizens can be trusted to conduct these measurements.	II	−1	+1	0
22	With a short training, citizens will be able to conduct measurements for the water authority.	IV	2	+1	2
23	Citizen Science is an interesting way to give meaning to the concept of citizen participation.	III	+3	+1	+3
24	Citizen Science is necessary, because it helps to decrease the awareness gap between citizens and the water authority.	III	+2	−3	+2
25	By using Citizen Science, the water authority shows that it is keeping pace with the times.	II	+1	−1	0
26	An important advantage of Citizen Science is that it reduces citizen's resistance to projects.	III	0	0	+1
27	One can connect with and involve another part of the audience using Citizen Science.	III	+2	0	+3
28	As long as Citizen Science is not included in the policy at the top levels, the water authority should not invest in it.	II	−3	−3	−2
29	It is a major bottleneck to create support within the water authority for the deployment of Citizen Science.	II	−1	0	−1
30	The water authority will benefit from using Citizen Science in conducting its tasks, because fewer (financial) resources are available.	II	0	−1	0
31	The conservative character of my organization is a major bottleneck for Citizen Science.	II	−1	−1	−2
32	The organization is not equipped to work with large groups of citizen scientists.	II	0	+3	0
33	My organization has no capacity to work with all this data.	II	−2	−1	−2
34	The water authority should incorporate in its policy how to deploy and stimulate Citizen Science more.	II	+2	−1	+1
35	If citizens are structurally contributing, they should be compensated for that.	I	0	0	0
36	If citizens collect data for the water authority, they should have a say in the measures taken afterwards.	III	−2	−4	−3
37	Citizens often have local knowledge and the water authority should use this knowledge.	IV	+3	+4	+4
38	Citizen Science is important, because it gives insight into the problems that citizens are concerned with.	IV	+1	0	+1
39	Citizens should have insight in the most recent information of the water quality that is available with the water authority.	IV	+1	+1	+2
40	If you provide citizens with a reference framework, they themselves can validate their data.	IV	0	−3	−3
41	I do not want citizens to interfere with our work.	II	−4	−1	−4
42	The water authority should maintain control of conducting measurements, since the water authority is indeed responsible.	IV	−2	+3	+2
43	I think the creation of Citizen Science does not fall within the tasks of the water authority.	II	−2	−1	−2
44	I do not have a full image of what is possible with Citizen Science.	II	−1	0	−1
45	An important caveat is that citizens will expect that their measurements will have a direct influence on policy.	III	0	+2	+1
46	Citizens cannot be motivated to participate in such projects for a long period.	I	−1	+2	−1

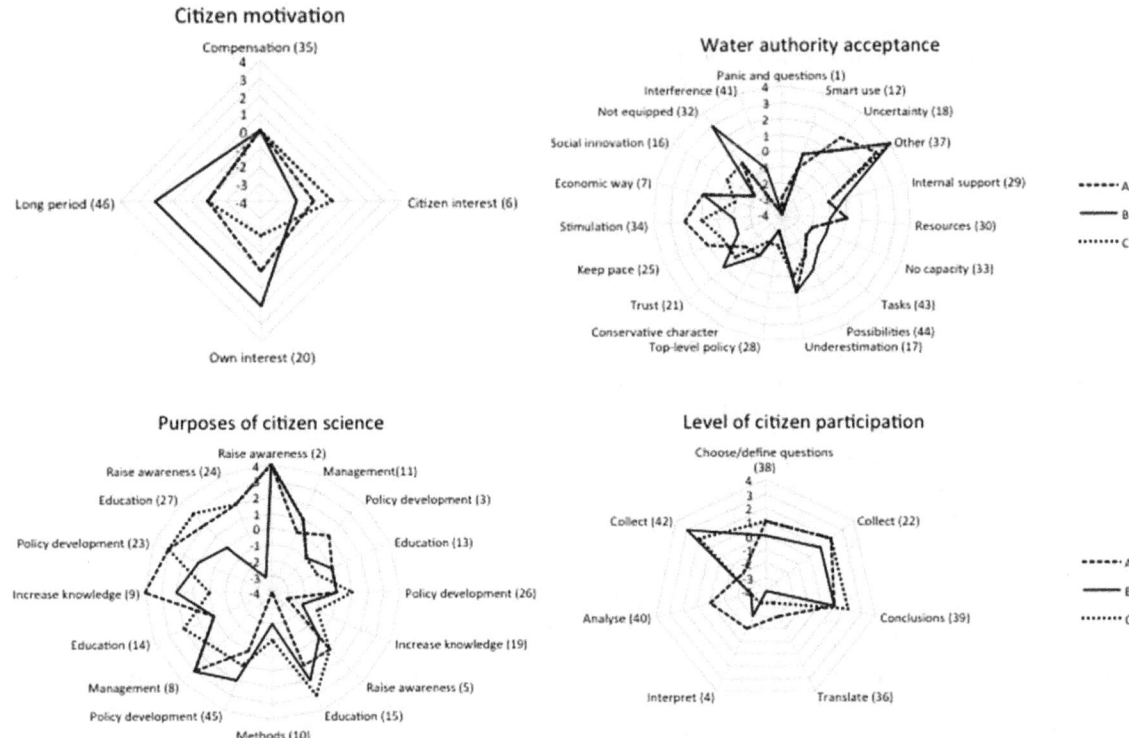

Figure 5. Theoretical item score for a person whose viewpoint is 100 % Factor A, B or C. Statements are clustered per theme (citizen motivation, water authority acceptance, purposes and level) in four radar charts. Numbers in brackets refer to the full statement given in Table 1.

ready measured by other parties" – Q25. They believe that citizens, if provided with a reference framework, can validate their own data (40: 0) "If they know what to do with it [the results], they can translate it to their environment" – Q11.

People with Viewpoint A think of citizen science as an interesting way to give meaning to the concept of citizen participation (23: +3) and decrease the gap between the water authority and citizens (24: +2). They are least (compared to Viewpoints B and C) afraid that citizens will expect their contribution to have a direct impact on policy (45: 0) and they do not think citizens should get this influence (36: −2). "You should prevent that, because manipulation [of results] is evident" – Q9. They consider citizen science to be a solution when it comes to explaining why you undertake certain measures (3: +1). This group further feels that citizen science will show that the water authority is keeping pace with the times (25: +1), although it is not a priority. Giving citizens' insight into water quality will not lead to unnecessary questions and panic (1: −3). "Those questions will come, but you should not be afraid, not afraid to say that you do not know everything" – Q21. They do not fear citizen interference with their work (41: −4).

3.2 Viewpoint B: "The water authority in control"

The six people loading significantly on Factor B (see Table 2) form a mixture of advisors, policy advisors and field staff. Five out of six are male and four of them are middle aged. They work at three different water authorities; two people work outside the Randstad. Four people work at water authorities that have recently (after 2005) gone through an organizational reform. All work in an area with a high flood risk.

People with Viewpoint B consider citizen science important for increasing water awareness (2: +4). In contrast to Viewpoint A, they think that the water authority should not incorporate citizen science as part of its policy (34: −1). They feel they do not have a full idea (yet) of what is possible with citizen science (44: 0). However, they do think the water authority should invest in citizen science, even if it is not yet included in top level policies (28: −3). They do fear that citizens cannot be motivated for long-term participation (46: +2) and will not participate unless participation is in their own interest (20: +2). This group is more concerned than the other two groups that the creation of a support base within the organization will be a bottleneck (29: 0), and they are convinced that their organization is not (yet) equipped to work with large groups of citizen scientists (32: +3).

Table 2. Final factor loadings after rotation. Sorts in bold indicate that a person's loading exceeds the significant factor loading (SFL) and a person's viewpoint thus resembles Viewpoint A, B or C. Also included are characteristics of respondents: water authority (Identification (ID), elevation above sea level (El), located in Randstad (RS) and recently reformed (since 2005)) and individual characteristics (sex, age and function group).

	Factors			Water authority				Individual respondent		
Sort	A	B	C	ID	El.	RS	RR	Sex	Age	Function group
1	**0.74**	0.18	−0.01	A	−	Yes	Yes	F	45–65	Politician
2	**0.49**	0.18	0.35	A	−	Yes	Yes	M	20–45	Policy advisor
3	0.35	**0.55**	0.18	A	−	Yes	Yes	M	45–65	Policy advisor
4	**0.49**	0.36	−0.03	B	−	Yes	No	F	20-45	Advisor (water quality/ecology)
5	**0.69**	−0.07	0,29	C	+	No	Yes	F	20-45	Policy advisor
6	**0.71**	0.09	0.32	C	−	Yes	No	M	20–45	Policy advisor
7	0.45	**0.62**	0.12	D	−	Yes	No	M	45–65	Field staff
8	0.31	**0.46**	**0.43**	D	−	Yes	No	F	20–45	Advisor (ecological monitoring)
9	**0.79**	0.19	0.06	D	−	Yes	No	F	20–45	Policy advisor
10	−0.05	0.22	**0.66**	D	−	Yes	No	M	45–65	Advisor (ecology)
11	**0.64**	−0.17	0.03	C	+	No	Yes	M	45–65	Policy advisor
12	**0.72**	0,12	**0.49**	C	+	No	Yes	M	45–65	Policy advisor
13	**0.75**	0.01	−0.12	C	+	No	Yes	M	20–45	Advisor (innovation)
14	**0.69**	−0.06	**0.41**	C	+	No	Yes	F	45–65	Field staff
15	**0.84**	−0.06	0.30	E	+	No	Yes	F	45–65	Policy advisor
16	**0.64**	0.33	**0.42**	E	+	No	Yes	F	20–45	Advisor (permits and assessment)
17	**0.58**	−0.05	0.25	E	+	No	Yes	M	45–65	Advisor (water system)
18	−0.04	**0.55**	0.24	F	−	No	Yes	M	45–65	Policy advisor
19	**0.69**	0.01	0.17	F	−	No	Yes	M	20–45	Hydrologist
20	0.32	**0.43**	−0.13	F	−	No	Yes	M	45–65	Policy advisor
21	**0.73**	−0.09	0.32	F	−	No	Yes	M	45–65	Policy advisor
22	**0.59**	0.25	0.04	F	−	No	Yes	F	20–45	Hydrologist
23	**0.82**	0.17	0.01	G	−	Yes	Yes	F	20–45	Advisor (innovation)
24	**0.74**	0.04	−0.12	G	−	Yes	Yes	M	20–45	Advisor (innovation)
25	**0.45**	0.00	0.23	G	−	Yes	Yes	M	20–45	Ecologist
26	**0.80**	−0.06	−0.07	G	−	Yes	Yes	M	45–65	Hydrologist
27	**0.85**	0.15	0.23	G	−	Yes	Yes	M	45–65	Policy advisor
28	**0.54**	0.20	0.27	H	−	Yes	No	M	45–65	Policy advisor
29	0.34	−0.03	0.34	H	−	Yes	No	M	45–65	Project leader
30	0.30	**0.39**	−0.09	A	−	Yes	Yes	M	20–45	Policy advisor
31	0.27	0.23	**0.43**	B	−	Yes	No	M	45–65	Policy advisor
32	**0.77**	0.21	0.10	B	−	Yes	No	F	20–45	Advisor (water quality)
33	**0.79**	−0.22	0.34	B	−	Yes	No	F	20–45	Policy advisor
EV[1]	12.55	2.29	2.58							
Var[2]	38 %	7 %	8 %							

People with Viewpoint B believe that local knowledge will be valuable for the water authority (37: +4), as the citizen "knows his own environment better than we do, on the small scale. We only have the broad overview in a large area" – Q20. They strongly believe that the water authority needs to maintain control of monitoring, because it has the final responsibility (42: +3). "I know what should be done with the data in the end. If we leave it to volunteers in this case, you have no reassurance on what comes when" – Q3. Citizen science allows for the collection of more measurements (9: +2) by conducting them more frequently (8: +3). They think citizens will be able to conduct measurements after they receive a short training (22: +1), but they do not expect that citizens will be able to validate their own data if provided with a reference framework (40: −3). "If [data] quality is important to you, I am not sure whether citizens can do this" – Q18. They further question whether all citizens can be trusted in doing measurements (21: +1). "If the citizen does not have personal interest, you have to wait to see what happens. Then he will think: I do not feel like it, I do not have time" – Q20. This reflects their belief that most citizens would not be interested in participating (6: −2). This group does not believe the water authority needs citizen science to help fulfil its tasks to compensate for fewer financial resources (30: −1).

People with this viewpoint are least convinced that citizen science will involve another part of the public (27: 0) or that it is an interesting way to give meaning to citizen participation (23: 1). Moreover, they believe that citizen science should not be used to decrease the gap between citizens and water authorities (24: −3) or to show that the water authority is keeping pace with the times (25: −1). "If this is your reason, I think it is rather cheap" – Q18. If citizens start collecting data for the water authority, this group strongly feels that they should not be given more influence over measures (36: −4), but they do fear that citizens will think that their work will influence policy directly (45: +2). "Citizens, I would almost say per definition, cannot do that [balance interests], they just want to do what they want" – Q30. People in this group do not fear questions or panic from citizens (1: −4). "If you are so suspicious towards your citizens, you have to question your role as government" – Q3.

3.3 Viewpoint C: "Education and sharing local knowledge"

The six people significantly loading on Factor C (see Table 2) are a mix of advisors, policy advisors and field staff. Four people are middle aged and three of them are male. They work at four different water authorities, three respondents within and three outside the Randstad. Three people work at a water authority that has recently (after 2005) gone through an organizational reform. Three out of six people with Viewpoint C work in an area with a lower flood risk.

People with Viewpoint C think that citizen science is important, because it contributes to increasing water awareness (2: +4). The most important purpose for this group is to teach people something about their environment (15: +3), and especially schools are considered to be a good target audience (14: +2). "It is a good way to keep them [students] engaged" – Q10. They think that citizens will understand what they measure, even though water quality is an abstract concept (4: −3). "I think this is an offensive comment toward the citizens, as if they are stupid" – Q10. They believe it is possible to teach people something within a short period of time (13: −1) and they find it difficult to think of reasons why people would not be interested in water quality (6: 0) compared to the other two groups. In contrast to Viewpoint B, they do think that citizens will participate, even if participation does not directly serve their own interests (20: −2). "I participate as a citizen in a sort of science project, I do not do that for my own benefit, but because I like it and want to contribute. I think I am not the only one" – Q10. Also, the conservative character of water authorities is not considered a major bottleneck (31: −2).

They feel that the water authority should use the local knowledge that citizens have (37: +4), but they consider citizen science to be merely a social innovation, rather than a way to collect useful data, compared to the other viewpoints (16: 0). This is reflected in their relatively small support of

the idea that citizen science will allow for collecting large amounts of data (9: 0) and for conducting measurements more frequently (8: +1). "It is mainly supportive material and not a replacement of existing sources, because it is invalidated and uncertified information. I do not think that will fit" – Q31. They strongly reject the view that citizens should not interfere with their work (41: −4), although they believe the water authority should stay in control (42: +2). "In my opinion information is essential for policy to be good, [so] I think they should be collected by a professional" – Q10. People in this group believe that citizens will not be able to validate their own data (40: −3).

People with Viewpoint C consider citizen science to be a good way to bind and involve another part of the audience (27: +3), to decrease the gap between citizens and water authorities (24: +2) and, to a lesser extent, to reduce citizens' resistance to projects (26: +1). A caveat could be that citizens will expect their measurements to have a direct influence on policy (45: +1), even though they should not be given a say in the measures taken afterwards (36: −3). "For me these are two separated tracks. [...] they have this influence via the representatives that they can elect for the board" – Q31. Citizens should be given insight into the most recent information about water quality that is available with the water authority (39: +2). "I believe that citizens and everyone have the right to get information from us" – Q10. These people strongly disagree that providing citizens with insight into water quality will lead to unnecessary panic and questions (1: −4).

4 Discussion

This study identified three different viewpoints with regard to citizen science derived with Q methodology. Participants sorted statements about four themes described in the Introduction and Method sections. In this discussion we first (Sect. 4.1) reflect on these four themes and relate the results to the literature, and second (Sect. 4.2) discuss the limitations of the research and make recommendations for future research.

4.1 Reflection on the four themes

The statements contributed to one of the four themes: (I) citizen motivation, (II) acceptance of citizen science at the water authority, (III) purposes and (IV) level of citizen participation. Several statements related to perceived citizen motivation by the participants (Theme I in Table 1 and the upper-left radar chart in Fig. 5) and acceptance of citizen science at the water authority (Theme II and the upper-right radar chart in Fig. 5). Trust in citizens' motivations and commitment ranged from low in Viewpoint B to high in Viewpoint C. The assessment of high, low or medium was based on the attitude that emerged from the ranking of statements in Themes I and

Table 3. Summary of the perception of the level of citizen motivations to contribute (Theme 1) and the level of acceptance of citizen science at the water authority (Theme II) for Viewpoints A, B and C.

	A	B	C
Theme 1: Citizen motivations to contribute	Medium	Low	High
Theme 2: Acceptance of citizen science at the water authority	High	Medium	High

Table 4. Overview of Theme III (purposes for citizen science) as supported in Viewpoints A, B and C. Purposes clearly indicated are marked with an X. (x) indicates that support for this purpose is not convincing.

	Viewpoint A	Viewpoint B	Viewpoint C
Increase knowledge	X	X	X
Improve methods	X		
Raise awareness	X	X	X
Improve management	X	(x)	
Public education			X
Policy development	X		(x)

II. Viewpoint C clearly had the most positive attitude towards citizens' motivations and B the least positive. For Theme II the distinction was less clear, although all viewpoints express a rather positive attitude. All three viewpoints are rather positive about implementing citizen science (perhaps due to some volunteer bias discussed in Sect. 4.2), although Viewpoints A and B are concerned with, respectively, the image of the water authority, and with the organizational capacity and a lack of internal support. Table 3 summarizes these findings. Table 4 summarizes the results with regard to the support per viewpoint for the purposes listed in the Introduction (Theme III). The same has been done for the roles that citizens can have (Theme IV) in Table 5 according to the classification of Bonney et al. (2009).

All viewpoints embrace data collection by citizens, i.e. contributory projects, and none support co-created projects (Table 5). Viewpoint A is optimistic towards citizen participation in the analysis of the data (see Statement 40), suggesting a potential for collaborative projects. Viewpoints B and C are wary of involving citizens in these steps of the research process. None of the viewpoints supports statements related to co-created projects. The following explanations are illustrative of the reluctance of participants to involve citizens in topic selection for monitoring (Statement 38): "There can be [topics] which we think they are important, while citizens do not find it important in the end" (Q24, Viewpoint A) and "In that case you should answer all [these questions of citizens] and I think our organisation is not equipped at the moment"

(Q18, Viewpoint B). Regarding the translation of results to action (Statement 36), participants said "They [citizens] can only focus on the problems in their direct environment, but not on the implications for a wider area" (Q19, Viewpoint A) and "I would not go that far" (Q18, Viewpoint B). Participants mentioned external causes such as the legal obligations a water authority has regarding the use of standardized methods and reports for water quality monitoring. This is consistent with previous conclusions about the responsibility of the water authority in relation to acceptance of citizen science (Wehn and Evers, 2014). Other participants mentioned internal causes, such as the existing procedures for citizens to influence decision-making, including complaint procedures and the water authority general elections once every 4 years.

There appears to be a mismatch between the intentions of the participants with respect to the purposes that citizen science should support and the expressed level of trust in citizens as reflected by the envisioned participation level. Especially people with Viewpoint C believe citizens have noble motivations, and they trust the citizens to a great extent. This trust is not reflected in the roles they envision for citizens, which is limited to data collection. The same goes for Viewpoint A. People with this viewpoint believe citizen science can serve many purposes (see Table 4). However, the envisioned role for citizens is limited to data collection and analysis. A lack of trust in citizens, few intentions to use the citizen scientists' data or a lack of support for higher levels of participation might conflict with citizens' motivations as described in the Introduction. A relation of mutual trust is required as the basis for effective citizen science projects and prolonged contributions by citizens (Rotman et al., 2012). Viewpoint B reveals distrust in the commitment of citizens, citizens' intentions to participate and their capacity (see Statements 20, 21, 22, 40 and 45). Another important motivation for citizens is the provision of feedback on how the data are used (e.g. Bonney et al., 2009; Rotman et al., 2012; Roy et al., 2012), which is particularly in contrast to Viewpoint C. Viewpoint C focuses on the goals of education (see Statements 14 and 15), with little emphasis on the actual use of the data (see Statements 8 and 9). Concerns regarding data quality may linger if recent development in systematic modelling and data assimilation approaches (e.g. Clark et al., 2015; Shoaib et al., 2016; Hut et al., 2015) would be adopted at the water authorities in the Netherlands. Such frameworks would allow systematic tracing of the propagation of uncertainty of citizen data in decision support tools and may identify opportunities for citizen science in model structure determination, which can be the largest source of uncertainty (Shoaib et al., 2016). Ottinger (2010) stressed the need for standardized methods in citizen science and legal embedding to increase the acceptance of the data and actual use of citizen science in policy-making. In the Netherlands systematic data handling and standardization may pave the way for using citizen science for public participation, in line with the vision of the Dutch water authorities governing body (UvW, 2015b).

Table 5. Overview of theme IV (level of citizen participation), based on Bonney et al. (2009). Supported activities are marked with an X. An (x) indicates the activities are sometimes assigned to citizens.

	Contributory projects	Collaborative projects	Co-created projects	Viewpoint A	Viewpoint B	Viewpoint C
Choose or define question(s) for study			X			
Design data collection methodologies		(x)	X			
Collect samples and/or record data	X	X	X	X	X	X
Analyse samples and data	(x)	X	X	X		
Interpret data and raw conclusions		(x)	X			
Disseminate conclusions/translate results into action	(x)	(x)	X			
Discuss results and ask new questions			X			

4.2 Research limitations and recommendations

Q methodology is an abductive research approach (Watts and Stenner, 2012), which means that we tried to understand and explain the data rather than describe it or test a hypothesis. This approach is subjective in nature. Researcher bias was reduced where possible as presented in the framework in Fig. 3. By collecting statements from various sources, reasonable saturation in the statements that formed the discourse was achieved and confirmed by the fact that no clearly new statements arose from the post-sorting interviews. The second sampling strategy (see Stage 4 in the Method section) recruited five participants that were expected to have different viewpoints. Two of them had Viewpoint B and one had Viewpoint C, thus broadening the scope of viewpoints. Still, the voluntary nature of the participation might have attracted participants with a positive attitude towards citizen science. On several occasions the research was presented to water authority employees. They were surprised that all viewpoints were relatively positive about citizen science, because they would have believed to have colleagues who were indeed more sceptical than these viewpoints suggest.

To further reduce researcher bias, we asked participants whether they recognized themselves in the assigned viewpoint described in an early draft of the text presented in the Results section. Fifteen participants responded, 13 of whom fully identified with their viewpoint, because they agreed with the viewpoints' main assertions. Two respondents were in doubt, due to overlap between Viewpoints A and C. The correlation between Factors A and C was 0.43, which indeed indicates that they are interrelated and overlap. Typically, correlations above 0.39 are considered significant (Watts and Stenner, 2012). Still, Factor A and Factor C are considered sufficiently different to regard them as the basis for separate viewpoints, particularly given the (in our opinion crucially) different opinion on involving citizens in data collection. Correlations between Factors A and B and Factors B and C were lower and not significant: 0.26 and 0.35, respectively. A mechanism to further reduce bias, beyond the mechanisms in Fig. 3, could be to execute stages such as statement definition, interviews and transcript analysis by independently

working researchers. However, this was not feasible within the scope of this research.

The importance of post-sorting interviews, as stated by Gallagher and Porock (2010), was recognized in this work. In particular, statements with multiple fragments per viewpoint illustrated this. For example, the interview fragments for Statement 36 revealed a difference in the reasoning to exclude citizens from decision-making on measures to realize policy goals. Participants with Viewpoint A feared manipulation of results, while participants with Viewpoint B emphasized the responsibility of the water authority to take an informed decision and balance conflicting interests. Hence, reformulation and/or splitting of Statement 36 may be considered if the set of statements was adopted for future research. Four participants with Viewpoint A literally recalled citizens' unfamiliarity with the tasks of the water authorities in response to Statement 2. Time limitations of the interviews resulted in an unequal distribution of interview fragments over statements and viewpoints (see Fig. 4). A higher coverage of interview fragments and more participants, particularly participants loading significantly on Viewpoint B and C, would likely have resulted in a more consistent image and more understanding of the participant's underlying reasoning. Future research should consider more time for post-sorting interviews or organizing group discussions.

The viewpoints identified in this study are expected to be representative of water authorities in the Netherlands. Only 8 out of 24 Dutch regional water authorities were included, so additional viewpoints may be found if the study would be repeated with participants from the remaining water authorities. We consider this unlikely as none of the selection criteria (see Stage 3 in the method section) were found to influence the results. Flood risk, age (years since the last organizational reform) and location were incorporated as characteristics that might influence participant's viewpoints. All viewpoints represented a combination of age and location; thus, there seems to be no relationship between these characteristics and the results. Sea level might, as all people with Viewpoint B were working at an authority responsible for an area below sea level. Job type did not influence the results; gender might, as almost all people with Viewpoint B were male. This study cannot make any claims regarding the influence of flood risk or gender, but this is a promising direction for further research.

A quantitative follow-up study can be used to determine the distribution of viewpoints across water authorities or within one organization. Such a study would further justify generalization of the results to the Dutch water authorities and would allow one to check how widespread the overall positive attitude towards citizen science found in this research actually is. This article presents results for practitioners in the Netherlands, but we encourage others to repeat the study, particularly in other countries facing low citizen water awareness as described by the OECD (OECD, 2011). The developed set of statements can serve as a basis for such a study as they are not unique to the Dutch situation.

5 Conclusion

This study contributes to understanding about government practitioners' acceptance and perception of citizen science. A Q methodological approach was applied to identify the viewpoints of practitioners on citizen science, in the case of water quality monitoring at Dutch regional water authorities. Water authority employees sorted a set of 46 statements related to citizen science. Three factors were identified in a factor analysis and translated into corresponding viewpoint narratives.

The first viewpoint, Viewpoint A, is named "Citizen participation for data application". People with Viewpoint A see more opportunities than challenges when it comes to citizen science. They see applications in practical use of the data, but also in the active engagement of people. The second viewpoint, Viewpoint B, is named "Water authority in control". People with this viewpoint see a potential for data contributions by citizens in an illustrative way, but are concerned with challenges in organizational capacity, expectation management and motivating citizens as well. The third viewpoint, Viewpoint C, is named "Education and local knowledge". People with this viewpoint focus on educational goals, such as teaching people about their environment and getting schools involved. They consider data applicability of secondary importance, although the data can be used illustratively.

The outcomes of this study provide strong indications that practitioners at the Dutch water authorities welcome citizen science. These practitioners further believe citizen science can contribute to bridging the awareness gap as identified by the OECD (2014) and the Dutch water authorities (UvW, 2015a). All three viewpoints are positive towards citizen science in the form of contributory projects in which citizens collect data. People with Viewpoint A support collaborative citizen science as well, but none of the viewpoints support co-created projects between citizens and water authorities. Interviews identified low expectations in citizens' motivations and capacities as underlying causes of this low support

for higher levels of citizen involvement. This may jeopardize the much-needed trust in the relationship between citizens and practitioners. Although participants recognized the potential of citizen science to change governance structures, the design of citizen science projects in the Netherlands is not expected to move beyond contributory projects in the near future.

Acknowledgements. We thank S. Abu Shoaib and an anonymous referee for their valuable comments on the discussion paper.

Edited by: S. Illingworth

References

Bonney, R., Cooper, C. B., Dickinson, J., Kelling, S., Phillips, T., Rosenberg, K. V., and Shirk, J.: Citizen Science: A Developing Tool for Expanding Science Knowledge and Scientific Literacy, BioScience, 59, 977–984, doi:10.1525/bio.2009.59.11.9, 2009.

Brown, S. R.: Political subjectivity: Applicaitons of Q methodology in political science, New Haven, CT, Yale University Press, 1980.

Buytaert, W., Zulkafli, Z., Grainger, S., Acosta, L., Alemie, T. C., Bastiaensen, J., De Bièvre, B., Bhusal, J., Clark, J., Dewulf, A., Foggin, M., Hannah, D., Hergarten, C., Isaeva, A., Karpouzoglou, T., Pandeya, B., Paudel, D., Sharma, K., Steenhuis, T., Tilahun, S., Van Hecken, G., and Zhumanova, M.: Citizen science in hydrology and water resources: opportunities for knowledge generation, ecosystem service management, and sustainable development, Front. Earth Sci., 2, 1–21, doi:10.3389/feart.2014.00026, 2014.

Chandler, D. and Kapelner, A.: Breaking monotony with meaning: Motivation in crowdsourcing markets, J. Econ. Behav. Organ., 90, 123–133, doi:10.1016/j.jebo.2013.03.003, 2013.

Clark, M. P., Nijssen, B., Lundquist, J. D., Kavetski, D., Rupp, D. E., Woods, R. A., Freer, J. E., Gutmann, E. D., Wood, A. W., Brekke, L. D., Arnold, J. R., Gochis, D. J., and Rasmussen, R. M.: A unified approach for process-based hydrologic modeling: 1. Modeling concept, Water Resour. Res., 51, 2498–2514, doi:10.1002/2015WR017200, 2015.

Cohn, J. P.: Citizen Science?: Can Volunteers Do Real Research?, BioScience, 58, 192–197, doi:10.1641/B580303, 2008.

Conrad, C. C. and Hilchey, K. G.: A review of citizen science and community-based environmental monitoring: Issues and opportunities, Environ. Monit. Assess., 176, 273–291, doi:10.1007/s10661-010-1582-5, 2011.

Cools, M., Moons, E., Janssens, B., and Wets, G.: Shifting towards environment-friendly modes: Profiling travelers using Q-methodology, Transportation, 36, 437–453, doi:10.1007/s11116-009-9206-z, 2009.

Cooper, C., Dickinson, J., Phillips, T., and Bonney, R.: Citizen Science as a Tool for Conservation in Residential Ecosystems, Ecol. Soc., 12, 11, doi:10.5751/ES-02197-120211, 2007.

Fraternali, P., Castelletti, A., Soncini-Sessa, R., Vaca Ruiz, C., and Rizzoli, A. E.: Putting humans in the loop: Social computing for Water Resources Management, Environ. Modell. Softw., 37, 68–77, doi:10.1016/j.envsoft.2012.03.002, 2012.

Gallagher, K. and Porock, D.: The use of interviews in Q methodology: card content analysis, Nurs. Res., 59, 295–300, 2010.

Gharesifard, M. and Wehn, U.: To share or not to share: Drivers and barriers for sharing data via online amateur weather networks, J. Hydrol., 535, 181–190, doi:10.1016/j.jhydrol.2016.01.036, 2016.

Hobbs, S. J. and White, P. C. L.: Motivations and barriers in relation to community participation in biodiversity recording. J. Nat. Conserv., 20, 364–373, doi:10.1016/j.jnc.2012.08.002, 2012.

Hollow, B., Roetman, P. E. J., Walter, M., and Daniels, C. B.: Citizen science for policy development: The case of koala management in South Australia. Environ. Sci. Policy, 47, 126–136, doi:10.1016/j.envsci.2014.10.007, 2015.

Hut, R., Amisigo, B. A., Steele-Dunne, S., and van de Giesen, N.: Reduction of Used Memory Ensemble Kalman Filtering (RumEnKF): A data assimilation scheme for memory intensive, high performance computing, Adv. Water Resour., 86, 273–283, doi:10.1016/j.advwatres.2015.09.007, 2015.

Land-Zandstra, A. M., Devilee, J. L. A., Snik, F., Buurmeijer, F., and van den Broek, J. M.: Citizen science on a smartphone: Participants' motivations and learning, Public Understanding of Science (Bristol, England), November, doi:10.1177/0963662515602406, 2015.

Macknick, J. E. and Enders, S. K.: Transboundary Forestry and Water Management in Nicaragua and Honduras: From Conflicts to Opportunities for Cooperation, Journal of Sustainable Forestry, 31, 376–395, doi:10.1080/10549811.2011.588473, 2012.

OECD: Water Governance in OECD Countries: A Multi-level Approach, OECD Studies on Water series, OECD Publishing: Organisation of Economic Cooperation and Development, doi:10.1787/9789264119284-en, 2011.

OECD: Water Governance in the Netherlands, OECD Publishing, doi:10.1787/9789264102637-en, 2014.

Ottinger, G.: Buckets of resistance: Standards and the effectiveness of citizen science, Sci. Technol. Hum. Val., 35, 244–270, 2010.

Raadgever, G. T., Mostert, E., and van de Giesen, N. C.: Identification of stakeholder perspectives on future flood management in the Rhine basin using Q methodology, Hydrol. Earth Syst. Sci., 12, 1097–1109, doi:10.5194/hess-12-1097-2008, 2008.

Raddick, M. J., Bracey, G., Gay, P. L., Lintott, C. J., Murray, P., Schawinski, K., Szalay, A. S., and Vandenberg, J.: Galaxy Zoo: Exploring the Motivations of Citizen Science Volunteers, Astronomy Education Review, 9, 010103-1, doi:10.3847/AER2009036, 2010.

Rogstadius, J., Kostakos, V., Kittur, A., Smus, B., Laredo, J., and Vukovic, M.: An Assessment of Intrinsic and Extrinsic Motivation on Task Performance in Crowdsourcing Markets, Fifth International AAAI Conference on Weblogs and Social Media, (Gibbons 1997), 321–328, 2011.

Rotman, D., Preece, J., Hammock, J., Procita, K., Hansen, D., Parr, C., Lewis, D., and Jacobs, D.: Dynamic Changes in Motivation in Collaborative Citizen-Science Projects, Proceedings of the ACM 2012 Conference on Computer Supported Cooperative Work – CSCW'12, 217–226, doi:10.1145/2145204.2145238, 2012.

Roy, H. E., Pocock, M. J. O., Preston, C. D., Roy, D. B., Savage, J., Tweddle, J. C., and Robinson, L. D.: Understanding Citizen Science & Environmental Monitoring, Final Report on behalf of UK-Environmental Observation Framework, 170, http://nora.nerc.ac.uk/20679/ (last access: 31 August 2016), 2012.

Shoaib, S. A., Marshall, L., and Sharma, A.: A metric for attributing variability in modelled streamflows, J. Hydrology, 541, 1475–1487, doi:10.1016/j.jhydrol.2016.08.050, 2016.

Silvertown, J.: A new dawn for citizen science, Trends Ecol. Evol., 24, 467–471, doi:10.1016/j.tree.2009.03.017, 2009.

Snik, F., Rietjens, J. H. H., Apituley, A., Volten, H., Mijling, B., Di Noia, A., Heikamp, S., Heinsbroek, R. C., Hasekamp, O. P., Smit, J. M., Vonk, J., Stam, D. M., Van Harten, G., De Boer, J., and Keller, C. U.: Mapping atmospheric aerosols with a citizen science network of smartphone spectropolarimeters, Geophys. Res. Lett., 41, 7351–7358, doi:10.1002/2014GL061462, 2014.

Tielrooij, F.: Aders omgaan met water: waterbeleid voor de 21e eeuw [Dealing differently with water: water policy for the 21st century], Ministry of Public Works, Transport and Water Management, Den Haag, 2000.

Tulloch, A. I. T., Possingham, H. P., Joseph, L. N., Szabo, J., and Martin, T. G.: Realising the full potential of citizen science monitoring programs, Biol. Conserv., 165, 128–138, doi:10.1016/j.biocon.2013.05.025, 2013.

UvW: Visie openbaar bestuur "waterbestuur dat werkt" [Vison on governance: "water management that works"], Unie van Waterschappen [Dutch Water Authorities], Den Haag, https://www.uvw.nl/wp-content/uploads/2015/07/Waterbestuur-dat-werkt-2015.pdf (last access: 12 March 2016), 2015a.

UvW: Waterbeheer doen we samen: waterschappen voor de burger [We do water management together: water authorities for citizens], Unie van Waterschappen [Dutch Water Authorities], Den Haag, www.uvw.nl/wp-content/uploads/2015/07/Waterschappen-voor-de-burger-2015.pdf (last access: 12 March 2016), 2015b.

Van der Auweraert, A.: The Science Communication Escalator, in: 2nd International Living Knowledge Conference Seville, Spain, 3–5 February 2005, 237–241, 2005.

Van Exel, J. and De Graaf, G.: Q methodology?: A sneak preview, Soc. Sci., 2, 1–30, 2005.

Varner, J.: Scientific Outreach: Toward Effective Public Engagement with Biological Science, BioScience, 64, 333–340, doi:10.1093/biosci/biu021, 2014.

Watts, S. and Stenner, P.: Doing Q methodological research: theory, method and interpretation, SAGE Publications, London, 2012.

Wehn, U. and Evers, J.: Citizen observatories of water: Social Innovation via eParticipation?, 2nd International Conference on ICT for Sustainability, (Ict4s), 10., http://www.atlantis-press.com/php/download_paper.php?id=13419 (last access: 29 August 2016), 2014.

Wehn, U., Rusca, M., Evers, J., and Lanfranchi, V.: Participation in flood risk management and the potential of citizen observatories: A governance analysis, Environ. Sci. Policy, 48, 225–236, doi:10.1016/j.envsci.2014.12.017, 2015.

Weng, Y. C.: Contrasting visions of science in ecological restoration: Expert-lay dynamics between professional practitioners and volunteers, Geoforum, 65, 134–145, doi:10.1016/j.geoforum.2015.07.023, 2015.

9

Long-term temporal trajectories to enhance restoration efficiency and sustainability on large rivers

David Eschbach[1,a], **Laurent Schmitt**[1], **Gwenaël Imfeld**[2], **Jan-Hendrik May**[3,b], **Sylvain Payraudeau**[2], **Frank Preusser**[3], **Mareike Trauerstein**[4], and **Grzegorz Skupinski**[1]

[1]Laboratoire Image, Ville, Environnement (LIVE UMR 7362), Université de Strasbourg, CNRS, ENGEES, ZAEU LTER, Strasbourg, France
[2]Laboratoire d'Hydrologie et de Géochimie de Strasbourg (LHyGeS UMR 7517), Université de Strasbourg, CNRS, ENGEES, Strasbourg, France
[3]Institute of Earth and Environmental Sciences, University of Freiburg, Freiburg, Germany
[4]Institute of Geography, University of Bern, Bern, Switzerland
[a]current address: Sorbonne Université, CNRS, EPHE, UMR 7619 Metis, 75005 Paris, France
[b]current address: School of Geography, University of Melbourne, Melbourne, Australia

Correspondence: David Eschbach (eschbach.pro@gmail.com)

Abstract. While the history of a fluvial hydrosystem can provide essential knowledge on present functioning, historical context remains rarely considered in river restoration. Here we show the relevance of an interdisciplinary study for improving restoration within the framework of a European LIFE+ project on the French side of the Upper Rhine (Rohrschollen Island). Investigating the planimetric evolution combined with historical high-flow data enabled us to reconstruct pre-disturbance hydromorphological functioning and major changes that occurred on the reach. A deposition frequency assessment combining vertical evolution of the Rhine thalweg, chronology of deposits in the floodplain, and a hydrological model revealed that the period of incision in the main channel corresponded to high rates of narrowing and lateral channel filling. Analysis of filling processes using Passega diagrams and IRSL dating highlights that periods of engineering works were closely related to fine sediment deposition, which also presents concomitant heavy metal accumulation. In fact, current fluvial forms, processes and sediment chemistry around Rohrschollen Island directly reflect the disturbances that occurred during past correction works, and up to today. Our results underscore the advantage of combining functional restoration with detailed knowledge of the past trajectory to (i) understand the functioning of the hydrosystem prior to anthropogenic disturbances, (ii) characterize the human-driven morphodynamic adjustments during the last 2 centuries, (iii) characterize physico-chemical sediment properties to trace anthropogenic activities and evaluate the potential impact of the restoration on pollutant remobilization, (iv) deduce the post-restoration evolution tendency and (v) evaluate the efficiency and sustainability of the restoration effects. We anticipate our approach will expand the toolbox of decision-makers and help orientate functional restoration actions in the future.

1 Introduction

During the last 2 centuries, numerous engineering works (e.g. channelization or damming), aimed at flood control, navigation improvement, expansion of agriculture or hydropower production, have altered the functioning of European large rivers (Brookes, 1988; Petts et al., 1989; Kondolf and Larson, 1995), including aquatic and riparian habitats and biodiversity (Bravard et al., 1986; Amoros and Petts, 1993; Dynesius and Nilsson, 1994). In order to balance these impacts by recovering fluvial processes (Knighton, 1984; Bravard et al., 1986; Naiman et al., 1993, 1988; Corenblit et al., 2007; Hering et al., 2015) and ecosystem services (Loomis et al., 2000; Acuña et al., 2013; Large and

Gilvear, 2015), an increasing number of restoration projects have been carried out over recent decades (Kondolf and Micheli, 1995; Wohl et al., 2005). In Europe, this trend has been supported by the Water Framework Directive (IKSR-CIPR-ICBR, 2005; WFD, 2000). Restoration activities progressively target hydromorphological processes and functioning rather than "static" fluvial forms (Jenkinson et al., 2006; Beechie et al., 2010; Arnaud et al., 2015; Jones and Johnson, 2015). Numerous studies have shown that current river functioning results from complex long-term trajectories driven by natural and anthropogenic factors at different spatio-temporal scales (e.g. Brown, 1997; Bravard and Magny, 2002; Gregory and Benito, 2003; Starkel et al., 2006; Ziliani and Surian, 2012). These trajectories provide a relevant basis to infer future trends and management principles (Bravard, 2003; Sear and Arnell, 2006; Dufour and Piégay, 2009; Fryirs et al., 2012). In order to understand the complete range of functional changes, a comprehensive understanding of human-driven channel adjustments over recent centuries therefore appears crucial in river restoration. Despite significant research efforts over the last 2 decades, integrating pluri-secular temporal trajectories into restoration projects remains an exception (Sear and Arnell, 2006; Fryirs et al., 2012).

In large modified hydrosystems such as the Upper Rhine River, the current lateral extent of the floodplain results from past disturbances that occurred during engineering works (Herget et al., 2005, 2007). Hydromorphological dynamics, chemical pollution and depositional processes were strongly impacted by diking along the main channel and disconnection of lateral channels since the beginning of the 19th century (Tümmers, 1999). Several studies have focused on the storage and remobilization of heavy metals (Schulz-Zunkel and Krueger, 2009; Ciszewski and Gryar, 2016; Falkowska et al., 2016; Grygar and Popelka, 2016) and/or organic pollutants in major floodplains worldwide (Lair et al., 2009; Zimmer et al., 2010; Berger and Schwarzbauer, 2016). Most studies concerned with the Rhine focused on the industrialized Lower Rhine region including the Rhine–Meuse delta (Evers et al., 1988; Middelkoop, 2000; Gocht et al., 2001; de Boer et al., 2010). In comparison, the Upper Rhine region is 2 times less contaminated than the lower parts of the Rhine with respect to total concentrations of both polychlorinated dibenzo(p)dioxins (PCDDs) and polychlorinated dibenzofurans (PCDFs) in sediments and tracers of industrial activity (Evers et al., 1988). However, contamination histories are still not well known and reference studies considering functioning or disturbance histories are missing.

Studying past functioning and disturbance histories as keys to understanding current forms and processes of floodplains can provide insights into current functioning and hydromorphological sensitivity to changes (Kondolf and Larson, 1995; Mika et al., 2010). Historical hydromorphological adjustments should be evaluated on an accurate spatial scale and at a high temporal resolution to identify past evolutionary processes and causal relationships (Bogen et al., 1992;

Horowitz et al., 1999). Interdisciplinary and retrospective studies, however, rarely combine different data sources to obtain a comprehensive view of functional changes (Bravard and Bethemont, 1989; Trimble and Cooke, 1991; Lawler, 1993; Gurnell et al., 2003; James et al., 2009; Rinaldi et al., 2013; Lespez et al., 2015). Furthermore, historical studies rarely consider sediment dating and pollution (Bogen et al., 1992; Garban et al., 1996; Horowitz et al., 1999; Woitke et al., 2003).

Within the framework of a functional geomorphological restoration project on Rohrschollen Island, we embarked on an interdisciplinary and retrospective pluri-secular study to provide a holistic understanding of the functional temporal trajectory of the fluvial hydrosystem, rather than to determine reference states. We hypothesized that long-term temporal trajectories allow us to identify driving factors, amplitude and response time of disturbances, and to provide key information to manage functional restoration actions, notably in terms of potential benefits and limits. This approach has the potential to accompany actions of functional restoration in order to maximize efficiency and sustainability, and infer future evolutionary trends. To test these hypotheses, this study combines horizontal (planimetric) and vertical (thalweg evolution, filling chronology) dynamics to (i) show the functioning of the hydrosystem prior to anthropogenic disturbances, (ii) characterize the human-driven morphodynamic adjustments during the last 2 centuries, including sediment transport and deposition processes as well as geochronology, (iii) assess physio-chemical sediment properties (e.g. heavy metals and organic contaminant concentrations) to trace anthropogenic activities and evaluate the potential impact of the restoration on pollutant remobilization (Middelkoop, 2000; Fedorenkova et al., 2013; IKSR-CIPR-ICBR, 2014), (iv) deduce post-restoration evolutionary trends and (v) propose an operational outlook to improve efficiency and sustainability of the restoration of Rohrschollen Island, and by extension of other river restoration projects (Sear et al., 1994; Grabowski and Gurnell, 2016).

2 Study area

With a total length of 1250 km and a drainage basin of about 185 000 km^2, the Rhine is the third largest river of Europe. Located between Basel and Bingen (Fig. 1a), the Upper Rhine Graben is 35–50 km wide and 310 km long. Hydrology in the southern part of this sector is characterized by a nivo-glacial regime and a mean discharge of 1059 m^3 s^{-1} (1891–2011; Basel gauging station; Uehlinger et al., 2009). Slope decrease and inherited geomorphological factors explain the longitudinal evolution of the channel pattern from braiding to anastomozing and meandering (Carbiener, 1983; Schmitt et al., 2016; Fig. 1b). Since the middle of the 19th century, three successive engineering works modified drastically the hydrosystem: (i) the correction stabilized the main

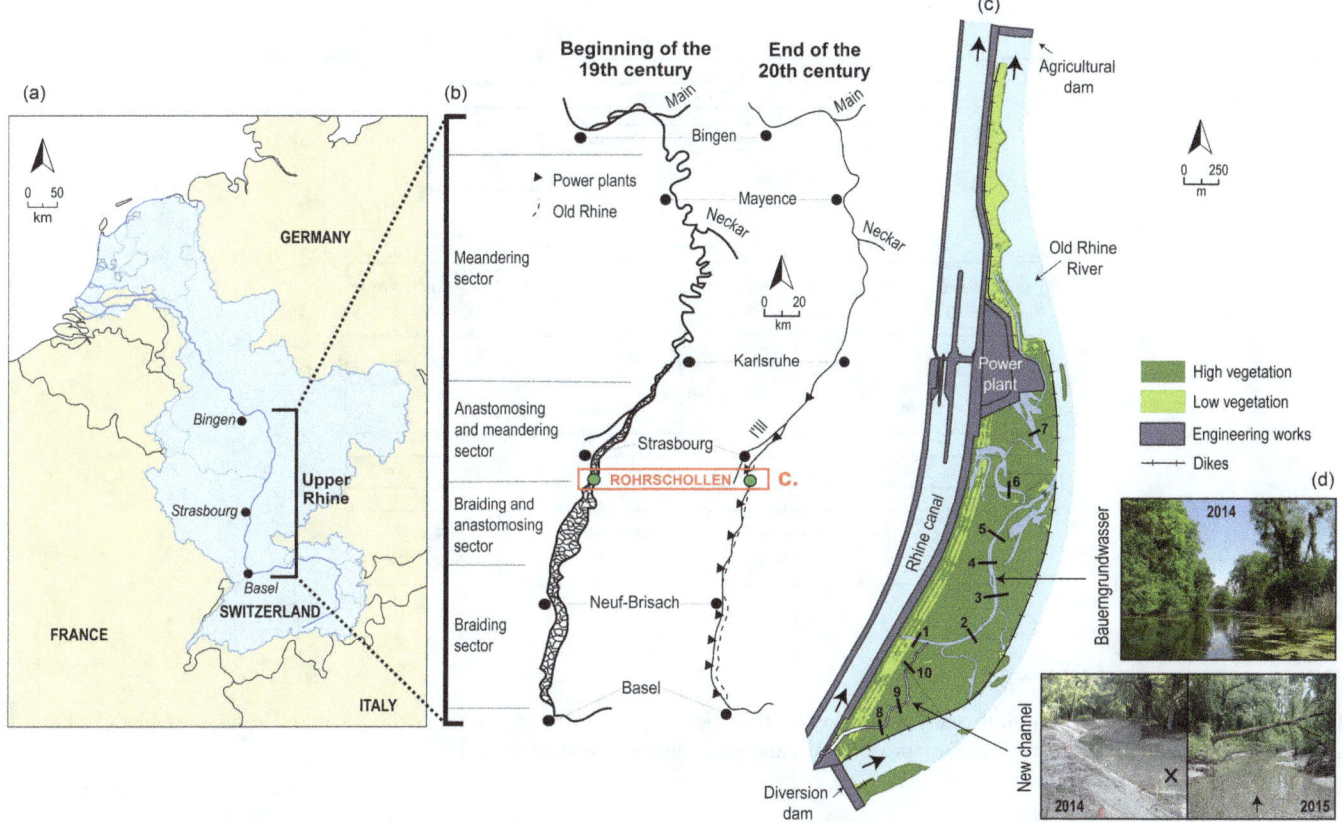

Figure 1. (a) Location of the Upper Rhine Graben, **(b)** channel pattern sectorization and evolution from the 18th century to present (Schmitt et al., 2009), **(c)** location of the study site and map of Rohrschollen Island, and **(d)** pictures of the Bauerngrundwasser and evolution of the new channel.

channel between two artificial banks and the floodplain between two high-flow dikes, (ii) the regularization consisted in building alternative in-channel groyne fields to improve navigation and (iii) the canalization by-passed the corrected main channel in many areas south of Strasbourg. Nowadays, the river consists of a single channel which is locally by-passed by artificial canalized sections. North of Strasbourg, the canalization concreted the Rhine bed itself.

The artificial Rohrschollen Island is located 8 km southeast of the city of Strasbourg and owes its existence to the construction of a power plant in 1970. Before engineering works, it was a braiding and anastomozing fluvial hydrosystem. The island is enclosed by the Rhine canal to the west and the Old Rhine River to the east, which corresponds to the by-passed corrected Rhine (Fig. 1c). On the southern part, a diversion dam diverts up to 1550 $m^3\,s^{-1}$ for usage by the power plant. When the discharge is less than 1563 $m^3\,s^{-1}$, an instream discharge of 13 $m^3\,s^{-1}$ flows to the Old Rhine. On the northern part of the island, an agricultural dam built in 1984 maintains a constant water level in the Old Rhine at 140 m NN (NormalNull) in order to increase groundwater level for agricultural purposes. When floods exceed 2800 $m^3\,s^{-1}$ (2-year instantaneous flood), the agricultural

dam can be raised for flood retention (IKSR-CIPR-ICBR, 2012), but flooding remains static and only a part of the island is flooded. The island is crossed by an anastomozing channel (Bauerngrundwasser), which is disconnected from the Rhine Canal at its upstream extremity and connects to the Old Rhine further downstream (Fig. 1c, d). Further north an additional minor channel flows towards the Rhine Canal. The water level of the entire length of the Bauerngrundwasser is artificially maintained by the hydraulic backwater of the agricultural dam. Classified as a natural reserve since 1997, the island has recently been restored (European LIFE+ project) in order to recover typical alluvial processes and biodiversity, including dynamic floods, bedload transport, active morphodynamics, and hygrophilous tree species. To attain these objectives, a large floodgate was built in 2013 in the southern part of the island and a new upstream channel of 900 m length was excavated (Fig. 1c, d). The downstream end of this channel is connected to the Bauerngrundwasser channel. Water input from the flood gate ranges from 2 $m^3\,s^{-1}$ (when Q Rhine < 1550 $m^3\,s^{-1}$) to 80 $m^3\,s^{-1}$ (when Q Rhine > 1550 $m^3\,s^{-1}$). As the bankfull discharge of the new channel is 20 $m^3\,s^{-1}$, flooding in the island occurs when the discharge exceeds this threshold. A 3-year monitoring

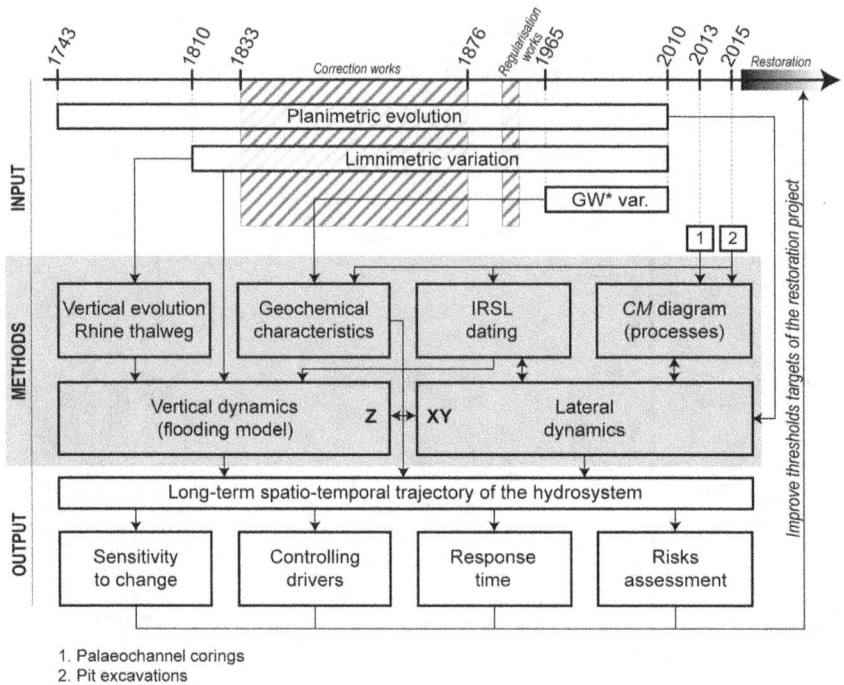

Figure 2. Methodological approach to analyse the long-term spatio-temporal trajectory of the hydrosystem by combining historical multi-source data with in situ morpho-sedimentary prospections and geochemical investigations.

showed that bedload transport and active lateral and vertical morphodynamics occur along the new channel (active bank erosion, formation of bars and logjams, enhancement of pool-riffle sequences, increase in groundwater–surface water exchange; Eschbach et al., 2017, 2018), but not along the Bauerngrundwasser, which is affected by the hydraulic backwater of the agricultural dam (Eschbach et al., 2017, 2018; see also the pictures of Fig. 1). Our study addresses embedded spatial scales: (i) the entire study site, which corresponds to the fluvial hydrosystem area around the natural reserve (about 1–3 km beyond the perimeter of the latter), (ii) Rohrschollen Island, which corresponds to the natural reserve area, (iii) seven transects on the Bauerngrundwasser used to characterize sediment transport and depositional processes, and (iv) two sediment pits excavated near the new and old channels in order to date sediment deposition and assess sediment pollution.

3 Material and methods

In this study, we have adopted an interdisciplinary approach that combines hydrological retrospective modelling with limnimetric, topographic (levelling and DEM) and hydrogeologic data as well as data on sediment filling processes, depositional chronology and geochemical characteristics (Fig. 2).

3.1 Planimetric analysis

The historical planimetric analysis covers a period of about 260 years (from 1743 to 2010) and was carried out in ArcMap (ESRI v.10.3) using six historical maps and two sets of aerial photographs. The map from 1828 compiled during the demarcation of the Franco-German border was georeferenced on the IGN BD ortho 2007 base map and used as a base layer for georeferencing of the 1743, 1778, 1838, 1872 and 1926 maps. Aerial photographs were georeferenced using the 2007 orthophotograph as a base layer. Fixed position objects such as churches, road crossings or bank protection structures were used as control points. Between 9 and 12 control points were selected for each historical map, and between 5 and 7 for the aerial photographs. Total root mean square error (RMSE) ranged from 0.94 to 25 m and increased with the distortion and the imprecision of the oldest maps, especially the 1743 and 1778 maps (Fig. 4c). However, the RMSE distortion is satisfactory considering the inherent relative imprecision of these maps and the difficulty in determining anchor points between old maps and 2007 orthophotography. Aerial photographs were selected at low-flow water level to enable comparison between morpho-ecological surfaces and active channel and gravel bar surfaces which are particularly sensitive to discharge variations (Rollet et al., 2014). Morpho-ecological units were then manually digitized at the detailed scale of 1 : 1000 to 1 : 2000 based on eight maps. According to the typology developed by Dufour (2006), 4 classes and 14

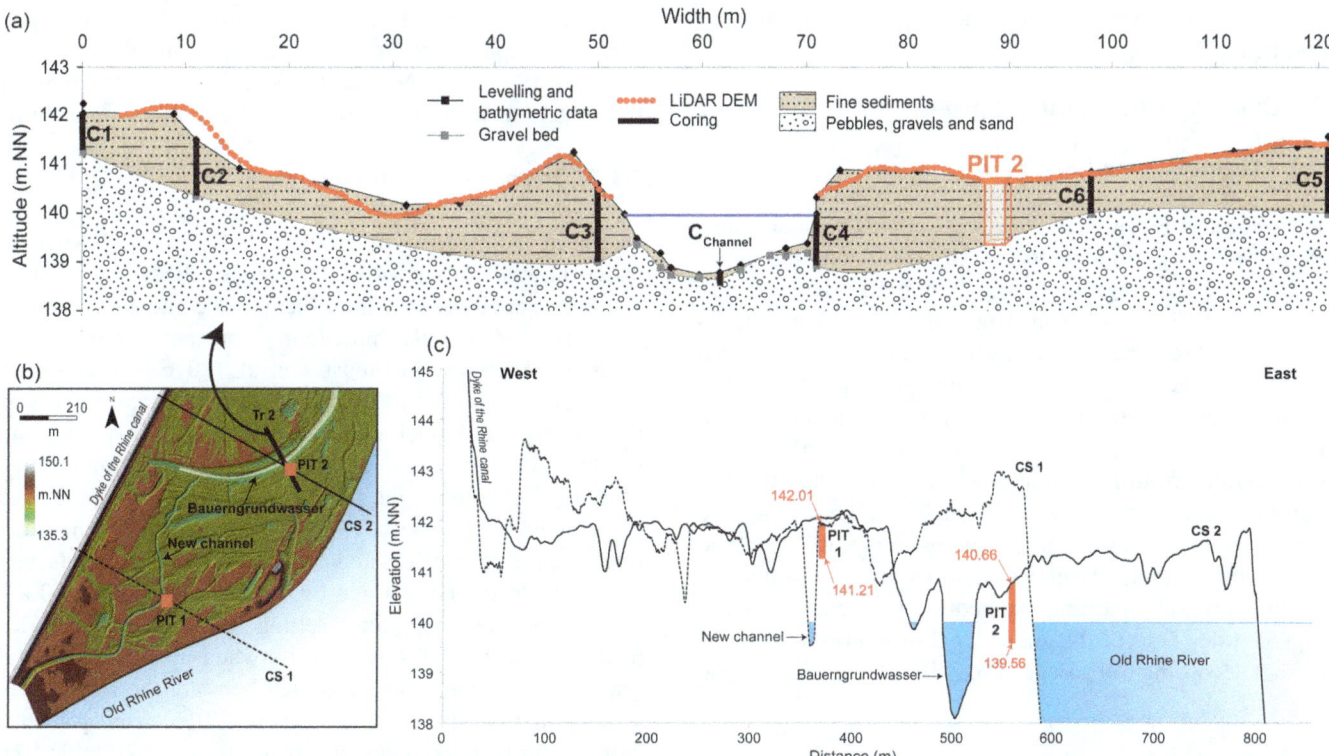

Figure 3. (a) Example of a transect (Tr 2) based on levelling and bathymetric data. Levelling data are compared to lidar DEM data. The location of corings and pit 2 is also given. **(b)** The location of Tr 2, the two excavation pits (red squares) and topographical profiles crossing the two pits. **(c)** Overlaying of the two topographical profiles.

subclasses were determined. Surface areas of each class and time slice were calculated to quantify the planimetric evolution. The results of the surface analyses were converted to a percentage ratio to facilitate regional interpretation of the morpho-ecological evolution. Two scales were considered: the study area of Rohrschollen Island with a total surface of 2181 ha and the area of the natural reserve with a total surface area of 314 ha.

3.2 Analysis of vertical data

3.2.1 Limnimetric and piezometric analysis

The vertical evolution of the Rhine thalweg was studied based on historical limnimetric data and bibliographic references (Bensing, 1966; Bull, 1885; CECR, 1978). Limnimetric data were compiled for low water discharge ($\sim 540\,\mathrm{m^3\,s^{-1}}$) at the Marlen gauging station (Kilometre Point 295; Jeanpierre, 1968; Felkel, 1969). The piezometric analysis was achieved using a German database (LUBW: Landesanstalt für Umwelt, Messungen und Naturschutz Baden-Württemberg) and the French regional groundwater database (APRONA, Association pour la PROtection de la Nappe phréatique de la plaine d'Alsace), by selecting datasets close to the Rohrschollen reach.

3.2.2 Palaeochannel corings

Seven coring transects were distributed along the Bauerngrundwasser to cover the different morphological characteristics of the study area. Six sediment cores per transect were hand-augered on both channel banks to measure the thickness of the post-correction deposits (Fig. 3a). From the 42 cores, a total of 81 sediment samples were taken at different depths of the filling layer. Two additional samples were extracted from the channel bottom at transects two and four with a piston sampler (e.g. Fig. 3a, sample named "C_{channel}").

3.2.3 Pit excavations

In addition to the transect-based prospection, two pits were excavated up to the gravel bottom, on the right bank of the two channels. Locations of the pits were determined by (i) identifying the main filling sectors revealed by old maps and the corings survey, and (ii) the proximity to the potential future erodible banks (concave banks; Fig. 3b). Stratigraphical units (SUs) were defined in the field on the basis of colour and textural differences. Two large topographical cross sections (600 m for pit 1 and 800 m for pit 2) intersecting both pits were extracted from the DEM to interpret the dynamics of fine sediment deposition in relation to the initial elevation of each pit, the thickness of the stratigraphical units and the flooding regime (Fig. 3b, c).

3.3 Characteristics of the sediments

3.3.1 Grain-size analysis

Depending on the width and thickness of the filling, 31 samples from three coring transects distributed along the entire length of the Bauerngrundwasser (2, 4 and 5 in Fig. 1) and two samples from the channel bottom (2 and 4 in Fig. 1) were selected for grain-size analysis. In addition, one sample per SU was taken from each excavated pit (Fig. 7). Munsell colour and qualitative SU description were completed in the field. Cumulative grain-size distribution and a sorting index were obtained from measurements with a Beckman Coulter laser diffraction particle size analyser. Then, the soil organic carbon ratio was determined by the loss on ignition method (375 °C during 16 h). We characterized transport and depositional processes by plotting the median (D_{50}) and the coarsest percentile (D_{99}) of the grain-size distributions in the CM diagram according to Passega (1964, 1977) and Bravard and Peiry (1999).

3.3.2 Geochemical and organic pollutant analyses

The 10 samples from the two pits (Fig. 7) were air-dried at 20 °C and sieved (< 2 mm). Dried sediments were pulverized (< 63 µm) using an agate disk mill prior to alkaline fusion and total dissolution by acids. Measurement of elemental concentrations was done as described previously (Duplay et al., 2014) by inductively coupled plasma atomic emission spectrometer and mass spectrometer analysis (ICP-AES and ICP-MS) using the geological standards BCR-2 (US Geological Survey, Reston, VA, USA) and SCL-7003 (Analytika, Prague, Czech Republic) for quality control. An enrichment factor (EF) was used to compare changes of Zn, Cr, Ni, Cu, Pb and Cd concentrations in the pit 1 and pit 2 profiles with the reference soil collected in the deepest SU of the considered pit:

$$\text{EF}_{\text{HM}} = \frac{\left(\dfrac{\text{HM}_{\text{sample}}}{\text{Ti}_{\text{sample}}}\right)}{\left(\dfrac{\text{HM}_{\text{reference}}}{\text{Ti}_{\text{reference}}}\right)}, \tag{1}$$

where HM and Ti are, respectively, the concentrations of the considered heavy metal and Ti ($\text{mg\,kg}^{-1}$ d.w.) in the SU sample of the pit and the reference soil. Concentrations of heavy metals were normalized relative to titanium (Ti) as a conservative element to limit EF variations due to local heterogeneities as it displays a low relative standard deviation (< 0.03) over both pits (Reimann and de Caritat, 2005).

Based on previous studies dedicated to the contamination of the Rhine sediments (Fedorenkova et al., 2013; van Helvoort et al., 2007), 38 legacy and modern organic pollutants (including 30 pesticides, the hexachlorobenzene

and 7 polychlorinated biphenyls) were analysed by liquid chromatography and gas chromatography mass spectrometry (LC-MS and GC-MS) with quantification limits ranging from 0.015 to 0.05 $\text{µg\,g}^{-1}$ (see Table S1 in the Supplement).

3.3.3 Depositional chronology

Dating of sediments sampled from both pits was carried out using infrared stimulated luminescence (IRSL), as no other alternative approach is achievable in this context (Preusser et al., 2016). The detailed procedures and methodological aspects are discussed in Preusser et al. (2016). In summary, IRSL (stimulated at 50 °C) of sand-sized feldspar grains was measured by applying both small-aliquot (ca. 100 grains) and single-grain techniques. When applying the Minimum Age Model (MAM) to single-grain data sets, the estimated ages coincide with the expected age of the sediment. At the same time, MAM ages calculated for the small-aliquot data sets overestimate the known age by up to 200 years (> 100 %). This is explained by partial resetting of the IRSL signal prior to deposition and masking effects when measuring several grains at the same time. Presented here are the results previously published by Preusser et al. (2016) from pit 1 together with additional samples taken from pit 2 (Table S1). For the new data, we again observed significant older ages when measuring several grains at the same time for most of the samples. Only for sample IDEX2-107 did multiple and single-grain approaches give the same result. However, for this sample overly small grains (100–150 µm) were measured using Risø single-grain discs. This results in accumulation of a few (ca. five) grains being measured at the same time and likely similar averaging effects as observed for larger aliquots. Hence, we consider these ages as maximum estimates.

3.3.4 Flooding frequency assessment

To determine the depositional chronology of the two pits and reconstruct flooding frequency over time, we developed a simple flooding model based on (i) the results of the vertical evolution of the corrected Rhine thalweg close to Rohrschollen Island, (ii) the thickness of fine sediments at both pits, and (iii) IRSL dating. Elevation of the pits at each time slice was determined by the gradual sediment accumulation and the corresponding elevation increase between the two adjacent IRSL dates. Active channel width in the vicinity of the pits was measured from the 1828, 1838, and 1872 maps. Flooding water depth was determined by subtracting the elevation of the pits from the thalweg elevation for each time slice (Fig. 5c). For the three dates corresponding to the 1828, 1838, and 1872 maps, we determined the maximum discharge using the hydrogram (Q_{max}) at Basel and the limnimetric data at the Kehl bridge, which is located about 7 km downstream from the study area. We calculated for these dates and the corresponding bankfull discharges the sections S (m^2), the mean water slopes I (m m^{-1}), the hydraulic radius Rh (m), and the roughness k (m$^{1/3}$ s^{-1}) close to the pits. This allowed us to estimate flood discharges ($Q_{flooding}$) for both pits during the entire period from 1828 to 1970 (end of the canalization) using the Manning–Strickler equation (Eq. 2):

$$Q_{\text{flooding}} \text{ pit}_x \text{ at date}_n =$$
$$S_{\text{date}_n} \times I_{\text{date}_n}^{1/2} \times \text{Rh}_{\text{date}_n}^{2/3} \times k. \qquad (2)$$

Finally, we compared these results with the limnimetric variations (historical hydrogram) to determine the frequency and the intensity of historical floods in each pit (Fig. S2 in the Supplement; Fig. 5c).

4   Results and discussion

4.1   Hydromorphological dynamics before the beginning of the correction works (up to ca. 1833)

Analysis of the three earliest historical maps (Fig. 4a, 1743, 1778 and 1828) at the scale of the entire study site documents the natural morphodynamics along the Upper Rhine before the beginning of the correction works. At that time, the Rhine was a wide and braiding channel system (width ranging from 500 to 1500 m) characterized by numerous in-channel gravel bars, which is consistent with the descriptions from Schäfer (1973) and Herget et al. (2005). Multiple anastomozing channels also existed along the floodplain (Carbiener, 1983), at a maximum distance of about 5 km from the thalweg. The braiding and anastomozing index ranged between 7.9 and 5.4 (Fig. 8). The period 1743–1828 is characterized by marked changes and strong channel shifting of about 1 to 2 km. Across the entire study area, gravel bar surface areas increased (+100 ha; +128 %), while vegetated areas changed only slightly (low vegetation: −45 ha; −18.4 %; high vegetation: +124 ha; +11 %; Fig. 4b). In 1828, more than 70 % of the present natural reserve area was occupied by the active channel (running water and gravel bars). The high morphodynamic activity may be caused by the high frequency of flood events (four 10-year floods from 1810 to 1828) characterizing the beginning of the 19th century (Fig. 5c). These dynamics could be an effect of the Little Ice Age (Martin et al., 2015; Schmitt et al., 2016), which may have had a considerable impact on discharge, bedload transport and flood regime intensifying lateral dynamics (Schirmer, 1988; Rumsby and Macklin, 1996). A similar phenomenon has been observed previously, for example, by Bonnefont and Carcaud (1997) on the River Moselle, or by Bravard (2003) for the Rhône Basin. However, this hypothesis has not been validated for the Rhine yet and awaits further testing (Wetter et al., 2011; Schmitt et al., 2016).

At the scale of the natural reserve, the reach was located on the left bank of the Rhine in 1743. It was almost completely occupied by the thalweg from 1778 to 1828 as it

shifted towards the western direction. According to the location of pit 1 (Fig. 4a, central bar on the 1828 map) and the depositional history deduced from historical maps, the resetting of the IRSL signal in basal sediments of pit 1 resulted from this major lateral migration (Preusser et al., 2016). In accordance with the IRSL dates, deposition of fine-grained sediments in the lower part of pit 1 took place between 1778 and 1806 (Fig. 5a) after the Rhine thalweg had moved over the area. The maximum age of the investigated sediments in pit 1 is therefore 238–210 years (i.e. 1777–1805; Preusser et al., 2016). The 1828 map in Fig. 4a shows that pit 2 is located in the main channel and accumulation of fine sediments must have commenced after this time. Despite local diking on the floodplain and across some lateral channels as revealed by the planimetric analyses (essentially in 1828), it seems that fluvial morphodynamics and lateral channel shifts before 1828 were not influenced by anthropogenic disturbances.

4.2   Hydromorphological disturbances during the correction works (1833–1876)

4.2.1   Main channel adjustments

The correction works commenced between 1828 and 1838 (around 1833) and induced major changes to the hydrosystem (Tulla, 1825; Herget et al., 2005). A detailed map based comparison of the 1828 and 1838 maps shows that the left bank was eroded in the southern part of the study area (maximum of about 50 m) after 1828 and before (possibly during?) the building of the *perpendicular in-channel dike* (Fig. 4a, 1838). This is in agreement with the current position of the upstream part of the Bauerngrundwasser (Fig. 9, period A). Subsequently, the Rhine thalweg was artificially shifted towards a western direction by the *right bank dike*, and then to the east by the *perpendicular in-channel dike* and the *high-flow dike*. A large flood ($Q \sim 3800$ m^3 s^{-1}; 10-year floods; Fig. 5c) occurred in 1831, which was thoroughly documented due to the important damages it caused (Champion, 1863). We hypothesize that this event intensified the morphological adjustments during this period. In 10 years (1828–1838) the surface of running water decreased by 20 ha (5 %) in the entire study area, while the surface area of stagnant water increased correspondingly (plus 75 ha equal to 85 %).

From 1840 onwards, the corrected Rhine channel began to incise (1 cm year^{-1} on average; Fig. 5a) in response to channel narrowing (250 m wide), slope increase and bank stabilization (Fig. 4c, 1872). In addition, areas of gravel bars, flowing water and low vegetation surfaces decreased (by 110 ha = 70 %; 194 ha = 50 %; 121 ha = 67 %, respectively), while areas of high vegetation and agricultural surfaces increased (by 218 ha = 18 %; 183 ha = 700 %, respectively; Fig. 4b). This is interpreted as the consequence of sediment deposition in the disconnected parts of the main channel, inducing channel narrowing, and forest and agricul-

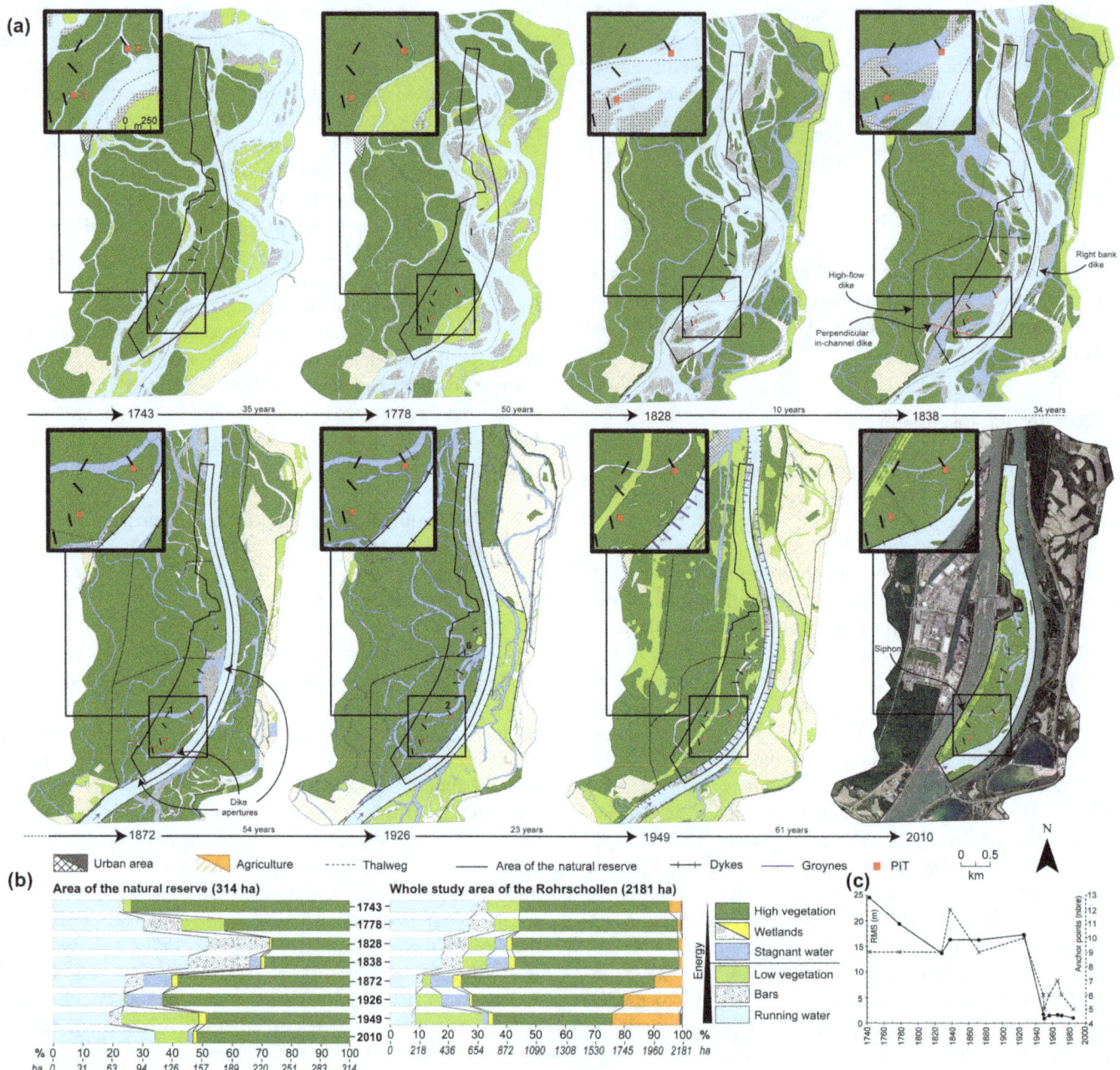

Figure 4. (a) Diachronic evolution of the whole study area. The black boxes are zooms (scale 1 : 2) of the surroundings of the two pits. **(b)** Surface evolutions of the morpho-ecological units in the natural reserve (1743–2010) and in the whole study area (1743–1949). **(c)** Number of anchor points used to georeference old maps and aerial photographs (dotted line) and values of RMSEs (continuous line).

ture expansion, as intended by the correction works (Bernhardt, 2000). These types of morphodynamic disturbances driven by channel correction were also observed by David et al. (2016) on the River Garonne, Magdaleno et al. (2012) on the River Ebro and Habersack et al. (2013) on the River Danube.

4.2.2 Lateral channel adjustments and filling processes

At the scale of the natural reserve, from 1828 to 1838, the braiding and anastomozing index decreased from 5.36 to 2.45, surface areas of running water decreased (by 21 ha = 13 %), and stagnant water and high-vegetation areas increased. In particular, vegetation populated central bars (Fig. 4a). This general trend became even more marked during the 1838–1872 period (Fig. 4b; the braiding and anas-

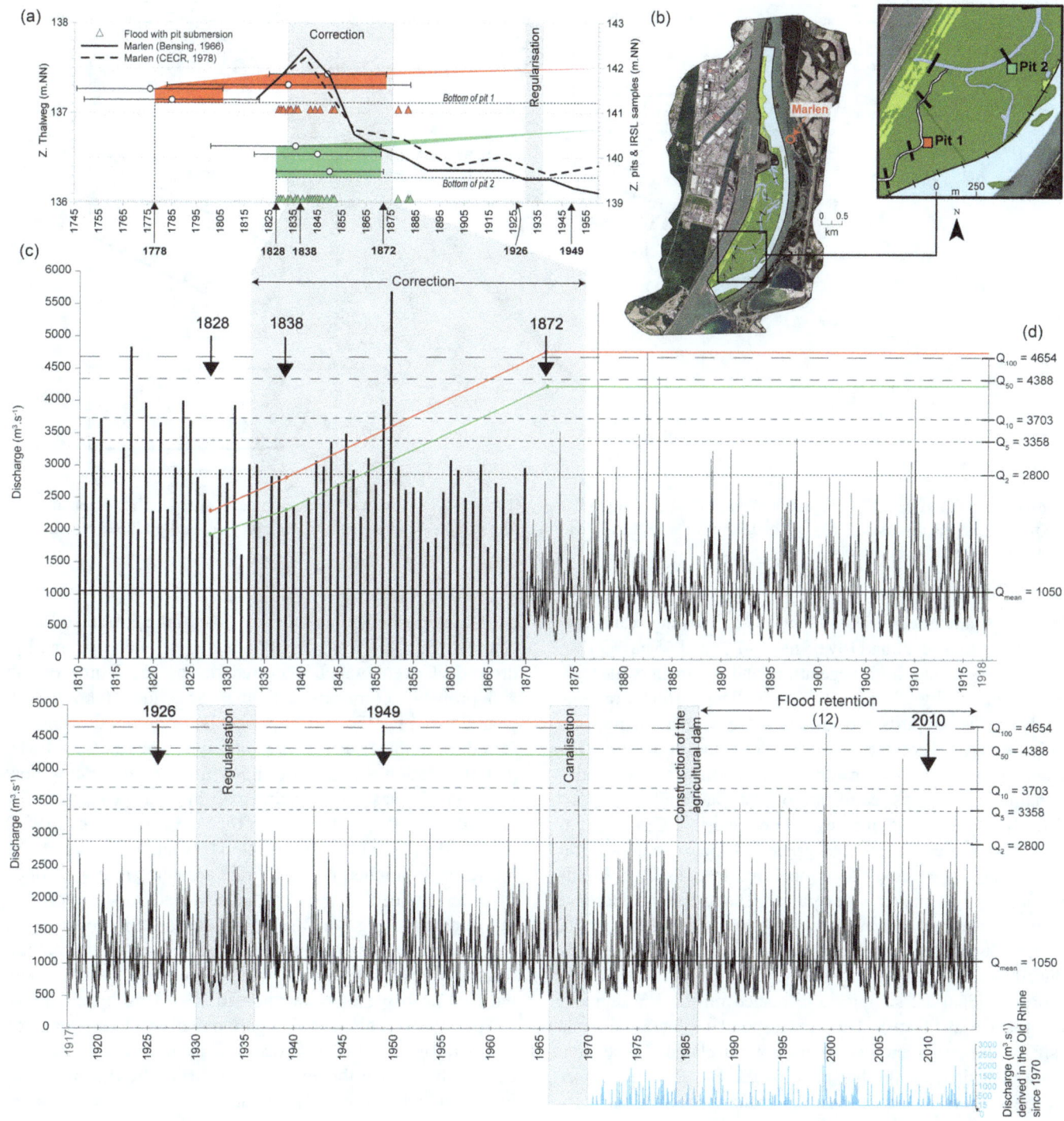

Figure 5. (a) Vertical evolution of the Rhine thalweg from 1815 to 1960 based on low water levels ($Q \sim 540\,\mathrm{m}^3\,\mathrm{s}^{-1}$) recorded at the Marlen gauging station (Bull, 1885; Bensing, 1966; CECR, 1978). Vertical evolution of the pits linked to the age–depth model and number of floods which attained each pit (triangle), **(b)** location of the two pits and of the ancient gauging station of Marlen, and **(c)** discharge of the Rhine at the gauging station of Basel. The period 1810–1870 corresponds to maximum instantaneous annual flows. The period 1870–2015 corresponds to the highest mean daily flow (OFEV: Office Fédéral de l'EnVironnement). The dates with arrows correspond to old maps or aerial photographs (see Fig. 4). The red and green lines correspond to the submersion discharges for pits 1 and 2, respectively. **(d)** Flood return periods at Basel between 1870 and 2015 ($\mathrm{m}^3\,\mathrm{s}^{-1}$; adjustment of Pearson III, OFEV).

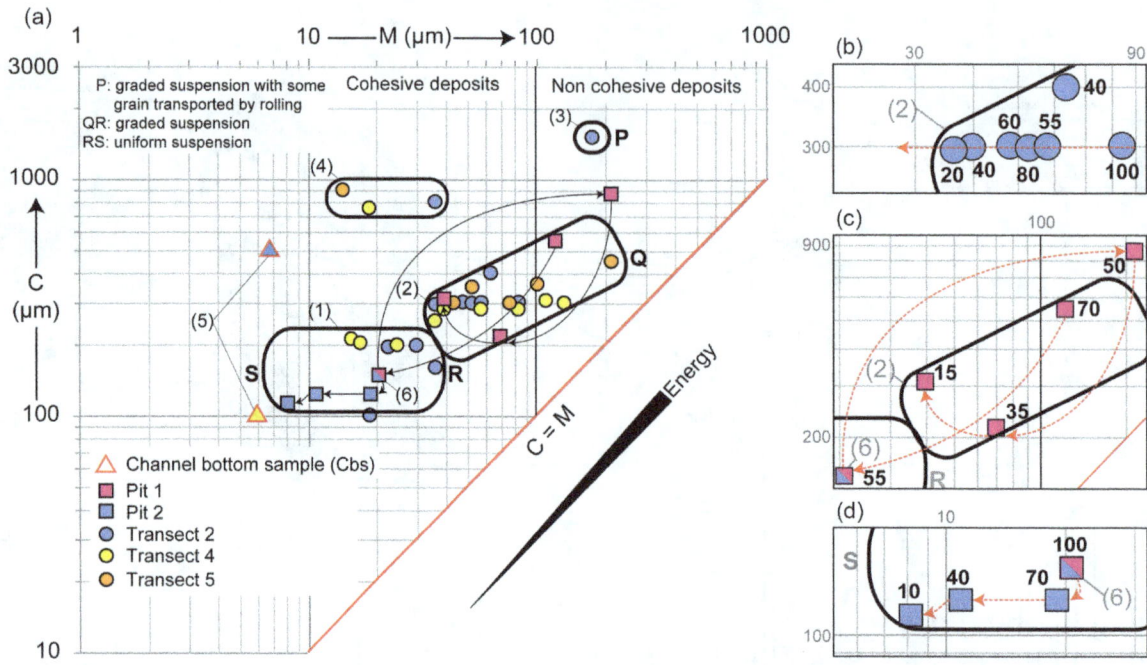

Figure 6. (a) Floodplain CM diagram and patterns according to Passega (1964, 1977). Numbers between brackets help in the presentation of the results and the discussion; **(b)** isolated samples with the indication of the sample depths for transect 2, pit 1 **(c)** and pit 2 **(d)**. Red arrows show the deposition chronology.

tomozing index decreased to 1 in 1872), running water and gravel bar areas declined (by 65 ha = 47 % and 45 ha = 71 %, respectively), and high-vegetation and stagnant water areas increased (by 90 ha = 101 % and 19 ha = 166 %, respectively). This supports the idea that the sequential impacts induced by the correction works along the floodplain (i.e. channel narrowing, expansion of vegetation) were especially dynamic in the upstream part of the Bauerngrundwasser. The channel filling dynamics are detailed in a CM diagram (Fig. 6a), which shows that graded suspension deposition (QR segment) of sandy loam occurred at transect 2. In addition, Fig. 6a shows a general and concomitant decrease in grain size and sediment sorting from the bottom to the surface, which reveals a decline in flood energy likely induced by channel diversion following the correction. Furthermore, residual T2 samples are located on the RS segment (uniform suspension). They correspond to fine sediments (silt) with sorting and depths are not correlated. These have been deposited on the left bank of the Bauerngrundwasser in low-turbulence conditions likely due to site-specific factors such as topography or vegetation, whose general importance has been underlined by Bravard et al. (2014), Toonen et al. (2015) and Riquier et al. (2015). Such kinds of depositional filling processes in newly by-passed channels have also been documented by Passega (1964, 1977) and Bravard and Peiry (1999).

The analysis of the evolution of the middle part of the natural reserve from 1838 to 1872 revealed that a large gravel

bar was deposited behind the left dike of the corrected Rhine channel (Fig. 4a). This clearly resulted from an extreme hydrological event which occurred during the Rhine diking and probably corresponds to the 1852 flood (above 300-year flood; Figs. 5c and 9, period B). This flood event, referred to as the "*flood of the century*" (6.63 m at the Basel gauging station, i.e. 5.78 m higher than the mean low-flow water level), has been documented in detail, especially by Wittmann (1859), Champion (1863), Eisenmenger (1907) and Pardé (1928). After this flood event, embankment of the Rhine continued but local dike apertures remained open in order to feed some old channels and to enhance filling dynamics (Fig. 4a, 1872). In addition, the downstream part of the Bauerngrundwasser (middle part of the natural reserve) presented an important connection to the Rhine, until its probable disconnection in 1876 (Fischbach, 1878; Casper, 1959). This specific hydrological condition impacted the depositional filling processes, as shown by the processes observed at the downstream part of the Bauerngrundwasser which are mainly characterized by graded suspension (Fig. 6a; T4 and T5 transects). Indeed, the CM diagram shows that energy is higher at T5 than at T4, although T4 is located upstream. Furthermore, the complexity is reinforced by the absence of upward fining in grain size at the surface as shown in the upstream part of the Bauerngrundwasser (Fig. 6b). Similar to the T2 samples, the position of the other T4 samples in the CM pattern corresponds to the mean level of turbulence in the uniform suspension (RS segment) and are located on the left bank (Fig. 6a). These T4 samples depend on the great distance from the main channel, as previously observed (Bravard and Peiry, 1999). Generally, the disconnection of the Bauerngrundwasser with the Rhine was reinforced by the total closure of the dike apertures after 1876 (Fig. 4a; 1872, 1926).

In Fig. 6a, square tag 6 corresponds to two overlapped samples from pit 1 – SU 3 – and pit 2 – SU 2 (Fig. 7). These samples are composed of the same grain-size characteristics. Square tag 6 is reported in Fig. 6c, d. Planimetric results combined with the CM diagram and IRSL ages imply that cohesive sediments were likely deposited during the same flood, probably the 1831 flood. Pit 1 is located on a former large gravel bar in 1828 (Fig. 4b, 1828). According to our chronology, pit 1 – SU 2 – was already deposited at this time (Preusser et al., 2016). This SU is composed of silt deposited during low-energy conditions (tag 6; Fig. 6c). It means that pit 1 likely consists of two depositional filling periods: the first started before 1828 (probably in 1778 for SUs 2, 3 and 4; Fig. 7), while the second started with the beginning of the correction works (for SUs 5 and 6; Fig. 7; Preusser et al., 2016). Conversely, pit 2 is located on the 1828 main channel (Fig. 4b, 1828), which became an overbank area characterized by low-energy depositional environments in close vicinity to the main channel in 1838 (Fig. 4a, 1838). Thus, if any deposits of fine sediments existed in 1828, the filling period in pit 2 began after 1828, as also shown by the IRSL ages (min: 1828; Fig. 5a), and was relatively regular. The end of the depositional filling periods at both pits occurred around 1872 as shown by the IRSL ages and is reflected in a grain-size refinement and sorting decrease (Fig. 5a). It coincides with a clear increase in vegetation areas shown by the 1872 map (Fig. 4a, 1872). Depositional filling differences (periods and processes) between the two pits are mainly controlled by the elevation and the location of the pits in the floodplain (Fig. 3c), which in turn determine the frequency of flooding (Fig. 5a, c) and mean sedimentation rates (from 0.9 cm year^{-1} in pit 1 and 6 cm year^{-1} in pit 2; Figs. 5a and 7). This may also explain the heterogeneity of depositional filling processes at pit 1 (Fig. 6c), in contrast to pit 2, where grain size generally decreases with increasing elevation; Fig. 6d). This phenomenon has been observed for disconnected lateral channels by several studies (e.g. Bravard et al., 1986; Hooke, 1995; Riquier et al., 2015).

4.2.3 Geochemistry of the sediment filling

Geochronological data combined with geochemical data confirmed changes in the hydrosystem and sediment deposition dynamics of Rohrschollen from the beginning of the correction works. Quartz (SiO$_2$) is the dominant mineral (ranging from 54 to 63 %; Fig. 7) and does not show any particular trend down the profile in pit 1 and pit 2. MnO, TiO$_2$ and P$_2$O$_5$ are least abundant in both pits. This shows that the patterns of mineral composition are not directly related

to grain size and likely reflect a common sediment source to both pits (Grygar and Popelka, 2016). However, lower ratios of Al / Si, especially in the lower sediment layers of SUs 2 and 3 in both pits 1 and 2, reflect a general low clay content with dominant sandy (pit 1) or silty (pit 2) sediment textures. This suggests relatively weak soil development and chemical weathering. The uppermost sediment layers of pit 1 and pit 2 differed in organic carbon (C) (16.4 and 27.0 g kg^{-1}, respectively), while organic matter content and organic carbon gradually decrease with depth in both pits. This suggests different temporal trajectories of deposition of organic-rich sediments on the two sites, i.e. the gravel bar (pit 1) and the Bauerngrundwasser (pit 2) until 1838.

Regarding pollution histories, polycyclic aromatic hydrocarbons (PAH), polychlorinated biphenyls (PCB), hexachlorocyclohexanes (HCH) and pesticides (Table S1) could not be detected in both pits. This is in agreement with the proposed chronology of pits filling because the first massive use of modern organic pollutants in the whole Rhine catchment gradually increased in the Rhine after 1940, followed by a strong reduction of these pollutants due to the Rhine Action Plan since 1970 (Evers et al., 1988; Middlekoop, 2000; Gocht et al., 2001). Similarly, the normalized REE patterns did not show any significant changes with depth in both pits (Fig. S3). Massive use of REE started after 2000 (Klaver et al., 2014), when sedimentation rates were low on Rohrschollen compared to the period during and immediately after the correction works (Fig. 5).

To evaluate the potential metal pollution resulting from anthropogenic activity, the relative enrichment in minor element concentrations in the sediments was evaluated in the pits accounting for the geochemical background. Depth distribution of minor elements in both pits follows a Sr > Ba > Zr > Cr > REE > Zn pattern. Heavy metal concentrations in pit 1 and pit 2 range from 31 to 196 mg kg^{-1} for Zn, 64 to 91 mg kg^{-1} for Cr, 10 to 46 mg kg^{-1} for Cu, 6 to 42 mg kg^{-1} for Ni, 12 to 35 mg kg^{-1} for Pb, and 0.23 to 1.24 mg kg^{-1} for Cd. Cu, Ni, and Cr enrichment factors in pit 1 and pit 2 remained within the natural background (i.e. normalized concentration profiles showed no enrichment in either profile). This suggests low industrial use of these heavy metals upstream from the study site in the 19th century, which is in agreement with previous results (Middlekoop, 2000; Gocht et al., 2001).

In contrast, Zn enrichment factors decrease with depth in pit 2 (from about 4 to 1), which probably reflects anthropogenic inputs of Zn in SUs 4 to 6 (up to 196 mg kg^{-1} in SU 6). This is in agreement with previous observations (Ciszewski and Gryar, 2016) insofar as metal pollution is expected to be greater at a shorter distance from the main channel thalweg (as for pit 2 until 1838) or from secondary channels that are hydrologically connected (as for the Bauerngrundwasser after 1838 for pit 2), compared to a site that is further and less frequently flooded (as for pit 1 after 1828). From about 1860 to the early 1930s, heavy metal concentra-

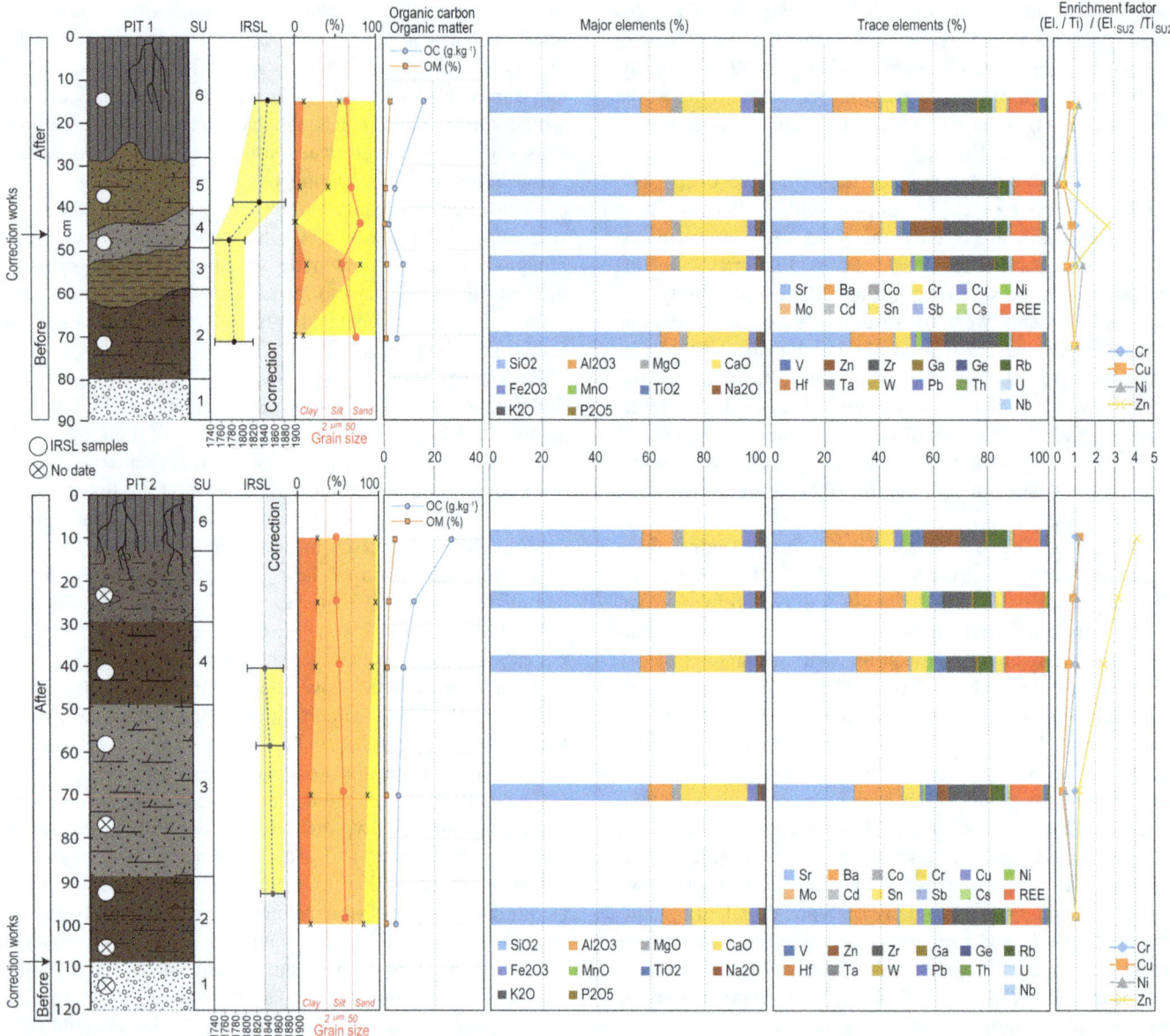

Figure 7. Stratigraphical log of the two pits including datings, organic contents, and sedimentological and geochemical results.

tions in the Rhine gradually increased in relation to the progressive industrialization (Middlekoop, 2000). Heavy metal enrichment, which is particularly marked for Zn, is therefore likely caused by changes in both heavy metal pollution of the Rhine over the past centuries and sedimentation rates. Although the main depositional filling at pit 2 ended around 1872, it was still flooded at least four times after this date (1876, 1881, 1882 and probably 1910), whereas pit 1 was more rarely and intensively flooded (1876 and probably 1881; Fig. 5c). Moreover, flow patterns during flooding from 1872 onwards likely controlled sedimentation rates of heavy metal-bound to suspended solids. Overall, our results suggest that Zn may be a proxy of anthropogenic deposition histories

in the Rhine floodplain, as previously shown in large fluvial systems (Grygar and Popelka, 2016; Lintern et al., 2016).

In addition, water table fluctuation and regular water saturation by flooding may have changed Zn speciation and mobilization in SUs 1–3 of pit 2, whereas pit 1 remained non-saturated (Fig. 8). In the period between floods, under non-saturated conditions and in the presence of oxygen, oxidation of metal sulfur, organic carbon and Fe–Mn oxyhydroxide in the upper SU may release heavy metals. High flood frequency possibly increased vertical transport of suspended solids and chemical re-distribution in pit 2, which smoothed changes in Zn concentrations with depth (Middelkoop, 2000). During flood events, heavy metals can thus not only move between the SU, but also be partly transported into the river water in association with suspended solids, by local surface erosion (Tao et al., 2005). This emphasizes that, in a restoration context, Zn may be mobilized and transported into the Rhine by groundwater table elevation (Ciszewski and Gryar, 2016) linked to flooding frequency increase as well as by bank and surface erosion.

4.3 Adjustments since the end of the correction works (after 1876)

Results of the CM pattern (Fig. 6a) show one sample in group 3 and three samples in group 4 that correspond to the left bank of the Bauerngrundwasser (surface layer). This last depositional period is linked to several large floods (1876, 1882; >300- and 100-year return periods, respectively; Fig. 5c) of high transport energy leading to suspension (Fig. 6a, group 4) and rolling processes (Fig. 6a, group 3-P). The 1876 flood, which was the major hydrological event during this period, fed the Bauerngrundwasser from upstream. The flood induced relative high-flow conditions with increased sediment transport capacities, which caused grain-size decrease along the channel. This process was particularly pronounced in the coarser P sample, which was deposited on the left bank of the channel close to T2 (Fig. 4a, 1926). As identified on the 1926 map, the flood probably opened a dike near the upstream part of the Bauerngrundwasser (Fig. 4a), which increased flow through the channel. During the same period (1872–1926), the last major planimetric and geomorphic changes occurred downstream from the Bauerngrundwasser close to transect 6 (Fischbach, 1878). In this area, the 1876 flood breached the downstream part of the high-flow dike (Fig. 4a, 1872, 1926), and widened and accentuated channel bends (Fig. 9, period C). The present morphology of the Bauerngrundwasser results from this event. After this flood, all dike apertures were closed, thereby reinforcing the disconnection between the floodplain and the main channel. This situation was exacerbated by the progressive incision of the corrected Rhine channel (Casper, 1959). Gravel bar surfaces were progressively covered by high vegetation (more than 60 % of the natural reserve area in 1926; Fig. 4b).

The regularization works (1930–1936) induced a second phase of incision (Casper, 1959; Marchal and Delmas, 1959), accentuating the gradual conversion of the Bauerngrundwasser into a wetland (+7 ha). At the same time gravel bars appeared on the groyne fields which extended into the corrected Rhine channel (+13 ha between 1926 and 1949; Fig. 4b). In 1970, this Rhine channel was by-passed by the Rhine canal on which a power plant was constructed. A continuous discharge of up to $1550 \, \mathrm{m^3 \, s^{-1}}$ is diverted towards the Rhine canal, altering the hydrology of the Old Rhine River drastically (Fig. 5c, blue hydrogram). Therefore, groundwater level was lowered by about 0.8 m (Fig. 8). In 1984, the construction of the agricultural dam raised and stabilized the water level of the Old Rhine River at 140.00 m

NN, and the groundwater level around 139.60–140.00 m NN, while the amplitude of groundwater fluctuation decreased suddenly from 1.5 to 0.4 m (Fig. S3). The entire Bauerngrundwasser was influenced by this backwater effect, which increased the stagnant water surface (+7 ha in 2010; Fig. 4b). To avoid drying of the Bauerngrundwasser during low-flow and drought periods along the Old Rhine River, an input of $1.5 \, \mathrm{m^3 \, s^{-1}}$ from the Rhine canal feeds the channel by a siphon (Fig. 4b, 2010). This explains the specific modern dynamics of sediment suspension at the bottom of the Bauerngrundwasser (group 5; Fig. 6a). Energy decreases along the channel as shown by the differences between T2-Cbs ($C = 500$; $M = 7$) and T4-Cbs ($C = 100$; $M = 6$). Similar dynamics were also observed by Peiry (1988) and underscore that in-channel deposition and filling was a current process until the start of the restoration.

4.4 Using Rhine long-term trajectory to enhance efficiency and sustainability: learning from the past to infer restoration guidelines

Combining spatial and temporal scales by overlapping multiple data sources in an interdisciplinary approach appears of crucial importance to (i) determine pre-disturbance dynamics of the hydrosystem, (ii) improve the understanding of the history of the hydrosystem in fine spatio-temporal scales (Brierley and Fryirs, 2008; Belletti et al., 2014; Bouleau and Pont, 2014), and (iii) provide key information to manage restoration projects in efficient and sustainable ways. Indeed, these results allow validation of our three working hypotheses.

4.4.1 Long-term temporal trajectories allow us to identify the driving factors, amplitude and response time of disturbances

The pre-disturbance functioning of the Rhine hydrosystem in our study area, as identified by historical map analysis, and sedimentological and geochronological data, was characterized by a high-energy depositional environment with active braiding–anastomozing, lateral channel mobility and important surface areas of gravel bars and pioneer vegetation. In part, this functioning has been targeted by recent restoration efforts. More specifically, the restoration induced, in the new channel, the recovery of bedload transport, lateral and vertical dynamics, as well as groundwater–surface water exchanges. Our results also highlight that the system has been drastically disturbed from the beginning of the correction works. In addition, the flood regime is an additional and crucial driving factor, inducing important and rapid changes in morphodynamics and sediment deposition, especially in low-elevation areas where floods are relatively frequent, long and intense (palaeochannels; e.g. pit 2). The hydrosystem remained morphologically sensitive to floods throughout the duration of the correction works. From the end of the correction, changes were less marked and also related to a decrease

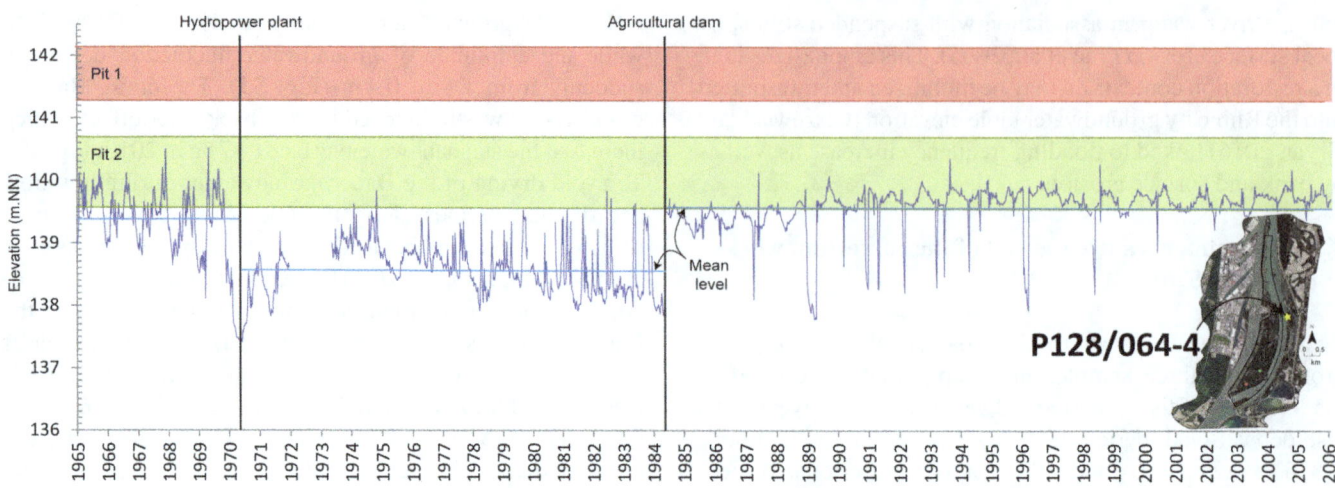

Figure 8. Hydrogeological variation between 1965 and 2006. After 1984, the lowest peaks correspond to periods during which the Old Rhine is empty.

in flood energy and frequency. The volume of fine sediment deposition is higher in the upstream part of the Bauerngrundwasser, with a maximum thickness of 1.8 m and a volume of about 400 000 m^3 (estimated by combining diachronic planimetric analysis with the thickness of the sediment layer close to the Bauerngrundwasser). A volume of about 800 000 m^3 of fine sediments with a mean thickness of about 0.4 m was deposited in the entire Rohrschollen Island up to today. This highlights the fact that impacts of correction works and further engineering works are irreversible because the removal of very large amounts of fine sediments seems unthinkable or extremely difficult. Furthermore, the strong hydrological alteration by the canalization works makes the functional alteration of the hydrosystem irreversible as well. In this constrained context, the main challenge of the restoration was to recover processes as dynamic floods (on the whole island) and an active morphodynamic gravel bed channel in a relatively restricted environment (new channel; see also below). On the basis of an environmental monitoring conducted during 3 years after the end of the restoration works, it appears that these restoration objectives are attained (Eschbach et al., 2017, 2018) and that the restoration choices were relevant.

4.4.2   Assessing potential benefits and limits of the restoration

Fine sediments are mainly located along the Bauerngrundwasser and represent a limitation for restoration purposes, because of the risk of their remobilization from both banks and the channel bottom. As shown by Richards and Bacon (1994) and Boulton et al. (1998), fine sediments (e.g. sand) are a limiting factor for many aquatic biological species. This point is important because fine sediments were relatively scarce in the braiding–anastomozing Rhine hydrosystem prior to the initiation of correction works (Ochsenbein, 1966). Furthermore, another risk concerns potential re-

mobilization of pollutants bound to fine sediments. This risk has also been identified and characterized on the Rhône River by Desmet et al. (2012), Provansal et al. (2012) and Bravard and Gaydou (2015). However, backwater effects induced by the agricultural dam control the water level on the Bauerngrundwasser, even during ecological floods (partly), and thus limit bank erosion. Thus, in the specific case of Rohrschollen Island, both risks are drastically lowered by this local hydraulic constraint. This demonstrates once again the relevancy of the principles of Rohrschollen's restoration.

Conversely, the new channel dug on a large former inchannel gravel bar (Fig. 1d) exposes a thinner layer of fine sediments along its banks, which are mainly composed of coarse sediments (gravels, pebbles) inherited from the precorrection Rhine. The backwater effect of the agricultural dam does not affect this channel (except the 100 downstream meters; Eschbach et al., 2017). Consequently, the restoration of this channel induces a recovery of bedload dynamics, lateral channel mobility, and morphodynamic and habitat diversification (Eschbach et al., 2018), which notably stimulate key processes such as downwelling/upwelling hydrological exchanges (Eschbach et al., 2017).

4.4.3   Provide key information to manage functional restoration actions in order to maximize efficiency and sustainability, and infer future evolutionary trends

Efficiency and sustainability are key issues in restoration projects (Bouleau and Pont, 2014; Loomis et al., 2000). For management strategies, knowledge of temporal trajectories is relevant for targeting pre-disturbance processes (Cairns, 1991) and performing hydromorphological process-based restorations (Mika et al., 2010; Rinaldi et al., 2015). It also allows us to identify floodplain areas with high hydromorphological functional potentials, i.e. sectors with thin lay-

Concerning the floodplain compartment, the sedimentation rate is relatively low on Rohrschollen Island (about $0.1\,\mathrm{cm\,year^{-1}}$), but it is higher (by about 1 order of magnitude) in areas where flood intensities and frequencies have been less impacted than Rohrschollen Island, along the non-canalized section of the Upper Rhine (Carbiener et al., 1993; Dister et al., 1990; Frings et al., 2014) and in the Rhine delta (Hudson et al., 2008). This floodplain geomorphological evolution reduces flood retention capacities. It probably will require in the future innovative flood management strategies (Hudson et al., 2008) that may notably be based on floodplain artificial excavations of fine sediments (which could be made more difficult if sediments are polluted) and/or natural fine sediment removal by the restoration of active bank erosion in lateral channels. This also should allow recovery of coarse sediments in the floodplains, a texture which is currently lacking. The long-term (> 100 years) sustainability of the hydro-geomorphological management of the Upper Rhine appears as a key question which is becoming more and more important and is made complex by possible fine sediment pollution.

Morphodynamic (and ecological) fluvial functionality requires relatively high, intense and long flood events superimposed on the mean hydrological regime (Bayley, 1991; Dister, 1992; Gurnell et al., 2012). This is the case on Rohrschollen Island, but following the first flooding events after restoration, a further question arises, among others: how to manage such a highly dynamic environment in the medium/long term? For example, it will probably be necessary to balance a relative sediment deficit in the upstream section of the new channel by artificial gravel augmentations, in the next years/decades (Eschbach et al., 2018).

The sustainability of functional restoration efforts and their management remains an open question: on which timescale, for which compartment and over which spatial scales should they be considered? In this context, it appears crucial to continue the post-restoration monitoring over short (3–5 years) and median (> 5–10 years) timescales to evaluate restoration success (Kondolf and Micheli, 1995; Palmer et al., 2005; Jähnig et al., 2011). This opens up avenues for developing integrative methodological approaches to improve pre-restoration knowledge and to implement post-restoration monitoring and modelling, both in the frame of fluvial hydrosystem temporal trajectories.

5 Conclusions

In this study we show the relevance of considering temporal trajectories in process-based river restoration. An interdisciplinary approach deployed at different spatio-temporal scales has been developed by combining planimetric data with sed-

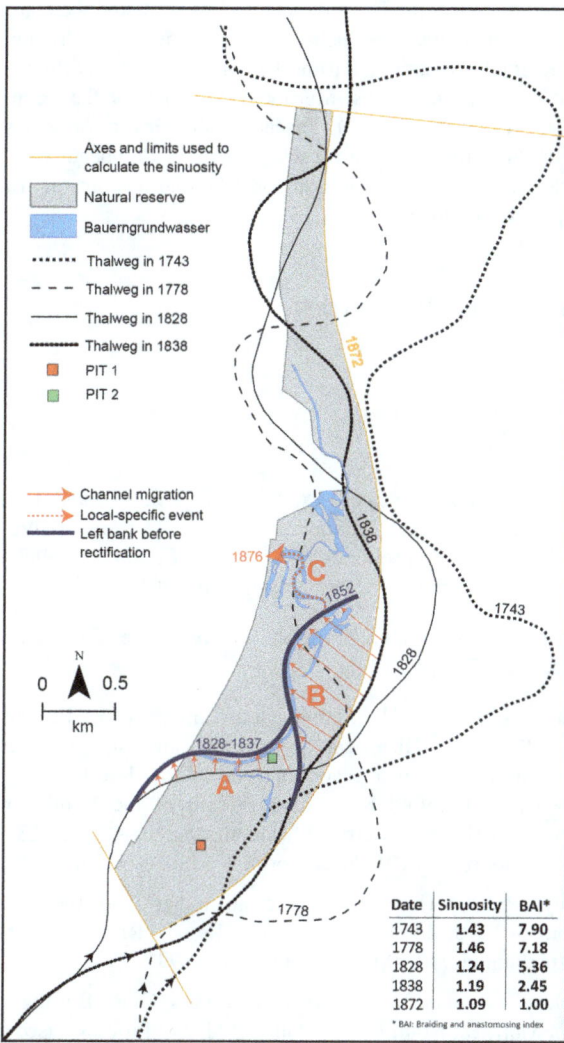

Figure 9. Rhine thalweg evolution from 1743 to the end of the correction works in 1876. (A) Area of lateral channel mobility during the 1828–1837 period. (B) Area of lateral channel mobility during the 1851–1852 flood. (C) Area impacted by the 1876 flood. BAI is a *Braiding and Anastomosing Index* which corresponds to the mean number of these two types of channels (channels showing stagnant water have been excluded).

ers of fine sediments located outside palaeochannels, notably on former gravel bars, as is the case on Rohrschollen Island (new channel). In such geomorphological areas, where the efficiency of restored lateral channels may be the highest, managers are encouraged to excavate new channels and enhance morphodynamics by floods (as has been performed on Rohrschollen Island), which may even erode self-formed lateral channels in some cases, rather than reconnecting filled palaeochannels directly. However, the latter restoration measure has dominantly been carried out on large rivers including the Rhine, which raises the question regarding its wider relevance and sustainability (Schmitt et al., 2009, 2012; see also below).

imentological, chemical and geochronological analysis, as well as a hydrological model. Prior to anthropogenic disturbances, the hydrosystem was mostly characterized by a high-energy depositional environment of braiding channels with high lateral mobility and important surfaces of gravel bars and pioneer vegetation. Correction works provoked a drastic temporal trajectory change, by intensifying filling of fine and polluted (Zn) sediments in palaeochannels and decreasing flood frequency, though some intense floods occurred. In contrast, the floodplain recorded lower deposition rates by quasi-unpolluted sediments. More recently, canalization resulted in very low sedimentation rates but strong hydrological and hydrogeological disturbances.

Our results highlight potential risks that restoration projects may face and need to mitigate along large rivers, for example the removal of fine and potentially polluted sediments by reactivating erosion/deposition processes in former channels. On Rohrschollen Island, this risk is reduced by the backwater effect of the agricultural dam which limits lateral erosion in the palaeochannel. By contrast, floodplain areas outside palaeochannels show thin layers of fine sediments and appear more relevant to restore dynamic lateral channels. Managers may benefit from excavating new channels in such areas, as has been performed on Rohrschollen Island. They are even encouraged to develop self-erosion of lateral channels by dynamic floods.

Finally, this research underscores the necessity to base functional river restorations on an interdisciplinary knowledge of hydrosystem past trajectories to maximize restoration efficiency and sustainability. On a practical level, we recommend managers conduct such studies in geomorphological restoration projects, even if they are less detailed than our study presented in this article, for example in the case of financial constraints. They should at least be based on (i) planimetric analysis (old maps and photographs), (ii) sedimentological prospection (hand auger) combined with both a lidar DEM analysis and a rapid study of former large floods and (iii) a physico-chemical analysis of sediments, especially these filling palaeo-channels.

Competing interests. The authors declare that they have no conflict of interest.

Acknowledgements. This study has been funded by the European Community (LIFE08 NAT/F/00471), the City of Strasbourg, the University of Strasbourg (IDEX-CNRS 2014 MODELROH project), the French National Center for Scientific Research (CNRS), the ZAEU (Zone Atelier Environnementale Urbaine – LTER), the Water Rhine-Meuse Agency, the DREAL Grand Est, the Région Alsace, the Département du Bas-Rhin and the company Électricité de France. We acknowledge Arthur Zimmermann for the GIS implementation, Jérôme Houssier and Erni Dillmann for use of the historical maps, Martine Trautmann for the grain-size analyses (EOST-UMS 830), Pascal Finaud-Guyot for reviewing the flooding frequency assessment, Claire Rambeau for field assistance, and Ferréol Salomon and Christian Damm for fruitful scientific discussions.

Edited by: Hubert H. G. Savenije

References

Acuña, V., Díez, J. R., Flores, L., Meleason, M., and Elosegi, A.: Does it make economic sense to restore rivers for their ecosystem services?, J. Appl. Ecol., 50, 988–997, https://doi.org/10.1111/1365-2664.12107, 2013.

Amoros, C. and Petts, G. E. (Eds.): Hydrosystèmes fluviaux, Masson, Paris, 301 pp., 1993.

Arnaud, F., Piégay, H., Schmitt, L., Rollet, A. J., Ferrier, V., and Béal, D.: Historical geomorphic analysis (1932–2011) of a by-passed river reach in process-based restoration perspectives: The Old Rhine downstream of the Kembs diversion dam (France, Germany), Geomorphology, 236, 163–177, https://doi.org/10.1016/j.geomorph.2015.02.009, 2015.

Bayley, P. B.: The flood pulse advantage and the restoration of river-floodplain systems, Regul. River., 6, 75–86, https://doi.org/10.1002/rrr.3450060203, 1991.

Beechie, T. J., Sear, D. A., Olden, J. D., Pess, G. R., Buffington, J. M., Moir, H., Roni, P., and Pollock, M. M.: Process-based principles for restoring river ecosystems, BioScience, 60, 209–222, 2010.

Belletti, B., Rinaldi, M., Buijse, A. D., Gurnell, A. M., and Mosselman, E.: A review of assessment methods for river hydromorphology, Environ. Earth Sci., 73, 2079–2100, https://doi.org/10.1007/s12665-014-3558-1, 2014.

Bensing, W.: Gewässerkundliche Probleme beim Ausbau des Oberrheins (Problèmes hydrologiques lies à l'aménagement du cours supérieur du Rhin), Deutsche Gewässerkundliche Mitteilungen, 85–102, 1966.

Bernhardt, C.: Die Rheinkorrektion. Der Bürger im Staat, 50, 76–81, 2000.

Berger, M. and Schwarzbauer, J.: Historical Deposition of Riverine Contamination on Terrestrial Floodplains as Revealed by Organic Indicators from an Industrial Point Source, Water Air Soil Poll., 227, https://doi.org/10.1007/s11270-015-2708-8, 2016.

Bogen, J., Bolviken, B., and Ottesen, R. T.: Environmental studies in Western Europe using overbank sediment, Erosion and Sediment Transport Monitoring Programmes in River Basins, Proceedings of the Oslo Symposium, August 1992, 317–325, 1992.

Bonnefont, J.-C. and Carcaud, N.: Le comportement morphodynamique de la Moselle avant ses aménagements/The morphodynamic behaviour of Moselle river before its harnessings, Géomorphologie, 3, 339–353, https://doi.org/10.3406/morfo.1997.932, 1997.

Bouleau, G. and Pont, D.: Les conditions de référence de la directive cadre européenne sur l'eau face à la dynamique des hydrosystèmes et des usages, Natures Sciences Sociétés, 22, 3–14, https://doi.org/10.1051/nss/2014016, 2014.

Boulton, A. J., Findlay, S., Marmonier, P., Stanley, E. H., and Valett, H. M.: The functional significance of the hyporheic zone in streams and rivers, Annu. Rev. Ecol. Syst., 29, 59–81, 1998.

Bravard, J.-P.: Dynamiques à long terme des systèmes écologiques ou de l'Eden impossible à la gestion de la variabilité, edited by: Lévêque, C. and Van Der Leeuw, S., Quelles Natures Voulons-Nous?, Elsevier, Paris, 133–139, 2003.

Bravard, J.-P. and Bethemont, J.: Cartography of rivers in France, in: Historical Change of Large Alluvial Rivers: Western Europe, edited by: Petts, G. E., United States, John Wiley & Sons Ltd, 1989.

Bravard, J.-P. and Gaydou, P.: Historical Development and Integrated Management of the Rhône River Floodplain, from the Alps to the Camargue Delta, France, in: Geomorphic Approaches to Integrated Floodplain Management of Lowland Fluvial Systems in North America and Europe, edited by: Hudson, P. F. and Middelkoop, H., Springer New York, New York, NY, 289–320, 289–320, 2015.

Bravard, J.-P. and Magny, M.: Les fleuves ont une histoire. Paléo-environnement des rivières et des lacs français depuis 15 000 ans, Errance édition, Errance, Paris, 2002.

Bravard, J.-P. and Peiry, J.-L.: The CM pattern as a tool for the classification of alluvial suites and floodplains along the river continuum, Geological Society, London, Special Publications, 163, 259–268, 1999.

Bravard, J.-P., Amoros, C., and Pautou, G.: Impact of Civil Engineering Works on the Successions of Communities in a Fluvial System: A Methodological and Predictive Approach Applied to a Section of the Upper Rhône River, France, Oikos 47, 92, https://doi.org/10.2307/3565924, 1986.

Bravard, J.-P., Goichot, M., and Tronchère, H.: An assessment of sediment-transport processes in the Lower Mekong River based on deposit grain sizes, the CM technique and flow-energy data, Geomorphology, 207, 174–189, https://doi.org/10.1016/j.geomorph.2013.11.004, 2014.

Brierley, G. J. and Fryirs, K. A.: River Futures: An Integrative Scientific Approach to River Repair, Island Press, 325 pp., 2008.

Brookes, A. B.: Channelized rivers: Perspectives for environmental management, Chichester, UK, 1988.

Brown, A. G.: Alluvial Geoarchaeology: Floodplain Archaeology and Environmental Change, Cambridge University Press, 404 pp., 1997.

Bull, F.: Technisch-statistische Mittheilungen über die Stromverhältnisse des Rheins längs des elsass-lothringischen Gebietes (Atlas), in: Ministerium für elsass-lothringen, Strassburg, edited by: Schmidt, C. F., 193 pp., 1885.

Carbiener, R.: Le grand ried central d'Alsace: écologie et évolution d'une zone humide d'origine fluviale rhénane, Bulletin Ecologie, 14, 249–277, 1983.

Carbiener, P., Carbiener, R., and Vogt, H.: Relations entre topographie, nature sédimentaire des dépôts et phytocénose dans le lit alluvial majeur sous forêt du Rhin dans le fossé rhénan : Forêt de la Sommerlet (commune d'Erstein), Revue de géographie de l'Est, 4, 207–312, 1993.

Cairns, J.: The status of the theoretical and applied science of restoration ecology, Environmental professional, 13, 186–194, 1991.

Casper, M.: L'aménagement du parcours allemand du Rhin : mesures exécutées, en cours et en projet, La Houille Blanche, 2, 229–248, https://doi.org/10.1051/lhb/1959037, 1959.

CECR: Ministère de l'Agriculture, Service Régional d'Aménagement des Eaux d'Alsace, Rapport finale de la Commission d'Etude des Crues du Rhin, Colmar, 87 pp., 1978.

Champion, M.: Les inondations en France depuis le VIème siècle jusqu'à nos jours, Dunod, Paris, 501 pp., 1863.

Ciszewski, D. and Grygar, T. M.: A Review of Flood-Related Storage and Remobilization of Heavy Metal Pollutants in River Systems, Water Air Soil Poll., 227, https://doi.org/10.1007/s11270-016-2934-8, 2016.

Corenblit, D., Tabacchi, E., Steiger, J., and Gurnell, A. M.: Reciprocal interactions and adjustments between fluvial landforms and vegetation dynamics in river corridors: A review of complementary approaches, Earth-Sci. Rev., 84, 56–86, https://doi.org/10.1016/j.earscirev.2007.05.004, 2007.

David, M., Labenne, A., Carozza, J.-M., and Valette, P.: Evolutionary trajectory of channel planforms in the middle Garonne River (Toulouse, SW France) over a 130-year period: Contribution of mixed multiple factor analysis (MFAmix), Geomorphology, 258, 21–39, https://doi.org/10.1016/j.geomorph.2016.01.012, 2016.

De Boer, J., Dao, Q. T., van Leeuwen, S. P. J., Kotterman, M. J. J., and Schobben, J. H. M.: Thirty-year monitoring of PCBs, organochlorine pesticides and tetrabromodiphenylether in eel from The Netherlands, Environ. Pollut., 158, 1228–1236, https://doi.org/10.1016/j.envpol.2010.01.026, 2010.

Desmet, M., Mourier, B., Mahler, B. J., Van Metre, P. C., Roux, G., Persat, H., Lefèvre, I., Peretti, A., Chapron, E., Simonneau, A., Miège, C., and Babut, M.: Spatial and temporal trends in PCBs in sediment along the lower Rhône River, France, Sci. Total Environ., 433, 189–197, https://doi.org/10.1016/j.scitotenv.2012.06.044, 2012.

Dister, E.: La maîtrise des crues par la renaturaliasation des plaines alluviales du Rhin supérieur, Espaces naturels rhénans – Bulletin de la société industrielle de Mulhouse 1, 10, 1992.

Dister, E., Gomer, D., Obrdlik, P., Petermann, P., and Schneider, E.: Water mangement and ecological perspectives of the upper rhine's floodplains, River Res. Appl., 5, 1–15, 1990.

Dufour, S.: Contrôles naturels et anthropiques de la structure et de la dynamique des forêts riveraines des cours d'eau du bassin rhodanien (Ain, Arve, Drôme et Rhône), Université Jean Moulin Lyon 3, 244 pp., 2006.

Dufour, S. and Piégay, H.: From the myth of a lost paradise to targeted river restoration: forget natural references and focus on human benefits, River Res. Appl., 25, 568–581, https://doi.org/10.1002/rra.1239, 2009.

Duplay, J., Semhi, K., Errais, E., Imfeld, G., Babcsanyi, I., and Perrone, T.: Copper, zinc, lead and cadmium bioavailability and retention in vineyard soils (Rouffach, France): The impact of cultural practices, Geoderma, 230–231, 318–328, https://doi.org/10.1016/j.geoderma.2014.04.022, 2014.

Dynesius, M. and Nilsson, C.: Fragmentation and flow regulation of river systems in the northern third the world, Science, 266, 753–762, 1994.

Eisenmenger, G.: Études sur l'évolution du Rhin et du système hydrographique Rhénan et propositions données par la faculté, Université de Paris, 512 pp., 1907.

Eschbach, D., Piasny, G., Schmitt, L., Pfister, L., Grussenmeyer, P., Koehl, M., Skupinski, G., and Serradj, A.: Thermal-infrared remote sensing of surface water-groundwater exchanges in a restored anastomosing channel (Upper Rhine River, France), Hydrol. Process., 31, 1113–1124, https://doi.org/10.1002/hyp.11100, 2017.

Eschbach, D., Grussenmeyer, P., Schmitt, L., Koehl, M., and Guillemin, S.: Combining geodetic and geomorphological methods to monitor restored dynamic mid-sized channels, Geomorphology, in review, 2018.

Evers, E. H. G., Ree, K. C. M., and Olie, K.: Spatial variations and correlations in the distribution of PCDDs, PCDFs and related compounds in sediments from the river Rhine – Western Europe, Chemosphere, 17, 2271–2288, https://doi.org/10.1016/0045-6535(88)90140-3, 1988.

Falkowska, L., Reindl, A. R., Grajewska, A., and Lewandowska, A. U.: Organochlorine contaminants in the muscle, liver and brain of seabirds (Larus) from the coastal area of the Southern Baltic, Ecotox. Environ. Safe., 133, 63–72, https://doi.org/10.1016/j.ecoenv.2016.06.042, 2016.

Fedorenkova, A., Vonk, J. A., Breure, A., Hendriks, A. J., and Leuven, R.: Tolerance of native and non-native fish species to chemical stress: a case study for the River Rhine, Aquat. Invasions, 8, 231–241, https://doi.org/10.3391/ai.2013.8.2.10, 2013.

Felkel, K.: Die erosion des oberrhein zwischen Basel und Karlsruhe, GWF (Wasser – Abwasser), 30, 801–810, 1969.

Fischbach, G.: Compte rendu des séances de la société – Séance ordinaire du 3 juillet 1878 présidée par M. de Turckheim, Société des sciences, agriculture et arts de la basse-Alsace – Bulletin Trimestriel – Tome XII – 3ième fascicule, Travaux sur les eaux entre Erstein et Strasbourg présentés par M. Schanté, 118 pp., 1878.

Frings, R. M., Gehres, N., Promny, M., Middelkoop, H., Schüttrumpf, H., and Vollmer, S.: Today's sediment budget of the Rhine River channel, focusing on the Upper Rhine Graben and Rhenish Massif, Geomorphology, 204, 573–587, https://doi.org/10.1016/j.geomorph.2013.08.035, 2014.

Fryirs, K., Brierley, G. J., and Erskine, W. D.: Use of ergodic reasoning to reconstruct the historical range of variability and evolutionary trajectory of rivers, Earth Surf. Proc. Land., 37, 763–773, https://doi.org/10.1002/esp.3210, 2012.

Garban, B., Ollivon, D., Carru, A. M., and Chesterikoff, A.: Origin, retention and release of trace metals from sediments of the river seine, Water Air Soil Poll., 87, 363–381, https://doi.org/10.1007/BF00696848, 1996.

Gocht, T., Moldenhauer, K.-M., and Püttmann, W.: Historical record of polycyclic aromatic hydrocarbons (PAH) and heavy metals in floodplain sediments from the Rhine River (Hessisches Ried, Germany), Appl. Geochem., 16, 1707–1721, https://doi.org/10.1016/S0883-2927(01)00063-4, 2001.

Grabowski, R. C. and Gurnell, A. M.: Using historical data in fluvial geomorphology, in: Tools in Fluvial Geomorphology, edited by: Kondolf, G. M. and Piégay, H., John Wiley & Sons, Ltd., 21 pp., 2016.

Gregory, K. J. and Benito, G.: Paleohydrology: Understanding Global Change, University of Southampton, 396 pp., 2003.

Grygar, T. and Popelka, J.: Revisiting geochemical methods of distinguishing natural concentrations and pollution by risk elements in fluvial sediments, J. Geochem. Explor., 170, 39–57, https://doi.org/10.1016/j.gexplo.2016.08.003, 2016.

Gurnell, A. M., Peiry, J.-L., and Petts, G. E.: Using Historical Data in Fluvial Geomorphology, in: Tools in Fluvial Geomorphology, edited by: Kondolf, G. M. and Piégay, H., John Wiley & Sons, Ltd, Hoboken, NJ, USA, 77–101, 2003.

Gurnell, A. M., Bertoldi, W., and Corenblit, D.: Changing river channels: The roles of hydrological processes, plants and pioneer fluvial landforms in humid temperate, mixed load, gravel bed rivers, Earth-Sci. Rev., 111, 129–141, https://doi.org/10.1016/j.earscirev.2011.11.005, 2012.

Habersack, H., Jäger, E., and Hauer, C.: The status of the Danube River sediment regime and morphology as a basis for future basin management, International Journal of River Basin Management, 11, 153–166, https://doi.org/10.1080/15715124.2013.815191, 2013.

Herget, J., Bremer, E., Coch, T., Dix, A., Eggenstein, G., and Ewald, K.: Engineering Impact on River Channels in the River Rhine Catchment (Menschliche Eingriffe in Flussläufe im Einzugsgebiet des Rheins), Erdkunde, 59, 294–319, 2005.

Herget, J., Dikau, R., Gregory, K. J., and Vandenberghe, J.: The fluvial system – Research perspectives of its past and present dynamics and controls, Geomorphology, 92, 101–105, https://doi.org/10.1016/j.geomorph.2006.07.034, 2007.

Hering, D., Aroviita, J., Baattrup-Pedersen, A., Brabec, K., Buijse, T., Ecke, F., Friberg, N., Gielczewski, M., Januschke, K., Köhler, J., Kupilas, B., Lorenz, A. W., Muhar, S., Paillex, A., Poppe, M., Schmidt, T., Schmutz, S., Vermaat, J., Verdonschot, P. F. M., Verdonschot, R. C. M., Wolter, C., and Kail, J.: Contrasting the roles of section length and instream habitat enhancement for river restoration success: a field study of 20 European restoration projects, J. Appl. Ecol., 52, 1518–1527, https://doi.org/10.1111/1365-2664.12531, 2015.

Hooke, J. M.: River channel adjustment to meander cutoffs on the River Bollin and River Dane, northwest England, Geomorphology, 14, 235–253, 1995.

Horowitz, A., Meybeck, M., Idlafkih, Z., and Biger, E.: Variations in trace element geochemistry in the Seine River Basin based on Foodplain deposits and bed sediments, Hydrol. Process., 13, 1329–1340, 1999.

Hudson, P. F., Middelkoop, H., and Stouthamer, E.: Flood management along the Lower Mississippi and Rhine Rivers (The Netherlands) and the continuum of geomorphic adjustment, Geomorphology, 101, 209–236, https://doi.org/10.1016/j.geomorph.2008.07.001, 2008.

IKSR-CIPR-ICBR: État des lieux DCE, partie A, Rapport soumis à la Commission européenne sur les résultats de l'état des lieux établi conformément à la directive 2000/60/CE du Parlement européen et du Conseil du 23 octobre 2000, 2005.

IKSR-CIPR-ICBR: Plan d'action contre les inondations 1995–2010 : objectifs opérationnels, mise en œuvre et résultats – Bilan synthétique, Coblence, 2012.

IKSR-CIPR-ICBR: Mise en œuvre du Plan de gestion des sédiments Rhin, Commission Internationale pour la Protection du Rhin (CIPR), ISBN-13: 978-3-941994-54-6, 2014.

Jähnig, S. C., Lorenz, A. W., Hering, D., Antons, C., Sundermann, A., Jedicke, E., and Haase, P.: River restoration success: a question of perception, Ecological society of America, 21, 2007–2015, 2011.

James, L. A., Singer, M. B., Ghoshal, S., and Megison, M.: Historical channel changes in the lower Yuba and Feather Rivers, California: Long-term effects of contrasting river-management strategies, in: Geological Society of America Special Papers, edited by: James, L. A., Rathburn, S. L., and Whittecar, G. R., Geological Society of America, 451, 57–81, https://doi.org/10.1130/2009.2451(04), 2009.

Jeanpierre, D.: L'érosion des fonds : conséquence secondaire de la protection de la plaine d'Alsace par l'endiguement longitudinal du Rhin, in: Société Hydrotechnique de France, Presented at the Xième journées de l'hydraulique, Paris, 8 pp., 1968.

Jenkinson, R. G., Barnas, K. A., Braatne, J. H., Bernhardt, E. S., Palmer, M. A., and Allan, J. D.: Stream restoration databases and case studies: a guide to information resources and their utility in advancing the science and practice of restoration, Restor. Ecol., 14, 177–186, 2006.

Jones, C. J. and Johnson, P. A.: Describing Damage to Stream Modification Projects in Constrained Settings, J. Am. Water Resour. As., 51, 251–262, https://doi.org/10.1111/jawr.12248, 2015.

Knighton, A. D.: Fluvial forms and processes, Edward Arnold Editions, London, 218 pp., 1984.

Kondolf, G. M. and Larson, M.: Historical channel analysis and its application to riparian and aquatic habitat restoration, Aquat. Conserv., 5, 109–126, https://doi.org/10.1002/aqc.3270050204, 1995.

Kondolf, G. M. und Micheli, E. R.: Evaluating stream restoration projects, Environ. Manage., 19, 1–15, https://doi.org/10.1007/BF02471999, 1995.

Lair, G. J., Zehetner, F., Khan, Z. H., and Gerzabek, M. H.: Phosphorus sorption-desorption in alluvial soils of a young weathering sequence at the Danube River, Geoderma, 149, 39–44, https://doi.org/10.1016/j.geoderma.2008.11.011, 2009.

Large, A. R. G. and Gilvear, D. J.: Using Google Earth, a virtual-globe imaging platform, for ecosystem services-based river assessment, River Res. Appl., 31, 406–421, https://doi.org/10.1002/rra.2798, 2015.

Lawler, D. M.: The measurement of river bank erosion and lateral channel change: A review, Earth Surf. Proc. Land., 18, 777–821, https://doi.org/10.1002/esp.3290180905, 1993.

Lespez, L., Viel, V., Rollet, A. J., and Delahaye, D.: The anthropogenic nature of present-day low energy rivers in western France and implications for current restoration projects, Geomorphology, 251, 64–76, https://doi.org/10.1016/j.geomorph.2015.05.015, 2015.

Lintern, A., Leahy, P. J., Heijnis, H., Zawadzki, A., Gadd, P., Jacobsen, G., Deletic, A., and Mccarthy, D. T.: Identifying heavy metal levels in historical flood water deposits using sediment cores, Water Res., 105, 34–46, https://doi.org/10.1016/j.watres.2016.08.041, 2016.

Loomis, J., Kent, P., Strange, L., Fausch, K., and Covich, A.: Measuring the total economic value of restoring ecosystem services in an impaired river basin: results from a contingent valuation survey, Ecol. Econ., 33, 103–117, 2000.

Magdaleno, F., Anastasio Fernández, J., and Merino, S.: The Ebro River in the 20th century or the ecomorphological transformation of a large and dynamic Mediterranean channel, Earth Surf. Proc. Land., 37, 486–498, https://doi.org/10.1002/esp.2258, 2012.

Marchal, M. and Delmas, G.: L'aménagement du Rhin à courant libre du Bâle à Lauterbourg, La Houille Blanche, 177–202, https://doi.org/10.1051/lhb/1959032, 1959.

Martin, B., Drescher, A., Fournier, M., Guerrouah, O., Giacona, F., Glaser, R., Himmelsbach, I., Holleville, N., Riemann, D., Schonbein, J., Vitoux, M.-C., and With, L.: Les évènements extrêmes dans le fossé rhénan entre 1480 et 2012. Quels apports pour la prévention des inondations?, La Houille Blanche, 82–93, https://doi.org/10.1051/lhb/20150023, 2015.

Middelkoop, H.: Heavy-metal pollution of the river Rhine and Meuse floodplains in the Netherlands, Neth. J. Geosci., 79, 411–427, 2000.

Mika, S., Hoyle, J., Kyle, G., Howell, T., Wolfenden, B., Ryder, D., Keating, D., Boulton, A., Brierley, G., Brooks, A. P., Fryirs, K., Leishman, M., Sanders, M., Arthington, A., Creese, R., Dahm, M., Miller, C., Pusey, B., and Spink, A.: Inside the "black box" of river restoration: using catchment history to identify disturbance and response mechanisms to set targets for process-based restoration, Ecol. Soc., 15, 20 pp., 2010.

Naiman, R. J., Décamps, H., Pastor, J., and Johnston, C. A.: The Potential Importance of Boundaries of Fluvial Ecosystems, J. N. Am. Benthol. Soc., 7, 289–306, https://doi.org/10.2307/1467295,1988.

Naiman, R. J., Decamps, H., and Pollock, M.: The Role of Riparian Corridors in Maintaining Regional Biodiversity, Ecol. Appl., 3, 209–212, https://doi.org/10.2307/1941822, 1993.

Ochsenbein, G.: La végétation sur les bords du Rhin, Saisons d'Alsace, 383–398, 1966.

Palmer, M. A., Bernhardt, E. S., Allan, J. D., Lake, P. S., Alexander, G., Brooks, S., Carr, J., Clayton, S., Dahm, C. N., Follstad Shah, J., Galat, D. L., Loss, S. G., Goodwin, P., Hart, D. D., Hassett, B., Jenkinson, R., Kondolf, G. M., Lave, R., Meyer, J. L., O'Donnell, T. K., Pagano, L., and Sudduth, E.: Standards for ecologically successful river restoration: Ecological success in river restoration, J. Appl. Ecol., 42, 208–217, https://doi.org/10.1111/j.1365-2664.2005.01004.x, 2005.

Pardé, M.: Périodicité des grandes inondations et crues exceptionnelles, Revue de géographie alpine, 16, 499–519, https://doi.org/10.3406/rga.1928.4457, 1928.

Passega, R.: Grain Size Representation by Cm Patterns as a Geological Tool, J. Sediment. Res., 34, 830–847, 1964.

Passega, R.: Significance of CM diagrams of sediments deposited by suspensions, Sedimentology, 24, 723–733, https://doi.org/10.1111/j.1365-3091.1977.tb00267.x, 1977.

Peiry, J.-L.: Approche géographique de la dynamique spatio-temporelle des sédiments d'un cours d'eau intra-montagnard : l'exemple de la plaine alluviale de l'Arve (Haute-Savoie), Lyon 3, 1988.

Petts, G. E., Moeller, H., and Roux, A. L. (Eds.): Historical change of large alluvial rivers: Western Europe, John Wiley and Sons, New York, NY (US), 1989.

Preusser, F., May, J.-H., Eschbach, D., Trauerstein, M., and Schmitt, L.: Infrared stimulated luminescence dating of 19th century fluvial deposits from the upper Rhine River, Geochronometria, 43, 131–142, https://doi.org/10.1515/geochr-2015-0045, 2016.

Provansal, M., Raccasi, G., Monaco, M., Robresco, S., and Dufour, S.: La réhabilitation des marges fluviales, quel intérêt, quelles contraintes? Le cas des annexes fluviales du Rhône aval, Méditerranée, 118, 85–94, 2012.

Reimann, C. and de Caritat, P.: Distinguishing between natural and anthropogenic sources for elements in the environment: regional geochemical surveys versus enrichment factors, Sci. Total Environ., 337, 91–107, https://doi.org/10.1016/j.scitotenv.2004.06.011, 2005.

Richards, C. and Bacon, K. L.: Influence of fine sediment on macroinvertebrate colonization of surface and hyporheic stream substrates, The Great Basin Naturalist, 54, 106–113, 1994.

Rinaldi, M., Surian, N., Comiti, F., and Bussettini, M.: A method for the assessment and analysis of the hydromorphological condition of Italian streams: The Morphological Quality Index (MQI), Geomorphology, 180–181, 96–108, https://doi.org/10.1016/j.geomorph.2012.09.009, 2013.

Rinaldi, M., Surian, N., Comiti, F., and Bussettini, M.: A methodological framework for hydromorphological assessment, analysis and monitoring (IDRAIM) aimed at promoting integrated river management, Geomorphology, 251, 122–136, https://doi.org/10.1016/j.geomorph.2015.05.010, 2015.

Riquier, J., Piégay, H., and Šulc Michalková, M.: Hydromorphological conditions in eighteen restored floodplain channels of a large river: linking patterns to processes, Freshwater Biol., 60, 1085–1103, https://doi.org/10.1111/fwb.12411, 2015.

Rollet, A. J., Piégay, H., Dufour, S., Bornette, G., and Persat, H.: Assessment of consequences of sediment deficit on a gravel river bed downstream of dams in restoration perspectives: application of a multicriteria, hierarchical and spatially explicit diagnosis, River Res. Appl., 30, 939–953, https://doi.org/10.1002/rra.2689, 2014.

Rumsby, B. T. and Macklin, M. G.: River response to the last neoglacial (the "Little Ice Age") in northern, western and central Europe, Geological Society, London, Special Publications, 115, 217–233, https://doi.org/10.1144/GSL.SP.1996.115.01.17, 1996.

Schäfer, W.: Der Oberrhein, sterbende Landschaft, Natur und Museum, 1–29, 1973.

Schirmer, W.: Holocene valley development on the upper Rhine and Main, in: Lake, Mire and River Environments, edited by: Lang, R. and Schlijchter, C., Balkema, Rotterdam, 153–160, 1988.

Schmitt, L., Trémolières, M., Blum, C., Dister, E., and Pfarr, U.: 30 years of restoration works on the two sides of the Upper Rhine River: feedback and future challenges, Int. Conf. Integrative Sciences and Sustainable Development of Rivers, Lyon, France, 101–103, 2012.

Schmitt, L., Houssier, J., Martin, B., Beiner, M., Skupinski, G., Boës, E., Schwartz, D., Ertlen, D., Argant, J., Gebhardt, A., Schneider, N., Lasserre, M., Trintafillidis, G., and Ollive, V.: Paléo-dynamique fluviale holocène dans le compartiment sud-occidental du fossé rhénan (France), Revue Archéologique de l'Est 42è supplément, 15–33, 2016.

Schmitt, L., Lebeau, M., Trémolières, M., Defraeye, S., Coli, C., Denny, E., Dillinger, M., Beck, T., Dor, J.-C., Gombert, P., Gueidan, A., Manné, S., Party, J.-P., Perrotey, P., Piquette, M., Roeck,

U., Schnitzler, A., Sonnet, O., Vacher, J.-P., Vauclin, V., Weiss, M., Zacher, J.-N., and Wilms, P.: Le "polder" d'Erstein : objectifs, aménagements et retour d'expérience sur cinq ans de fonctionnement et de suivi scientifique environnemental (Rhin, France), Ingénieries no. spécial, 67–84, 2009.

Schulz-Zunkel, C. and Krueger, F.: Trace Metal Dynamics in Floodplain Soils of the River Elbe: A Review, J. Environ. Qual., 38, 1349, https://doi.org/10.2134/jeq2008.0299, 2009.

Sear, D. A. and Arnell, N. W.: The application of palaeo-hydrology in river management, Catena, 66, 169–183, https://doi.org/10.1016/j.catena.2005.11.009, 2006.

Sear, D. A., Darby, S. E., Thorne, C. R., and Brookes, A. B.: Geomorphological approach to stream stabilization and restoration: Case study of the Mimmshall brook, hertfordshire, UK, Regul. River., 9, 205–223, https://doi.org/10.1002/rrr.3450090403, 1994.

Starkel, L., Soja, R., and Michczyńska, D. J.: Past hydrological events reflected in Holocene history of Polish rivers, CATENA, Past Hydrological Events Related to Understanding Global Change, 66, 24–33, https://doi.org/10.1016/j.catena.2005.07.008, 2006.

Tao, F., Jiantong, L., Bangding, X., Xiaoguo, C., and Xiaoqing, X.: Mobilization potential of heavy metals: A comparison between river and lake sediments, Water Air Soil Poll., 161, 209–225, 2005.

Toonen, W. H. J., Winkels, T. G., Cohen, K. M., Prins, M. A., and Middelkoop, H.: Lower Rhine historical flood magnitudes of the last 450 years reproduced from grain-size measurements of flood deposits using End Member Modelling, CATENA, 130, 69–81, https://doi.org/10.1016/j.catena.2014.12.004, 2015.

Trimble, S. W. and Cooke, R. U.: Historical Sources for Geomorphological Research in the United States, The Professional Geographer, 43, 212–228, https://doi.org/10.1111/j.0033-0124.1991.00212.x, 1991.

Tulla, J. G.: Ueber die Rectification des Rheins, von seinem Austritt aus der Schweiz bis zu seinem Eintritt in Hessen, Müller's, Karlsruhe, 1825.

Tümmers, H. J.: Der Rhein: ein europäischer Fluß und seine Geschichte, 2. Überarbeitete und aktualisierte Auflage, Beck, München, 481 pp., 1999.

Uehlinger, U., Wantzen, K. M., Leuven, R. S. E. W., and Arndt, H.: The Rhine river basin, in: Rivers of Europe, edited by: Tockner, K., Academic Press, Amsterdam, London, 199–245, 2009.

van Helvoort, P.-J., Griffioen, J., and Hartog, N.: Characterization of the reactivity of riverine heterogeneous sediments using a facies-based approach; the Rhine–Meuse delta (The Netherlands), Appl. Geochem., 22, 2735–2757, https://doi.org/10.1016/j.apgeochem.2007.06.016, 2007.

Wetter, O., Pfister, C., Weingartner, R., Luterbacher, J., Reist, T., and Trösch, J.: The largest floods in the High Rhine basin since 1268 assessed from documentary and instrumental evidence, Hydrolog. Sci. J., 56, 733–758, https://doi.org/10.1080/02626667.2011.583613, 2011.

WFD: 2000/60/EC, Directive 2000/60/EC of the European parliament and of the council of 23 October 2000 establishing a framework for Community action in the field of water policy, 2000.

Wittmann, J.: Chronik der niedrigsten Wasserstände des Rheins vom Jahre 70, Chr. Geb. bis 1858 und Nachrichten über die im Jahre 1857 – 58 im Rheinbette von der Schweiz bis nach Holland zu Tage gekommenen Alterthümer und Merkwürdigkeiten, insbesondere die damals sichtbaren Steinpfeilerreste der ehemaligen festen Brücke bei Mainz und die unfern dieser Stadt im Rheinstrome gemachten Entdeckungen, Seifert, Mainz, 142 pp., 1859.

Wohl, E., Angermeier, P. L., Bledsoe, B., Kondolf, G. M., MacDonnell, L., Merritt, D. M., Palmer, M. A., Poff, N. L., and Tarboton, D.: River restoration, Water Resour. Res., 41, W10301-1–W10301-12, https://doi.org/10.1029/2005WR003985, 2005.

Woitke, P., Wellmitz, J., Helm, D., Kube, P., Lepom, P., and Litheraty, P.: Analysis and assessment of heavy metal pollution in sus-

pended solids and sediments of the river Danube, Chemosphere, 51, 633–642, https://doi.org/10.1016/S0045-6535(03)00217-0, 2003.

Ziliani, L. and Surian, N.: Evolutionary trajectory of channel morphology and controlling factors in a large gravel-bed river, Geomorphology, 173–174, 104–117, 2012.

Zimmer, D., Kiersch, K., Jandl, G., Meissner, R., Kolomiytsev, N., and Leinweber, P.: Status Quo of Soil Contamination with Inorganic and Organic Pollutants of the River Oka Floodplains (Russia), Water Air Soil Poll., 211, 299–312, https://doi.org/10.1007/s11270-009-0301-8, 2010.

10

Definition of efficient scarcity-based water pricing policies through stochastic programming

H. Macian-Sorribes[1], **M. Pulido-Velazquez**[1], **and A. Tilmant**[2]

[1]Research Institute of Water and Environmental Engineering (IIAMA), Universitat Politècnica de València, Valencia, Spain
[2]Department of Civil and Water Engineering, Université Laval, Québec City, Québec, Canada

Correspondence to: H. Macian-Sorribes (hecmasor@upv.es)

Abstract. Finding ways to improve the efficiency in water usage is one of the most important challenges in integrated water resources management. One of the most promising solutions is the use of scarcity-based pricing policies. This contribution presents a procedure to design efficient pricing policies based on the opportunity cost of water at the basin scale. Time series of the marginal value of water are obtained using a stochastic hydro-economic model. Those series are then post-processed to define step pricing policies, which depend on the state of the system at each time step. The case study of the Mijares River basin system (Spain) is used to illustrate the method. The results show that the application of scarcity-based pricing policies increases the economic efficiency of water use in the basin, allocating water to the highest-value uses and generating an incentive for water conservation during the scarcity periods. The resulting benefits are close to those obtained with the economically optimal decisions.

1 Introduction

One of the main challenges in integrated water resources management (IWRM) is improving the efficiency in water usage while balancing it with equity. Given that in the majority of the developed world the building of new water supply systems has well passed its zenith, water management strategies are now devoted to achieve better operating policies. Several criteria can be considered when designing a policy for water allocation: flexibility in the allocation, security of tenure for the users, real cost recovery, predictability of its performance, fairness and acceptability (Dinar et al., 2007). Each system has a unique configuration and, in consequence,

a unique combination of factors that lead to an adequate management policy.

There are four major water allocation mechanisms: public water allocation, water markets, user-based allocation and marginal cost pricing. Public water allocation provides an adequate treatment of water as a public good, allows for the development of large-scale infrastructures often beyond the private investment capacity, and focuses on equity issues and non-economic objectives. However, it usually fails in achieving optimal economic performance, leads to water prices which are below the water value, and provides no incentive to water saving and efficient use (Meinzen-Dick and Mendoza, 1996). Water markets encourage both sellers and buyers to use it efficiently, provide flexible allocation mechanisms and allow considering the real value of the employed resource. On the contrary, unique characteristics of water can turn markets into a bad allocation mechanism if externalities are not adequately considered (Garrick et al., 2009). User-based allocation, in which water users regulate water resources by themselves, is especially suited for local needs in water management and is likely to be accepted by the users. However, it may be inadequate in inter-sectorial allocation, requiring also a very transparent structure (Dinar et al., 2007).

Finally, marginal cost pricing provides a theoretically adequate way to consider water values in allocation, encourages users to save it and puts water in its most valuable uses, leading to efficient allocations. It also can play a major role in the long-term planning and conservation of water supplies, delaying the need of capacity expansions and offering higher economic returns while holding rationing requirements (Gysi and Loucks, 1971). However, marginal cost pricing would require estimating the non-accounting opportunity costs involved in water allocation (Griffin, 2001). Calculating the marginal value of water is challenging as it varies in space and time according to supply–demand imbalances, requires adequate monitoring, and has some difficulties to deal with equity when water prices are beyond what lower-value users can afford (Dinar et al., 2007). Moreover, administrative constraints on price charges can limit their benefits (Dandy et al., 1984). In Europe, the EU Water Framework Directive (European Commission, 2000) calls for the implementation of new pricing policies that assure the contribution of water users to the recovery of the cost of water services (financial instrument) while providing adequate incentives for an efficient use of water (economic instrument). Not only financial costs should be recovered but also environmental and resource (opportunity) costs. This issue has been addressed through the use of hydro-economic models as tools able to couple physical and economic water resource aspects (Heinz et al., 2007; Pulido-Velazquez et al., 2008, 2013; Riegels et al., 2013; Ward and Pulido-Velazquez, 2008).

A pricing policy is efficient, according to economic theory, if the prices charged correspond to the marginal cost of water. Therefore, it must take into account supply costs, opportunity costs and externalities (Rogers et al., 2002). Measuring the opportunity costs of scarce water is difficult: since water markets are usually absent or ineffective, scarcity values are not reflected in the water prices. Given that opportunity cost depends on the alternative uses, an integrated basin-wide approach is needed to simultaneously account for all major competing water uses in the basin (Rogers et al., 2002; Pulido-Velazquez et al., 2013). The assessment of these opportunity costs requires a systems approach and a proper method to estimate the value of water across the different users (Young, 2005; Pulido-Velazquez et al., 2008). If pricing policies reflect the entire basin-wide marginal opportunity costs, then they will act as an economic instrument for efficient water resources management, modifying the demand–supply interaction by acting on the demand side and supporting water allocation to the most valuable users.

The marginal resource opportunity cost (MROC), or marginal value of water, can be defined as the benefits that would have been obtained at one location and one time if the available resource at that location and time had been increased by one unit (Pulido-Velazquez et al., 2008, 2013; Tilmant et al., 2008, 2014). MROC can be derived from hydro-economic models. Pulido-Velazquez et al. (2013) developed a method to obtain scarcity-based pricing policies using MROC values, in which the time series of MROC obtained after running a hydro-economic model are post-processed to derive step pricing policies whose performance can be simulated using a decision support system (DSS) shell. However, in those studies pricing policies were based on either priority-based simulation (which are not representing an optimal policy) or deterministic hydro-economic optimization, with the inherent limitation of the perfect foresight (the optimization algorithm knows future flows in advance

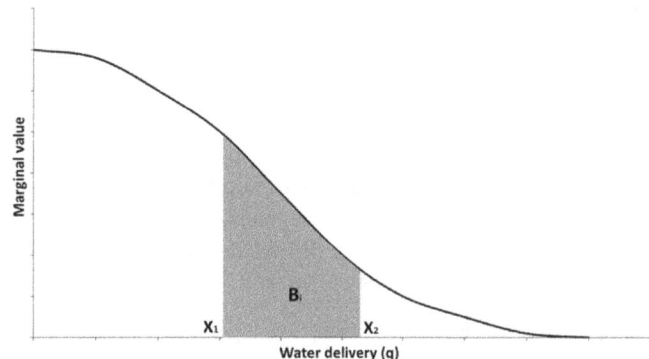

Figure 1. Benefits from an increase of water delivery from x_1 to x_2.

and, in consequence, it has an unrealistic advantage that diminishes the applicability of the results) (Labadie, 2004).

The main purpose of this paper is to propose a method for the design of scarcity-based water pricing policies based on the MROC derived from a stochastic hydro-economic model. With stochastic programming procedures, uncertainty is taken into account in the optimization process. Therefore, it removes the effect that the "perfect foresight" phenomenon causes in the marginal values, which are flattened across time and thus lose an important part of their short-term variability. The marginal values obtained using stochastic programming are representative of an optimal policy while reflecting the future uncertainties in the system's inflows. After describing the method to obtain the MROC values, we propose a method for the definition of a stochastic-programming-based water pricing policy. Finally, a case study is developed to prove and illustrate the methodology using a hydro-economic simulation model of the Mijares River basin system (Spain). Pricing policies are applied in this paper exclusively as economic instruments whose purpose is achieving an efficient use of water. Financial issues are not addressed.

2 Method and materials

2.1 Assessment of the marginal resource opportunity cost (MROC)

For a specific water demand, the benefit obtained by the user, B_i, given a change in water delivery level from x_1 to x_2, can be calculated by integrating the demand curve (D_i) (Fig. 1):

$$B_i = \int_{x_1}^{x_2} D_i(q)\mathrm{d}q. \tag{1}$$

Similarly, for a given location L and time t, the benefit B_t achieved by a change in its state $s_{L,t}$ (water availability) from x_1 to x_2 can be calculated integrating the marginal water value (or MROC) function:

$$B_t = \int_{x_1}^{x_2} \mathrm{MROC}_{L,t}(s_{L,t})\mathrm{d}s. \tag{2}$$

The MROC can be defined as the derivative of the benefit function with respect to the system state. Therefore, if the MROC integration obtains the system-wide benefits, the MROC can be calculated as

$$\mathrm{MROC}_{L,t} = \frac{\mathrm{d}B_t(s_{L,t})}{\mathrm{d}s}. \tag{3}$$

The MROC value for a specific location and time can be estimated (1) under a simulation approach, as the benefits obtained by an increase of one unit in the available resource at that location and time (Pulido-Velazquez et al., 2008, 2013); and (2) under an optimization approach, as the shadow value, dual variable or Lagrange multiplier associated with the mass-balance equation at the desired place and the specified time (Pulido-Velazquez et al., 2008, 2013; Tilmant et al., 2008).

2.2 MROC assessment through stochastic programming

Stochastic programming (SP) procedures are powerful and useful methodologies to derive optimal management of water systems with uncertain inputs (Tejada-Guibert et al., 1993). Various SP algorithms are available. Among them, stochastic dynamic programming (SDP) has been widely used in water resources management because (1) it is able to handle non-linearities in the objective function in an efficient way, (2) the inflow uncertainty representation is clear and simple, and (3) it treats the decision-making process sequentially, as done in real-life operation (Labadie, 2004). The SDP algorithm solves the Bellman's recursive equation as follows:

$$F_t(S_t, Q_t) \tag{4}$$
$$= \max_{D_t}\left[B_t(S_t, Q_t, D_t) + E_{Q_{t+1}|Q_t}\{F_{t+1}(S_{t+1}, Q_{t+1})\}\right],$$

where F_t is the total benefit function, S_t the current (time t) system state vector, Q_t the current inflow vector, D_t the decision made at time step t, B_t the immediate benefit function, $E_{Q_{t+1}|Q_t}$ the expectation operator between the current and future inflows, and F_{t+1} the future benefit function or benefit-to-go function.

In the SDP method, the state variables S_t and Q_t are discretized over all the state space forming a grid, allowing only transitions between grid points. The expectation operator is then defined by using a Markov chain that relates the current hydrological state Q_t to all the possible future states Q_{t+1} through a set of transition probabilities.

With the application of Eq. (4), the optimal policies $D_t(S_t, Q_t)$ and benefit-to-go function $F_t(S_t, Q_t)$ are calculated at the grid points. Then, interpolation methodologies

can be applied to obtain the optimal policies $D_t^*(S_t, Q_t)$ and the optimal benefits $F_t^*(S_t, Q_t)$ over the entire state space. An alternative is to use a reoptimization approach as in Tejada-Guibert et al. (1993). With this approach, the Bellman function is implemented forward with the SDP-derived benefit-to-go functions as inputs.

$$F_t(S_t, Q_t) \tag{5}$$
$$= \max_{D_t}\left[B_t(S_t, Q_t, D_t) + \sum_q \left\{p_{p,q}^t \cdot F_{t+1}^*(S_{t+1}, Q_{t+1})\right\}\right],$$

where S_t and Q_t are the simulated system state (storage) and inflows at stage t, and $p_{p,q}^t$ is the transition probability (Markov Chain) between inflow class p at time stage t and inflow class q at time stage $t+1$. The S_{t+1} and Q_t values are not subjected to a discrete grid. The reoptimization provides time series of allocation decisions and the corresponding λ values associated with the system's nodes, which correspond to the MROC.

2.3 From MROC values to pricing policies

The results given by the SDP algorithm are the optimal allocation policies, benefits and MROC values at each point of the discrete mesh. Those values vary with the time stage of the year, storages and inflows. A pricing scheme based on those values would be in theory the most efficient. Highly variable prices are normal in hydropower production, in which deregulated electricity markets' prices and demands vary even during the same day and, in consequence, hydropower producers need to make decisions on very short notice, independently of previous choices. However, this situation is distinctly different in consumptive demands, especially in irrigated agriculture. The majority of farmers make most of their decisions in an annual or inter-annual basis (area to be irrigated, cropping pattern and so on), where in-year choices are dependent on decisions in previous time stages. Farmers act as risk-averse decision-makers, since errors in the expectations of crop prices, input costs and water deliveries can cause significant economic losses. For those reasons, a pricing policy based on the raw MROC values would introduce too much uncertainty in the water price and thus in the agricultural sector. On the other hand, the pricing schemes derived from MROC values were conceived as the basis for a process involving discussion, negotiation and approval of a certain simple pricing policy with certain consensus among the stakeholders. As a result, the raw MROC values previously obtained have to be post-processed in order to transform them into simpler a priori scarcity-based pricing policies, so that the rule can be negotiated and known beforehand by everybody, allowing farmers to react accordingly with a more predictable price. Several operations must be carried out to transform the time series of MROC into a step pricing policy depending on the system state variables (t, S_t, Q_t), in which a step function defines the price to be applied each time period. Those operations can be summarized as MROC values of aggregation/disaggregation, MROC statistical analysis, and step pricing policy construction. Although the SDP method was used to obtain the MROC time series, the operations explained below can be used regardless of the algorithm employed (another stochastic one such as SDDP, deterministic optimization or simulation) to provide MROC time series.

The aggregation/disaggregation of the MROC time series previously obtained is required in order to derive pricing functions at a certain spatial and temporal scale. Regarding the spatial dimension of the intended pricing policy, different pricing schedules for raw water in different zones in the system will better capture the MROC spatial variability. However, the complexity of pricing policies will probably imply greater implementation difficulties. With regard to the temporal scale, as stated earlier, pricing policies varying at a lower time resolution (seasonal or monthly) are more accurate than annual ones, although they might also face more implementation problems and higher uncertainty in future prices. Defining a general procedure to aggregate/disaggregate MROC time series is difficult, since it depends on the desired pricing policy features and each system's unique features. An example of the aggregation/disaggregation process for the specific features of the desired pricing policy is shown in the case study section.

Once the aggregated MROC values are obtained, their cumulative probability distribution can be determined. Several characteristic values can then be chosen using different percentiles of the cumulative probability distribution. Those characteristic values can be used to estimate the MROC–state relationship by (1) sorting the time series of state variables obtained with SDP according to their respective aggregated MROC values, (2) selecting the MROC–state pairs in which the MROC value was a characteristic one, and (3) organizing the results in the form of state–MROC steps. To sum up, the method presented in this paper can be divided in the following steps:

1. definition of the main pricing policy features;

2. development of a hydro-economic stochastic programming model of the system;

3. determination of MROC (marginal water values or λ-values) time series at the reference nodes (e.g., main reservoirs);

4. aggregation/disaggregation of MROC time series to calculate the aggregated MROC values;

5. development of a statistical analysis over the aggregated MROC values to obtain their cumulative probability distribution;

6. building of k steps by

(a) choose k different cumulative probability values (characteristic values),

(b) sorting according to the aggregated MROC values the system state values obtained in the stochastic programming run,

(c) obtaining, for each characteristic value, the system states associated with it,

(d) summarizing all the possible state values associated with each characteristic value in the form of steps,

7. definition of several step pricing policies based on the obtained steps.

Pricing policies can be simulated to assess their performance and to compare them to the SDP results and to other alternatives such as different operating rules. In case the pricing policies' performance is found to be inadequate, the process must be restarted: the pricing policies' features are reassessed and the build-up and analysis stages must be redone. The most straightforward way to determine its adequacy is to quantify the forgone benefits that the users would be willing to accept as counterpart of using a simpler pricing policy. It is impossible to establish a unique threshold value since it totally depends on the system features. An alternative approach, employed in the case study of this paper, is to compare the performance of the pricing policy with the one achieved by the optimal operating rules expressed by the SDP results. In that way, a pricing policy could be considered as adequate as long as it obtains similar economic returns than to for the optimal policy.

2.4 Case study: Mijares River basin (Spain)

The Mijares River basin is located in eastern Spain (Fig. 2). It is characterized by the existence of several relevant water springs at its headwater (Mas Royo and Babor), the implementation of conjunctive-use water strategies to improve water management (Andreu and Sahuquillo, 1987), and the existence of an allocation framework accepted by all the users (SCRM, 1974). Regulated by the Arenós (93 Mm³) and Sichar (49 Mm³) reservoirs, surface water is mostly devoted to agricultural purposes (mainly orange trees), with groundwater as complementary or substitutive resource, while urban demands are entirely supplied using groundwater. There are 10.499 ha irrigated exclusively by surface water and 11.622 ha irrigated by surface and groundwater.

The Mijares River simplified flow network is shown in Fig. 3. Although the groundwater supply is significant in the lower basin (Plana de Castellón aquifer), it has not been explicitly represented in the optimization model, as there is no hydraulic connection between the river and the aquifer (disconnected aquifer). Upstream, stream–aquifer interaction is implicit in the inflow (discharge) time series. Seepage equations are also added in certain lower reaches of the river. Consequently, the demands supplied entirely by groundwater

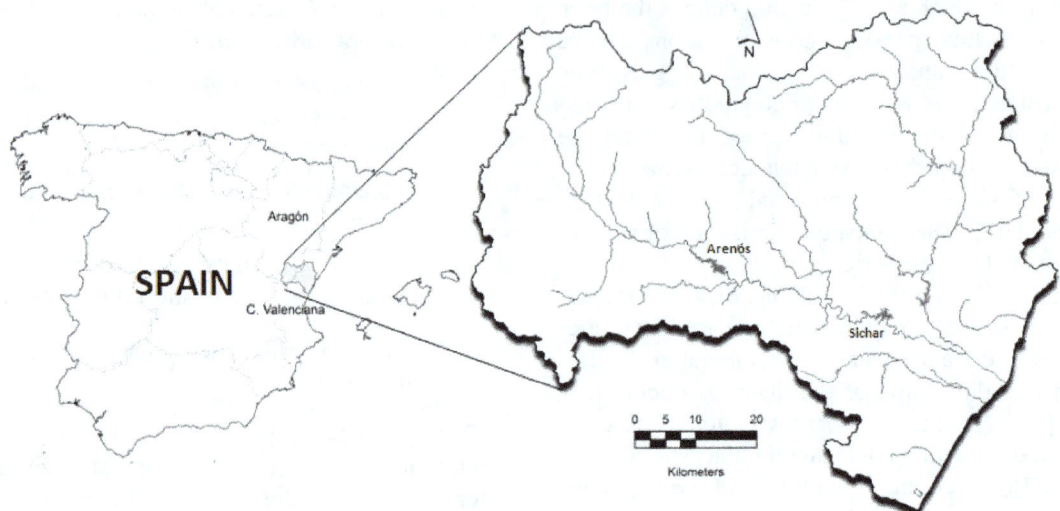

Figure 2. Mijares River basin location (eastern Spain).

Table 1. Characteristic values of elements of the Mijares River network.

Element	Characteristic value
Arenós reservoir	93 Mm^3 capacity
Sichar reservoir	49 Mm^3 capacity
Upper basin inflow	138 Mm^3 annual discharge
Middle basin inflow	55 Mm^3 annual discharge
Traditional irrigation district	83.5 Mm^3 annual demand
MC canal irrigation district	7.6 Mm^3 annual demand
CC100 canal irrigation district	16.3 Mm^3 annual demand
CC220 canal irrigation district	11.9 Mm^3 annual demand
Minimum flow downstream Sichar	0.2 Mm^3 annual requirement

have not been considered, and the mixed-supplied demands have been reduced by an amount equivalent to its groundwater supply. The characteristics of each element are shown in Table 1.

Current water management agreements give priority to the supply to the traditional irrigation district (ID), which has been using water since the 13th century, over the remaining IDs (established in mid 20th century). In year 1970, before the construction of the Arenós dam (with public funding), an agreement was signed between users to regulate the use of the Sichar reservoir (funded by the traditional ID) (SCRM, 1974). That agreement established a monthly storage limit for the Sichar reservoir below which only the traditional ID can be supplied (see Fig. 4). That agreement continued to be applied after construction of the Arenós reservoir, but referred to the total system storage (Arenós and Sichar).

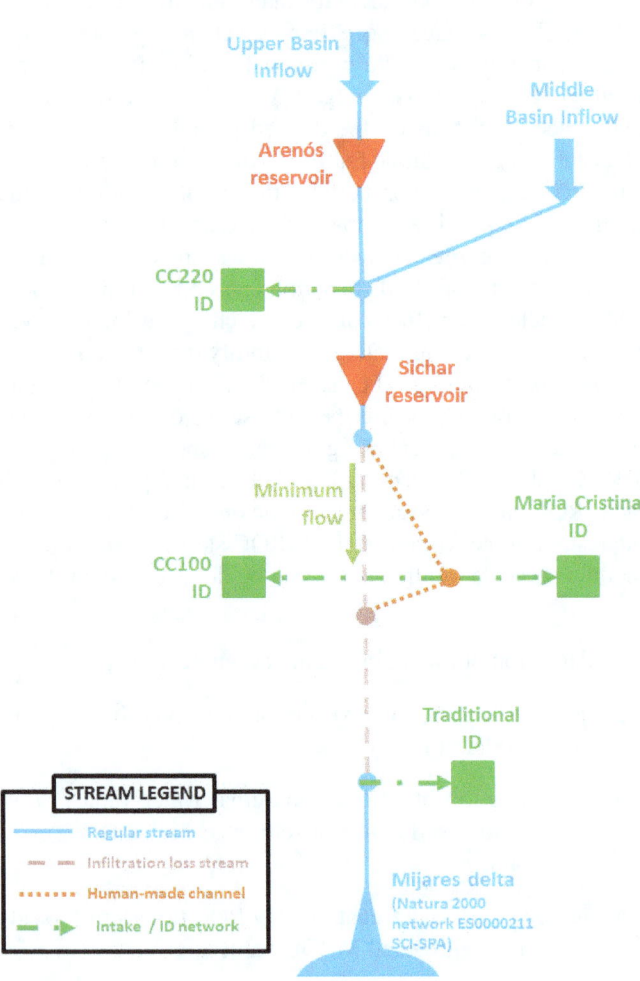

Figure 3. Mijares River network schematic.

2.5 SDP hydro-economic model of the Mijares River

The SDP hydro-economic model comprises all the elements previously described and depicted in Fig. 4. A monthly time step was used. The hydrologic variables $(q_t, t = 1, \dots, 12)$ were discretized into four equally likely intervals per sub-basin, each one represented by a characteristic value. Water demand curves are derived from Alvarez-Mendiola (2012). The minimum flow requirement has been considered as a constraint. A lag-1 Markov chain captures the temporal persistence found in the inflow data. The discrete storage classes adopted were 13 (Arenós) and 7 (Sichar). Minimum flows, demand curves, evaporation and infiltration losses, stream capacities and benefits (obtained as the sum of integrations under all the demand curves) are also taken into account in the model. The model was built using a generalized SDP algorithm developed using GAMS software (Macian-Sorribes and Pulido-Velazquez, 2014). This model was optimized, for an infinite horizon, taking target storages as decision variables.

3 Results

3.1 SDP-obtained benefits, policies and MROC values

The monthly policies and benefits obtained depend on a vector consisting of four variables: Arenós storage, Sichar storage, upper basin inflow and middle basin inflow. The optimal decisions obtained with the algorithm followed the classic "rule of thumb" of reservoirs in series devoted to water supply – fill the upper reservoirs first, and empty the lower reservoirs first (Lund and Guzman, 1999) – as the results empty first Sichar (the lower reservoir) and fill first Arenós (the upper reservoir). In addition, traditional ID users are subject to greater water deficits compared to the other ones, inverting the current criteria, caused by the river seepage in the lower Mijares streams.

A reoptimization procedure was applied to obtain the time series of MROC values at the Arenós and Sichar reservoirs, depicted compared with the sum of storages in Fig. 5. The plots show the same values during most of the historical time series. The slight differences between them found in certain time stages correspond to the opportunity cost of the CC220 ID delivery. Water values increase between 1977 and 1986, a period that corresponds to the largest drought suffered in the Mijares River basin. The average MROC value is equal to EUR $0.15 \, \mathrm{m}^{-3}$, ranging from EUR 0 to $0.68 \, \mathrm{m}^{-3}$.

3.2 Pricing policies in the Mijares River basin

Regarding the aggregation/disaggregation of the MROC time series at the Arenós and Sichar reservoirs, the pricing policy used was defined at basin-wide scale. This decision has been made considering the proximity of the intakes for the demands and the possibility of releasing water from the two reservoirs to satisfy almost all of them. The chosen temporal scale for the pricing policy was annual, with the same pricing policy for all the months. For simplicity, the state variable for defining the pricing schedule was the sum of the storage in the Arenós and Sichar reservoirs, without considering the corresponding monthly inflow. That departs from the SDP formulation but is consistent with the current management policies, based exclusively on storages. The aggregation operation driven by these features was simply a non-weighted average of the MROC values at the Arenós and Sichar reservoirs, as the MROC values are almost coincident for both reservoirs.

Figure 6 shows the MROC cumulative probability distribution. To establish pricing policies, we sampled the 5th (EUR $0 \, \mathrm{m}^{-3}$), 25th (EUR $0.06 \, \mathrm{m}^{-3}$), 50th (EUR $0.13 \, \mathrm{m}^{-3}$), 75th (EUR $0.24 \, \mathrm{m}^{-3}$) and 95th (EUR $0.51 \, \mathrm{m}^{-3}$) percentiles. The MROC–storage pairs were then organized in intervals (as depicted in Fig. 7). Each interval or step represents the range of storage values associated with that MROC.

The previous steps were used to define the pricing policies. Firstly, the storage space was divided into intervals of $25 \, \mathrm{Mm}^3$. A price was then defined for each interval as either the minimum or the maximum or the average over the MROC values associated with the steps found within the interval. As a result, a set of 15 pricing policies was obtained. Figure 7 shows some of them, corresponding to policies regarding maximum between steps (pricing policy 1), average (pricing policy 2) and minimum (pricing policy 3). The remaining pricing policies were based on different combinations between prices obtained in the first three.

3.3 Pricing policy performance by hydro-economic modeling

Each pricing policy was simulated for the 1940–2009 period with a hydro-economic simulation model, previously built using MatLab (Macian-Sorribes, 2012), whose features are identical to the SDP one. This model implements the network shown in Fig. 3 with the corresponding element features (storage capacity, historical monthly inflows, seepage losses equations, etc.), the current demand priority scheme (first the traditional ID, then the rest), and the current system operation scheme (first fill Arenós, first empty Sichar and avoid as much as possible the streams subjected to seepage losses). More details can be found in Macian-Sorribes (2012). This simulation model calculates at each month the price that corresponds to the available storage, redefines water demands using the demand curves, and then allocates resources using the system's river network and infrastructure. Simulation results are then analyzed and compared to the performances obtained with both current and SDP-derived policies (Table 2). Figure 8a shows the time series of benefits resulting from SDP-derived policies (the optimal policies obtained from the SDP once interpolated as suggested

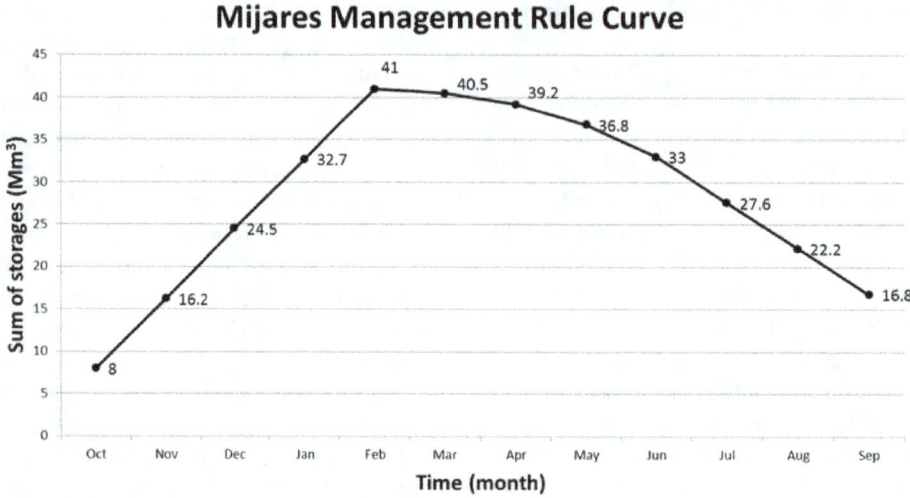

Figure 4. Current management rule curve established in the Mijares River basin.

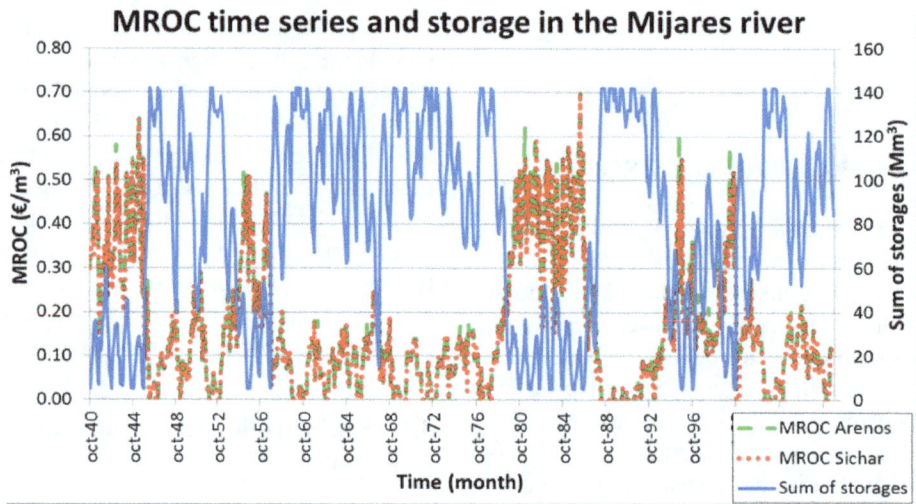

Figure 5. MROC time series and storages in the Mijares River.

by Tejada-Guibet in 1993), current management rules and the best pricing policies for the 1940–2009 period.

Regarding Table 2 and Fig. 8a, only slight differences can be found between policies. All pricing policies increase the economic results of current management policies by around EUR 0.70 million per year, being similar to the ones obtained with the direct use of the SDP policies. For that reason, we consider those pricing policies to be adequate, as it was not necessary to test complex ones. This situation is caused by the natural robustness of the Mijares River water system and by the homogeneity of the cropping pattern (mainly citrus crops, mostly oranges) found in the basin. The improvement caused by pricing policies is due to temporal reallocations: the prices hedge the immediate supplies to allow for greater deliveries in the next months. In that way, the deficits and their induced scarcity costs are distributed over

several months of slight delivery reductions rather than a single large deficit. As the income losses are non-linear with respect to the deliveries, that deficit distribution improves the total economic return for the system. Despite having the same global benefits, the way they are distributed among the users' changes for all the pricing policies tested; thus it is necessary to take them into account when deciding which one to be implemented.

Focusing on the most severe historical drought faced by the Mijares Basin, from year 1977 to 1986 (Table 2, Fig. 8b), the differences on benefits between the current management and the SDP results are higher (around EUR 1.10 million per year), indicating that SDP-derived policies hedge available resources better against the drought events. To sum up, the pricing policy application resulted in greater benefits. Especially in drought situations, the adoption of these strategies

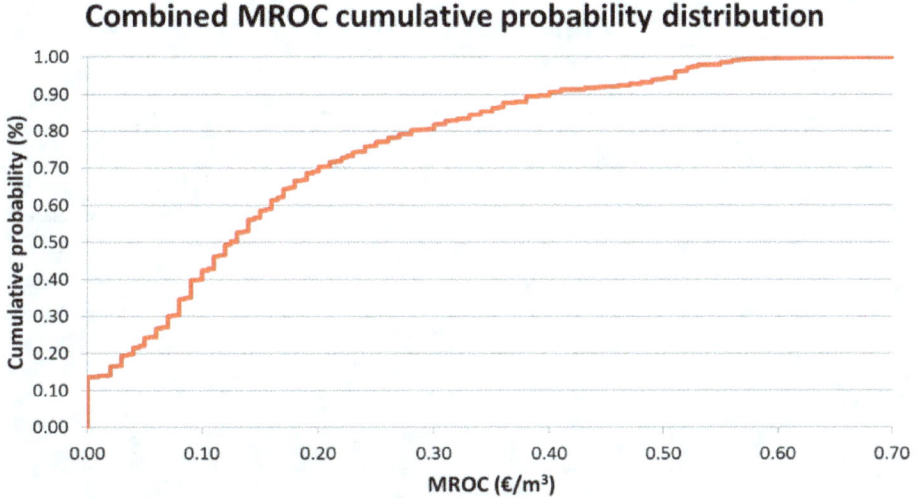

Figure 6. Combined MROC cumulative probability distribution.

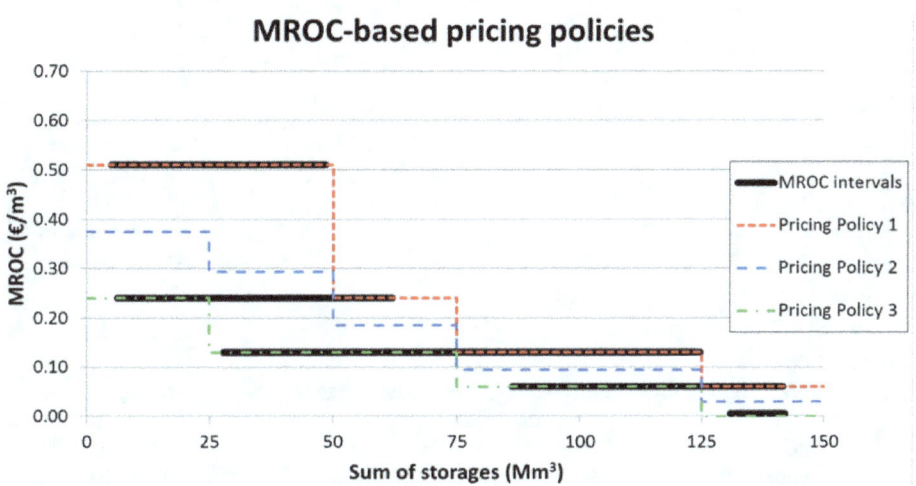

Figure 7. MROC-based pricing policies.

Table 2. Benefits for the 1940–2009 and 1977–1986 periods with stochastic optimization (SDP), current management rules and pricing policies.

Simulation	Traditional M EUR	MC M EUR	CC100 M EUR	CC220 M EUR	Total M EUR
1940–2009 benefits per demand and total					
SDP	44.49	4.14	8.56	6.56	63.75
Current policies	46.31	3.60	7.42	5.73	63.06
Pricing policy 10	44.99	4.06	8.29	6.47	63.81
Pricing policy 11	45.00	4.05	8.29	6.46	63.81
Pricing policy 12	45.05	4.04	8.27	6.44	63.81
1977–1986 benefits per demand and total					
SDP	35.97	3.22	6.80	5.07	51.05
Current policies	42.05	1.69	3.52	2.68	49.93
Pricing policy 4	37.11	3.06	6.09	4.86	51.12
Pricing policy 5	37.11	3.06	6.09	4.86	51.12

would lead to a greater economic performance and to a more efficient water use.

4 Discussion and conclusions

This paper presents a method to design an efficient scarcity-based pricing policy based on marginal water values (MROC) derived from stochastic programming. The method is applied to a case study, the Mijares River basin, in Spain. The results show that the benefits from the application of the resulting pricing policies are close to those obtained by the optimal SDP policy for both the entire historical hydrological data series and the drought conditions. By pricing marginal water opportunity costs, water would be reallocated to the highest-valued uses, significantly increasing the total net benefit of water use in the basin (by EUR 0.75 million per year).

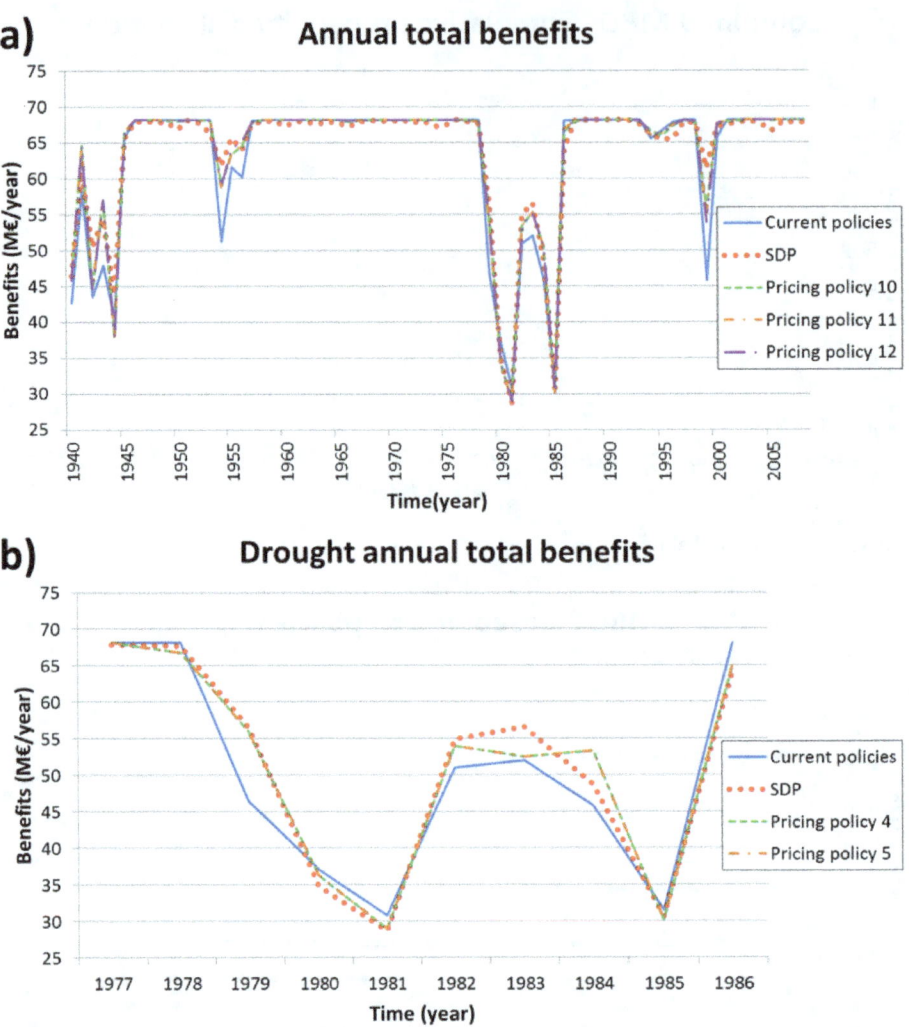

Figure 8. Annual total benefits comparison for the 1940–2009 period (**a**) and for the 1977–1986 drought (**b**).

The reason why a simple pricing policy is able to achieve a similar performance as that of a complex optimal operating rule in this case study is due to the in-year time pattern possessed by this policy: the majority of the MROC values that determined the water prices for the lower storage levels correspond to start-of-refill ones, while the MROC values associated with high storage levels are start-of-drawdown ones. For that, the prices triggered vary across time in accordance to the refill–drawdown cycle of the system, reproducing in some way the water value annual cycle.

Given the uncertainties associated with the inputs of the model, the predictions concerning the pricing policy performance are therefore uncertain. The most important source of uncertainty are the demand curves, since they directly affect the MROC values and the reliability of the simulated performance of a pricing policy. Given the strong influence of the demand curves in the results, demand curves should be properly estimated and tested. The robustness in the estimation of the demand curves will be subject to the availability

of the proper information for the economic characterization of the water uses in the basin as well as the suitability of the method used in the definition of those curves. This could be a limitation in the applicability of the method to certain cases. The resulting pricing policies should be in any case regarded just as a starting point for a negotiation process involving the users and policymakers to determine the final prices to be charged for water abstraction. On the other hand, the pricing policies defined in this paper are conceived exclusively as economic instruments for achieving an economically efficient use of water. Financial issues (such as revenue sufficiency and cost recovery) and other goals of pricing policies like equity and environmental sustainability should be considered as well.

Unlike the method proposed in Pulido-Velazquez et al. (2013), this one uses a stochastic programming approach instead of deterministic programming or simulation. It also employs a different method to derive the pricing policies based on the MROC and state time series.

The MROC values measure the opportunity cost associated with water use. Therefore, in order to determine the final prices charged to the users, the cost recovery component of the supply costs (operation, maintenance and capital charges) and the environmental externalities should be added (Rogers et al., 2002). The main objective in the design of the pricing policies discussed here focuses on the use of water prices as economic instrument for an efficient management of the interaction between supply and demand. The role of pricing for the cost recovery of water services (pricing as financial instruments) will require a complementary analysis.

Comparing pricing policies with water markets, both will be theoretically valid approaches for enhancing economic efficiency in water allocation in the system. Nowadays in Spain water markets are allowed by law but, in practice, only in a few occasions have they been operative and never in this system (Palomo-Hierro et al., 2015). Factors like high transaction costs, farmers' reluctance to participate, low physical connectivity, etc., often prevent more transfers. While the experience and literature on water markets is more abundant, water pricing is clearly underused regarding its potential for dealing with water scarcity. Despite its limitations, drawbacks, barriers and issues for its implementation, water pricing offers some interesting features: contributes to match supply and demand, generates revenues, and maintains customer choices (against command and control policies). On the other hand, the river basin authority holds the formal control of the system, which is essential for addressing environmental requirements, third party effects and so on.

Regarding the established methodology and the case study, several conclusions can be drawn.

1. Stochastic programming is a useful tool for estimating optimal policies and MROC time series under hydrological uncertainty. These time series capture and summarize the overall performance of the optimization policies and can be therefore used to assess pricing policies able to be applied at the basin scale.

2. Pricing policies defined using MROC data series, after statistical analysis and step building, are adequate to enhance a system's global economic efficiency. They establish a univocal relationship between the system state (storages and inflows) and a water price based on the marginal value of water in a reservoir, linking the price concept to the MROC one.

3. Participatory framework processes might be desirable to define the features and characteristics that the pricing policies should have, in order to find as much consensus as possible for its implementation.

4. The proposed methodology aims at designing efficient pricing policies. Other issues should be incorporated in the design of a final pricing policy, such as cost recovery of financial costs related to water services and of environmental cost (externalities), as well as equity issues and other social objectives (eg. rural development and environmental protection).

5. Pricing policy is one of the economic policy instruments that can be implemented to adapt individual decisions to collective goals. We can also apply a mix of them (water markets, pollution taxes, etc.) in order to better reach the social and environmental targets in the management of water resource systems.

Acknowledgements. This study has been partially funded by the European Union's Seventh Framework Program (FP7) ENHANCE (number 308.438). In addition, the authors acknowledge the editor and reviewers for their helpful and constructive comments.

Edited by: F. Pappenberger

References

Alvarez-Mendiola, E.: Diseño de una política eficiente de precios del agua integrando costes de oportunidad del recurso a escala de cuenca, PhD dissertation, Universitat Politècnica de València, Valencia, Spain, 2012 (in Spanish).

Andreu, J., and Sahuquillo, A.: Efficient aquifer simulation in complex systems, J. Water Res. Pl.-ASCE, 113, 110–129, 1987.

CHJ: Esquema provisional de Temas Importantes, Ministerio de Medio Ambiente y Medio Rural y Marino, Confederación Hidrográfica del Júcar, Valencia, Spain, 2009 (in Spanish).

Dandy, G. C., McBean, E. A., and Hutchinson, B. G.: A model for constrained optimum water pricing and capacity expansion, Water Resour. Res., 20, 511–520, 1984.

Dinar, A., Rosegrant, M. W., and Meinzen-Dick, R.: Water Allocation Mechanisms – Principles and Examples, Agriculture and Natural Resources Department, World Bank, Washington, DC, USA, 2007.

European Commission: Directive 2000/60/Ec of the European Parliament and of the Council, of 23 October 2000, establishing a framework for community Action in the Field of Water Policy, Official Journal of the European Communities (OJL), 327, 1–73, 2000.

Fisher, F., Huber-Lee, A., and Amir, I.: Liquid Assets: An Economic Approach for Water Management and Conflict Resolution in the Middle East and Beyond, RFF Press, Washington, DC, USA, 2005.

Garrick, D., Siebentritt, M. A., Aylward, B., Bauer, C. J., and Purkey, A.: Water markets and freshwater ecosystem services: policy reform and implementation in the Columbia and Murray-Darling Basins, Ecol. Econ., 69, 366–379, 2009.

Griffin, R. C.: Effective water pricing, J. Am. Water Resour. As., 37, 1335–1347, 2001.

Griffin, R. C.: Water Resource Economics: The Analysis of Scarcity, Policies, and Projects, The MIT Press, Cambridge, USA, 402 pp., 2006.

Gysi, M. and Loucks, D. P.: Some long run effects of water-pricing policies, Water Resour. Res., 7, 1371–1382, 1971.

Harou, J. J., Pulido-Velazquez, M., Rosenberg, D. E., Medellín-Azuara, J., Lund, J. R., and Howitt, R. E.: Hydro-economic models: concepts, design, applications, and future prospects, J. Hydrol., 375, 627–643, 2009.

Heinz, I., Pulido-Velazquez, M., Lund, J., and Andreu, J.: Hydroeconomic modeling in river basin management: implications and applications for the European Water Framework Directive, Water Resour. Manag., 21, 1103–1125, 2007.

Howe, C. W., Schurmeier, D. R., and Shaw Jr., W. D.: Innovative approaches to water allocation: the potential for water markets, Water Resour. Res., 22, 439–445, 1986.

Johansson, R. C., Tsur, Y., Roe, T. L., Doukkali, R., and Dinar, A.: Pricing irrigation water: a review of theory and practice, Water Policy, 4, 173–199, 2002.

Karamouz, M., Houck, M. H., and Delleur, J. W.: Optimization and simulation of multiple reservoir systems, J. Water Res. Pl.-ASCE, 118, 71–81, 1992.

Kelman, K., Stedinger, J. R., Cooper, L. A., Hsu, E., and Yuan, S.-Q.: Sampling stochastic dynamic programming applied to reservoir operation, Water Resour. Res., 26, 447–454, 1990.

Labadie, J. W.: Optimal operation of multireservoir systems: state-of-the-art review, J. Water Res. Pl.-ASCE, 130, 93–111, 2004.

Lund, J. R. and Guzman, J.: Derived operating rules for reservoirs in series or in parallel, J. Water Res. Pl.-ASCE, 125, 143–153, 1999.

Macian-Sorribes, H.: Utilización de Lógica Difusa en la Gestión de Embalses, Master Thesis dissertation, Universitat Politècnica de València, Valencia, Spain, 2012 (in Spanish).

Macian-Sorribes, H. and Pulido-Velazquez, M.: Hydro-economic optimization under inflow uncertainty using the SDP-GAMS generalized tool, in: Evoling Water Resources Systems: Understanding, Predicting and Managing Water-Society Interactions, IAHS Press, Wallingford, UK, 410–415, 2014.

Massarutto, A.: El precio de agua: herramienta básica para una política sostenible del agua?, Ingeniería del Agua, 10, 293–326, 2003 (in Spanish).

Meinzen-Dick, R. and Mendoza, M.: Alternative water allocation mechanisms indian and international experiences, Econ. Polit. Weekly, 31, A25–A30, 1996.

Mousavi, J. J., Ponnambalam, K., and Karray, F.: Reservoir operation using a dynamic programming fuzzy rule-based approach, Water Resour. Manag., 19, 655–672, 2005.

Nandalal, K. D. W. and Bogardi, J. J.: Dynamic Programming Based Operation of Reservoirs: Applicability and Limits, Cambridge University Press, Cambridge, UK, 144 pp., 2007.

Palomo-Hierro, S., Gomez-Limon, J. A., and Riesgo, L.: Water Markets in Spain: Performance and Challenges, Water 2015, 7, 652–678, 2015.

Pulido-Velazquez, M., Jenkins, M., and Lund, J. R.: Economic values for conjunctive use and water banking in southern California, Water Resour. Res., 40, W03401, doi:10.1029/2003WR002626, 2004.

Pulido-Velazquez, M., Andreu, J., Sahuquillo, A., and Pulido-Velazquez, D.: Hydro-economic river basin modelling: the application of a holistic surface-groundwater model to assess opportunity costs of water use in Spain, Ecol. Econ., 66, 51–65, 2008.

Pulido-Velazquez, M., Alvarez-Mendiola, E., and Andreu, J.: Design of efficient water pricing policies integrating basinwide resource opportunity costs, J. Water Res. Pl.-ASCE, 139, 583–592, 2013.

Riegels, N., Pulido-Velazquez, M., Doulgeris, C., Sturm, V., Jensen, R., Moller, F., and Bauer-Gottwein, P.: Systems analysis approach to the design of efficient water pricing policies under the EU Water Framework Directive, J. Water Res. Pl.-ASCE, 139, 574–582, 2013.

Rogers, P., de Silva, R., and Bhatia, R.: Water is an economic good: How to use prices to promote equity, efficiency and sustainability, Water Policy, 4, 1–17, 2002.

SCRM: Convenio de Bases para la Ordenación de las Aguas del río Mijares, Ministerio de Obras Públicas, Transportes y Medio Ambiente, Confederación Hidrográfica del Júcar, Valencia, Spain, 50 pp., 1974 (in Spanish).

Stedinger, J. R., Sule, B. F., and Loucks, D. P.: Stochastic dynamic programming models for reservoir operation optimization, Water Resour. Res., 20, 1499–1505, 1984.

Tejada-Guibert, J. A., Johnson, S. A., Stedinger, J. R.: Comparison of two approaches for implementing multireservoir operating policies derived using stochastic dynamic programming, Water Resour. Res., 29, 3969–3980, 1993.

Tilmant, A. and Kelman, R.: A stochastic approach to analyze trade-offs and risks associated with large-scale water resources systems, Water Resour. Res., 43, W06425, doi:10.1029/2006WR005094, 2007.

Tilmant, A., Pinte, D., and Goor, Q.: Assessing marginal water values in multipurpose multireservoir systems via stochastic programming, Water Resour. Res., 44, W12431, doi:10.1029/2008WR007024, 2008.

Tilmant, A., Goor, Q., and Pinte, D.: Agricultural-to-hydropower water transfers: sharing water and benefits in hydropower-irrigation systems, Hydrol. Earth Syst. Sci., 13, 1091–1101, doi:10.5194/hess-13-1091-2009, 2009.

Tilmant, A., Arjoon, D., and Fernandes Marques, G.: Economic value of storage in multireservoir systems, J. Water Res. Pl.-ASCE, 140, 375–383, 2014.

US Army Corps of Engineers (USACE) Hydrologic Engineering Center (HEC): Developing Seasonal and Long-Term Reservoir System Operation Plans Using HEC-PRM, US Army Corps of Engineers, Hydrologic Engineering Center, Davis, California, USA, 1996.

Ward, F. and Pulido-Velazquez, M.: Efficiency, equity and sustainability in a water quantity-quality optimization model in the Rio Grande basin, Ecol. Econ., 66, 23–37, 2008.

Wurbs, R. A.: Reservoir-system simulation and optimization models, J. Water Res. Pl.-ASCE, 119, 455–472, 1993.

Young, R. A.: Water economics, in: Water Resources Handbook, McGraw-Hill, New York, NY, USA, 3.1–3.57, 1996.

Young, R. A.: Determining the Economic Value of Water: Concepts and Methods, RFF Press, Washington, DC, USA, 375 pp., 2005.

11

Links between the Big Dry in Australia and hemispheric multi-decadal climate variability – implications for water resource management

D. C. Verdon-Kidd[1], A. S. Kiem[2], and R. Moran[2]

[1]Environmental and Climate Change Research Group, School of Environmental and Life Sciences, University of Newcastle, Callaghan, Australia

[2]Water Group, Department of Environment and Primary Industries, Victoria, Australia

Correspondence to: D. C. Verdon-Kidd (danielle.verdon@newcastle.edu.au)

Abstract. Southeast Australia (SEA) experienced a protracted drought during the mid-1990s until early 2010 (known as the Big Dry or Millennium Drought) that resulted in serious environmental, social and economic effects. This paper analyses a range of historical climate data sets to place the recent drought into context in terms of Southern Hemisphere inter-annual to multi-decadal hydroclimatic variability. The findings indicate that the recent Big Dry in SEA is in fact linked to the widespread Southern Hemisphere climate shift towards drier conditions that began in the mid-1970s. However, it is shown that this link is masked because the large-scale climate drivers responsible for drying in other regions of the mid-latitudes since the mid-1970s did not have the same effect on SEA during the mid- to late 1980s and early 1990s. More specifically, smaller-scale synoptic processes resulted in elevated autumn and winter rainfall (a crucial period for SEA hydrology) during the mid- to late 1980s and early 1990s, which punctuated the longer-term drying. From the mid-1990s to 2010 the frequency of the synoptic processes associated with elevated autumn/winter rainfall decreased, resulting in a return to drier than average conditions and the onset of the Big Dry. The findings presented in this paper have marked implications for water management and climate attribution studies in SEA, in particular for understanding and dealing with "baseline" (i.e. current) hydroclimatic risks.

1 Introduction

1.1 The Big Dry and other protracted droughts in Australia's history

Australia, with its naturally highly variable climate, is no stranger to drought conditions. For example, the Federation Drought ($\sim$ 1895–1902) was associated with drought conditions covering the majority of the eastern two-thirds of Australia (Verdon-Kidd and Kiem, 2009a). From 1937 to 1945 Southeast Australia (SEA) was subjected to another multi-year drought, known as the World War II Drought, while more recently the Big Dry (or Millennium Drought) affected SEA during the mid-1990s through to early 2010 and resulted in a marked reduction in rainfall and runoff (NWC, 2006; Murphy and Timbal, 2008; Verdon-Kidd and Kiem, 2009a, b; Kiem and Verdon-Kidd, 2010; Gallant et al., 2012). In addition to the multi-year droughts mentioned, a number of shorter, equally intense droughts have also occurred during SEA's instrumental history (e.g. 1914–1915, 1965–1968 and 1982–1983).

In terms of annual rainfall deficits the Big Dry has been shown to be more severe in parts of SEA than the earlier multi-year droughts for durations of 3–19 years (CSIRO, 2012); however, the Federation and World War II droughts were more widespread (Verdon-Kidd and Kiem, 2009a).

The Big Dry was also characterised by a lack of high 1-day rainfall totals (Murphy and Timbal, 2008; Verdon-Kidd and Kiem, 2009a) and wet months (CSIRO, 2012), which is consistent with a reduction in the amount of rainfall associated with cut-off low pressure systems over this period (Pook et al., 2006) and an absence of persistent pre-frontal troughs that aid the penetration of rain-producing cold fronts into SEA (Verdon-Kidd and Kiem, 2009b; Alexander et al., 2010). There is considerable debate about the causes of the Big Dry. For example, Verdon-Kidd and Kiem (2009a) attribute the rainfall deficits primarily to the El Niño–Southern Oscillation (ENSO) and the Southern Annular Mode (SAM), while van Dijk et al. (2013) estimate that the latter part of the drought (2001–2009) was driven by ENSO with a small contribution by the Interdecadal Pacific Oscillation (IPO). Others (e.g. Timbal et al., 2010) attribute the drought to an intensification of the Subtropical Ridge.

In 2010/11, a strong La Niña combined with warm sea surface temperatures (SSTs) off northwestern Australia (i.e. a negative Indian Ocean Dipole (IOD) event) resulted in wet austral spring/summer conditions across much of SEA, providing relief from the extended drought. A second La Niña event followed in 2011/12 which resulted in average to above-average rainfall across much of SEA from mid-spring to early autumn, and further replenished water storages.

1.2   The Big Dry in the context of other Southern Hemisphere droughts

It has been suggested that the Big Dry in SEA has similarities to the extended dry spell that began in the late 1960s/mid-1970s in southwest Western Australia (SWWA) (e.g. Hope et al., 2009; IOCI, 2002) and has continued to the present time, with a possible intensification occurring from the mid-1990s (Hope et al., 2006). Indeed, both regions exhibit a common winter-maximum rainfall regime, with peak rainfall occurring from May to October, and rainfall variability (on a synoptic to interannual scale) in the two regions is significantly related, with rainfall during May, June and July being significantly correlated (Hope et al., 2009). However, while both regions have experienced reduced cool season rainfall, the timing with respect to the start of the decreased rainfalls is not consistent. For example, in SWWA average winter (June–August) rainfall totals have decreased by approximately 20 % since the 1970s, while in Victoria the decreased rainfall trend occurred mostly from the mid-1990s, predominantly during late autumn and early winter (a decrease of approximately 20 % during the 1997–2010 period, relative to the historic average) (e.g. Murphy and Timbal, 2008; Hope et al., 2009; Verdon-Kidd and Kiem 2009a; Gallant et al., 2012). The decreased rainfalls in SWWA have been linked to a reduction in winter storm formation, with the growth rate of the leading storm track modes affecting southern Australia being more than 30 % lower during the period 1975–1994 compared to the period 1949–1968 (Frederiksen et al., 2005,

2007). The differences in timing of the decreases between SWWA and SEA make it unclear as to whether the post-mid-1970s SWWA climate shift is related to or independent of the mid-1990s to early 2010 SEA Big Dry.

The mid-1970s corresponds to a period of change in ocean–atmospheric processes of the Pacific Ocean (i.e. ENSO and the Interdecadal Pacific Oscillation (IPO)), which have resulted in more frequent droughts for much of eastern Australia (Power et al., 1999; Kiem and Franks, 2004; Verdon and Franks, 2006). There is also evidence that generalised warming across the Indian Ocean since the mid-1970s may be linked to the decreased rainfalls in SWWA (Verdon and Franks, 2005; Samuel et al., 2006). The mid-1940s and mid-1920s relate to periods of significant climatic shifts in the Pacific and Indian Ocean regions, highlighting the fact that the mid-1970s climate shift is not an isolated event (Mantua et al., 1997; Power et al., 1999; Kiem et al., 2003; Verdon and Franks, 2006). It has also been shown that inter-annual to multi-decadal variability not only affects Australia, with numerous studies identifying similar dry and wet epochs in other regions of the Southern Hemisphere. For example, a regime shift in New Zealand's climate around 1976 has also been identified (Salinger and Mullan, 1999), with the 1976–1994 period characterised by annual rainfall decreases in the north of the North Island, and increases in much of the South Island, except the east. An earlier shift in New Zealand's climate was also observed in the early 1950s, where the period 1951–1975 was characterised by increased rainfall in the north of the North Island, particularly in autumn, with rainfall decreases in the southeast of the South Island, especially in summer (Salinger and Mullan, 1999). In southern Africa, overall moist conditions were reported between 1960 and 1970 and since then a switch to drier conditions has been noted (Ngongondo, 2006). Western Africa (in the Northern Hemisphere), which regularly experiences protracted drought conditions, has also experienced an extreme drought known as "the Sahel Drought" beginning in the 1970s (Hulme, 1992), continuing to the present. Recent research by Van Ommen and Morgan (2010) reported a significant inverse correlation between the records of precipitation at Law Dome, East Antarctica and SWWA over the instrumental period. In particular, since the mid-1970s rainfall has increased at Law Dome, while rainfall has decreased in SWWA.

The apparent connection between rainfall in SWWA and East Antarctica, along with the similar timing of climate shifts in southern Africa and New Zealand, provides evidence to suggest synchronicity in climate shifts occurs in the Southern Hemisphere (a concept that has recently been analysed by Verdon-Kidd and Kiem, 2013). These observations also highlight the presence of climate phases/cycles that operate over multi-decadal timescales. Yet it is still unclear how/if the Big Dry in SEA is related to these hemispheric changes.

1.3 Managing the Big Dry and future water availability of the SEA region

Resource managers in the water and agricultural sectors across SEA have a long history of dealing with climate variability. During the early to mid-1990s, at both the national and state level, significant effort went into ensuring drought preparedness and the development of effective drought response strategies. However, the unprecedented severity and duration of the Big Dry "stress-tested" these processes. In terms of the water entitlement, and water planning and management framework in Victoria, existing drought response processes served well during the early part of the drought. However, 2006 saw the lowest flows on record across most of the state and, from 2006 to 2009, additional contingency measures were required in some situations to ensure that essential water needs were met. These measures included water carting, groundwater bores, pumping from dead storage, qualification of rights to water, and changes to water sharing arrangements and water trading rules. In addition, the Victorian Government and water corporations invested in infrastructure and in water efficiency and conservation measures to augment supplies (DSE, 2011).

Recent research conducted under the South Eastern Australian Climate Initiative indicates that, while the Big Dry has broken, rainfall deficits during the cool season (April–October), which is the traditional filling season for water storages, have tended to persist (CSIRO, 2012). This persistent rainfall deficit has been shown to be associated with changes in the global atmospheric circulation via the expansion of the Hadley circulation (estimated at 50 km per decade) and associated increase in pressure in the subtropical ridge, resulting in mid-latitude storm tracks being "pushed" further south (CSIRO, 2012; Whan et al., 2014). Cai et al. (2012) also related the observed expansion of the Hadley Cell to a poleward progression of the tropical belt and subtropical dry zone, which they claim can explain most of the southeastern Australian rainfall decline. The expansion of the Hadley Cell can only be reproduced by global climate models when human influences (in the form of greenhouse gases, aerosols and stratospheric ozone) are included, leading to the assertion that the expansion is at least partially attributed to anthropogenic climate change (CSIRO, 2012; Lucas et al., 2012). However, the exact cause of the expansion is highly debated. The width of the Hadley circulation is determined by the tropical tropopause height, pole–equator temperature difference and the global mean radiative equilibrium temperature (Hu and Fu, 2007). Each anthropogenic factor (i.e. greenhouse gases, aerosols and stratospheric ozone) impacts differently upon these dynamics – making it difficult to determine the relative importance of each. Model projections indicate that these trends are likely to continue, although the models tend to underestimate observed trends (e.g. Lu et al., 2007; CSIRO, 2012). There are therefore significant uncertainties about the timing and magnitude of associated reductions in mid-latitude rainfall. The overall implication, nevertheless, is that the traditional filling season for water supply systems across most of SEA, which historically was considered to run from about May to November, may not be as reliable in the future. Rather, replenishment of storages may in the future be more dependent on warm season rainfall events which, in turn, will primarily depend on the status of the El Niño–Southern Oscillation (ENSO), the IOD and the Southern Annular Mode (SAM) (Verdon-Kidd and Kiem, 2009b; Gallant et al., 2012). However, future changes in these key climate influences (in particular ENSO) are also not certain, and it is unclear to what extent continuing reductions in cool season rainfall are likely to be offset by higher warm season rainfalls (which also suffer from higher evaporative losses). Without this understanding it is difficult to plan for future water availability in the region. For example, do we need to change the "baseline" estimation of climate to reflect an overall "new" drier climate state?

In the light of the experiences during the Big Dry and the significant uncertainties currently associated with future climate, adaptive management principles have been built into the water management and planning framework across SEA, and water trading arrangements have been modified to assist in maintaining reliable water supplies in the face of a variable and changing climate (see for example McMahon et al., 2008; DSE, 2011; www.nccarf.edu.au/content/robust-optimization-urban-drought).

For example, in Victoria, prior to the Big Dry, the framework for maintaining an acceptable reliability of water supplies required water corporations to develop long-term strategies (aimed at balancing supply and demand over the next 50 years) and complementary short-term drought response plans (aimed at managing short-term deficits in supply) for each of the supply systems under their control. The flow scenarios considered over the 50-year time frame included the continuation of historic conditions and low, medium and high climate change flow projections out 50 years. These strategies were reviewed and updated every 5 years. As a consequence of the Big Dry, the possibility of the conditions of the drought returning in the immediate future, and persisting, is an additional scenario that is considered in long- and short-term planning processes. Given that flow reductions during the Big Dry exceeded medium to high climate change projections out to 2060, this means that a wide range of possible futures are considered. Water corporations are now also required to develop (in November every year) annual water security outlooks over a 1–5 year time frame using a range of possible future flow scenarios and, in the light of an associated risk assessment, make decisions as to whether to implement any of the short- or long-term options to reduce demand and/or augment supplies. Water trading arrangements have also been modified to allow users to take allocations that are unused at the end of a season into the following season. This provides all water users – irrigators, urban water corporations, and environmental managers – with greater flexibility to manage their own water availability. All these measures help ensure an appropriate and timely response to variable and changing climatic conditions.

1.4 Objectives

This paper aims to establish if the Big Dry that impacted SEA from the mid-1990s to 2010 is related to the 1970s climate shift experienced in many regions of the Southern Hemisphere (Sect. 3), and in particular the step change in climate observed in SWWA (Sect. 4). The synoptic processes that contribute to the differences observed in SWWA and SEA since the 1970s are explored in Sect. 5, while the role of remote large-scale drivers are analysed in Sect. 6. Implications of the findings for water resource management in the region are analysed and discussed in Sect. 7 of the paper.

2 Data

2.1 Gridded climate data

The daily NCEP/NCAR (National Centres for Environmental Prediction and the National Centre for Atmospheric Research) Reanalysis global gridded data sets (available 1948–present), from the US National Oceanic and Atmospheric Administration (NOAA, www.esrl.noaa.gov/psd/), are used to identify various synoptic patterns that influence the rainfall of SEA and to develop an understanding of Southern Hemisphere climatology during different climate epochs. The NCEP/NCAR Reanalysis data is derived from a global spectral model with a grid resolution of 2.5 degree latitude ×2.5 degree longitude global (144 × 73 grids). As with all reanalysis data, this data has various limitations, particularly in the Southern Hemisphere where historical recorded data tends to be sparse. Trenberth and Guillemot (1998) provide a review of limitations associated with the NCEP/NCAR reanalysis data, in particular the uncertainties inherent in the atmospheric moisture representation.

Gridded monthly precipitation data for Australia was obtained from the Australian Water Availability Project (AWAP), a joint initiative of the Bureau of Meteorology (BoM) and the Commonwealth Scientific and Industrial Research Organisation (CSIRO). In the daily/monthly AWAP data set the observed daily/monthly rainfall from gauges within the BoM gauging network (i.e. up to approximately 7500 gauges, both open and closed) are decomposed into a monthly average and associated anomaly (Jones et al., 2009). The daily/monthly anomalies are interpolated using the Barnes successive correction technique, and the monthly climatological averages are interpolated using three dimensional smoothing splines (Jones et al., 2009). The rainfall grids are produced by multiplying the monthly climate average grids and daily/monthly anomaly grids. An unexplained microscale variance term is used in AWAP to allow for observational or measurement error, such that exact reproduc-

Figure 1. Location of **(a)** monthly rainfall stations (red circles) and **(b)** eight high-quality daily rainfall stations in SEA (blue diamonds). Note monthly station data varies in length with all stations containing data from at least 1920 to 2009 (with the exception of two stations from Argentina that include data from 1931 onwards, which were included in order to improve the spatial coverage of the east coast of South America).

tion of gauged values at each gauge location is not expected (Jones et al., 2009). AWAP rainfall grids are freely available from 1900 onwards at http://www.bom.gov.au/jsp/awap/.

2.2 Station-based rainfall data

The dry conditions during the Big Dry were largely confined to regions south of 30 degrees (Verdon-Kidd and Kiem, 2009a). Therefore, in order to provide a spatial/geographical context to the recent dry conditions in SEA, and to ensure limitations of gridded rainfall data sets are not skewing our results (see Tozer et al., 2012), station-based rainfall data from regions south of 30 degrees (shown in Fig. 1a) from the following data sets was also used:

– *Australia* – Monthly station-based rainfall data was obtained from the Australian Bureau of Meteorology (www.bom.gov.au/climate/data/weather-data.shtml). Data chosen for analysis was at least 95 % complete and covered the period 1900–2009. Daily rainfall data was also obtained for eight stations located in SEA (see Fig. 1b) in order to further investigate the daily characteristics of the monthly rainfall totals (i.e. number of rain days, daily rainfall magnitude etc.). The stations

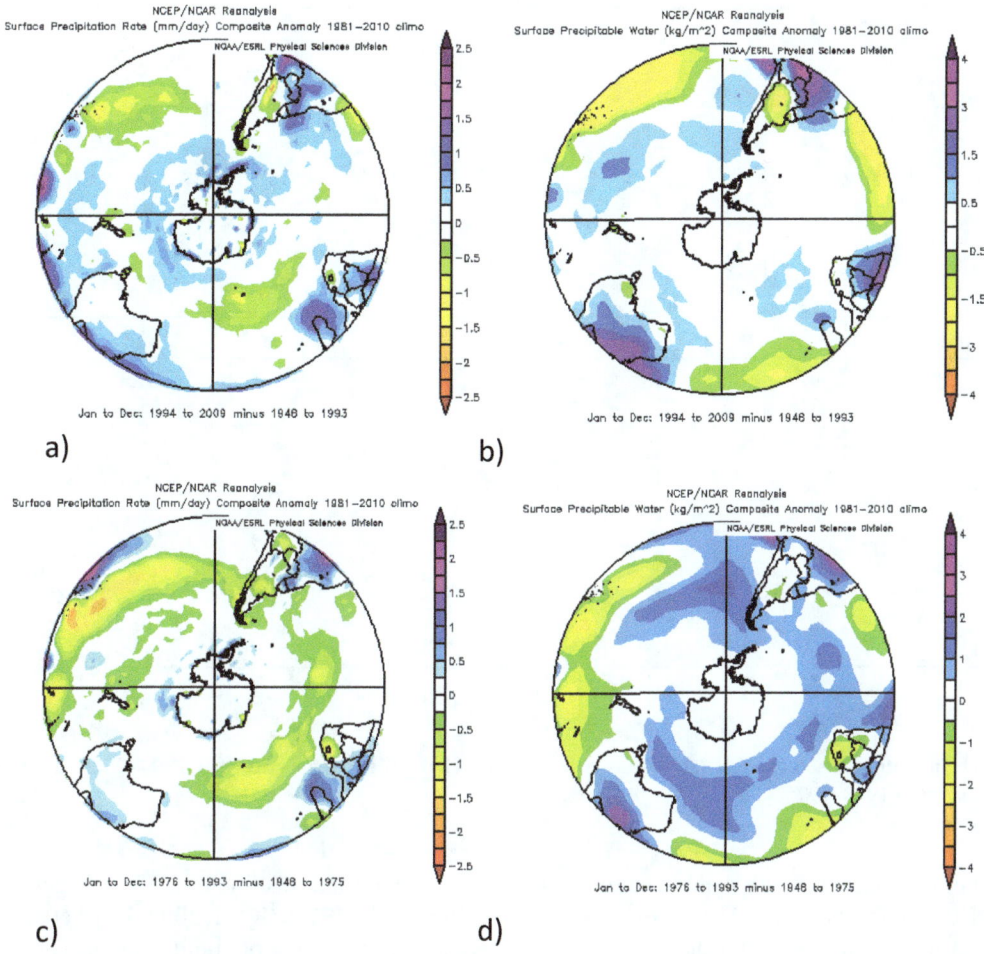

Figure 2. Difference in annual (**a**) surface precipitation rate (1994–2009 compared to 1948–1993), (**b**) precipitable water (1994–2009 compared to 1948–1993), (**c**) surface precipitation rate (1976–1993 compared to 1948–1975) and (**d**) precipitable water (1976–1993 compared to 1948–1975).

chosen for analysis have sufficiently long records (i.e. at least 100 years) and minimal missing data (at least 99 % complete).

– *New Zealand* – Daily station data was obtained from the National Climate Database of the National Institute of Water and Atmospheric Research (NIWA, http://cliflo.niwa.co.nz/). Data chosen for analysis was at least 95 % complete and covered the period 1920–present. An additional station was chosen for analysis (located at Reefton on the west coast of New Zealand) that contained daily data from 1948 onwards in order to sufficiently represent the west coast of New Zealand, since the climatology there differs significantly from the east coast. Daily data was aggregated to monthly data prior to analysis.

– *South America and southern Africa* – Monthly rainfall data for South America and southern Africa was obtained from the Global Historical Climatol-

ogy Network (GHCN, www.ncdc.noaa.gov/oa/climate/ghcn-monthly/index.php). The historical GHCN data has previously undergone rigorous quality assurance review (including pre-processing checks on source data, time series checks that identify spurious changes in the mean and variance, spatial comparisons that verify the accuracy of the climatological mean and the seasonal cycle, and neighbour checks that identify outliers from both a serial and a spatial perspective). GHCN-Monthly is used operationally by NCDC to monitor long-term trends in temperature and precipitation. It has also been employed in several international climate assessments, including the Arctic Climate Impact Assessment, and the "State of the Climate" report published annually by the Bulletin of the American Meteorological Society.

Despite the level of quality assurance, the rainfall data available through the GHCN for South America and southern Africa is not as complete as data from Australia and New Zealand, with the more recent data being particularly sparse,

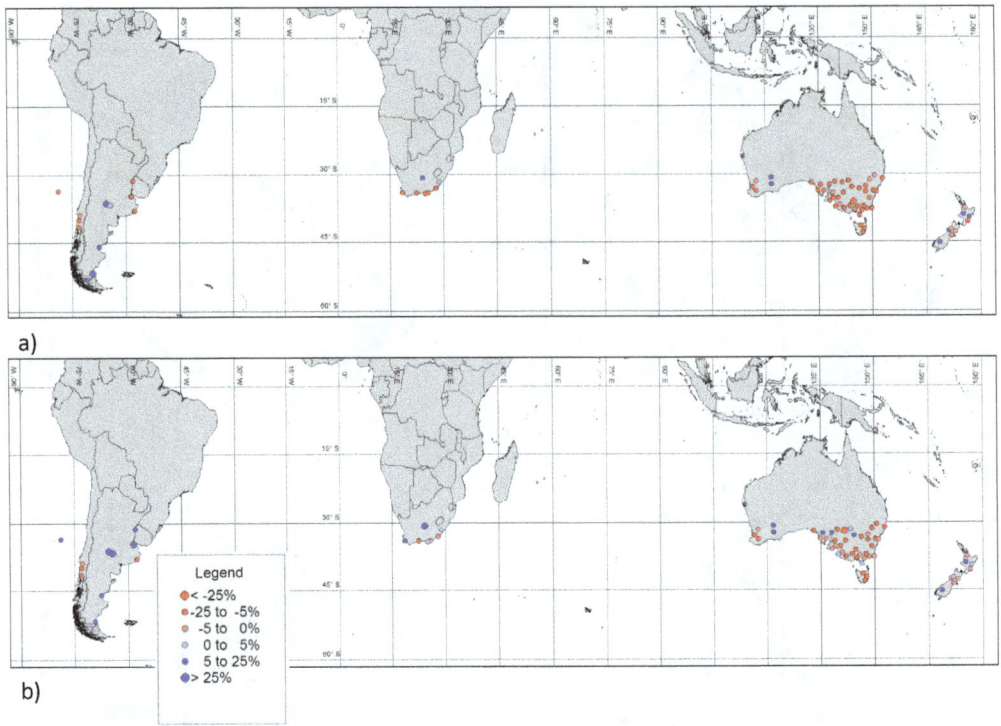

Figure 3. Difference in station-based rainfall (shown as a percentage change) for the period **(a)** 1994–2009 compared to 1948–1993, **(b)** 1976–1993 compared to 1948–1975.

especially in southern Africa and South America. Given that the main focus of this study is on the 15 years up to 2009 (i.e. period covered by the Big Dry) and that only stations with complete data over this period were included in our analysis, the spatial coverage of the station-based rainfall data is poor in some regions (see Fig. 1).

3 Setting the context of recent dry conditions in SEA in terms of Southern Hemisphere variability

While it is well known that SEA experienced a prolonged drought from the mid-1990s to early 2010 (e.g. NWC, 2006; Murphy and Timbal, 2008; Verdon-Kidd and Kiem, 2009a, b; Kiem and Verdon-Kidd, 2010; Gallant et al., 2012), it is unclear if other regions of similar latitude also experienced a step change in climate (either too dry or wet) at this time. The NCEP/NCAR Reanalysis data sets described in Sect. 2.1 were used to study this as a first step towards identifying possible climate linkages in the Southern Hemisphere during this period. Figure 2a and b show, for the Southern Hemisphere, the annual difference in surface precipitation rate and precipitable water, respectively, during the period 1994–2009 compared to the period 1948–1993 (note that 1994 was used as a starting year for the drought as per the findings of Kiem and Verdon-Kidd (2010); 1948 was used as the start of the analysis period, as this is restricted by the NCEP/NCAR data set). As discussed in the Introduction, it is well known

that the rainfall decline in SWWA began as early as the late 1960s/mid-1970s (IOCI, 2002; Hope et al., 2009), which corresponds to a time of significant changes in the climate system in the Southern Hemisphere. Therefore, it is possible that any drying trends observed from 1994–2009 are part of a continuation from an earlier step change in climate. To examine this, the precipitation rate and precipitable water analysis was also repeated for the period 1976–1993 (i.e. the period prior to the mid-1990s step change in SEA) as shown in Fig. 2c and d, respectively. The year 1976 was chosen as the start year of this analysis, as this corresponds to a statistically significant change in ocean/atmosphere processes and a reorganisation of the climate system (Mantua et al., 1997).

Tozer et al. (2012), and other references reviewed within, demonstrate that gridded rainfall data sets often do not capture extreme events particularly well (due to smoothing during the interpolation process, inadequate representation of topographical effects, uncertainties associated with remotely sensed data, density of available station data that varies in space and time, etc.). Further, as mentioned in Sect. 2.2, rainfall data observations for the Southern Hemisphere tend to be sparser than the Northern Hemisphere, particularly for regions such as South America, southern Africa and Antarctica. Therefore, in order to ground truth the observations made using the gridded data set, the station-based monthly rainfall data (described in Sect. 2.2) was also used to compare rainfall during the period 1994–2009 with the period

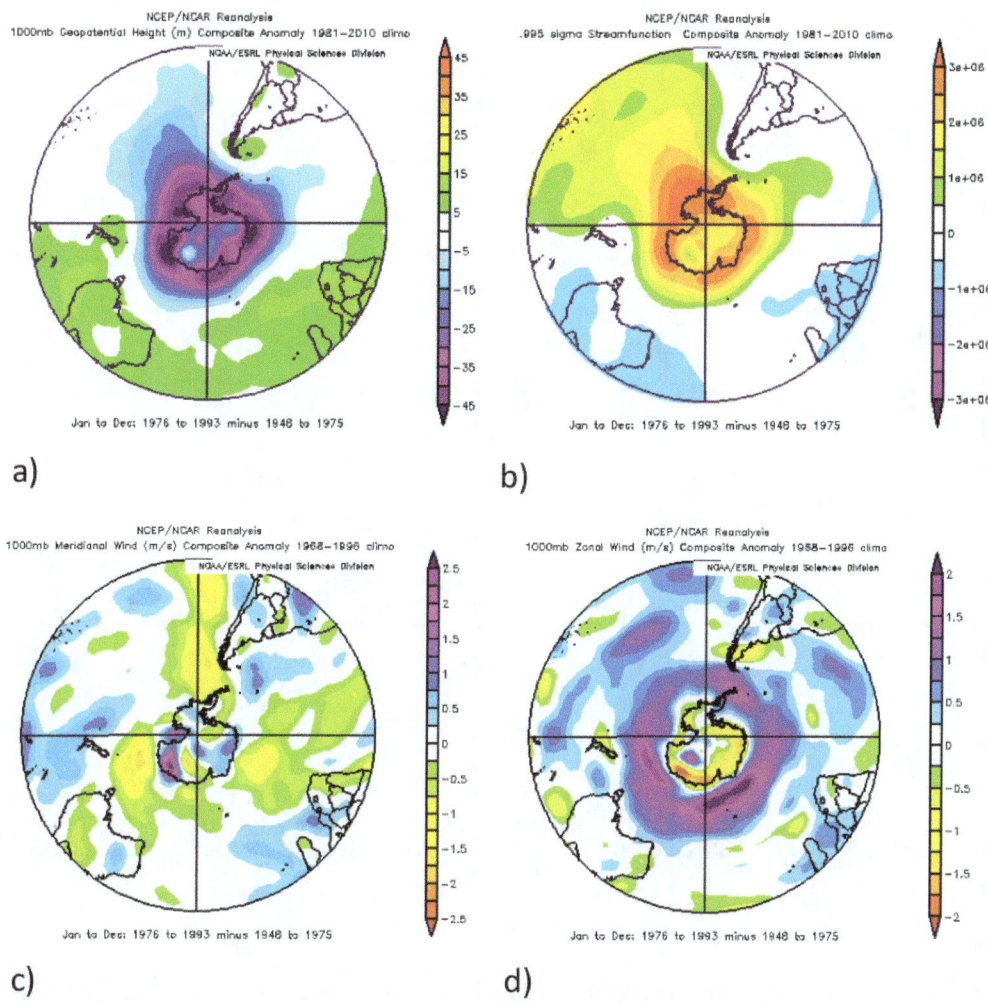

Figure 4. Difference in annual (**a**) geopotential height, (**b**) streamfunction, (**c**) meridional and (**d**) zonal wind anomalies during the period 1976–1993 compared to 1948–1975.

1948–1993 (the period which the NCEP/NCAR data covers) and the period 1976–1993 to 1948–1975 (to investigate the possibility of an earlier step change). The results are shown in Fig. 3.

Figure 2a and b show that, for the period 1994–2009, there was a clear increase in the surface precipitation rate and precipitable water relative to 1948–1993 in the northwest of Australia, the northeast and northwest of South America and the northwest of southern Africa. Figure 2a and b also show a corresponding decrease in precipitation during this period in the southwest and southeast of Australia, New Zealand and central South America. Further, the station-based data (Fig. 3a) support the findings above, providing multiple lines of evidence (from three data sets) that, since the mid-1990s, there appears to be a southward migration of rain-bearing systems in the Southern Hemisphere, resulting in a dry band across approximately 30–45° S (consistent with the findings of CSIRO, 2012).

The analysis presented here also suggests that the trends in precipitation across the Southern Hemisphere observed in the post-1994 analysis are likely to be part of a longer-term trend beginning around the mid-1970s, given the similarity in the anomaly patterns (Figs. 2c, 2d and 3b) between the two time slices (for regions other than SEA). Figure 4a clearly shows a reduction in geopotential height since the mid-1970s over Antarctica (indicating an increase in conditions conducive to storms) and a corresponding increase in geopotential height over Australia, southern Africa and New Zealand (associated with clear weather). Similarly, Fig. 4b shows an intensification of streamfunction over Antarctica and the southern oceans and a decrease over Australia and southern Africa, further evidence of a southward migration of storm tracks over this period. Figure 4b and c confirm that there has also been a corresponding decrease in meridional winds south of 30 degrees and an increase in zonal winds since the mid-1970s. These observations are consistent with the findings

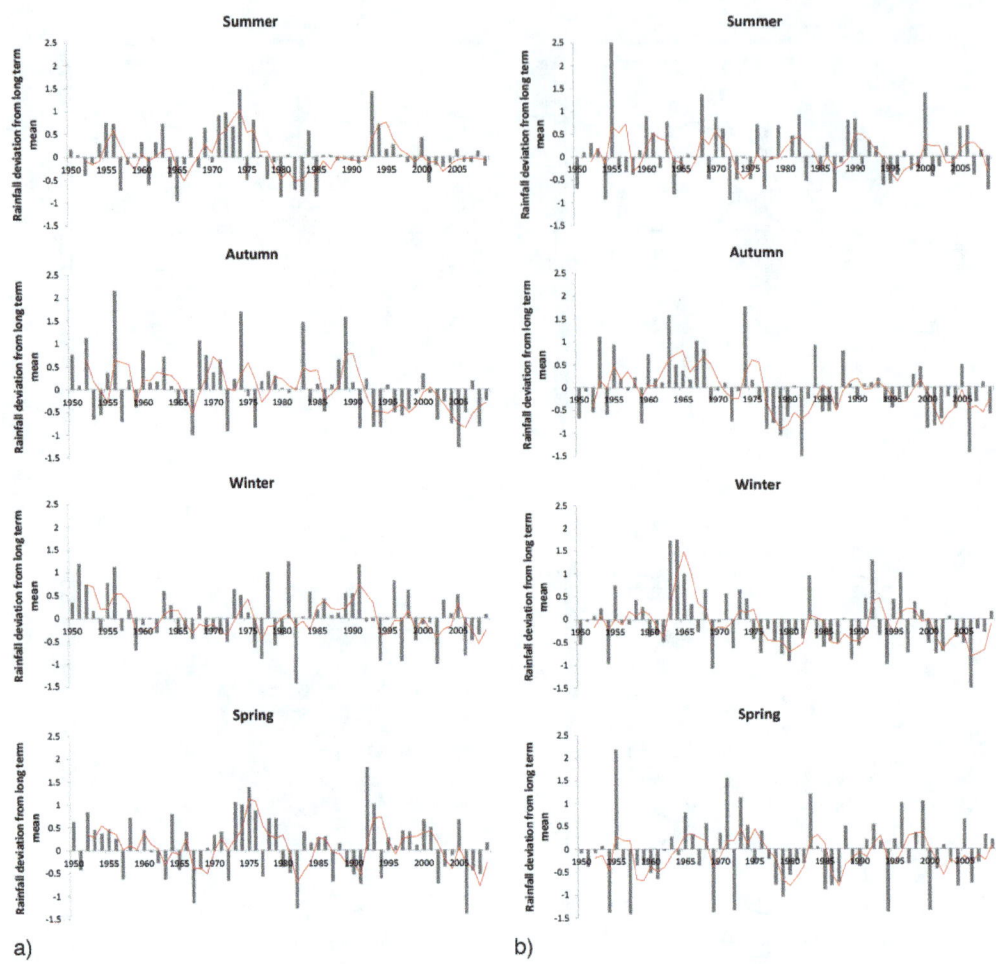

Figure 5. Rainfall deviation from the long-term mean (1900–2009) for **(a)** all SEA (left panel) and **(b)** all SWWA stations (right panel) combined. Note anomalies displayed from 1950 onwards.

of Frederickson et al. (2005, 2007) regarding a reduction in winter storm formation in the three decades from 1969.

4 Comparison of rainfall anomalies in SEA and SWWA since the 1970s climate shift

A comparative analysis between SEA and SWWA rainfall was carried out using the station-based rainfall data as described in Sect. 2.2. Normalised rainfall anomalies for the period 1900–2009 were combined and the time series plotted for all SEA stations and all SWWA stations for each season (Fig. 5).

Figure 5 shows that, following below-average summer rainfall and average autumn rainfall in the late 1970s in SEA, a series of eight consecutive back-to-back wetter than average winters occurred between 1984 and 1991 (actually 1983 is also slightly wetter than average, but is not considered as "wet" in this analysis), along with a series of wetter than average autumns in 1988, 1989 and 1990. Therefore, while the dominant mid-latitude trend had been towards dry since

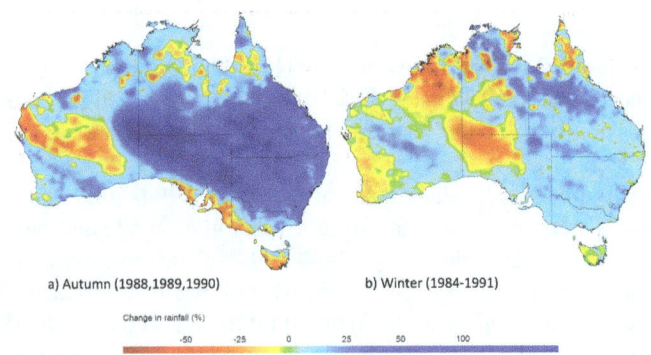

Figure 6. Difference in **(a)** autumn and **(b)** winter rainfall during "wet seasons" compared to the long-term mean (1900–2009).

the mid-1970s, other factors acted to offset the trend for SEA during these wet autumn/winter seasons. Importantly, in terms of winter rainfall, the series of eight back-to-back wet events is unprecedented in the 100-year instrumental record. It can also be seen from Fig. 5 that the mid-1980s to early

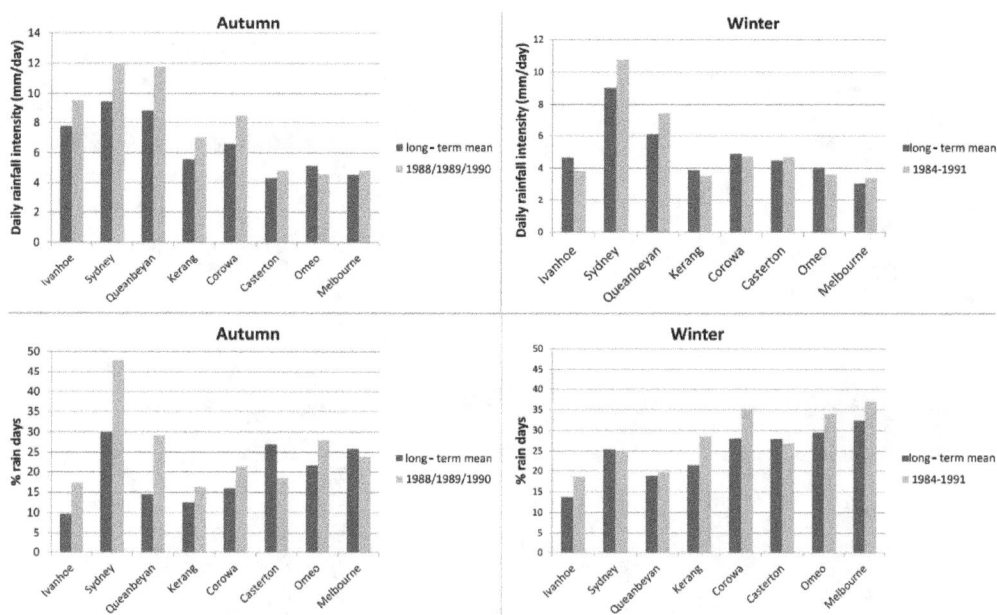

Figure 7. Daily rainfall intensity and percentage of rain days for the eight selected high-quality rainfall stations in SEA during wet years compared to long-term mean (1900–2009).

1990s autumn and winter rainfall relief did not occur simultaneously in SWWA; rather these seasons have consistently been dry since the mid-1970s for this region. The results presented here also indicate that, based on the station data used in our analysis, the cool season drying trend in SWWA occurred starting from the mid-1970s (rather than the late 1960s as reported by Hope et al., 2009).

The spatial nature of these wetter than average autumn and winter conditions is analysed in Fig. 6. In this figure "wet seasons" are defined as 1988–1990 for autumn and 1984–1991 for winter (based on the analysis presented above).

Figure 6a shows that the enhanced autumn rainfall during the period 1988–1990 occurred in a broad northwest to southeast band across the continent. Not all SEA regions experienced this elevated autumn rainfall; however, as shown in Fig. 6a, southern regions of South Australia, Victoria and Tasmania actually experienced a decrease in rainfall during this time. Figure 6b shows that the elevated winter rainfall during the 1984–1991 period penetrated all of SEA (note SWWA did not receive this winter relief and hence the overall drying trend initiated in the mid-1970s continued without interruption in this region).

Thus far the rainfall has been analysed in terms of seasonal totals, however, changes in daily rainfall statistics (i.e. changes to the frequency, intensity, duration and/or sequencing of rainfall events) have important hydrological implications, as this will influence soil moisture and, in turn, the runoff generated. Therefore, daily characteristics of the seasonal rainfall during the elevated rainfall seasons identified above were analysed using the station-based daily rainfall data (refer to Sect. 2.2 and Fig. 1b). Figure 7 shows the daily

rainfall intensity and percentage of rain days (defined as any day with rainfall greater than 1 mm) for the eight daily rain stations during autumn (1988/1989/1990) and winter (1984–1991) compared to the long-term mean.

As shown in Fig. 7, the autumns of 1988–1990 were associated with both an increase in rainfall intensity and number of rain days for those stations located in the region where seasonal rainfall totals were shown to be elevated in Fig. 6. The stations in the region where autumn rainfall decreases were experienced (Casterton and Melbourne) either show no change, or a decrease in rainfall intensity and/or rain days during autumn. Overall, there is no evidence of increased rainfall intensity during the winters of 1984–1991 (with the exception of Queanbeyan and Sydney). The elevation in winter rainfall during this period is thus primarily a result of an increase in the number of rain days (i.e. the increase in rainfall appears to be due to the fact that it rained more often rather than due to an increase in daily intensity).

5 Synoptic processes driving increases in autumn and winter rainfall during the mid-1980s to the early 1990s in SEA

5.1 Identification of synoptic types using self-organising mapping

Given that the majority of enhanced autumn/winter rainfall is due to increased rain days (with some evidence of elevated rainfall intensity for autumn), persistent climate systems are most likely responsible for the elevated rainfall occurrence. Therefore, we have used a process to identify daily

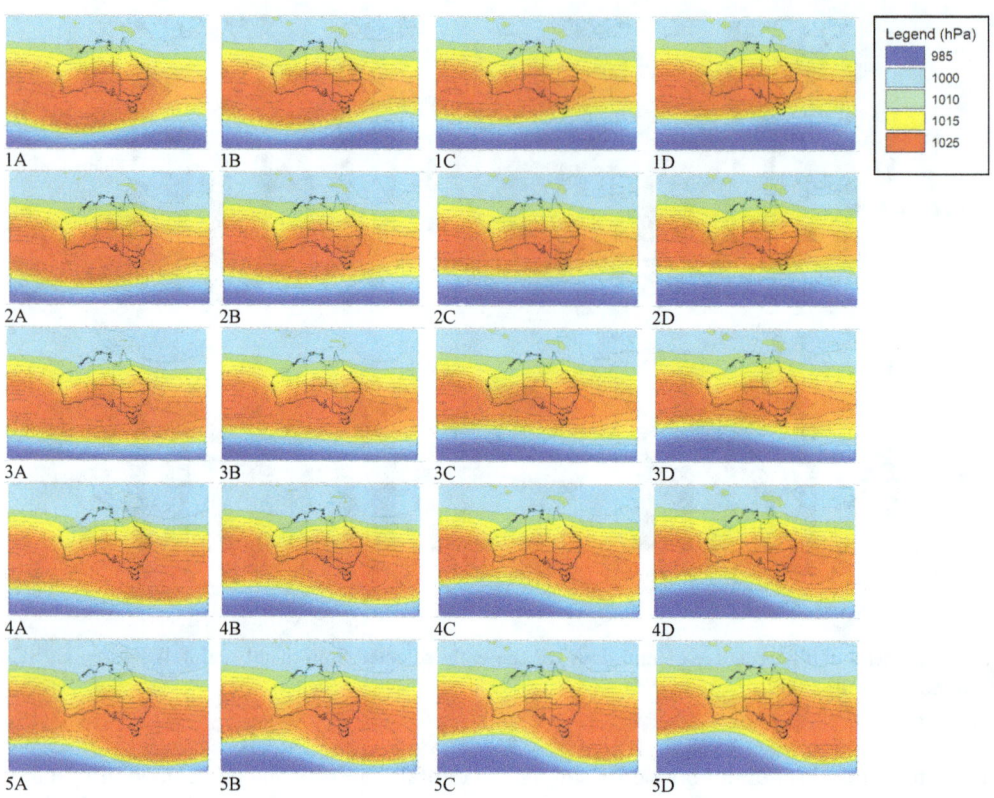

Figure 8. The 20 autumn synoptic types identified using SOM based on data from 1948–2009.

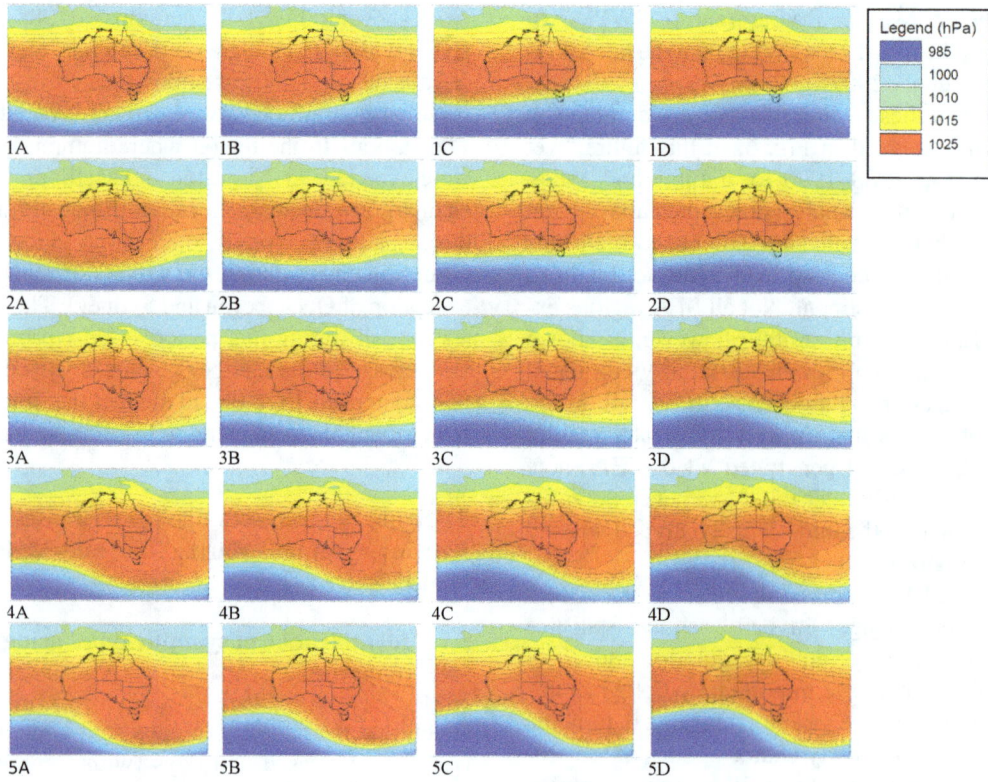

Figure 9. The 20 winter synoptic types identified using SOM based on data from 1948–2009.

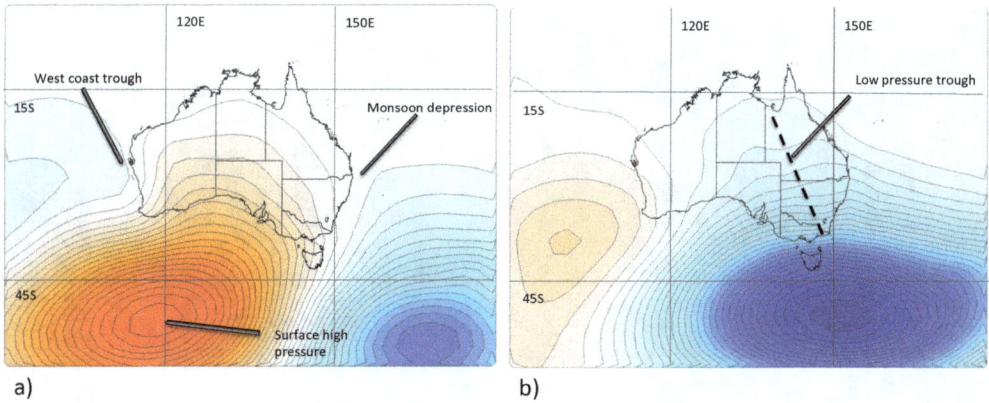

Figure 10. (**a**) Synoptic type 1A expressed as an anomaly from the autumn mean. (**b**) Synoptic type 1D expressed as an anomaly from the winter mean. Note blue (red) indicates lower (higher) than average pressure.

synoptic systems that occurred during the period of elevated autumn/winter rainfall which can then be compared to the long-term climatology. The method by which the synoptic systems have been identified is known as a self-organising map (SOM). An SOM is a non-linear neural network classification technique developed to recognise relevant structures in complex, high-dimensional data via an unsupervised learning and self-adaptation process (Cavazos et al., 2002). SOMs have been described as less complex, more robust and less subjective than more traditional techniques, including cluster analysis and principal component analysis, which are commonly used to identify synoptic patterns (Hewitson and Crane, 2002). SOMs are essentially a mapping of many vectors onto a two-dimensional array of representative nodes (in this case synoptic types) via an unsupervised learning algorithm. The SOM methodology has been shown to be successful in identifying key regional synoptic patterns that drive local climate in other regions of the world (e.g. Cavazos, 2000; Cavazos et al., 2002; Hewitson and Crane, 2002; Hope et al., 2006; Reusch et al., 2007; Verdon-Kidd and Kiem, 2009b). Importantly, the SOM methodology is less subjective than other forms of pattern recognition, and the non-linear approach lends itself to regions where local climate is constantly changing due to large-scale climate variability.

Daily global sea level pressure (SLP) data for the years 1948–2009 was used to develop the SOM (see discussion of the NCEP/NCAR Reanalysis data in Sect. 2.1). This data set has been widely used in similar studies (e.g. Cavazos, 2000; Cavazos et al., 2002; Hope et al., 2006) and is considered to be the best SLP data available for the study region and type of analysis (see Hope et al., 2006 for a detailed discussion). In order to study the regional-scale synoptic systems that are important for SEA, a subset of the global SLP data was extracted to carry out the SOM. This region was chosen so as to capture the synoptic patterns that influence autumn and winter rainfall in SEA. Two separate SOMs were generated – one for autumn (MAM) and another for winter (JJA). The

location of the SLP field used in this analysis is 90–180° E, 0–60° S.

The size of the SOM array directly influences the range of synoptic patterns represented. A number of array sizes were trialled in order to determine the optimum number of synoptic patterns. It was determined that a 3×4 SOM (i.e. 12 types) was not large enough to adequately identify the subtle differences between types that are likely to be important in generating rainfall; however these subtleties were found to be captured by a 4×5 SOM (i.e. 20 types). Larger array sizes (e.g. a 5×6 SOM) resulted in further refinement of the transitionary synoptic types (resulting in very discrete differences between types), yet did not alter the extreme types. In addition, there was no improvement in the mean error per sample (calculated as the average Euclidian distance between the input vector and the synoptic types it best matches) by increasing the size of the SOM array beyond 20 types. Given these findings, a 4×5 SOM was chosen for the synoptic typing performed in this study – this array size satisfactorily captures extreme types and a range of synoptic patterns with sufficient differences observed between types.

Twenty synoptic types (using a 4×5 grid) were generated using the daily SLP data, as shown in Fig. 8 (autumn) and Fig. 9 (winter). By virtue of the method, similar types are clustered together in the SOM, with the most dissimilar types located at the far corners of the SOM map.

A range of synoptic systems for both seasons were identified using the SOM methodology. In general, the top half of the SOM represents systems with a westerly flow (since winds move in the direction of high to low pressure), while the bottom half of the SOM represents easterly flow. Systems with a westerly flow are more likely to bring rain into eastern Australia (particularly west of the Great Dividing Range); however, coastal stations may benefit from easterly systems if they are laden with moist air (for example, a "black noreaster"). The westerly systems are more likely to deliver lower intensity, longer duration events, while the easterly

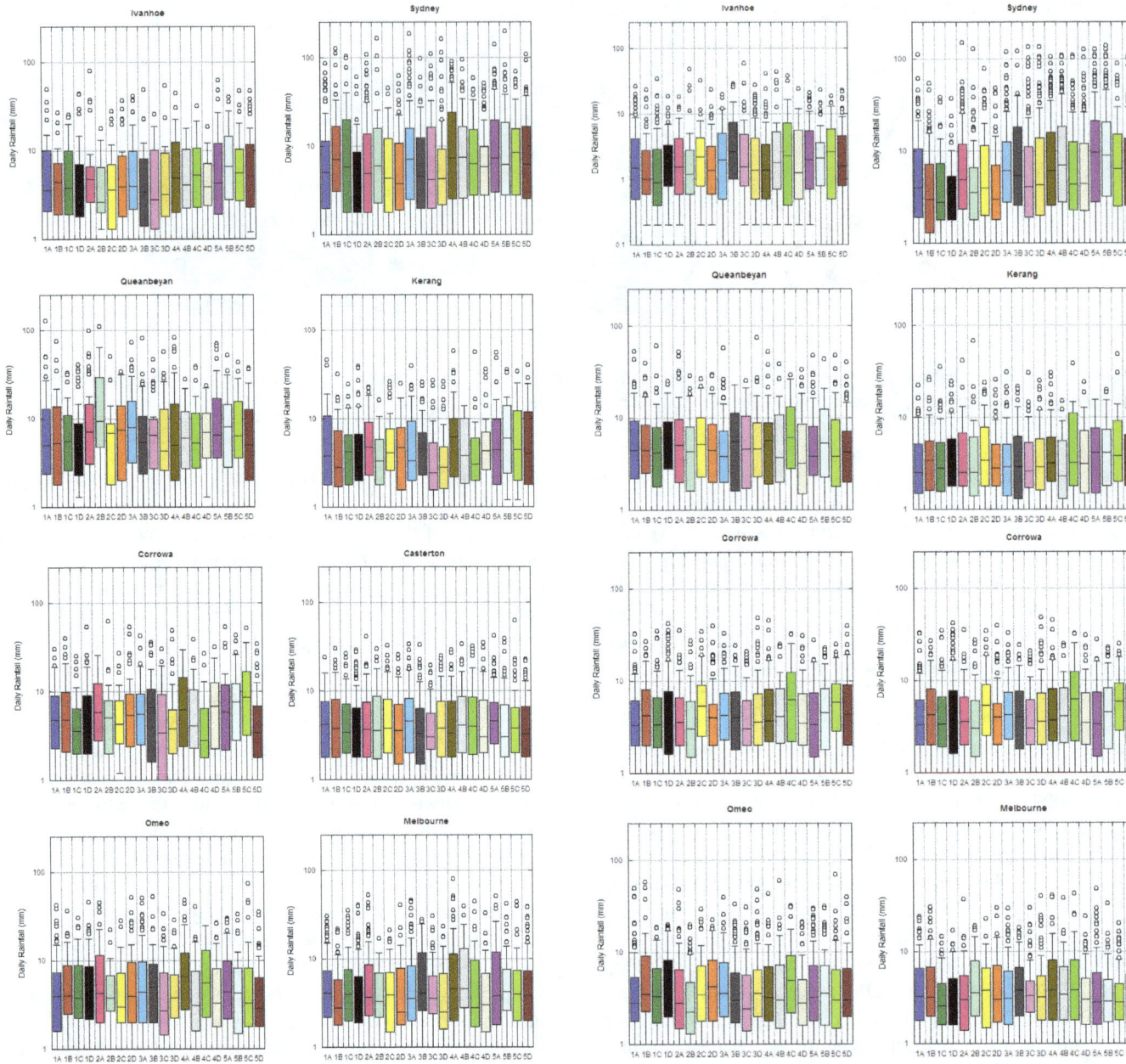

Figure 11. Daily rainfall distributions associated with each autumn synoptic type.

Figure 12. Daily rainfall distributions associated with each winter synoptic type.

systems are more likely to deliver rainfall in shorter, more intense storm events (from cut-off lows, east coast lows, etc.). The types located along Row 3 of the SOM (Figs. 8 and 9) show a southward retraction of the Subtropical Ridge. This is likely to result in reduced rainfall across SEA (a more summer-like pattern) due to the surface high pressure being located over southern Australia (Sturman and Tapper, 2004).

Autumn synoptic types located in the top left-hand corner (1A, 1B, 2A) of Fig. 8 show a monsoon depression over eastern Australia, as seen by the strong "dip" in the isobars over eastern Australia (note also the low pressure trough over

western Australia, known as the "west coast trough" that occurs at this time of year and results in dry conditions for SWWA). To aid in interpretation of the SOM results, these features are highlighted in Fig. 10a, which displays synoptic type 1A as a seasonal anomaly. Monsoon depressions often have their origin in the northwest of the continent and move southeast, bringing warm moist air that often results in prolonged rainfall (particularly west of the Great Dividing Range). This type of system was responsible for the breaking of the severe 1982/83 drought (Tapper and Hurry, 1996), and more recently was the primary weather system that resulted

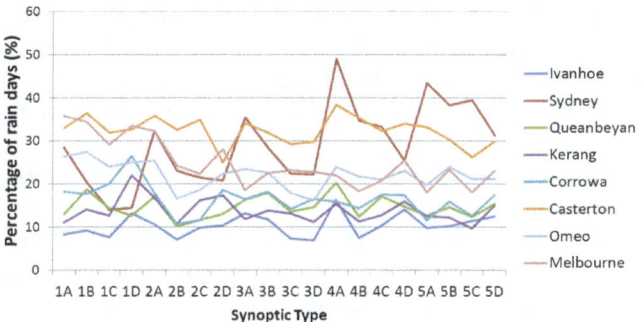

Figure 13. Percentage of rain days associated with each autumn synoptic type.

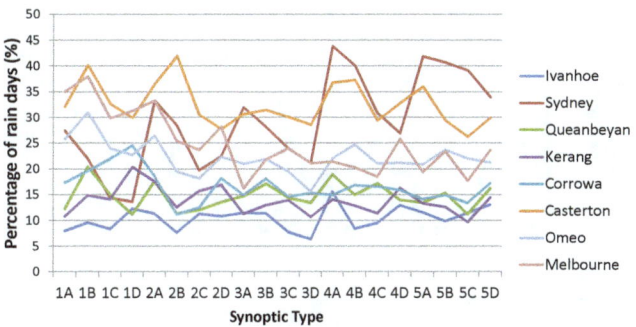

Figure 14. Percentage of rain days associated with each winter synoptic type.

in elevated rainfalls in 2010/11. The strong surface high pressure system located to the south of Australia in the "Bight" (also highlighted in Fig. 10a) is likely to block the propagation of the rain-bearing systems into the southern regions, and can even lead to a "cold outbreak" in southern Victoria and South Australia (Tapper and Hurry, 1996). Occasionally, east coast lows can also form out of these monsoon depressions, resulting in heavy recorded rainfall for east coast stations.

Similar synoptic patterns to those observed in autumn were identified for winter. However, some key differences were observed. In particular, as shown in Fig. 9, winter types 1C, 1D and 2D display a low pressure trough stretching from the northwest through to the southeast of the continent (note this feature is also highlighted as a seasonal anomaly in Fig. 10b to aid interpretation). The low pressure trough represents conditions suitable for cloud band development, a system common in winter and spring and less common in autumn. The "northwest cloud bands", as they are known, often result in substantial rainfall across SEA in the winter and spring seasons. Similar to low pressure trough systems, the type of rainfall associated with northwest cloud bands tends to be of a lower intensity and longer duration than isolated low pressure systems (such as east coast lows, Tasman lows, etc.).

5.2 Rainfall statistics associated with the 20 autumn and winter synoptic types

The daily rainfall associated with each autumn and winter synoptic type is shown in Fig. 11 (autumn) and Fig. 12 (winter). Note that days with zero rain were removed in this analysis in order to demonstrate how much rain is associated with each type when that type results in rainfall. The percentage of rain days associated with each synoptic type is shown in Fig. 13 (autumn) and Fig. 14 (winter).

From Figs. 11 and 12 it is clear that daily rainfall distributions vary markedly for different synoptic types at the same site (and for the same synoptic type across the different sites). For example, the most intense rainfall in autumn is associ-

ated with type 5B at Ivanhoe and Kerang, 4A at Sydney and Omeo, 2B at Queanbeyan, 5C at Corowa, 3A/4C at Casterton and 4B at Melbourne. These types are clustered in the bottom left corner of the SOM in Fig. 8 (strong easterly flow).

There is also a clear trend in the number of rain days associated with each synoptic type, as demonstrated by Figs. 13 and 14. For those stations located in the northern part of the study region (e.g. Sydney and Ivanhoe) there is a general increase in rain days in both autumn and winter for those synoptic types located in the bottom half of the SOM, while for stations located in the south of the study region (e.g. Corowa, Casterton, Omeo and Melbourne) the opposite is true (i.e. a greater number rain days occur for synoptic types located in the top half of the SOM). This indicates that the more southern stations benefit most from westerly flow, while those located further north benefit from easterly systems. The four stations located in the south of the study region are within the "Victorian" region as defined by Timbal et al. (2010) which, according to the authors, is most strongly influenced by local mean sea level pressure in autumn and winter, with the Southern Annular Mode (i.e. westerly flow) also playing a significant role in winter. The stations in our study that are located further north correspond to the regions defined as "Central" and "Eastern" by Timbal et al. (2010) which, according to the authors, are most strongly influenced by the Pacific driver (i.e. ENSO) which affects easterly systems.

5.3 Relative occurrence of synoptic types during the wetter than average autumns of 1988–1990 and winters of 1984–1991

Next we investigate how frequently each synoptic type occurred during the wetter than average autumns (1988–1990) and winters (1984–1991) in order to determine which (if any) types were responsible for the elevated rainfall during these seasons. The relative percentage of each synoptic type occurring during the wetter than average autumns of 1988–1990 and winters of 1984–1991 are shown in Fig. 15 compared to the long-term climatology (based on the period 1948–2009).

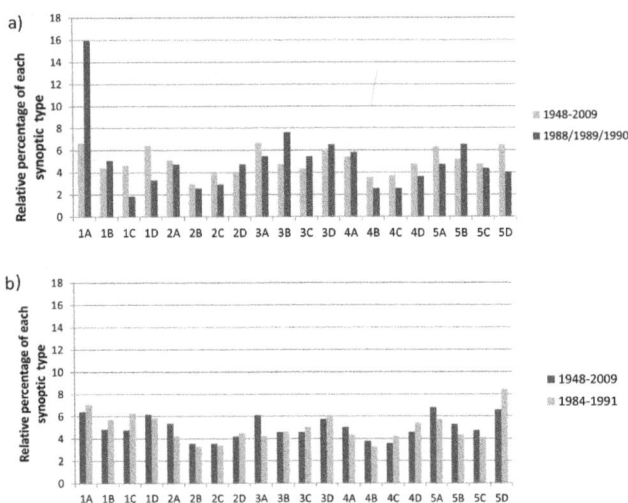

Figure 15. Relative percentage (frequency of occurrence) for each synoptic type during (**a**) autumn 1988–1990 and autumn 1948–2009, (**b**) winter 1984–1991 and winter 1948–2009.

Figure 15a shows a large increase in type 1A during the wet autumns of 1988–1990 (result statistically significant at < 10 % (p value 0.07) using a Student's t test), representing a monsoon depression over eastern Australia. As noted previously, this system often results in substantial rainfall across SEA (however, the presence of a west coast trough is associated with dry conditions in SWWA). There is also a general increase in synoptic types representing westerly flow (as opposed to easterly flow).

Based on Fig. 15b there appears to be two separate weather systems that were more common than usual during the wet winter period of 1984–1991. Increases in synoptic types located along the first row of the SOM indicate that, as for autumn, monsoon trough systems over the east coast were more frequent, along with cloud band development. These systems are also associated with strong westerly flow. In addition, an increase in types located in the bottom right corner of the SOM (representing systems with a very strong easterly flow) would have resulted in rainfall being generated from the Pacific in the form of cut-off lows, which are closed low pressure systems that have become completely displaced from the basic westerly current flowing across Australia's southern oceans (Tapper and Hurry, 1996). While these increases were not found to be statistically significant using a Student's t test, a small increase in "wet" synoptic types may have a larger impact on the seasonal rainfall totals.

An assumption that has been made thus far is that the likelihood of rainfall for a given synoptic type is stationary and consistent in time. In order to analyse the validity of this assumption, the percentage change in the likelihood of rainfall occurrence (i.e. a day with rain > 1 mm) for a given synoptic type during the wet autumns of 1988/1989/1990 compared to the long-term climatology was calculated (Table 1). Similarly, results for winter (1984–1991 compared to the long-term mean) are shown in Table 2. Note that a small deviation is not unexpected given the reduced sample size of daily synoptic types during the "wet" seasons compared to the long-term climatology.

Table 2 shows that the likelihood of rain for a given synoptic type during winter is fairly consistent for the time periods being analysed. However, in the autumns of 1988–1990 there does appear to be a trend towards more rain days for a given synoptic type (up to 50 % more in some cases), as shown in Table 1. An initial investigation into climate variables other than sea level pressure has highlighted some interesting findings that may help explain why some stations received so much rain during this period. For example, Fig. 16 shows precipitable water anomalies, SST anomalies, scalar wind speed and outgoing longwave radiation, during the wet autumns of 1988–1990.

Figure 16a shows that a pool of precipitable water was centred over southern Queensland, New South Wales and northern Victoria during the autumns of 1988–1990, which is likely to have increased the chance of rain for a given synoptic system. Another reason for an increased chance of rain in autumn (for Sydney in particular) may be due to warmer than average SSTs (shown in Fig. 16b) off the west coast of Australia feeding moisture to the westerly winds. Thirdly, the outgoing longwave radiation (Fig. 16c) is particularly low along eastern Queensland and New South Wales during this time. Low longwave radiation is typically associated with increased storm activity and subsequent rainfall. Therefore, both the moisture source and the atmospheric processes that actually deliver the rainfall were available during this time.

While some preliminary insights have been gained, the reason why Sydney received so much rainfall during the autumns of 1988–1990 requires further investigation. Indeed it is likely that local climate phenomena not studied here may have played a role (i.e. phenomena that are too small to be captured based on the synoptic types using NCEP/NCAR grid resolution).

6 Role of large-scale climate phenomena in driving enhanced autumn and winter rainfall during the mid-1980s to the early 1990s in SEA

Numerous studies have demonstrated that the four most influential climate modes on SEA's climate are

– El Niño–Southern Oscillation (ENSO; Chiew et al., 1998; Kiem and Franks, 2001; Verdon et al., 2004) – ENSO is represented by the Oceanic Niño Index (ONI) from the United States National Oceanic and Atmospheric Administration (NOAA) Climate Prediction Centre (CPC) (www.cpc.ncep.noaa.gov). The ONI is a 3-month running mean of ERSST.v3b SST anomalies in the Niño 3.4 region, centred on 30-year base periods updated every 5 years. For historical purposes, cold (La

Table 1. Percentage change in the likelihood of a rain day (greater than 1 mm) for a given synoptic type during the autumns of 1988–1990 compared to the 1948–2009 period (substantial deviations from the long-term mean (> 20 % or < −20 %) are highlighted in italic).

Type	Ivanhoe	Sydney	Queanbeyan	Kerang	Corowa	Casterton	Omeo	Melbourne
1A	5	10	8	3	7	−10	3	−6
1B	19	8	10	0	−11	−8	1	8
1C	*52*	*46*	*46*	7	*40*	8	16	−9
1D	−2	*30*	−2	0	−4	−10	−3	−11
2A	5	−2	14	−2	−10	*−20*	−2	−2
2B	7	*34*	18	−11	−11	−4	−17	*−24*
2C	15	−9	13	*21*	1	−10	6	3
2D	−3	18	−5	−2	4	−17	1	3
3A	−6	*25*	−3	1	−10	*−27*	−3	1
3B	−7	14	6	−4	−4	−13	−4	−8
3C	*26*	11	6	13	19	*−23*	*22*	17
3D	10	*22*	8	11	11	−13	12	11
4A	−10	*38*	11	−9	−3	−7	−5	−16
4B	*50*	*37*	*30*	3	*43*	*−21*	*21*	−4
4C	−10	10	−17	−13	−3	−4	−7	−7
4D	6	*25*	5	4	3	6	17	−5
5A	12	*21*	*23*	2	3	−12	16	3
5B	1	17	−3	4	1	*−30*	−2	−18
5C	*22*	2	*29*	15	*37*	−10	*20*	−1
5D	6	*41*	3	12	10	*−30*	*24*	*22*

Table 2. Percentage change in the likelihood of a rain day (greater than 1 mm) for a given synoptic type during winter 1984–1991 compared to the 1948–2009 period (substantial deviations from the long-term mean (> 20 %) are highlighted in italic).

Type	Ivanhoe	Sydney	Queanbeyan	Kerang	Corowa	Casterton	Omeo	Melbourne
1A	−1	−2	1	−3	8	−10	2	1
1B	5	7	−2	5	5	−8	2	7
1C	0	8	−11	−1	−7	−8	−3	9
1D	4	−3	−3	0	0	2	−5	−5
2A	1	−6	−10	6	2	−11	*−22*	6
2B	−4	−4	−9	−4	−1	−13	1	4
2C	2	13	12	−2	2	−14	5	12
2D	1	3	4	12	14	6	*21*	−2
3A	11	10	−1	9	15	−16	4	−4
3B	8	−5	3	13	4	*−24*	13	11
3C	11	4	−4	12	11	0	11	5
3D	0	−10	0	11	10	1	2	12
4A	−2	3	−2	3	−1	0	2	12
4B	5	4	−10	−3	3	−6	−2	3
4C	8	−5	−7	10	9	7	−1	6
4D	16	−2	4	19	19	−12	18	18
5A	4	3	0	4	−1	2	3	1
5B	9	−3	0	−6	5	0	0	1
5C	4	12	5	−3	6	−5	0	−1
5D	4	1	1	7	8	−12	5	5

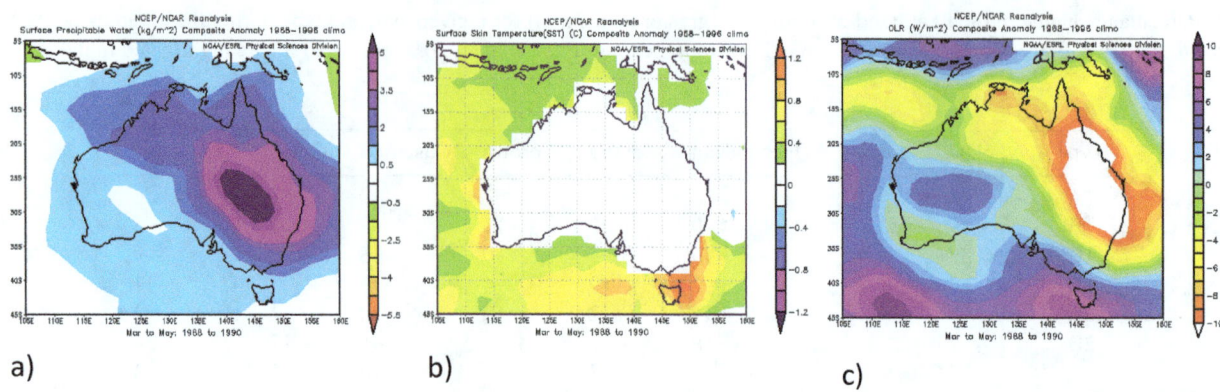

Figure 16. (a) Precipitable water anomalies, **(b)** SST anomalies, **(c)** outgoing longwave radiation, during the autumns of 1988–1990.

Niña) and warm (El Niño) episodes are defined when the threshold (±0.5°) is met for a minimum of 5 consecutive over-lapping seasons.

– Interdecadal Pacific Oscillation (IPO; Power et al., 1999; Kiem et al., 2003; Verdon et al., 2004; Power and Colman, 2006) – Both the raw and the smoothed time series of Power et al. (1999) are used in this study in order to identify epochs of positive and negative IPO.

– Indian Ocean Dipole (IOD; Saji et al., 1999; Ashok et al., 2003; Verdon and Franks, 2005) – SST anomalies to the northwest of Australia are used here rather than the IOD, as previous studies have shown that warming to the northwest of Australia (i.e. the eastern pole of the IOD) is the most important for cloud band development and is strongly related to above-average rainfall in SEA (e.g. Verdon and Franks, 2005), while the need for an anomaly further west has not been demonstrated. In fact, the poles of the IOD are not negatively correlated as one would expect (Dommenget and Latif, 2001; Gallant et al., 2012) and intermittent decoupling of the east and west pole of the IOD can lead to false classification of events.

– Southern Annular Mode (SAM; Thompson and Wallace, 2000; Thompson et al., 2000; Ho et al., 2012) – The positive phase of SAM has been associated with reduced winter rainfall in SEA (e.g. Risbey et al., 2009) and also reduced autumn SEA rainfall via a reduction in frontal systems (e.g. Verdon-Kidd and Kiem, 2009b. Nicholls, 2009). In this study the NOAA CPC version of the AAO is used when it exists (i.e. from 1979 onwards) and the Thompson and Wallace (2000) AAO data is used prior to that (1948 to 1978). Overlapping periods (1979 to 2002) of the two versions of the AAO were compared and the difference found to be negligible ($R^2 = 0.95, N = 288$).

It is well known that the 1970s change in climate observed in the Southern Hemisphere (and parts of the Northern Hemi-

sphere) corresponds to changes in both ENSO and IPO (e.g. Mantua et al., 1997; Power et al., 1999; Kiem and Franks, 2004; Verdon and Franks, 2006). Changes in Indian Ocean SSTs also occurred during this time (Verdon and Franks, 2005; Samuel et al., 2006). It has also been established that these large-scale climate modes modulate the frequency and timing of synoptic systems in Australia, at least at a monthly timescale (Verdon-Kidd and Kiem, 2009b). The question remains, however, whether these modes either collectively or individually can explain the increase in rainfall observed during autumn and winter of the late 1980s/early 1990s. Table 3 shows the state of each of the large-scale climate drivers during this wetter than average autumn and winter period. Note that while typically an ENSO event does not establish itself until after autumn (and therefore is unlikely to be a driver of autumn rainfall), the La Niña of 1988 was particularly long lasting and extended through the following autumn and winter of 1989.

Based on the analysis presented in Table 3 it may appear that SSTs to the northwest of Australia were primarily responsible for the elevated rainfall; however, it must be noted that in fact this region has been warmer than average during the entire period from the mid-1970s (including during the recent drought where autumn and winter rainfalls have been lower than average). What is clear from the analysis presented in Table 3 is that, for every wet season identified (except winter 1987) at least one of the four large-scale climate drivers was in a "wet phase". This indicates that, while a single climate mode can sometimes dominate, it is more often the interaction between drivers that is most important (Kiem and Verdon-Kidd, 2010). It also appears that the SAM plays an important role in the autumn rainfall relief (in that SAM is not positive during any of the wet autumns) but that a negative SAM on its own is not usually enough to ensure wetter than average conditions. This possibly explains why previous studies which analyse correlation relationships between SEA rainfall and SAM on its own conclude that SAM has minimal influence in autumn (e.g. Cai and Cowan, 2008a, b; Risbey et al., 2009). Based on the findings presented here

Table 3. State of large-scale drivers during "wet seasons".

Autumn	ENSO	IPO (smooth)*	IPO (raw/annual)	Northwest of Australia SSTs	SAM
1988	Neutral	Positive	Positive	Warm SSTs	Negative
1989	La Niña	Positive	Negative	Warm SSTs	Neutral
1990	El Niño	Positive	Negative	Warm SSTs	Negative
Winter					
1984	Neutral	Positive	Negative	Neutral	Negative
1985	Neutral	Positive	Negative	Warm SSTs	Positive
1986	El Niño	Positive	Positive	Warm SSTs	Positive
1987	El Niño	Positive	Positive	Neutral	Neutral
1988	La Niña	Positive	Negative	Warm SSTs	Negative
1989	Neutral	Positive	Negative	Warm SSTs	Neutral
1990	El Niño	Positive	Negative	Warm SSTs	Neutral
1991	El Niño	Positive	Positive	Neutral	Negative

* Power et al. (1999a) applied a spectral filter with a 13-year cut-off to the raw IPO (i.e. annual value) to generate a smoothed (or slowly varying) IPO time series.

(also Kiem and Verdon-Kidd, 2010), we suggest that this is unlikely to be true, and that SAM does indeed play an important role in modulating autumn/winter rainfall in SEA; however, this modulating effect is dependent on the phase of the Pacific Ocean and Indian Ocean drivers (see Kiem and Verdon-Kidd, 2010 for more details).

Of particular interest is the winter of 1991, which was extremely wet and the only climate mode in a wet phase at the time was the SAM, demonstrating that SAM does indeed have an influence; however, this result also indicates that correlation studies of SAM and rainfall may be skewed by this one very wet winter and negative SAM. Also of interest is the winter of 1987 where ENSO was in a dry phase, and the Indian and Southern Modes were neutral (indicating that rainfall would be expected to be average to below average). The rainfall for this season was only slightly elevated (whereas the other seven winters were substantially above average). However, this result points to the fact that it is also possible that other climate drivers may have played a role that has not been considered here – for example upper atmosphere phenomena or small-scale synoptic systems.

7 Implications for water resource management in SEA

There is evidence that changes in the dominant Southern Hemisphere climate drivers experienced during the mid-1970s created a shift to drier than average conditions across most of the mid-latitude belt. However, a series of wetter than average winters in SEA during the 1984–1991 period (and three wet autumns), due to weather systems originating primarily from the tropical Indian Ocean (i.e. northwest cloud bands and monsoon depressions) prevented, delayed or interrupted this change in climate in SEA.

What would have happened to water resources and supplies in the region in the absence of these wet winters and autumns? Given the experience of SWWA and the findings presented here, it is suggested that this should be considered as a possible "scenario" in terms of water management planning, and indeed water corporations in Victoria are using an immediate return to the dry conditions of the Big Dry as a possible future scenario alongside climate change projections (DSE, 2011).

To demonstrate the importance of considering such a scenario, a simple analysis of the effects on soil moisture and streamflow has been carried out. To achieve this, the observed rainfall sequences were altered such that the elevated rainfall totals actually experienced during the winters of 1984–1991 were replaced by the long-term mean winter rainfall. Similarly, the autumns of 1988, 1989 and 1990 were replaced with the long-term autumn mean rainfall. This perturbed synthetic rainfall sequence was then used to calculate the monthly value of the Palmer Drought Severity Index (PDSI, Palmer, 1965), a physically based index of meteorological drought that takes into account precipitation, evapotranspiration and soil moisture conditions, all of which are determinants of agricultural drought (Alley, 1984). Droughts are classified based on PDSI values between -10 (extreme drought) and $+10$ (extremely wet). The calculation of the PDSI also requires maximum temperature data (which was sourced from the BoM for the locations at the rainfall gauges). Sheffield et al. (2012) highlighted significant issues with the PDSI in terms of overestimation of drought (albeit on a global scale) given that the index uses a simplified model of potential evaporation that responds only to changes in temperature (see Mishra and Singh, 2010 for a comprehensive review of drought indices). In the absence of local meteorological data required to calculate potential evaporation using

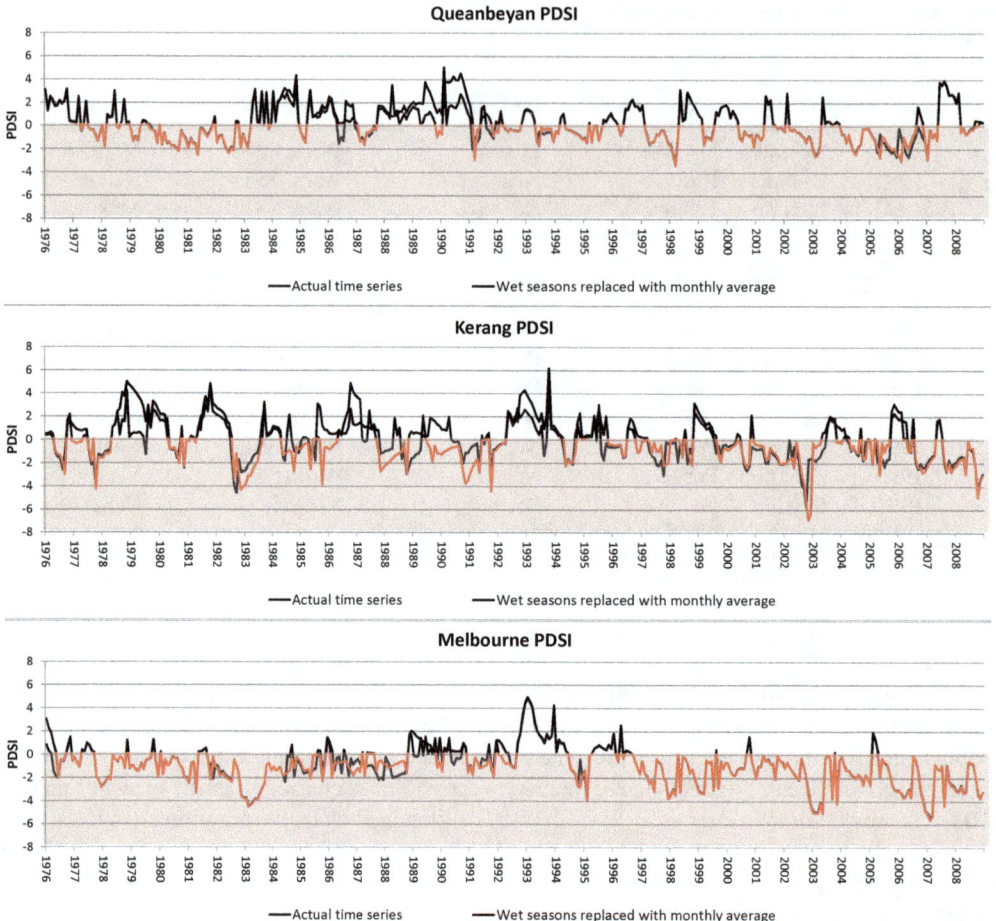

Figure 17. Monthly time series of PDSI from 1976 to 2009 at Queanbeyan, Kerang and Melbourne (red shading shows PDSI values that indicate dry conditions).

the Penman–Monteith method (a physically based estimate of potential evaporation), a second measure of drought, the Standardized Precipitation Index (SPI, McKee et al., 1993) was calculated to compare with the PDSI results. The SPI is a purely statistical measure of meteorological drought that requires only precipitation as input, where negative values of the SPI indicate dry conditions (values < −2 represent extreme drought). It is acknowledged that all indices used to measure drought are only proxies (with associated pros and cons); however, the analysis presented here is simply to demonstrate the potential changes in drought conditions that could have been experienced had the elevated rainfalls in autumn/winter of the later 1980s and early 1990s not occurred.

Figure 17 shows the time series of the PDSI for Queanbeyan, Kerang and Melbourne in SEA (see Figure 1b for locations) calculated using the instrumental record and then again using the "altered" rainfall sequence. Figure 18 shows the SPI time series for the same locations. Note drought indices are shown from 1976 only; however, the long-term indices were calculated over the entire instrumental record.

Figures 17 and 18 confirm that drought conditions (varying degrees of severity) associated with the Big Dry occurred between the mid-1990s through to 2009 at Kerang and Melbourne, while the drought appears to be less severe for Canberra (based on the PDSI and SPI). Based on both the PDSI and SPI time series for Melbourne (the most southerly station analysed here), the period from 1976 through to the late 1980s was also consistently dry, similar to SWWA.

It can be seen from Figs. 17 and 18 that replacing the wetter than average autumns and winters of the 1980s and early 1990s with average rainfall has an impact on the drought index time series at all stations for both indices to varying degrees. The largest impact on drought conditions was obtained at Kerang (the most inland station analysed here), particularly for the PDSI. It can be seen from Fig. 17 that, in the absence of wetter than average winters (1984–1991) and autumns (1988–1990), the return to "wet" soil moisture conditions (i.e. PDSI > 0.5) experienced between ∼1984–1991 at Kerang would not have occurred. Based on the recorded data (actual time series) dry (PDSI < −0.5) conditions prevailed 33% of the time between 1984–1991 compared to 55% of

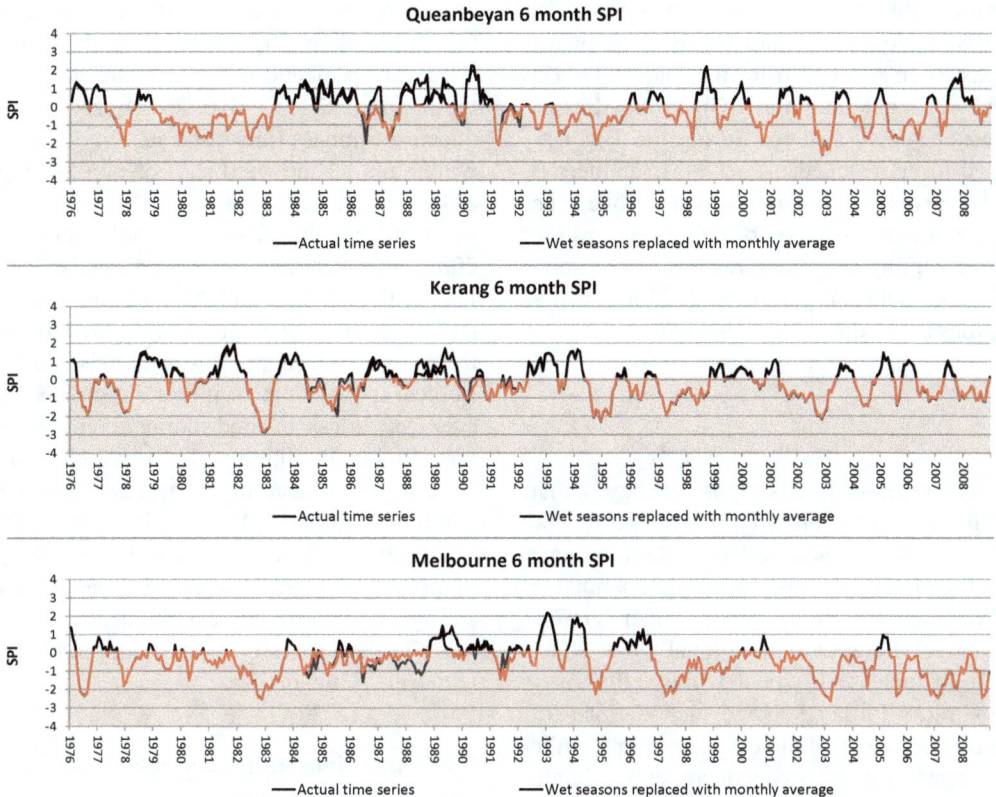

Figure 18. Monthly time series of SPI from 1976 to 2009 at Queanbeyan, Kerang and Melbourne (red shading shows SPI values that indicate dry conditions).

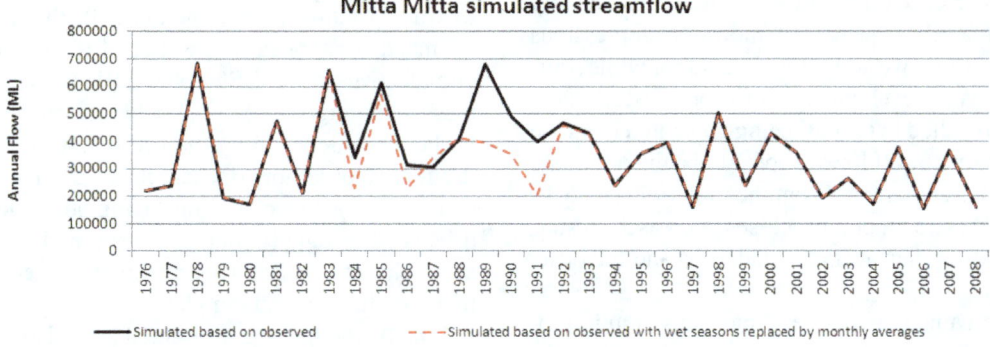

Figure 19. Monthly simulated streamflow for Mitta Mitta Creek located in Victoria, Australia.

the time for the time series with wet seasons replaced by averages. Indeed, if the rainfall had been average during these seasons, the entire period post-1982 would have been characterised by predominantly dry conditions (PDSI < -0.5).

The altered (i.e. wet winters and autumns replaced by the seasonal average) rainfall sequence for Omeo (see Figure 1 for location) was also run through a simple hydrological model (Kiem et al., 2007) to determine the impact this scenario would have on streamflow for the Mitta Mitta River (see Fig. 19). Mitta Mitta River feeds Dartmouth Dam, the largest capacity dam in Victoria, storing water for irrigation

and domestic and stock use in Victoria and New South Wales. The model simulates changes in the soil moisture storage deficit over a monthly period with fluxes in rainfall, evaporation and overflow represented. A rainfall–runoff regression converts the estimated soil moisture deficit to streamflow (overall R^2 of model $= 0.8$).

Figure 19 shows that the impacts of replacing the wet autumns/winters of the late 1980s and early 1990s with average conditions are magnified (compared to meteorological drought analysis) for streamflow. Rather than streamflow reductions occurring around the mid-1990s (as was experienced during the Big Dry), the streamflow simulation for Mitta Mitta River (Fig. 19) shows the possibility for streamflows to have been reduced from the mid-1980s (in the absence of the wet winter/autumns). Given the stresses placed on SEA water supply systems during the Big Dry (i.e. critical storage levels reached in ~2006 in numerous places) an extra 10 years of low streamflow would have been very challenging for water managers, bringing forward the need for the types of contingency measures that were implemented from 2006 and, potentially, further additional major system augmentations as the drought progressed.

8 Discussion

Based on the findings presented in this paper, we suggest that the Big Dry (or Millennium Drought) is in fact likely to be connected to the widespread Southern Hemisphere climate shift that began in the mid-1970s (also resulting in a notable decrease in SWWA winter rainfall). Since the mid-1970s there has been a general decrease in meridional winds south of 30 degrees and an increase in zonal winds, effectively shifting the storm tracks further south; hence stations located south of 45 degrees show an increase in rainfall, while mid-latitude stations have experienced a decrease in rainfall. However, the effects in SEA were shown to be masked due to smaller-scale synoptic processes which resulted in elevated autumn and winter rainfall during the mid- to late 1980s and early 1990s (a crucial period for SEA hydrology). The elevation in rainfall occurrence during this time was subsequently linked to an increased frequency of monsoon depressions and systems with strong westerly flow in autumn while, during winter, cloud band development and strong easterly flow (that would have resulted in rainfall being generated from the Pacific in the form of cut-off lows) appear to be the dominant drivers of the increased rainfall. From the mid-1990s to 2010 the frequency of these synoptic processes decreased, resulting in a return to drier than average conditions and the onset of the Big Dry.

While we have gained insight here into the synoptic processes responsible for the delayed response in the 1970s climate shift in SEA, the probability of this situation occurring needs to be established in order to properly quantify drought risk – and importantly, the chances of a drought being broken. An analysis of how the large-scale drivers (e.g. ENSO, IPO, IOD and SAM) may have influenced this period of elevated rainfall for SEA, but which was not experienced elsewhere around the Southern Hemisphere, was inconclusive. However, what was made clear from the analysis is the importance of the interaction between drivers that operate on an inter-annual to annual timescales, with the majority of elevated rainfall events occurring during periods where more than one climate mode was in its "wet" phase. It is also suggested that other large-scale processes may have played a role in the elevated rainfall period – for example upper at-

mosphere phenomena such as the seasonal longwave trough (which should be the subject of future research).

Current projections for SEA are for continued drying of the region (CSIRO, 2010, 2012) primarily as a result of reduced storminess (increased stability) in the mid-latitudes in winter and a southward shift of rain-bearing systems (Frederiksen et al., 2011), along with a projected weakening and poleward expansion of the Hadley circulation (Lu et al., 2007), the continuing observed trends (Lucas et al., 2012). During the 2010/11 and 2011/12 austral summer, rainfall totals were above average in SEA, breaking the long running hydrological drought. However, this rainfall was primarily tropical in origin and winter rainfall in SEA was still below average, with spring and summer being the greatest contributors to this rainfall relief. The wet spring/summer conditions were in part related to strong La Niña events in the Pacific combined, in 2010/11 only, with warm SSTs off northwestern Australia (i.e. negative IOD). Therefore, establishing the likelihood of repeat events (such as those that occurred in spring/summers of 2010/11 and 2011/12) in the next few years to decades is of crucial importance to water resource managers in the region.

9 Conclusions and recommendations

As discussed in Sect. 1.3, in the light of the experiences during the Big Dry and the significant uncertainties associated with future climate, changes have been made to the water management and planning framework across SEA, and also to water trading arrangements, with the aim being to better manage supplies in the face of a variable and changing climate (Moran and Sharples, 2011). The effectiveness of these planning processes would nevertheless be facilitated by better understanding the full risk profile associated with drought (that is how dry can it get and for how long). This can only be achieved by improving our understanding of the drivers of drought (both in the short- and long term) using all available information (instrumental, palaeoclimate, stochastic modelling, climate models etc.).

Finally, the findings presented here also have implications for climate change attribution studies, since many assume or aim to prove the Big Dry is due to increasing CO_2 emissions. However, if this climate shift is part of a much longer (and more widespread) change in climate since the mid-1970s, the "goal post" (in terms of timing) for studies examining the potential role of CO_2 (and other non-CO_2 drivers) is somewhat shifted.

Acknowledgements. This study was partially funded by the Victorian Department of Sustainability and Environment (DSE) (which has subsequently become the Department of Environment and Primary Industries (DEPI)).

Edited by: M. Werner

References

Alexander, L. V., Uotila, P., Nicholls, N., and Lynch, A.: A new daily pressure dataset for Australia and its application to the assessment of changes in synoptic patterns during the last century, J. Climate, 23, 1111–1126, doi:10.1175/2009JCLI2972.1, 2010.

Alley, W. M.: The Palmer Drought Severity Index: Limitations and Assumptions, J. Clim. Appl. Meteorol., 23, 1100–1109, 1984.

Ashok, K., Guan, Z., and Yamagata, T.: Influence of the Indian Ocean Dipole on the Australian winter rainfall, Geophys. Res. Lett., 30, 1821, doi:10.1029/2003GL017926, 2003.

Cai, W. and Cowan, T.: Dynamics of late autumn rainfall reduction over southeastern Australia, Geophys. Res. Lett., 35, L09708, doi:10.1029/2008GL033727, 2008a.

Cai, W. and Cowan, T.: Evidence of impacts from rising temperature on inflows to the Murray-Darling Basin, Geophys. Res. Lett., 35, L07701, doi:10.1029/2008GL033390, 2008b.

Cai, W., Cowan, T., and Thatcher, M.: Rainfall reductions over Southern Hemisphere semi-arid regions: the role of subtropical dry zone expansion, Scientific Reports, 2, doi:10.1038/srep00702, 2012.

Cavazos, T.: Using self-organizing maps to investigate extreme climate events: An application to wintertime precipitation in the Balkans, J. Climate, 13, 1718–1732, 2000.

Cavazos, T., Comrie, A. C., and Liverman, D. M.: Intraseasonal variability associated with wet monsoons in southeast Arizona, J. Climate, 15, 2477–2490, 2002.

Chiew, F. H. S., Piechota, T. C., Dracup, J. A., and McMahon, T. A.: El Niño Southern Oscillation and Australian rainfall, streamflow and drought – links and potential for forecasting, J. Hydrol., 204, 138–149, 1998.

CSIRO: Climate variability and change in south-eastern Australia: A synthesis of findings from Phase 1 of the South Eastern Australian Climate Initiative (SEACI), 2010.

CSIRO: Climate and water availability in south-eastern Australia: A synthesis of findings from Phase 2 of the South Eastern Australian Climate Initiative (SEACI), 2012.

DSE: Guidelines for the Development of a Water Supply-Demand Strategy, Department of Sustainability and Environment, August 2011, http://www.water.vic.gov.au/governance/water-corporations/water-supply-demand-strategy, 2011.

Dommenget, D. and Latif, M.: A cautionary note on the interpretation of EOFs, J. Climate, 15, 216–225, 2001.

Frederiksen, J. S. and Frederiksen, C. S.: Decadal Changes in Southern Hemisphere Winter Cyclogenesis, CSIRO Marine and Atmospheric Research Paper No. 002, 35 pp., 2005.

Frederiksen, J. S. and Frederiksen, C. S.: Inter-decadal changes in Southern Hemisphere winter storm track modes, Tellus, 59, 559–617, 2007.

Frederiksen C. S., Frederiksen, J. S., Sissons, J. M., and Osbrough, S. L.: Changes and projections in Australian winter rainfall and circulation: Anthropogenic forcing and internal variability, Int. J. Clim. Change Impacts Responses, 2, 143–162, 2011.

Gallant, A. J. E., Kiem, A. S., Verdon-Kidd, D. C., Stone, R. C., and Karoly, D. J.: Understanding hydroclimate processes in the Murray-Darling Basin for natural resources management, Hydrol. Earth Syst. Sci., 16, 2049–2068, doi:10.5194/hess-16-2049-2012, 2012.

Hewitson, B. C. and Crane, R. G.: Self-organizing maps: application to synoptic climatology, Clim. Res., 22, 13–26, 2002.

Ho, M., Kiem, A. S., and Verdon-Kidd, D. C.: The Southern Annular Mode: a comparison of indices, Hydrol. Earth Syst. Sci., 16, 967–982, doi:10.5194/hess-16-967-2012, 2012.

Hope, P. K., Drosdowsky, W., and Nicholls, N.: Shifts in the synoptic systems influencing southwest Western Australia, Clim. Dynam., 26, 751–764, 2006.

Hope, P. K., Timbal, B., and Fawcett, R.: Associations between rainfall variability in the southwest and southeast of Australia and their evolution through time, Int. J. Climatol., 30, 1360–1371, doi:10.1002/joc.1964, 2009.

Hu, Y. and Fu, Q.: Observed poleward expansion of the Hadley circulation since 1979, Atmos. Chem. Phys., 7, 5229–5236, doi:10.5194/acp-7-5229-2007, 2007.

IOCI: Climate variability and change in south west Western Australia, Indian Ocean Climate Initiative (IOCI) Panel, Perth, Australia, 2002.

Jones, D. A., Wang, W., and Fawcett, R.: High-quality spatial climate analyses for Australia, Aust. Meteorol. Oceanogr. J., 58, 233–248, 2009.

Kiem, A. S. and Franks, S. W.: On the identification of ENSO-induced rainfall and runoff variability: a comparison of methods and indices, Hydrol. Sci. J., 46, 715–727, 2001.

Kiem, A. S. and Franks, S. W.: Multi-decadal variability of drought risk – Eastern Australia, Hydrol. Process., 18, 2039–2050, 2004.

Kiem, A. S. and Verdon-Kidd, D. C.: Towards understanding hydroclimatic change in Victoria, Australia – preliminary insights into the "Big Dry", Hydrol. Earth Syst. Sci., 14, 433–445, doi:10.5194/hess-14-433-2010, 2010.

Kiem, A. S. and Verdon-Kidd, D. C.: Steps towards "useful" hydroclimatic scenarios for water resource management in the Murray-Darling Basin, Water Resour. Res., 47, W00G06, doi:10.1029/2010WR009803, 2011.

Kiem, A. S., Franks, S. W. and Kuczera, G.: Multi-decadal variability of flood risk, Geophys. Res. Lett., 30, 1035, doi:10.1029/2002Gl015992, 2003.

Kiem, A. S., Verdon, D. C., Hill, P. I., Payne, E., and Goodwin, I.: Seasonal forecasting of Victorian streamflows: Phase 1 – Identifying the drivers of Victoria's climate, Technical report prepared by Sinclair Knight Merz for the Victorian Department of Sustainability and Environment (Project: Seasonal forecasting of Victorian streamflows (VW03931)), 2007.

Lu, J., Vecchi, G. A., and Reichler, T.: Expansion of the Hadley cell under global warming, Geophys. Res. Lett., 34, L06805, doi:10.1029/2006GL028443, 2007.

Lucas, C., Nguyen, H., and Timbal, B.: An observational analysis of Southern Hemisphere tropical expansion, J. Geophys. Res., 117, D17112, doi:10.1029/2011JD017033, 2012.

Mantua, N. J., Hare, S. R., Zhang, Y., Wallace, J. M., and Francis, R. C.: A Pacific interdecadal climate oscillation with impacts on salmon production, B. Am. Meteorol. Soc., 78, 1069–1079, 1997.

McMahon, T. A., Kiem, A. S., Peel, M. C., Jordan, P. W., and Pegram, G. G. S.: A new approach to stochastically generating six-monthly rainfall sequences based on Empirical Model Decomposition, J. Hydrometeorol., 9, 1377–1389, 2008.

Mishra, A. K. and Singh, V. P.: A review on drought concepts, J. Hydrol., doi:10.1016/j.jhydrol.2010.07.012, 2010.

Moran, R. and Sharples, J.: Guidelines for the development of a water supply-demand strategy: Technical Supplement for Section 3.5 – Forecasting Supply, Department of Sustainability and Environment, Victoria, available at: http://www.water.vic.gov.au/_data/assets/pdf_file/0015/134061/Water-Supply-Demand-Strategy-Guidelines_technical-supplement.pdf, 2011.

Murphy, B. F. and Timbal, B.: A review of recent climate variability and climate change in southeastern Australia, Int. J. Climatol., 28, 859–879, 2008.

Ngongondo, C. S.: An analysis of long-term rainfall variability, trends and groundwater availability in the Mulunguzi river catchment area, Zomba mountain, Southern Malawi, Quaternary Int., 148, 45–50, 2006.

Nicholls, N.: Local and remote causes of the southern Australian autumn-winter rainfall decline, 1958–2007, Clim. Dynam., 34, 835–845, doi:10.1007/s00382-009-0527-6, 2009.

NWC: Australia's Water Supply Status and Seasonal Outlook, National Water Commission (NWC), Canberra, ACT, 2006.

Palmer, W. C.: Meteorological Drought, US. Weather Bureau, Research Paper No. 45, 58 pp., 1965.

Pook, M. J., McIntosh, P. C., and Meyers, G. A.: The synoptic decomposition of cool-season rainfall in the southeastern Australian cropping region, J. Appl. Meteorol. Clim., 45, 1156–1170, 2006.

Power, S. and Colman, A.: Multi-year predictability in a coupled general circulation model, Clim. Dynam., 26, 247–272, 2006.

Power, S., Casey, T., Folland, C., Colman, A., and Mehta, V.: Interdecadal modulation of the impact of ENSO on Australia, Clim. Dynam., 15, 319–324, 1999.

Reusch, D. B., Alley, R. B., and Hewitson, B. C.: North Atlantic climate variability from a self-organizing map perspective, J. Geophys. Res., 112, D02104, doi:10.01029/02006JD007460, 2007.

Risbey, J. S., Pook, M. J., McIntosh, P. C., Wheeler, M. C., and Hendon, H. H.: On the remote drivers of rainfall variability in Australia, Mon. Weather Rev., 137, 3233–3253, doi:10.1175/2009MWR2861.1, 2009.

Saji, N. H., Goswami, B. N., Vinayachandran, P. N., and Yamagata, T.: A dipole mode in the tropical Indian Ocean, Nature, 401, 360–363, 1999.

Salinger, M. J. and Mullan, A. B.: New Zealand climate: Temperature and precipitation variations and their links with atmospheric circulation 1930–1994, Int. J. Climatol., 19, 1049–1071, 1999.

Samuel, J. M., Verdon, D. C., Sivapalan, M., and Franks, S. W.: Influence of Indian Ocean sea surface temperature variability on Southwest Western Australia winter rainfall, Water Resour. Res., 42, W08402, doi:10.1029/2005WR004672, 2006.

Sheffield, J., Wood, E. F., and Roderick, M. L.: Little change in global drought over the past 60 years, Nature, 491, 435–438, doi:10.1038/nature11575, 2012.

Sturman, A. and Tapper, N.: The Weather and Climate of Australia and New Zealand, Oxford University Press, Melbourne, Victoria, Australia, 2004.

Tapper, N. and Hurry, L.: Australia's Weather Patterns – An Introductory Guide, Dellasta Pty Ltd, Mount Waverly, Victoria, 1996.

Thompson, D. W. J. and Wallace, J. M.: Annular modes in the extratropical circulation, part I: month-to-month variability, J. Climate, 13, 1000–1016, 2000.

Thompson, D. W. J., Wallace, J. M., and Hegerl, G. C.: Annular modes in the extratropical circulation, part II: Trends, J. Climate, 13, 1018–1036, 2000.

Timbal, B., Arblaster, J., Braganza, K., Fernandez, E., Hendon, H., Murphy, B., Raupach, M., Rakich, C., Smith, I., Whan, K., and Wheeler, M.: Understanding the anthropogenic nature of the observed rainfall decline across south-eastern Australia, The Centre for Australian Weather and Climate Research, Technical Report 026, 2010.

Tozer, C. R., Kiem, A. S., and Verdon-Kidd, D. C.: On the uncertainties associated with using gridded rainfall data as a proxy for observed, Hydrol. Earth Syst. Sci., 16, 1481–1499, doi:10.5194/hess-16-1481-2012, 2012.

Trenberth, K. E. and Guillemot, C. J.: Evaluation of the atmospheric moisture and hydrological cycle, Clim. Dynam., 14, 213–231, 1998.

Van Dijk, A. I. J. M., Beck, H. E., Crosbie, R. S., de Jeu, R. A. M., Liu, Y. Y., Podger, G. M., Timbal, B., and Viney, N. R.: The Millennium Drought in southeast Australia (2001–2009): Natural and human causes and implications for water resources ecosystems, economy, and society, Water Resour. Res., 49, 1040–1057, doi:10.1002/wrcr.20123, 2013.

Van Ommen, T. D. and Morgan, V.: Snowfall increase in coastal East Antarctica linked with southwest Western Australian drought, Nat. Geosci., 3, 267–272, doi:10.1038/NGEO761, 2010.

Verdon, D. C. and Franks, S. W.: Indian Ocean sea surface temperature variability and winter rainfall: Eastern Australia, Water Resour. Res., 41, W09413, doi:10.1029/2004WR003845, 2005.

Verdon, D. C. and Franks, S. W.: Long-term behaviour of ENSO: Interactions with the PDO over the past 400 years inferred from paleoclimate records, Geophys. Res. Lett., 33, L06712, doi:10.1029/2005GL025052, 2006.

Verdon-Kidd, D. C. and Kiem, A. S.: Nature and causes of protracted droughts in southeast Australia: Comparison between the Federation, WWII, and Big Dry droughts, Geophys. Res. Lett., 36, L22707, doi:10.1029/2009GL041067, 2009a.

Verdon-Kidd, D. C. and Kiem, A. S.: On the relationship between large-scale climate modes and regional synoptic patterns that drive Victorian rainfall, Hydrol. Earth Syst. Sci., 13, 467–479, doi:10.5194/hess-13-467-2009, 2009b.

Verdon-Kidd, D. C. and Kiem, A. S.: Quantifying drought risk in a non-stationary climate, J. Hydrometeorol., 11, 1020–1032, 2010.

Verdon-Kidd, D. C. and Kiem, A. S.: Synchronicity of historical dry spells in the Southern Hemisphere, Hydrol. Earth Syst. Sci., 18, 2257–2264, doi:10.5194/hess-18-2257-2014, 2014.

Verdon, D. C., Wyatt, A. M., Kiem, A. S., and Franks, S. W.: Multi-decadal variability of rainfall and streamflow – Eastern Australia, Water Resour. Res., 40, W10201, doi:10.1029/2004WR003234, 2004.

Whan, K., Timbal, B., and Lindsay, J.: Linear and nonlinear statistical analysis of the impact of sub-tropical ridge intensity and position on south-east Australian rainfall, Int. J. Climatol., 34, 326–342, doi:10.1002/joc.3689, 2014.

12

A new probability density function for spatial distribution of soil water storage capacity leads to the SCS curve number method

Dingbao Wang

Department of Civil, Environmental, and Construction Engineering, University of Central Florida, Orlando, Florida, USA

Correspondence: Dingbao Wang (dingbao.wang@ucf.edu)

Abstract. Following the Budyko framework, the soil wetting ratio (the ratio between soil wetting and precipitation) as a function of the soil storage index (the ratio between soil wetting capacity and precipitation) is derived from the Soil Conservation Service Curve Number (SCS-CN) method and the variable infiltration capacity (VIC) type of model. For the SCS-CN method, the soil wetting ratio approaches 1 when the soil storage index approaches ∞, due to the limitation of the SCS-CN method in which the initial soil moisture condition is not explicitly represented. However, for the VIC type of model, the soil wetting ratio equals the soil storage index when the soil storage index is lower than a certain value, due to the finite upper bound of the generalized Pareto distribution function of storage capacity. In this paper, a new distribution function, supported on a semi-infinite interval $x \in [0, \infty)$, is proposed for describing the spatial distribution of storage capacity. From this new distribution function, an equation is derived for the relationship between the soil wetting ratio and the storage index. In the derived equation, the soil wetting ratio approaches 0 as the storage index approaches 0; when the storage index tends to infinity, the soil wetting ratio approaches a certain value (≤ 1) depending on the initial storage. Moreover, the derived equation leads to the exact SCS-CN method when initial water storage is 0. Therefore, the new distribution function for soil water storage capacity explains the SCS-CN method as a saturation excess runoff model and unifies the surface runoff modeling of the SCS-CN method and the VIC type of model.

1 Introduction

The Soil Conservation Service Curve Number (SCS-CN) method (Mockus, 1972) has been popularly used for direct runoff estimation in engineering communities. Even though the SCS-CN method was obtained empirically (Ponce, 1996; Beven, 2012), it is often interpreted as an infiltration excess runoff model (Bras, 1990; Mishra and Singh, 1999). Yu (1998) showed that partial area infiltration excess runoff generation on a statistical distribution of soil infiltration characteristics provided a similar runoff generation equation to the SCS-CN method. Recently, Hooshyar and Wang (2016) derived an analytical solution for Richards' equation for ponded infiltration into a soil column bounded by a water table, and they showed that the SCS-CN method, as an infiltration excess model, is a special case of the derived general solution. The SCS-CN method has also been interpreted as a saturation excess runoff model (Steenhuis et al., 1995; Lyon et al., 2004; Easton et al., 2008). During an interview, Mockus, who developed the proportionality relationship of the SCS-CN method, stated that "saturation overland flow was the most likely runoff mechanism to be simulated by the method" (Ponce, 1996). Recently, Bartlett et al. (2016a) developed a probabilistic framework, which provides a statistical justification of the SCS-CN method and extends the saturation excess interpretation of the event-based runoff of the method.

Since the 1970s, various saturation excess runoff models have been developed based on the concept of probability distribution of soil storage capacity (Moore, 1985). TOP-MODEL is a well-known saturation excess runoff model based on spatially distributed topography (Beven and Kirkby, 1979; Sivapalan et al., 1987). To quantify the dynamic change of saturation area during rainfall events, the spatial variability of soil moisture storage capacity is described by a cumulative probability distribution function in the Xinanjiang model (Zhao, 1977; Zhao et al., 1992) and the variable infiltration capacity (VIC) model (Wood et al., 1992; Liang et al., 1994). The spatial distribution of storage capacity in these models is described by the generalized Pareto distribution, which has been used for catchment-scale runoff prediction and large-scale land surface hydrologic simulations. Bartlett et al. (2016b) proposed an event-based probabilistic storage framework for unifying TOPMODEL, the VIC type of model, and the SCS-CN method, and the framework includes a spatial description of the runoff concept of "prethreshold" and "threshold-excess" runoff (Bartlett et al., 2016a).

Even though the SCS-CN method has been interpreted as a saturation excess runoff model in the literature, there is a knowledge gap for the direct linkage between the SCS-CN method and the Xinanjiang and VIC type of model based on a probability distribution function for the spatial variability of soil water storage capacity. If the SCS-CN method is a saturation excess runoff model, is there a distribution function for soil water storage capacity which leads to the SCS-CN method? If yes, what is the probability density function (PDF)? This is an unsolved research question. The objective of this paper is to fill this knowledge gap, i.e., discovering the distribution function for soil water storage capacity which leads to the SCS-CN method. This is a procedure of inverse modeling, i.e., identifying the distribution function of the saturation excess runoff model for a known functional form of runoff generation.

Meanwhile, the identification of the new distribution function is intrigued by the linkage between the SCS-CN method and the Budyko equation (Budyko, 1974). By applying the generalized proportionality hypothesis from the SCS-CN method to mean annual water balance, Wang and Tang (2014) derived a one-parameter Budyko equation for the mean annual evaporation ratio (i.e., the ratio of evaporation to precipitation) as a function of the climate aridity index (i.e., the ratio of potential evaporation to precipitation). As an analogy to the Budyko framework, the SCS-CN method and the VIC type of model at the event scale can be represented by the relationship between the soil wetting ratio, defined as the ratio between soil wetting and precipitation, and the soil storage index, which is defined as the ratio between soil wetting capacity and precipitation. The representation of runoff generation in the Budyko type of framework facilitates the identification of the new distribution function for soil water storage capacity leading to the SCS-CN method.

The identified new distribution function for soil water storage capacity will unify the SCS-CN method and the VIC type of model. In Sect. 2, the SCS-CN method is presented in the form of the Budyko-type framework with two parameterization schemes. In Sect. 3, the VIC type of model is presented in the form of the Budyko-type framework. In Sect. 4, the SCS-CN method is then compared with the VIC type of

model from the perspectives of the number of parameters and boundary conditions (i.e., the lower and upper bounds of the soil storage index). In Sect. 5, the proposed new distribution function is introduced and compared with the generalized Pareto distribution of the VIC type of model, and a modified SCS-CN method considering initial storage explicitly is derived from the new distribution function. Conclusions are drawn in Sect. 6.

2 SCS curve number method

In this section, the SCS-CN method is described in the form of surface runoff modeling and then is presented for infiltration modeling in the Budyko-type framework. The initial storage at the beginning of a time interval (e.g., rainfall event) is denoted by S_0 (mm), and the maximum value of average storage capacity over the catchment is denoted by S_b (mm). The storage capacity for soil wetting for the time interval, S_p (mm), is computed by

$$S_p = S_b - S_0. \tag{1}$$

The total rainfall during the time interval is denoted by P (mm). Before surface runoff is generated, a portion of rainfall is intercepted by vegetation and infiltrates into the soil. This portion of rainfall is called initial abstraction or initial soil wetting denoted by W_i (mm). The remaining rainfall $(P - W_i)$ is partitioned into runoff and continuing soil wetting. This competition is captured by the proportionality relationship in the SCS-CN method:

$$\frac{W - W_i}{S_p - W_i} = \frac{Q}{P - W_i}, \tag{2}$$

where W (mm) is the total soil wetting, $W - W_i$ is continuing wetting and $S_p - W_i$ is its potential value, Q (mm) is surface runoff, and $P - W_i$ is the available water and interpreted as the potential value of Q. Since rainfall is partitioned into total soil wetting and surface runoff, i.e., $P = W + Q$, surface runoff is computed by substituting $W = P - Q$ into Eq. (2):

$$Q = \frac{(P - W_i)^2}{P + S_p - 2W_i}. \tag{3}$$

This equation is used for computing direct runoff in the SCS-CN method.

The SCS-CN method can also be represented in terms of the soil wetting ratio $\left(\frac{W}{P}\right)$. Substituting Eq. (3) into $W = P - Q$ and dividing P on both sides, the soil wetting ratio equation is obtained:

$$\frac{W}{P} = \frac{\frac{S_p}{P} - \frac{W_i^2}{P^2}}{1 + \frac{S_p}{P} - 2\frac{W_i}{P}}. \tag{4}$$

The climate aridity index is defined as the ratio between potential evaporation and precipitation. In the climate aridity

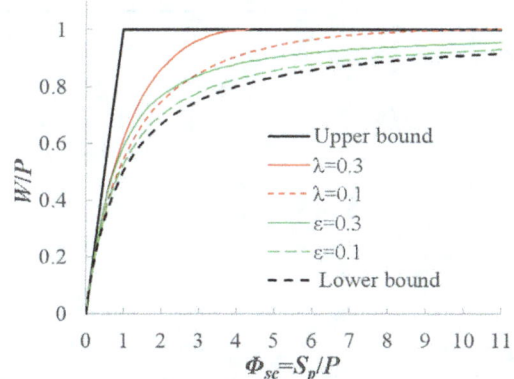

Figure 1. The wetting ratio $\left(\frac{W}{P}\right)$ versus the soil storage index $\left(\frac{S_p}{P}\right)$ from the SCS-CN method based on two parameterization schemes: $\lambda = \frac{W_i}{S_p - W_i}$ (scheme 1) and $\varepsilon = \frac{W_i}{W}$ (scheme 2).

index, both available water supply and water demand are determined by climate.

$$\Phi_{sc} = \frac{S_p}{P} \tag{5}$$

A similar dimensionless parameter for the ratio between the maximum soil storage capacity and mean rainfall depth of rainfall events was defined in Porporato et al. (2004). In the soil storage index, water demand is determined by soil and available water supply is determined by climate. Substituting Eq. (5) into Eq. (4), the soil wetting equation for the SCS-CN method is obtained:

$$\frac{W}{P} = \frac{\Phi_{sc} - \frac{W_i^2}{P^2}}{1 + \Phi_{sc} - 2\frac{W_i}{P}}. \tag{6}$$

There are two potential schemes for parameterizing the initial wetting in Eq. (6). As the first scheme, the initial wetting is usually parameterized as the ratio between initial wetting and storage capacity in the SCS-CN method. The detail of this scheme is described in Appendix A and plotted in Fig. 1. As we can see, the range of Φ_{sc} is dependent on the parameter $\lambda = \frac{W_i}{S_p - W_i}$.

In order to avoid the situation where the range of Φ_{sc} is dependent on the parameter λ, we can use the following parameterization scheme (Chen et al., 2013; Tang and Wang, 2017):

$$\varepsilon = \frac{W_i}{W}. \tag{7}$$

Substituting Eq. (7) into Eq. (6), we can obtain the following equation:

$$\frac{W}{P} = \frac{1 + \Phi_{sc} - \sqrt{(1 + \Phi_{sc})^2 - 4\varepsilon(2 - \varepsilon)\Phi_{sc}}}{2\varepsilon(2 - \varepsilon)}. \tag{8}$$

Equation (8) has the same functional form as the derived Budyko equation for the long-term evaporation ratio (Wang

and Tang, 2014; Wang et al., 2015). Equation (8) satisfies the following boundary conditions: $\frac{W}{P} \to 0$ as $\Phi_{sc} \to 0$ and $\frac{W}{P} \to 1$ as $\Phi_{sc} \to \infty$. Based on Eq. (7), the range of ε is [0, 1], and $\varepsilon = 1$ corresponds to the upper bound (Fig. 1). Equation (8) becomes Eq. (A3) as $\varepsilon \to 0$, and it is the lower bound. Figure 1 plots Eq. (8) for $\varepsilon = 0.1$ and 0.3. Due to the dependence of the range of Φ_{sc} on the parameter λ in the first parameterization scheme, the second parameterization scheme is focused on in the following sections.

In the SCS-CN method, the soil wetting ratio is a function of the soil storage index with a parameter for describing initial wetting. The average wetting capacity at the catchment scale is used for computing the soil storage index, but the spatial variability of wetting capacity is not represented in the SCS-CN method.

3 Saturation excess runoff model

The spatial variability of soil water storage capacity is explicitly represented in the saturation excess runoff models such as VIC and Xinanjiang. In these models, the spatial variation of the point-scale storage capacity (C) is represented by a generalized Pareto distribution:

$$F(C) = 1 - \left(1 - \frac{C}{C_m}\right)^\beta, \tag{9}$$

where $F(C)$ is the cumulative probability, i.e., the fraction of the catchment area for which the storage capacity is less than C (mm), and C_m (mm) is the maximum value of the point-scale storage capacity over the catchment. The water storage capacity includes vegetation interception, surface retention, and soil moisture capacity; β is the shape parameter of the storage capacity distribution and is usually assumed to be a positive number. β ranges from 0.01 to 5.0 as suggested by Wood et al. (1992). The storage capacity distribution curve is concave down for $0 < \beta < 1$ and concave up for $\beta > 1$. The average value of storage capacity over the catchment is equivalent to S_b in the SCS-CN method, and it is obtained by integrating the exceedance probability of storage capacity $S_b = \int_0^{C_m}(1 - F(x))\,\mathrm{d}x$:

$$S_b = \frac{C_m}{\beta + 1}. \tag{10}$$

Similarly, for a given C, the catchment-scale storage S (mm) can be computed as follows (Moore, 1985):

$$S = S_b\left[1 - \left(1 - \frac{C}{C_m}\right)^{\beta+1}\right]. \tag{11}$$

To derive the wetting ratio as a function of the soil storage index, the initial storage at the catchment scale is parameterized by the degree of saturation:

$$\psi = \frac{S_0}{S_b}. \tag{12}$$

Recalling Eq. (1) and the definition of the soil storage index (i.e., Eq. 5), we obtain

$$\frac{S_b}{P} = \frac{\Phi_{sc}}{1 - \psi}. \tag{13}$$

The value of C corresponding to the initial storage S_0 is denoted as C_0, and $S_0 = S_b \left[1 - \left(1 - \frac{C_0}{C_m} \right)^{\beta+1} \right]$ is obtained by substituting S_0 and C_0 into Eq. (11). When $P + C_0 \geq C_m$, each point within the catchment is saturated and soil wetting reaches its maximum value; i.e., $W = S_p$. Substituting $C_0 = C_m - C_m \left(1 - \frac{S_0}{S_b} \right)^{\frac{1}{\beta+1}}$ into $P + C_0 \geq C_m$, we obtain

$$\Phi_{sc} \leq b, \text{ where } b = (\beta + 1)^{-1}(1 - \psi)^{\frac{\beta}{\beta+1}}. \tag{14}$$

Therefore, this condition is equivalent to

$$\frac{W}{P} = \Phi_{sc} \text{ when } \Phi_{sc} \leq b. \tag{15}$$

Next, we will derive $\frac{W}{P}$ for the condition of $\Phi_{sc} > b$. The storage at the end of the modeling period (e.g., rainfall–runoff event) is denoted as S_1, which is computed by

$$S_1 = S_b \left[1 - \left(1 - \frac{P + C_0}{C_m} \right)^{\beta+1} \right]. \tag{16}$$

From Eq. (16) one obtains (see Appendix B for details)

$$\frac{W}{P} = \Phi_{sc} \left[1 - \left(1 - b\Phi_{sc}^{-1} \right)^{\beta+1} \right] \text{ when } \Phi_{sc} > b. \tag{17}$$

The limit of Eq. (17) for $\Phi_{sc} \to \infty$ can be obtained as follows (see Appendix C for details):

$$\lim_{\Phi_{sc} \to \infty} \frac{W}{P} = (1 - \psi)^{\frac{\beta}{\beta+1}}. \tag{18}$$

Equations (15) and (17) provide $\frac{W}{P}$ as a function of Φ_{sc} with two parameters (ψ and β). Figure 2 plots Eqs. (15) and (17) for $\psi = 0$ and 0.5 when $\beta = 0.2$ and 2. As we can see, $\frac{W}{P}$ decreases as β increases for given values of ψ and Φ_{sc}, and $\frac{W}{P}$ decreases as ψ increases for given values of β and Φ_{sc}, implicating that the soil wetting ratio decreases with the degree of initial saturation under a given the soil storage index.

4 Comparison between the SCS-CN model and the VIC type of model

The SCS-CN model with the parameterization of the ratio between initial wetting and total wetting is compared with the VIC type of saturation excess runoff model. In Sects. 2 and 3, we derived $\frac{W}{P}$ as a function of Φ_{sc} based on the SCS-CN method and the VIC type of model, which uses a generalized

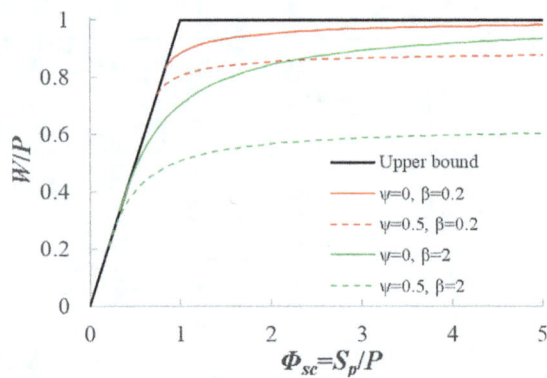

Figure 2. The impact of β and the degree of initial storage ($\psi = S_0 / S_b$) on the soil wetting ratio (W/P).

Pareto distribution to describe the spatial distribution of storage capacity. The SCS-CN method is a function of storage capacity S_p, but the VIC type of model is a function of storage capacity S_p and the degree of initial saturation $\frac{S_0}{S_b}$. As a result, the function of $\frac{W}{P} \sim \frac{S_p}{P}$ for the SCS-CN method has only one parameter (ε), but it has two parameters (β and ψ) for the VIC type of model.

Table 1 shows the boundary conditions for the relationships between $\frac{W}{P}$ and Φ_{sc} from the SCS-CN method and the VIC type of model. The lower boundary of the SCS-CN method with parameter ε is $\frac{W}{P} \to 0$ as $\Phi_{sc} \to 0$. However, for the VIC type of model, $\frac{W}{P} = \Phi_{sc}$ when $\Phi_{sc} \leq b$. For the SCS-CN method, W reaches its maximum (S_p) when rainfall reaches infinity, while for the VIC type of model, W reaches its maximum value (S_p) when rainfall reaches a finite number ($C_m - C_0$). In other words, for the SCS-CN method, the entire catchment becomes saturated when rainfall reaches infinity, while for the VIC-type model, the entire catchment becomes saturated when rainfall reaches a finite number.

As shown in Table 1, the upper boundary of the SCS-CN method (with parameter ε) is 1. However, for the VIC type of model, the upper boundary is $(1 - \psi)^{\frac{\beta}{\beta+1}}$ instead of 1. This is due to the effect of initial storage in the VIC type of model. When initial storage is 0 (i.e., $\psi = 0$), the wetting ratio $\frac{W}{P}$ for the VIC type of model has the same upper boundary condition as the SCS-CN method.

5 Unification of the SCS-CN method and the VIC type of model

Based on the comparison between the SCS-CN method and the VIC type of model, a new distribution function is proposed in this section for describing the spatial distribution of soil water storage capacity, which unifies the SCS-CN method and the VIC type of model. As discussed in Sect. 4, the upper boundary condition of the SCS-CN model (i.e., $\frac{W}{P} \to 1$ as $\Phi_{sc} \to \infty$) does not depend on the initial stor-

Table 1. The boundary conditions of the functions for relating the wetting ratio $\left(\frac{W}{P}\right)$ to the soil storage index (Φ_{sc}): (1) the SCS-CN method, (2) the VIC type of model, and (3) the modified SCS-CN method based on the proposed new distribution for the VIC type of model.

Surface runoff model	Parameters	Lower boundary condition	Upper boundary condition
SCS-CN, parameterization of initial wetting	S_p, ε	$\frac{W}{P} \to 0$ as $\Phi_{\text{sc}} \to 0$	$\frac{W}{P} \to 1$ as $\Phi_{\text{sc}} \to \infty$
Generalized Pareto distribution for storage capacity (VIC type of model)	C_{m}, β	$\frac{W}{P} = \Phi_{\text{sc}}$ when $\Phi_{\text{sc}} \leq b$	$\frac{W}{P} \to (1-\psi)^{\frac{\beta}{\beta+1}}$ as $\Phi_{\text{sc}} \to \infty$
Modified SCS-CN method based on the proposed distribution for storage capacity	S_b, a	$\frac{W}{P} \to 0$ as $\Phi_{\text{sc}} \to 0$	$\frac{W}{P} \to \dfrac{\sqrt{(m+1)^2-2am+a-m-1}}{a\sqrt{(m+1)^2-2am}}$ as $\Phi_{\text{sc}} \to \infty$

age. This upper boundary condition needs to be modified by including the effect of initial storage so that the limit of $\frac{W}{P}$ as $\Phi_{\text{sc}} \to \infty$ is dependent on the degree of initial saturation like the VIC type of model. However, the lower boundary condition of the VIC model needs to be modified so that the lower boundary condition follows that of $\frac{W}{P} \to 0$ as $\Phi_{\text{sc}} \to 0$ like the SCS-CN method. Through these modifications, the SCS-CN method and the VIC type of saturation excess runoff model can be unified from the functional perspective of the soil wetting ratio.

Based on the comparison one may have the following questions. (1) Can the SCS-CN method be derived from the VIC type of model by setting initial storage to 0? (2) If yes, what is the distribution function for soil water storage capacity? Once we answer these questions, a modified SCS-CN method considering initial storage explicitly can be derived as a saturation excess runoff model based on a distribution function of water storage capacity, and it unifies the SCS-CN method and the VIC type of model. In this section, a new distribution function is proposed for describing the spatial variability of soil water storage capacity, from which the SCS-CN method is derived as a VIC type of model.

5.1 A new distribution function

The probability density function (PDF) of the new distribution for describing the spatial distribution of water storage capacity is represented by

$$f(C) = \frac{(2-a)\mu^2}{\left[(C+\mu)^2 - 2a\mu C\right]^{3/2}},\qquad (19)$$

where C is the point-scale water storage capacity and supported on a positive semi-infinite interval ($C \geq 0$), a is the shape parameter and its range is $0 < a < 2$, and μ is the mean of the distribution (i.e., the scale parameter). Figure 3a plots the PDFs for five sets of shape and scale parameters. When $a \leq 1$, the PDF monotonically decreases with the increase of C; i.e., the peak of the PDF occurs at $C = 0$, while when

$a > 1$ the peak of the PDF occurs at $C > 0$ and the location of the peak depends on the values of a and μ. For comparison, Fig. 3b plots the PDF for the VIC model. As shown by the solid black curve in Fig. 3b, when $0 < \beta < 1$, $f(C)$ approaches infinity as $C \to C_{\text{m}}$. It is a uniform distribution when $\beta = 1$. The peak of the PDF occurs at $C = 0$ when $\beta > 1$. Therefore, the peak of the PDF for the VIC model occurs at $C = 0$ or C_{m}.

The cumulative distribution function (CDF) corresponding to the proposed PDF is obtained by integrating Eq. (19):

$$F(C) = 1 - \frac{1}{a} + \frac{C+(1-a)\mu}{a\sqrt{(C+\mu)^2 - 2a\mu C}}.\qquad (20)$$

Figure 4a plots the CDFs corresponding to the PDFs in Fig. 3a. For comparison, Fig. 4b plots the CDFs corresponding to the PDFs in Fig. 3b. The storage capacity distribution curve for the proposed distribution is concave up for $a \leq 1$ and S shaped for $a > 1$ (Fig. 4a), while the storage capacity distribution curve for the VIC model is concave up for $\beta > 1$ and concave down for $0 < \beta < 1$ (Fig. 4b). The S shape of the CDF (Fig. 4a) is more significant with a higher value of a (e.g., $a = 1.9$). For a smaller value of a, the difference between the new PDF and the VIC type of model becomes smaller. The proposed distribution can fit the S shape of the cumulative distribution for storage capacity which is observed from soil data (Huang et al., 2003), but the generalized Pareto distribution of the VIC type of model is not able to fit the S shape of the CDF.

5.2 Deriving the SCS-CN method from the proposed distribution function

The soil wetting and surface runoff can be computed when Eq. (20) is used to describe the spatial distribution of soil water storage capacity in a catchment. The average value of storage capacity over the catchment is the mean of the distri-

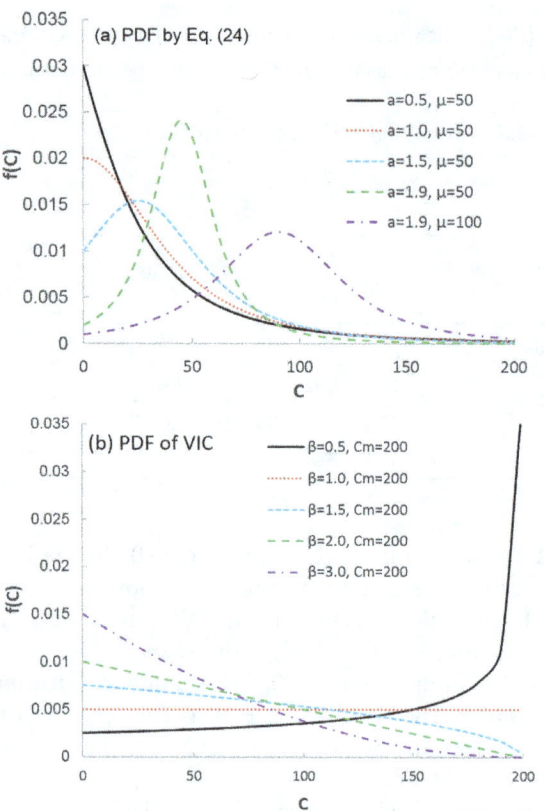

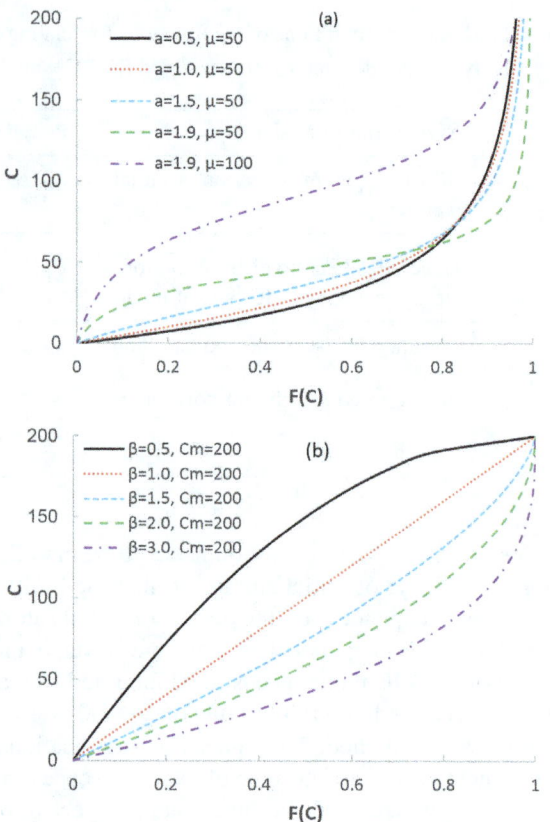

Figure 3. The probability density functions (PDFs) with different parameter values: **(a)** the proposed PDF represented by Eq. (24) and **(b)** the generalized Pareto distribution of the VIC model, i.e., Eq. (25).

Figure 4. The cumulative distribution functions (CDFs) with different parameter values: **(a)** the proposed distribution function represented by Eq. (26) and **(b)** the generalized Pareto distribution of the VIC model represented by Eq. (13).

bution:

$$\mu = S_b. \tag{21}$$

For a given C, the catchment-scale storage S can be computed by $S = \int_0^C [1 - F(x)]\,dx$ (Moore, 1985). From Eq. (20), we obtain

$$S = \frac{C + S_b - \sqrt{(C + S_b)^2 - 2aS_bC}}{a}. \tag{22}$$

For a rainfall–runoff event, the average initial storage at the catchment scale is denoted as S_0 and the corresponding value of C is denoted as C_0. Substituting S_0 and C_0 into Eq. (22), we obtain

$$m = \frac{\psi\,(2 - a\psi)}{2\,(1 - \psi)}, \tag{23}$$

where $\psi = \frac{S_0}{S_b}$ is defined in Eq. (12) and $m = \frac{C_0}{S_b}$.

The rainfall in the catchment is assumed to be spatially uniform and the rainfall depth is denoted as P. If the spatial distribution of rainfall is not uniform, the method is applied to sub-catchments where the effect of spatial variability of rainfall is negligible. The average storage at the catchment

scale after infiltration is computed by substituting $C = C_0 + P$ into Eq. (22):

$$S_1 = \frac{C_0 + P + S_b - \sqrt{(C_0 + P + S_b)^2 - 2aS_b(C_0 + P)}}{a}. \tag{24}$$

The soil wetting is computed as the difference between S_1 and S_0:

$$W = \frac{P + \sqrt{(C_0 + S_b)^2 - 2aS_bC_0} - \sqrt{(C_0 + P + S_b)^2 - 2aS_b(C_0 + P)}}{a}. \tag{25}$$

Dividing P on both sides of Eq. (25) and substituting $m = \frac{C_0}{S_b}$, we obtain

$$\frac{W}{P} = \frac{1 + \frac{S_b}{P}\sqrt{(m+1)^2 - 2am} - \sqrt{\left(1 + (m+1)\frac{S_b}{P}\right)^2 - 2am\left(\frac{S_b}{P}\right)^2 - 2a\frac{S_b}{P}}}{a}. \tag{26}$$

Substituting Eq. (13) into Eq. (26), we obtain

$$\frac{W}{P} =$$

$$\frac{1+\frac{\sqrt{(m+1)^2-2am}}{1-\psi}\Phi_{sc}-\sqrt{\left(1+\frac{m+1}{1-\psi}\Phi_{sc}\right)^2-2am\left(\frac{\Phi_{sc}}{1-\psi}\right)^2-\frac{2a}{1-\psi}\Phi_{sc}}}{a}. \quad (27)$$

Figure 5 plots Eq. (27) for $\psi = 0$, 0.4, and 0.6 when $a = 0.6$ and 1.8. As we can see, $\frac{W}{P}$ increases with a for given values of ψ and Φ_{sc}, and $\frac{W}{P}$ decreases with ψ for given values of a and Φ_{sc}, which is consistent with the VIC model and implicates that the soil wetting ratio decreases with the degree of initial saturation under a storage index. As shown in Fig. 5, Eq. (27) satisfies the lower boundary of the SCS-CN method and the upper boundary of the VIC model. Specifically, Eq. (27) satisfies the following boundary conditions (see Appendix D for details) shown in Table 1:

$$\lim_{\Phi_{sc}\to 0}\frac{W}{P}=0, \quad (28a)$$

$$\lim_{\Phi_{sc}\to\infty}\frac{W}{P}=\frac{\sqrt{(m+1)^2-2am}+a-m-1}{a\sqrt{(m+1)^2-2am}}. \quad (28b)$$

When the effect of initial storage is negligible (i.e., $\psi = 0$), $\frac{S_b}{P} = \Phi_{sc}$ from Eq. (13) and $m = 0$ from Eq. (23). Then, Eq. (27) becomes

$$\frac{W}{P}=\frac{1+\frac{S_b}{P}-\sqrt{\left(1+\frac{S_b}{P}\right)^2-2a\frac{S_b}{P}}}{a}. \quad (29)$$

Equation (29) is same as Eq. (8) with $a = 2\varepsilon(2-\varepsilon)$. We can obtain the following equation from Eq. (29) (see Appendix E for a detailed derivation):

$$\frac{Q}{P-\varepsilon W}=\frac{W-\varepsilon W}{S_b-\varepsilon W}, \quad (30)$$

where εW is defined as initial abstraction (W_i) in the SCS-CN method. Since $S_b = S_p$ when $\psi = 0$, Eq. (30) is same as Eq. (2), i.e., the proportionality relationship of the SCS-CN method.

Equation (27) is derived from the VIC-type model by using Eq. (20) to describe the spatial distribution of soil water storage capacity. From this perspective, Eq. (27) is a saturation excess runoff model. Since Eq. (27) becomes the SCS-CN method when initial storage is negligible, Eq. (27) is the modified SCS-CN method which considers the effect of initial storage on runoff generation explicitly. Therefore, the new distribution function represented by Eq. (20) unifies the SCS-CN method and the VIC type of model.

Bartlett et al. (2016a) developed an event-based probabilistic storage framework including a spatial description of prethreshold and threshold-excess runoff, and the framework has been utilized for unifying TOPMODEL, VIC, and SCS-CN (Bartlett et al., 2016b). The extended SCS-CN method (SCS-CNx) from the probabilistic storage framework is derived given the following assumptions: (1) the spatial distribution of rainfall is exponential, (2) the spatial distribution of the soil moisture deficit is uniform, and (3) the spatial distribution of storage capacity is exponential. When prethreshold

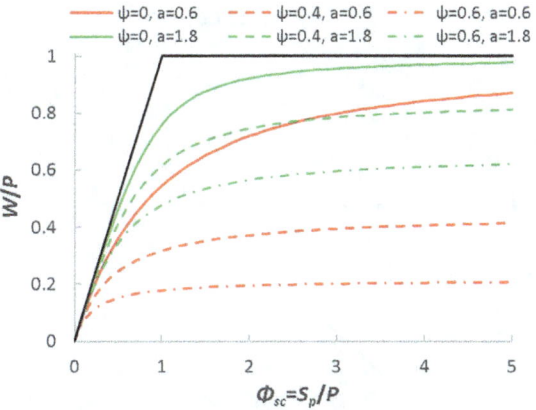

Figure 5. The effects of the degree of initial storage ($\psi = 0, 0.4$, and 0.6) and shape parameter ($a = 0.6$ and 1.8) on soil wetting in the modified SCS-CN method derived from the proposed distribution function for soil water storage capacity.

runoff is 0 (i.e., there is only threshold-excess or saturation excess runoff), the SCS-CNx method leads to the SCS-CN method without the initial abstraction term (i.e., there is no εW term in Eq. 30). In this paper, the new probability distribution function is used for storage capacity in the VIC model in which the spatial distribution of precipitation is assumed to be uniform. The obtained equation for saturation excess runoff leads to the exact SCS-CN method as shown in Eq. (30).

This research started with the following research question: if the SCS-CN method is a saturation excess runoff generation model, what is the distribution function of soil water storage capacity? Wang and Tang (2014) showed that Eq. (29) is derived from the proportionality relationship of the SCS-CN method, i.e., Eq. (30). From the comparison of boundary conditions between the SCS-CN method and the VIC type of model discussed in Sect. 4, it is observed that Eq. (29) does not include initial soil water storage, and the derived one from the distribution function will include initial soil water storage (e.g., Eq. 26). However, Eq. (29) can be viewed as the result of $S_0 = 0$, and W for Eq. (29) can be written as

$$W = \int_0^P [1 - F(x)]\,\mathrm{d}x. \quad (31)$$

From Eq. (29), one obtains

$$W = \frac{P + S_b - \sqrt{(S_b + P)^2 - 2aPS_b}}{a}. \quad (32)$$

Substituting Eq. (32) into Eq. (31), one obtains

$$\frac{P + S_b - \sqrt{(S_b + P)^2 - 2aPS_b}}{a} = \int_0^P [1 - F(C)]\,\mathrm{d}C. \quad (33)$$

Equation (20) is obtained from Eq. (33).

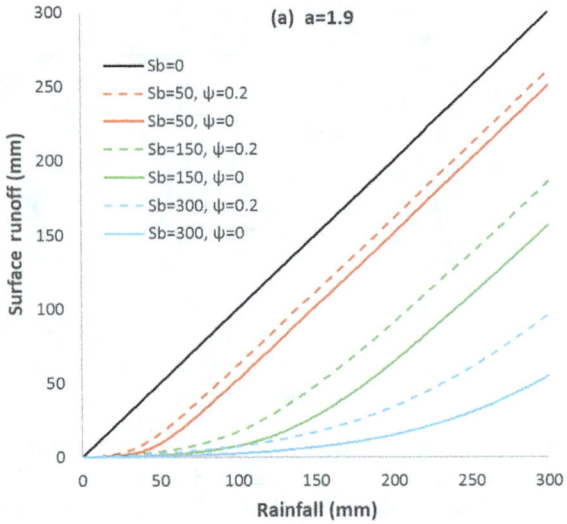

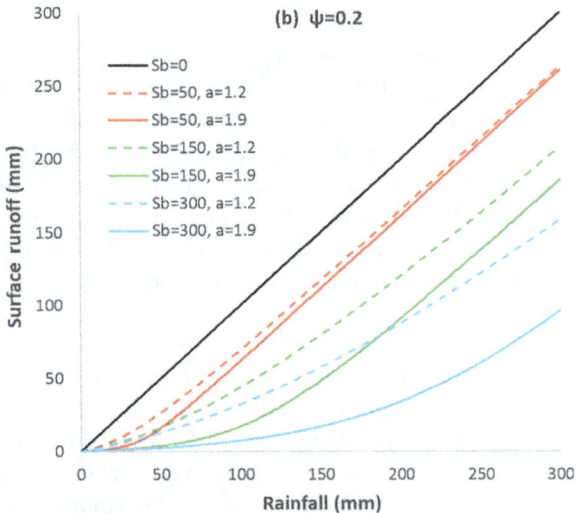

Figure 6. (a) The effects of average storage capacity and initial storage on the rainfall–runoff relation and **(b)** the effects of average storage capacity and shape parameter on the rainfall–runoff relation.

5.3 Surface runoff of the unified SCS-CN and VIC model

From the unified SCS-CN and VIC model (i.e., Eq. 26), surface runoff (Q) can be computed as

$$Q = $$

$$\frac{(a-1)\,P - S_b\sqrt{(m+1)^2 - 2am} + \sqrt{[P + (m+1)S_b]^2 - 2amS_b^2 - 2aS_bP}}{a}. \quad (34)$$

The parameter m is computed by Eq. (23) as a function of ψ and a. Equation (34) represents surface runoff as a function of precipitation (P), average soil water storage capacity (S_b), the shape parameter of the storage capacity distribution (a), and initial soil moisture (ψ). Fig. 6 plots Eq. (34) under different values of P, S_b, a, and ψ. Figure 6a shows the effects of S_b and ψ on the rainfall–runoff relationship

with a given shape parameter of $a = 1.9$. The solid lines show the rainfall–runoff relations with zero initial storage ($\psi = 0$) and the dashed lines show the rainfall–runoff relations with $\psi = 0.2$. Given the same amount of precipitation and storage capacity, wetter soil ($\psi = 0.2$) generates more surface runoff than drier soil ($\psi = 0$), and the difference of runoff is higher for watersheds with larger average storage capacity. Figure 6b shows the effects of S_b and a on the rainfall–runoff relationship with a given initial soil moisture ($\psi = 0.2$). The solid lines show the rainfall–runoff relations for $a = 1.9$ and the dashed lines show the rainfall–runoff relations for $a = 1.2$. As we can see, the shape parameter affects the runoff generation significantly for watersheds with larger average storage capacity.

In the SCS-CN method, surface runoff is computed as $Q = \frac{(P - 0.2S_b)^2}{P + 0.8S_b}$. The effect of initial soil moisture on runoff is considered implicitly by varying the curve number for normal, dry, and wet conditions depending on the antecedent moisture condition. In the unified SCS-CN model shown in Eq. (34), the effect of initial soil moisture is explicitly included through ψ, which is the ratio between average initial water storage and average storage capacity. In the SCS-CN method, the value of initial abstraction W_i is parameterized as a function of average storage capacity; i.e., $W_i = 0.2S_b$. In the unified SCS-CN model shown in Eq. (34), W_i is dependent on the shape parameter a. Therefore, the unified SCS-CN model extends the original SCS-CN method for including the effect of initial soil moisture explicitly and estimating the parameter for initial abstraction.

6 Conclusions

In this paper, the SCS-CN method and the saturation excess runoff models based on distribution functions (e.g., VIC model) are presented in terms of soil wetting (i.e., infiltration). Like the Budyko framework, the relationship between the soil wetting ratio and the soil storage index is obtained for the SCS-CN method and the VIC type of model. It is found that the boundary conditions for the obtained functions do not fully match. For the SCS-CN method, the soil wetting ratio approaches 1 when the soil storage index approaches infinity, and this is due to the limitation of the SCS-CN method; i.e., the initial soil moisture condition is not explicitly represented in the proportionality relationship. However, for the VIC type of model, the soil wetting ratio equals the soil storage index when the soil storage index is lower than a certain value, and this is due to the finite bound of the distribution function of storage capacity.

In this paper, a new distribution function, which is supported by $x \in [0, \infty)$ instead of a finite upper bound, is proposed for describing the spatial distribution of soil water storage capacity. From this new distribution function, an equation is derived for the relationship between the soil wetting ratio and the storage index, and this equation satisfies the following boundary conditions: when the storage index approaches 0, the soil wetting ratio approaches 0; when the storage index approaches infinity, the soil wetting ratio approaches a certain value (≤ 1) depending on the initial storage (e.g., at the beginning of a rainfall event, runoff is generated at the initially saturated areas, Yu et al., 2001; Gao et al., 2018). Meanwhile, the model becomes the exact SCS-CN method when initial storage is negligible. Therefore, the new distribution function for soil water storage capacity explains the SCS-CN method as a saturation excess runoff model and unifies the SCS-CN method and the VIC type of model for surface runoff modeling.

Future potential work could test the performance of the proposed new distribution function for quantifying the spatial distribution of storage capacity by analyzing the spatially distributed soil data. On the one hand, the distribution functions of the probability distributed model (Moore, 1985), VIC model, and Xinanjiang model could be replaced by the new distribution function and the model performance would be further evaluated. On the other hand, the extended SCS-CN method (i.e., Eq. 27), which includes initial storage explicitly, could be used for surface runoff modeling in the SWAT (Soil and Water Assessment Tool) model, and the model performance would be evaluated.

Appendix A

The potential for continuing wetting is called potential maximum retention and is denoted by $S_m = S_p - W_i$. S_m is computed as a function of curve number which is dependent on land use–land cover and soil permeability. The ratio between W_i and S_m in the SCS curve number method is denoted by $\lambda = \frac{W_i}{S_p - W_i}$, and then the ratio between initial soil wetting and storage capacity is computed by

$$\frac{W_i}{S_p} = \frac{\lambda}{1+\lambda}. \tag{A1}$$

The value of λ varies in the range of $0 \leq \lambda \leq 0.3$, and a value of 0.2 is usually used (Ponce and Hawkins, 1996). Substituting Eq. (A1) into Eq. (6) leads to

$$\frac{W}{P} = \frac{1 - \left(\frac{\lambda}{1+\lambda}\right)^2 \Phi_{sc}}{1 - \frac{2\lambda}{1+\lambda} + \Phi_{sc}^{-1}}. \tag{A2}$$

Equation (A2) is plotted in Fig. 1 for $\lambda = 0.1$ and 0.3. As we can see, the range of Φ_{sc} is dependent on the parameter λ. Since $W_i \leq P$, Φ_{sc} is in the range of $\left[0, 1 + \frac{1}{\lambda}\right]$. Equation (A2) satisfies the following boundary conditions: $\frac{W}{P} \to 0$ as $\Phi_{sc} \to 0$ and $\frac{W}{P} \to 1$ as $\Phi_{sc} \to \frac{\lambda+1}{\lambda}$. When $\lambda \to 0$, Eq. (A2) becomes

$$\frac{W}{P} = \frac{1}{1 + \Phi_{sc}^{-1}}. \tag{A3}$$

Equation (A3) is the lower bound for $\frac{W}{P}$ based on this parameterization scheme.

Appendix B

Substituting $W = S_1 - S_0$ into Eq. (16), wetting is computed by

$$W = S_b \left[1 - \left(1 - \frac{P + C_0}{C_m}\right)^{\beta+1}\right] - S_0. \tag{B1}$$

The following equation is obtained by dividing P on both sides of Eq. (B1):

$$\frac{W}{P} = \frac{S_b - S_0}{P} - \frac{S_b}{P}\left(1 - \frac{P + C_0}{C_m}\right)^{\beta+1}. \tag{B2}$$

Substituting $\frac{C_0}{C_m} = 1 - \left(1 - \frac{S_0}{S_b}\right)^{\frac{1}{\beta+1}}$ into Eq. (B2), we obtain

$$\frac{W}{P} = \frac{S_b - S_0}{P} - \frac{S_b}{P}\left(1 - \frac{P}{C_m} - \left[1 - \left(1 - \frac{S_0}{S_b}\right)^{\frac{1}{\beta+1}}\right]\right)^{\beta+1}. \tag{B3}$$

Substituting Eq. (10) into Eq. (B3), we obtain

$$\frac{W}{P} = \frac{S_b - S_0}{P} - \left(\left(\frac{S_b - S_0}{P}\right)^{\frac{1}{\beta+1}} - \frac{\left(\frac{S_b}{P}\right)^{-\frac{\beta}{\beta+1}}}{\beta+1}\right)^{\beta+1}. \tag{B4}$$

Substituting Eqs. (5) and (13) into Eq. (B4), we obtain

$$\frac{W}{P} = \Phi_{sc} - \left(\Phi_{sc}^{\frac{1}{\beta+1}} - \frac{\left(\frac{\Phi_{sc}}{1-\psi}\right)^{-\frac{\beta}{\beta+1}}}{\beta+1}\right)^{\beta+1}, \tag{B5}$$

which leads to

$$\frac{W}{P} = \Phi_{sc}\left[1 - \left(1 - b\Phi_{sc}^{-1}\right)^{\beta+1}\right], \tag{B6}$$

where b is defined in Eq. (14).

Appendix C

$$\lim_{\Phi_{sc}\to\infty}\frac{W}{P}=\lim_{\Phi_{sc}\to\infty}\Phi_{sc}\left[1-\left(1-b\Phi_{sc}^{-1}\right)^{\beta+1}\right] \tag{C1}$$

The right-hand side of Eq. (C1) is rewritten as

$$\lim_{\Phi_{sc}\to\infty}\Phi_{sc}\left[1-\left(1-b\Phi_{sc}^{-1}\right)^{\beta+1}\right]=$$
$$\lim_{\Phi_{sc}\to\infty}\frac{1-\left(1-b\Phi_{sc}^{-1}\right)^{\beta+1}}{\Phi_{sc}^{-1}}. \tag{C2}$$

Since $\lim_{\Phi_{sc}\to\infty}1-\left(1-b\Phi_{sc}^{-1}\right)^{\beta+1}=0$ and $\lim_{\Phi_{sc}\to\infty}\Phi_{sc}^{-1}=0$, we apply the L'Hospital's rule,

$$\lim_{\Phi_{sc}\to\infty}\frac{\left[1-\left(1-b\Phi_{sc}^{-1}\right)^{\beta+1}\right]'}{\left(\Phi_{sc}^{-1}\right)'}=$$
$$\lim_{\Phi_{sc}\to\infty}b(\beta+1)\left(1-b\Phi_{sc}^{-1}\right)^{\beta}. \tag{C3}$$

Since $\lim_{\Phi_{sc}\to\infty}\left(1-b\Phi_{sc}^{-1}\right)^{\beta}=1$, the limit for $\frac{W}{P}$ is obtained as follows:

$$\lim_{\Phi_{sc}\to\infty}\frac{W}{P}=b(\beta+1). \tag{C4}$$

Substituting Eq. (14) into Eq. (C4), we obtain

$$\lim_{\Phi_{sc}\to\infty}\frac{W}{P}=(1-\psi)^{\frac{\beta}{\beta+1}}. \tag{C5}$$

Appendix D

$$\lim_{\Phi_{sc}\to\infty}\frac{W}{P}=\lim_{\Phi_{sc}\to\infty}$$
$$\frac{1+\frac{\sqrt{(m+1)^2-2am}}{1-\psi}\Phi_{sc}-\sqrt{\left(1+\frac{m+1}{1-\psi}\Phi_{sc}\right)^2-2am\left(\frac{\Phi_{sc}}{1-\psi}\right)^2-\frac{2a}{1-\psi}\Phi_{sc}}}{a} \tag{D1}$$

Multiplying

$$1+\frac{\sqrt{(m+1)^2-2am}}{1-\psi}\Phi_{sc}$$
$$+\sqrt{\left(1+\frac{m+1}{1-\psi}\Phi_{sc}\right)^2-2am\left(\frac{\Phi_{sc}}{1-\psi}\right)^2-\frac{2a}{1-\psi}\Phi_{sc}} \tag{D2}$$

with the denominator and numerator of the right-hand side Eq. (D1) leads to

$$\lim_{\Phi_{sc}\to\infty}\frac{W}{P}=\frac{1}{a}\lim_{\Phi_{sc}\to\infty}$$
$$\frac{\frac{2\sqrt{(m+1)^2-2am}}{1-\psi}\Phi_{sc}-\frac{2(m+1)}{1-\psi}\Phi_{sc}+\frac{2a}{1-\psi}\Phi_{sc}}{1+\frac{\sqrt{(m+1)^2-2am}}{1-\psi}\Phi_{sc}+\sqrt{\left(1+\frac{m+1}{1-\psi}\Phi_{sc}\right)^2-2am\left(\frac{\Phi_{sc}}{1-\psi}\right)^2-\frac{2a}{1-\psi}\Phi_{sc}}}. \tag{D3}$$

Dividing Φ_{sc} in the denominator and numerator, we obtain

$$\lim_{\Phi_{sc}\to\infty}\frac{W}{P}=\frac{1}{a(1-\psi)}\lim_{\Phi_{sc}\to\infty}$$
$$\frac{2\sqrt{(m+1)^2-2am}-2(m+1)+2a}{\frac{1}{\Phi_{sc}}+\frac{\sqrt{(m+1)^2-2am}}{1-\psi}+\sqrt{\left(\frac{1}{\Phi_{sc}}+\frac{m+1}{1-\psi}\right)^2-2am\left(\frac{1}{1-\psi}\right)^2-\frac{2a}{(1-\psi)\Phi_{sc}}}}. \tag{D4}$$

Therefore, the limit of $\frac{W}{P}$ as $\Phi_{sc}\to\infty$ is

$$\lim_{\Phi_{sc}\to\infty}\frac{W}{P}=\frac{\sqrt{(m+1)^2-2am}+a-m-1}{a\sqrt{(m+1)^2-2am}}. \tag{D5}$$

Appendix E

Substituting $a=2\varepsilon(2-\varepsilon)$ into Eq. (29), one can obtain

$$\frac{W}{P}=\frac{1+\frac{S_b}{P}-\sqrt{\left(1+\frac{S_b}{P}\right)^2-4\varepsilon(2-\varepsilon)\frac{S_b}{P}}}{2\varepsilon(2-\varepsilon)}. \tag{E1}$$

Equation (E1) is the solution of the following quadratic function:

$$\varepsilon(2-\varepsilon)\left(\frac{W}{P}\right)^2-\left(1+\frac{S_b}{P}\right)\frac{W}{P}+\frac{S_b}{P}=0. \tag{E2}$$

Multiplying P^2 on both sides of Eq. (E2), Eq. (E2) becomes

$$\varepsilon(2-\varepsilon)W^2-(P+S_b)W+S_bP=0. \tag{E3}$$

Equation (E3) can be written as the following one:

$$\frac{P-W}{P-\varepsilon W}=\frac{W-\varepsilon W}{S_b-\varepsilon W}. \tag{E4}$$

Substituting $Q=P-W$ into Eq. (E4), we obtain the proportionality relationship of the SCS-CN method:

$$\frac{Q}{P-\varepsilon W}=\frac{W-\varepsilon W}{S_b-\varepsilon W}. \tag{E5}$$

Competing interests. The authors declare that they have no conflict of interest.

Acknowledgements. This research was funded in part under award CBET-1804770 from the National Science Foundation (NSF) and the United States Geological Survey (USGS) Powell Center Working Group Project "A global synthesis of land-surface fluxes under natural and human-altered watersheds using the Budyko framework". The authors would also like to thank the Associate Editor and three reviewers for their constructive comments and suggestions that have led to substantial improvements over an earlier version of the manuscript.

Edited by: Zhongbo Yu

References

Bartlett, M. S., Parolari, A. J., McDonnell, J. J., and Porporato, A.: Beyond the SCS-CN method: A theoretical framework for spatially lumped rainfall-runoff response, Water Resour. Res., 52, 4608–4627, https://doi.org/10.1002/2015WR018439, 2016a.

Bartlett, M. S., Parolari, A. J., McDonnell, J. J., and Porporato, A.: Framework for event-based semidistributed modeling that unifies the SCS-CN method, VIC, PDM, and TOPMODEL, Water Resour. Res., 52, 7036–7052, https://doi.org/10.1002/2016WR019084, 2016b.

Beven, K.: Rainfall-Runoff Modelling: The Primer, 2nd Edn., Wiley-Blackwell, Chichester, UK, 2012.

Beven, K. and Kirkby, M. J.: A physically based, variable contributing area model of basin hydrology, Hydrol. Sci. J., 24, 43–69, 1979.

Bras, R. L.: Hydrology: an introduction to hydrologic science, Addison Wesley Publishing Company, Reading, MA, 1990.

Budyko, M. I.: Climate and Life, 508 pp., Academic Press, New York, 1974.

Chen, X., Alimohammadi, N., and Wang, D.: Modeling interannual variability of seasonal evaporation and storage change based on the extended Budyko framework, Water Resour. Res., 49, 6067–6078, https://doi.org/10.1002/wrcr.20493, 2013.

Easton, Z. M., Fuka, D. R., Walter, M. T., Cowan, D. M., Schneiderman, E. M., and Steenhuis, T. S.: Re-conceptualizing the soil and water assessment tool (SWAT) model to predict runoff from variable source areas, J. Hydrol., 348, 279–291, 2008.

Gao, H., Birkel, C., Hrachowitz, M., Tetzlaff, D., Soulsby, C., and Savenije, H. H. G.: A simple topography-driven and calibration-free runoff generation module, Hydrol. Earth Syst. Sci. Discuss., https://doi.org/10.5194/hess-2018-141, in review, 2018.

Hooshyar, M. and Wang, D.: An analytical solution of Richards' equation providing the physical basis of SCS curve number method and its proportionality relationship, Water Resour. Res., 52, 6611–6620, https://doi.org/10.1002/2016WR018885, 2016.

Huang, M., Liang, X., and Liang, Y.: A transferability study of model parameters for the variable infiltration capacity land surface scheme, J. Geophys. Res., 108, 8864, https://doi.org/10.1029/2003JD003676, 2003.

Liang, X., Lettenmaier, D. P., Wood, E. F., and Burges, S. J.: A simple hydrologically based model of land surface water and energy fluxes for general circulation models, J. Geophys. Res.-Atmos., 99, 14415–14428, 1994.

Lyon, S. W., Walter, M. T., Gérard-Marchant, P., and Steenhuis, T. S.: Using a topographic index to distribute variable source area runoff predicted with the SCS curve – number equation, Hydrol. Process., 18, 2757–2771, 2004.

Mishra, S. K. and Singh, V. P.: Another look at SCS-CN method, J. Hydrol. Eng., 4, 257–264, 1999.

Mockus, V.: National Engineering Handbook Section 4, Hydrology, NTIS, available at: https://directives.sc.egov.usda.gov/OpenNonWebContent.aspx?content=18393.wba (last access: 19 December 2018), 1972.

Moore, R. J.: The probability-distributed principle and runoff production at point and basin scales, Hydrol. Sci. J., 30, 273–297, 1985.

Ponce, V.: Notes of my conversation with Vic Mockus, unpublished material, available at: http://mockus.sdsu.edu/ (last access: 29 September 2017) 1996.

Ponce, V. M. and Hawkins, R. H.: Runoff curve number: has it reached maturity?, J. Hydrol. Eng., 1, 9–20, 1996.

Porporato, A., Daly, E., and Rodriguez-Iturbe, I.: Soil Water Balance and Ecosystem Response to Climate Change, Am. Nat., 164, 625–632, 2004.

Sivapalan, M., Beven, K., and Wood, E. F.: On hydrologic similarity: 2. A scaled model of storm runoff production, Water Resour. Res., 23, 2266–2278, 1987.

Steenhuis, T. S., Winchell, M., Rossing, J., Zollweg, J. A., and Walter, M. F.: SCS runoff equation revisited for variable-source runoff areas, J. Irrig. Drain. Eng., 121, 234–238, 1995.

Tang, Y. and Wang, D.: Evaluating the role of watershed properties in long-term water balance through a Budyko equation based on two-stage partitioning of precipitation, Water Resour. Res., 53, 4142–4157, https://doi.org/10.1002/2016WR019920, 2017.

Wang, D. and Tang, Y.: A one-parameter Budyko model for water balance captures emergent behavior in Darwinian hydrologic models, Geophys. Res. Lett., 41, 4569–4577, https://doi.org/10.1002/2014GL060509, 2014.

Wang, D., Zhao, J., Tang, Y., and Sivapalan, M.: A thermodynamic interpretation of Budyko and L'vovich formulations of annual water balance: Proportionality hypothesis and maximum entropy production, Water Resour. Res., 51, 3007–3016, https://doi.org/10.1002/2014WR016857, 2015.

Wood, E. F., Lettenmaier, D. P., and Zartarian, V. G.: A land – surface hydrology parameterization with subgrid variability for general circulation models, J. Geophys. Res.-Atmos., 97, 2717–2728, 1992.

Yu, B.: Theoretical justification of SCS method for runoff estimation, J. Irrig. Drain. Eng., 124, 306–310, 1998.

Yu, Z., Carlson, T. N., Barron, E. J., and Schwartz, F. W.: On evaluating the spatial-temporal variation of soil moisture in the Susquehanna River Basin, Water Resour. Res., 34, 1313–1326, 2001.

Zhao, R.: Flood forecasting method for humid regions of China, East China College of Hydraulic Engineering, Nanjing, China, 1977.

Zhao, R.: The Xinanjiang model applied in China, J. Hydrol., 135, 371–381, 1992.

13

Sharing water and benefits in transboundary river basins

Diane Arjoon[1], Amaury Tilmant[1], and Markus Herrmann[2]

[1]Department of Civil Engineering and Water Engineering, Université Laval, Québec, Canada
[2]Department of Economics, Université Laval, Québec, Canada

Correspondence to: Amaury Tilmant (amaury.tilmant@gci.ulaval.ca)

Abstract. The equitable sharing of benefits in transboundary river basins is necessary to solve disputes among riparian countries and to reach a consensus on basin-wide development and management activities. Benefit-sharing arrangements must be collaboratively developed to be perceived not only as efficient, but also as equitable in order to be considered acceptable to all riparian countries. The current literature mainly describes what is meant by the term benefit sharing in the context of transboundary river basins and discusses this from a conceptual point of view, but falls short of providing practical, institutional arrangements that ensure maximum economic welfare as well as collaboratively developed methods for encouraging the equitable sharing of benefits. In this study, we define an institutional arrangement that distributes welfare in a river basin by maximizing the economic benefits of water use and then sharing these benefits in an equitable manner using a method developed through stakeholder involvement. We describe a methodology in which (i) a hydrological model is used to allocate scarce water resources, in an economically efficient manner, to water users in a transboundary basin, (ii) water users are obliged to pay for water, and (iii) the total of these water charges is equitably redistributed as monetary compensation to users in an amount determined through the application of a sharing method developed by stakeholder input, thus based on a stakeholder vision of fairness, using an axiomatic approach. With the proposed benefit-sharing mechanism, the efficiency–equity trade-off still exists, but the extent of the imbalance is reduced because benefits are maximized and redistributed according to a key that has been collectively agreed upon by the participants. The whole system is overseen by a river basin authority. The methodology is applied to the Eastern Nile River basin as a case study. The described technique not only ensures economic efficiency, but may also lead to more equitable solutions in the sharing of benefits in transboundary river basins because the definition of the sharing rule is not in question, as would be the case if existing methods, such as game theory, were applied, with their inherent definitions of fairness.

1 Introduction

With growing water scarcity, as a result of expanding population demand, environmental concerns and climate change effects, there is increased international recognition of the importance of cooperation for the effective governance of water resources. This is particularly evident in the case of transboundary river basins in which unidirectional, negative externalities, caused by the upstream regulation of the natural flow, often place some parties at a disadvantage and result in asymmetric relationships that add to the challenge of coordinating resource use (van der Zaag, 2007). There is a consensus among water professionals that the cooperative management of shared river basins should provide opportunities to increase the scope and scale of benefits (Phillips et al., 2006; Grey and Sadoff, 2007; Leb, 2015), stepping beyond the volumetric allocation of water that reduces negotiations between riparians to a zero-sum game. In their seminal paper, Sadoff and Grey (2002) discussed the types of benefits that river basins can provide, assuming cooperation: *benefits to the river* can result from sustainable cooperative management of the ecosystem; efficient, cooperative management and development of river flow can yield *benefits from the river* in the form of increased water quality, quantity and productivity; policy shifts away from riparian disputes/conflicts toward cooperative development can reduce costs of non-cooperation arising *because of the river*; and cooperation between riparian states can lead to economic, political and institutional integration, resulting in *benefits beyond the river*.

A large proportion of past research has focused mainly on the economic benefits of cooperation (benefits from the river). Focussing on benefits in strictly economic terms does not lessen the importance of benefits from other spheres (Qaddumi, 2008). An economic perspective, however, may be an effective method for encouraging cooperation because it may help riparian countries to realize win–win situations (Dombrowsky, 2009).

The traditional approach to estimating the economic benefits of cooperation relies on hydro-economic modeling (Arjoon et al., 2014; Jeuland et al., 2014; Tilmant and Kinzelbach, 2012; Teasley and McKinney, 2011; Whittington et al., 2005). These studies present various implementation strategies representing various levels of cooperation, but all show that there are significant economic benefits to be had through basin-wide cooperation. However, economic efficiency is not necessarily compatible with equitability due to the different production abilities of water users (Wang et al., 2003). Analytical methods, including game theory solutions such as the Shapley value (Jafarzadegan et al., 2013; Abed-Elmdoust and Kerachian, 2012) and bankruptcy theory (Sechi and Zucca, 2015; Mianabadi et al., 2014, 2015; Madani et al., 2014; Ansink and Weikard, 2012), have been examined for use in water allocation as equitable alternatives to the efficient economic allocation produced by hydro-economic models. Analytical methods were also used by van der Zaag et al. (2002), who looked at possible equitable criteria for sharing water and developed allocation algorithms to operationalize these, applying them to the Orange, Nile and Incomati rivers. It has been argued that the notion of equity, or fairness, involves a cultural component that should be incorporated into any type of water policy and, therefore, stakeholder involvement in decision-making is a significant determinant in the judgement of fairness (Syme et al., 1999; Asmamaw, 2015). The explicit provision of benefit-sharing arrangements that are collaboratively developed and, thus, perceived as fair, is therefore necessary to help solve disputes and to reach a consensus in transboundary river basin development and management activities (MRC Initiative on Sustainable Hydropower, 2011).

Increasingly, efforts are focussing on the sharing of benefits generated through cooperation in order to solve the problem of equitability. The rapidly growing body of literature on benefit sharing mainly describes what is meant by this in the context of transboundary river basins and discusses benefit sharing from a conceptual point of view (Suhardiman et al., 2014; Skinner et al., 2009; Qaddumi, 2008). This literature introduces and defines different approaches but falls short of providing practical institutional arrangements for the sharing of benefits. Recently, Ding et al. (2016) introduced a methodology to address the problem of water allocation in the Nile River through a revenue re-distribution mechanism that leads to a fairly allocated revenue for each water user based on the proportion of its contribution to the basin.

Analytical methods, such as game theory and related bankruptcy methods, may also be useful for determining ways to fairly allocate generated benefits. Game theory, which is the mathematical study of competition and cooperation, can provide a somewhat realistic simulation of the interest-based behavior of stakeholders (Madani, 2010). The framework that relates the preferences of players to the observable features of a game is the hypothesis that players care about nothing except their own payoffs (Hausman, 1999). Fair outcomes are captured in solution concepts such as the *core*, which selects the payoff allocations that give each group of individuals no less than their collective worth and the *Shapley value* in which payoffs are related to the marginal contributions of individuals to a coalition (de Clippel and Rozen, 2013). The aim of bankruptcy methods is to distribute an estate or asset among a group of creditors, all having a claim to the asset, where the sum of the creditors' claims is larger that the amount available to distribute (Herrero and Villar, 2001). An overview of bankruptcy rules has been presented by Thomson (2003, 2013). Each bankruptcy rule defines fairness based on the properties underlying the rule. The three most well-known bankruptcy rules (the proportional rule, the constrained equal awards rule and the constrained equal losses rule) all define equity through the *equal treatment of equals* requirement in which agents with identical claims should be treated the same[1]. In other words, agents with the same claim should receive the same compensation. The analysis and formulation of properties and principles of distribution rules, such as those in cooperative game theory and bankruptcy theory, are the object of the axiomatic method (Thomson, 2001).

The axiomatic method allows desirable properties to be translated into a sharing rule. If a particular rule has been adopted to solve a problem involving a group of agents, it is assumed that all agents have agreed on the properties that such a rule fulfills. The concept of fairness, then, can be embedded into a rule. The axiomatic approach is easily incorporated into negotiations because the axioms can be interpreted quite naturally as describing characteristics of a negotiation procedure (Ansink and Houba, 2014).

As discussed previously, the economically efficient allocation of water is not necessarily equitable. Axiomatic approaches, on the other hand, allow the characterization of an equitable distribution of welfare, but do not necessarily maximize the aggregated economic welfare over the basin. Institutional arrangements that ensure maximum economic welfare, as well as the equitable sharing of these benefits over the basin, are required.

---

[1]Equal treatment of equals is one of the properties upon which these bankruptcy rules are defined. For a complete discussion of all properties, refer to Thomson (2003, 2013).

In this study we define an institutional arrangement that distributes welfare in a river basin by maximizing the economic benefits of water use and then sharing these benefits in an equitable manner. The methodology relies on a pseudo-market arrangement in the form of a highly regulated market in which the behavior of water users is restrained to control externalities associated with water transfers and to ensure basin-wide coordination and enhanced efficiency. The term pseudo-market indicates that bulk water users are not free to choose how much water will be moved in the system. Freedom of contract and private property rights, which are necessary conditions for the existence of a market, are restrained, giving rise to a pseudo-market[2]. These restrictions are due to the flow characteristics of water and to the need to account for externalities and third-party effects, which can seldom be achieved within a traditional market.

The institutional arrangement described in this paper should encourage full cooperation between water users because it is intended as a replacement for traditional types of agreements on international river basins, which can lead to distrust and tension between riparian countries. What we present is an entirely different perspective that may help to avoid the pitfalls and limitations of current agreements.

In the following section, we describe this arrangement, which uses a hydro-economic model to determine the economically efficient allocation of water and a collaboratively developed sharing method for the equitable allocation of monetary benefits. Section 3 presents the application of this framework to the Eastern Nile River basin. Section 4 presents and discusses the results and Sect. 5 concludes the paper.

2 Methodology

In the proposed pseudo-market approach, a river basin authority (RBA) plays the role of water system operator, identifying economically efficient allocation policies that are then imposed on the agents (water users). The agents are charged for water use and these payments are redistributed to ensure equitability among the users. In this particular system, the mandate of the RBA consists of (1) collecting information on water use and productivity, (2) efficiently allocating water between the different agents in the system based on the information collected in the first step, (3) preserving the hydrologic integrity of the river basin, and (4) coordinating the collection and redistribution of the benefits associated with the optimal allocation policies.

2.1 Information collection

In this first step, the RBA collects information that is required to assess the demand curves, or at least the productivity (unit net benefit), of all users in the system, once at the beginning

of each year. The information must be validated to ensure that it is complete and reasonable since the economically efficient allocation of water in the next step depends on it. The collection of information can be the basis of a bidding process in which agents offer to buy water at a given price. In the case of irrigation agents, information such as crop area, crop type, yield, crop price and crop water requirement over a period can be used to determine the bid for each agent and, based on the bid information, the demand curve can be inferred using the residual imputation method (Pulido-Velazquez et al., 2008; Riegels et al., 2013). This method assumes that all input costs, except for the cost of water, are known. The water value is then imputed as the residual of the observed gross benefits after all non-water costs are subtracted (Young, 2005).

In order to control the declarations of agents in the agricultural sector, the RBA can use techniques such as remote sensing to validate land classification and cropping areas (Gallego et al., 2014; El-Kawy et al., 2011; Rozenstein and Karnieli, 2011). As an example, the European Union uses an Integrated Administration and Control System (IACS), which includes a land-parcel identification system (LPIS), to control declarations from farmers for financial aid grants (Oesterle and Hahn, 2004). The LPIS uses orthophotos to monitor the evolution of the land cover and the management of crops, and enables more accurate declarations by farmers.

In the case of hydropower, information regarding energy production and scheduling is important. For example, power plants might be offline for maintenance or might be obliged to generate a minimum amount of energy to meet their contractual commitments. Also, water use requirements such as environmental flow and minimum domestic use supply will be required.

The unconstrained or expected net benefits (ENB) for a water user are the consumer surplus (Fig. 1), which is the area under the demand curve above the price P_D. The surplus is the private user cost of water and corresponds to the willingness to pay for the last unit of water demanded in a situation where allocation is unconstrained. This area is made up of three regions (A, B and C) that will be discussed later.

2.2 Water allocation

Once water user information has been collected, allocation decisions are identified by matching demand with supply in a cost-efficient way, i.e., by giving priority of access to users with the highest productivity. In order to do this, an aggregation of the demand curve is carried out, which means that a distinction must be made between rival and non-rival water uses. When water users are not in competition for the same unit of water, non-rivalness is observed. For example, water flowing through a dam may be considered a non-rival water use since a unit of water released through one dam can be used downstream by another dam. In rival water use, units are consumed and are no longer available to other water users

---

[2]One could also argue that a true market is created by assuming that every agent agrees with, and respects, having to pay for water.

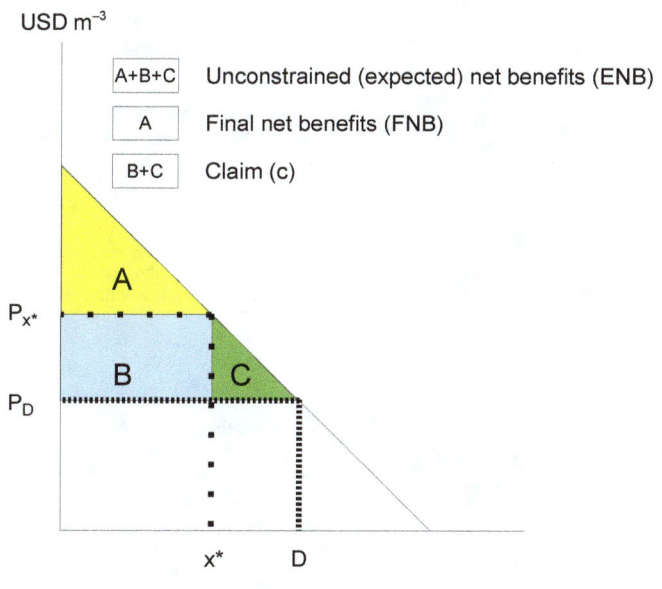

Figure 1. Demand curve. D: quantity of water demanded for a time period; x^*: quantity of water allocated for a time period; P: price of water.

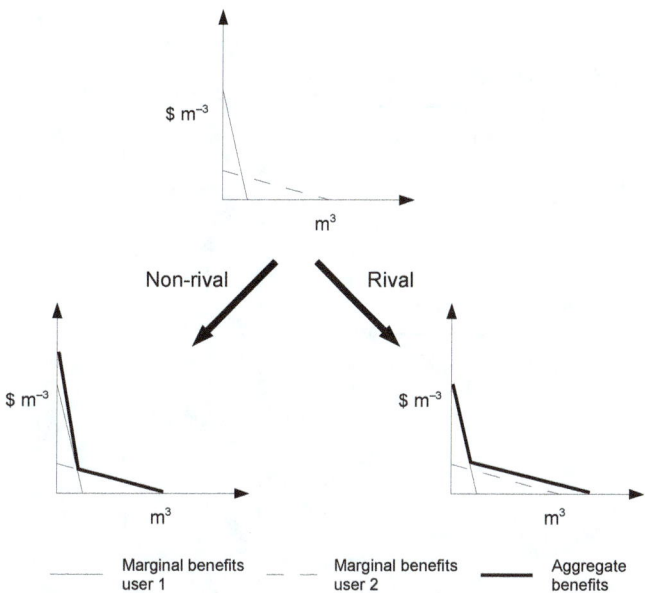

Figure 2. Aggregation of demand curves for rival and non-rival water uses for a given time period.

(for example, water lost to irrigation or water held in a reservoir during a period when it is required downstream for irrigation). In this case, the demand curves are summed horizontally (see Fig. 2). Rival water uses need to be coordinated to prevent conflicts. The decision to divert one additional unit of water to any rival use depends on the at-source value[3] of water for that use. If this value is larger than the at-source value of all downstream marginal users, then it will be diverted to the rival use. See Tilmant and Kinzelbach (2012) for a detailed description of rival and non-rival water uses. The value of the last unit of water at any site, then, is the sum of the marginal values of the non-rival users since the demand curves can be summed up vertically (see Fig. 2). This aggregation of the demand curve is done automatically in hydro-economic models. Hydro-economic models, then, can be used to determine the allocation of water between users at the same site and over a basin (comprising a number of sites) and to determine the marginal value of water and economic benefits at each site. A description of the mathematical formulation involved is given in the Appendix.

---

[3]The at-source value of water is observed at the location where bulk water is diverted. The at-site value corresponds to the value of water delivered to the users (for example, a farm at the end of a conveyance and distribution system). At-site water values are generally larger than at-source values because they include losses in the system and conveyance costs. In the study of intersectoral allocation choices, at-source water values should be used (Young, 2005).

2.3 Collection of bulk water charges

Based on the water allocation decisions and the corresponding water fluxes, pseudo-market transactions occur between the RBA and the water users. Users must pay the RBA for the water allocated to them. The cost of water is the marginal water value or shadow price (λ) calculated by the hydro-economic model at the site of water abstraction or use. Economic theory indicates that for efficient water allocation to occur, the price that users pay for the resource must be equal to the marginal value of still available opportunities of water use, which reflects the social cost of using water at a particular site. If the user pays less than this, the resource is overconsumed or overutilized, as no efficient rationing occurs. Conversely, a user price higher than the marginal value would result in underconsumption/underutilization.

The RBA charges for the water entering the system in order to cover the costs associated with its mandates (conservation, coordination, compensation). In the case of consumptive users, water is purchased from the RBA at the marginal water value (the value of a marginal unit of water) at the site of abstraction. Non-consumptive users buy inflow from the RBA at a price equal to the difference between the marginal value of water at the user site and the marginal value of water at the downstream site (Fig. 3). This bulk water charge system is based on a dynamic water accounting framework presented by Tilmant et al. (2015).

Payment for bulk water use has been addressed, recently, by the United Nations in their 2014 World Water Development Report (United Nations World Water Assessment Programme, 2014) in which they state that economic instruments such as markets for buying and selling a resource (such

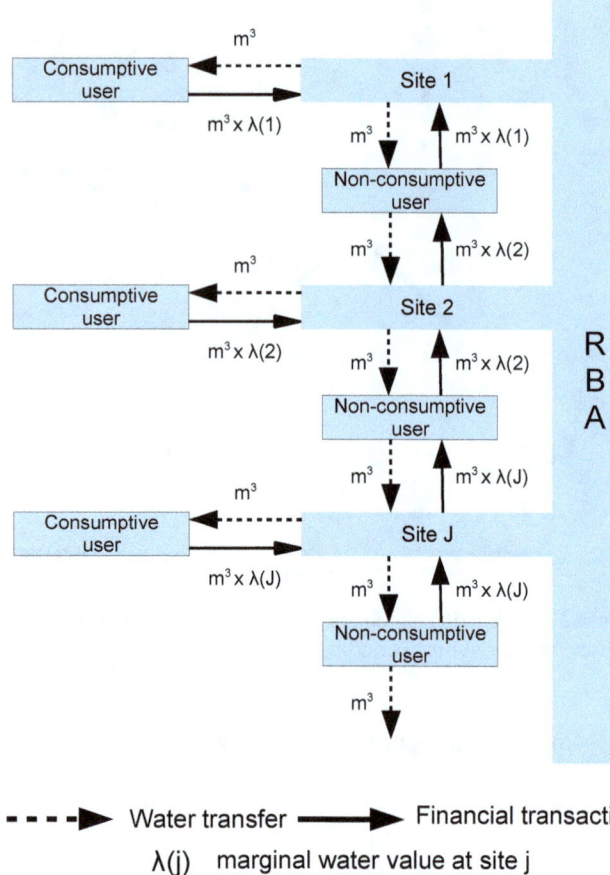

- - - ▶ Water transfer   ━━━▶ Financial transacti

$\lambda(j)$ marginal water value at site j

Figure 3. Collection of bulk water charges for a given time peric

as water) or the imposition of water use tariffs could cre incentives for more efficient use. And, in fact, payment for bulk water supply has been established in recent water laws in Zimbabwe, Tanzania and Mozambique (The World Bank, 2008).

Once transactions are collected by the RBA, water costs (CW) for each water user can be calculated along with the final net benefits (FNB), which are equivalent to the consumer surplus shown, in Fig. 1, as the area above the line P_{x*} (area A). Line P_{x*} is the social cost of water where x^* is the economically efficient water allocation.

The difference between the benefits expected by each agent (ENB) and the final net benefits received (FNB) is the amount an agent will claim for compensation in the next step (c) and is equal to the value of the externalities (B+C in Fig. 1). These claims are composed of the difference in water costs between the unconstrained water demand (D) and the actual water allocation (x^*), which is area B in the figure, and the cost of cooperation (CC), which is the loss in benefits due to the allocation of fewer resources than what was demanded (area C in the figure).

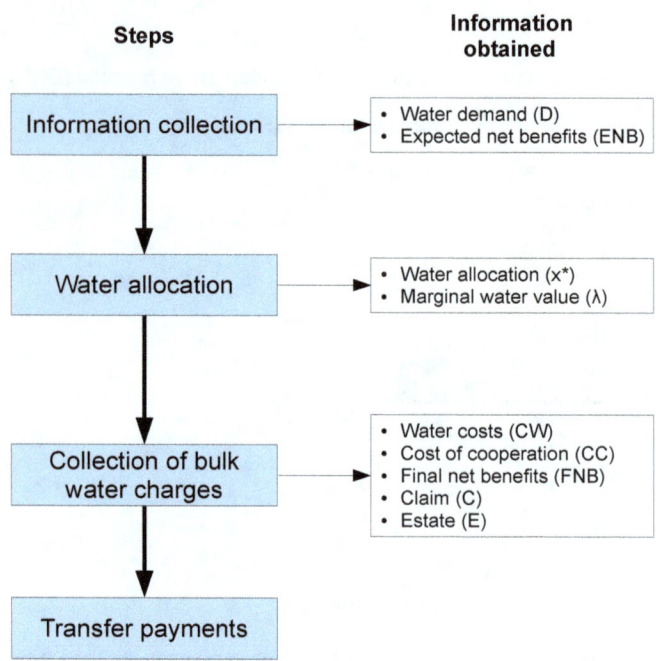

Figure 4. Flowchart of methodology including information obtained at each step.

2.4 Transfer payments

At this point in the methodology, the RBA has collected an amount of money, referred to as the *estate* (E), that can be shared among the water use agents. Using an axiomatic approach, a method of sharing this estate should be determined. The aim of the axiomatic approach is to find and capture the notion of fairness that water users could agree upon. The approach then sets out axioms (properties) that fairness should or should not satisfy. Finally, these properties are translated into a sharing rule that quantifies the particular definition of fairness. How the benefits are shared depends entirely on this definition as agreed to by water users. For example, a simple proportional sharing method may satisfy the properties of equity defined by the users, or an egalitarian method, or some other form of sharing may be required. Since each river basin will have a different definition of fairness (depending on conditions in the basin and the outcome of negotiations with the water users), each river basin will likely have its own unique sharing rule.

A flowchart of the complete methodology, including information obtained at each step, is shown in Fig. 4.

3 Case study

3.1 Eastern Nile River basin

The Eastern Nile River basin is used to illustrate the methodology described in the previous section. Covering an area of

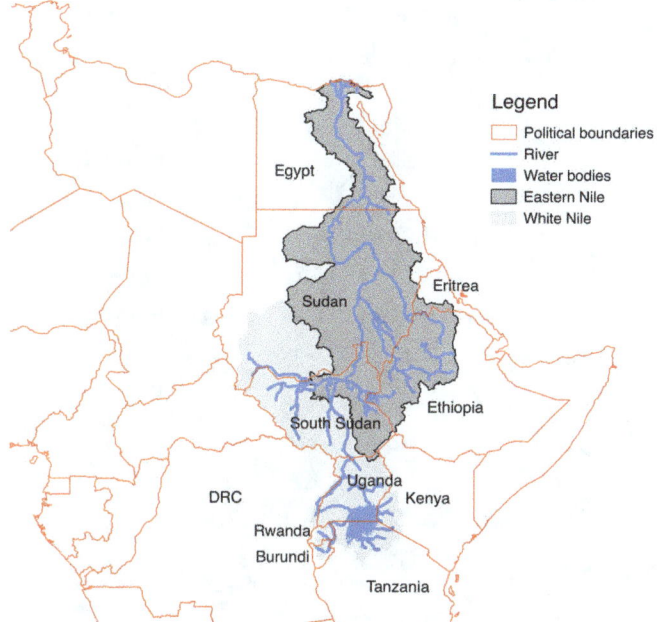

Figure 5. Eastern Nile River basin

approximately $330\,000\,\text{km}^2$ and with a length of $1529\,\text{km}$, the Blue Nile originates in the highlands of Ethiopia and flows into Sudan, where it joins the White Nile at Khartoum to form the Main Nile. The Main Nile then flows out of Sudan, into Egypt, and discharges into the Mediterranean Sea. The Eastern Nile River basin is composed of the Blue Nile, the Tekeze-Atbara, the Baro-Aboko-Sobat, the White Nile downstream from Malakal and the Main Nile sub-basins (Fig. 5).

The dominant uses of water in the Eastern Nile River basin are irrigated agriculture and hydropower generation, mostly in Sudan and Egypt. This is, however, likely to change in the near future with the completion of the Grand Ethiopian Renaissance Dam on the border of Ethiopia and Sudan.

There is a long history of unsuccessful negotiations over water allocation and development of Nile water resources. Attempts at cooperation and benefit sharing within the Eastern Nile basin go back to the early part of the 20th century. The 1929 Nile Waters Agreement between Sudan and Egypt prioritized Egyptian water needs and reportedly gave Egypt the right to veto future hydroelectric projects along the Nile (Brunnée and Toope, 2003). Sudan and Egypt subsequently replaced the 1929 treaty, in 1959, with the Agreement for the Full Utilization of the Nile Waters, which essentially allocated the entire flow of the Nile at the Aswan Dam to Sudan and Egypt. Unsurprisingly, this has caused regional tension with the other riparians, who invoke the Nyerere Doctrine[4],

and general principles of international water law, to contest the 1959 agreement and claim a share of the Nile waters.

In 1999 the Nile Basin Initiative (NBI) was undertaken with the goal being to adopt a comprehensive, permanent, legal and institutional agreement on the Nile River basin. So far there has been little success in negotiations leading to an agreement. However, a Cooperative Framework Agreement (CFA) was signed by a number of the Nile basin countries, with the notable exceptions of Egypt, Sudan and South Sudan.

Regional tensions have further complicated Nile cooperation efforts. For example, Ethiopia and Egypt have a long history of distrust and Egypt and Sudan, as well as Eritrea and Ethiopia, have long unresolved border disputes. Additionally, many Nile riparians have been broken by internal conflicts and instabilities that result in challenges to international relations.

In recent years, the construction of the Grand Ethiopian Renaissance Dam has been a source of concern and conflict among the three riparian countries. It should be noted, however, that in early 2015, Egypt, Sudan and Ethiopia signed an agreement on the declaration of principles with respect to the project.

It is pretty much agreed, at this point, that benefit sharing may offer a solution to the stalemate surrounding water use and allocation in the Eastern Nile River basin. While the concept of benefit sharing can be appreciated by most riparian countries, questions regarding methods of sharing benefits have emerged. The three Eastern Nile River basin countries need to, first and foremost, identify the bundle of benefits that can be generated, and then agree on a mechanism for sharing these (Tafesse, 2009)).

3.2 Information collection

Given the lack of accurate data with respect to irrigated agriculture in the Nile River basin, a net return of $0.05\,\text{USD}\,m^{-3}$ is chosen as in Whittington et al. (2005). For hydropower it is assumed that each MWh generated has an economic value averaging $80\,\text{USD}\,\text{MWh}^{-1}$ for firm power and $50\,\text{USD}\,\text{MWh}^{-1}$ for secondary power. These values are consistent with feasibility studies of hydroelectric dams in Ethiopia. Using these values the unconstrained ENB are determined for each water use agent as

$$\text{ENB}_j = D_j \cdot P_j, \tag{1}$$

where D_j is the unconstrained quantity of water demanded by agent j and P_j is its productivity. Note that the assumption is made that users do not currently pay for water.

The water demand for the irrigation agents is equal to the crop water demand. For the hydropower agents the water demand is equal to the amount that they are allocated in the next

---

[4]The Nyerere Doctrine of state succession, founded by the first President of Tanzania, states that a new nation should not be bound to international agreements dating back to colonial times and that

these agreements should be re-negotiated when a state becomes independent.

step. Since the allocation is economically efficient, the hydropower agents are assumed to be satisfied with the amount of water flowing through the turbines.

3.3 Water allocation

The stochastic multistage decision-making problem (Eqs. A1 to A4 defined in the Appendix) was solved using stochastic dual dynamic programming (SDDP). Details of this algorithm can be found in Goor et al. (2010) and in Tilmant and Kinzelbach (2012). The hydro-economic model of the Eastern Nile basin is based on the schematization shown in Fig. 6. In this study the assumption is made that the Grand Ethiopian Renaissance Dam (located at H8 in Fig. 6) is online. Allocation decisions are chosen to maximize expected net economic returns from irrigated agriculture and hydropower generation over a planning horizon of 10 years and for 30 hydrologic sequences (see Arjoon et al. (2014) for a description of the model).

Once the allocation decisions are determined, the actual gross benefits (GB) can be calculated as

$$GB_j = x_j^* \cdot P_j, \qquad (2)$$

where x_j^* is the water allocation decision for agent j. The difference between the ENB and GB is the cost of cooperation (CC) to the agent due to the efficient allocation of water. In other words, it is the difference between the amount of benefits the agent is expecting to get if their unconstrained water demand is met and the actual benefits the agent receives given the allocation decision, excluding water costs.

3.4 Collection of bulk water charges

The total of the transactions collected by the RBA (E), minus yearly operating expenses of 3 million USD, will be used to compensate the agents for a percentage of the benefits lost either through efficient allocation (cost of cooperation) or water costs. Operating expenses of 3 million USD yr^{-1} are in line with those published by power pools (Southern African Power Pool, 2009) and river commissions (Mekong River Commission, 2013).

Final net benefits for each agent can be calculated as

$$FNB_j = GB_j - CW_j, \qquad (3)$$

where CW_j is the cost of water for agent j.

3.5 Transfer payments

Once the final net benefits have been determined, transfer payments can be calculated for each agent. To do this, the total cost for each agent needs to be calculated, which will give the upper limit to the claim (c) of an agent to the estate.

Figure 7 shows the annual demand curve for an irrigation agent in this case study. In this study, we implicitly assume that the input demand is horizontal (perfectly elastic) with the

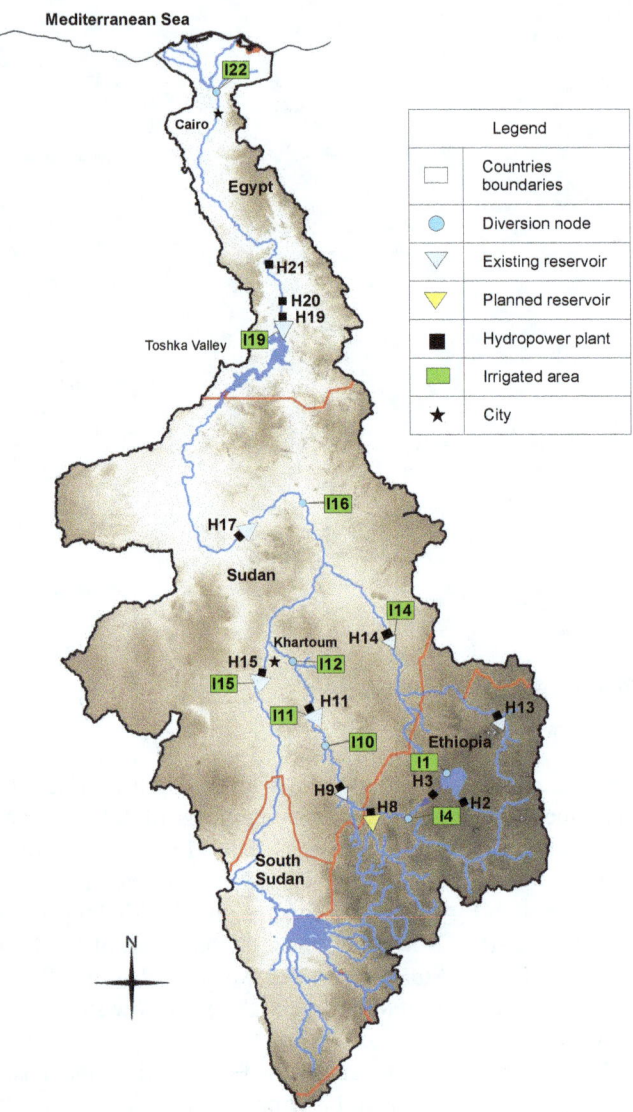

Figure 6. Model schematic of the Eastern Nile River basin. Irrigation agents (I) and hydropower agents (H) for this case study are shown. Note that the numbering is not consecutive because there are nodes that represent agents that are not part of the case study.

price (P) = marginal productivity. The area to the left of line D (comprising areas A, B and C) is the ENB (we see that the agent does not pay for water) resulting from unconstrained water use. When water is constrained, area A is the FNB. The claims (c) are divided into two parts: area B is the cost of water (CW) to the agent and area C is the cost of cooperation (CC) due to the efficient allocation of water. Area B also represents the amount of money that the RBA collects from this agent. As previously mentioned, for hydropower agents the water demand and the water allocation are equal; therefore, there is no cost of cooperation. The claim (c), then, for a hydropower agent, is the cost of water (CW). Over the whole basin the amount that the RBA collects (and is available for

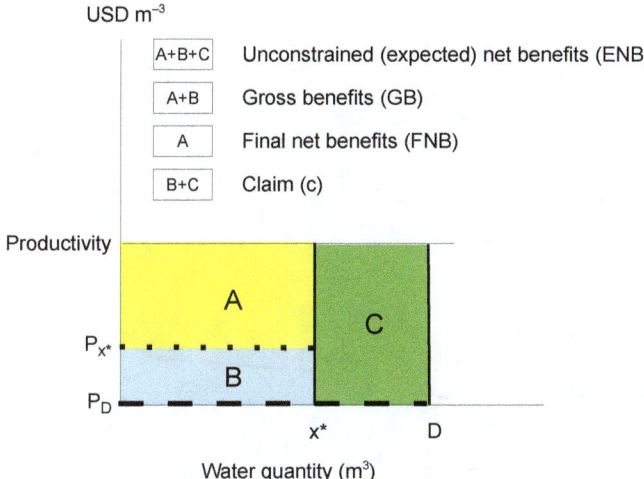

Figure 7. Demand curve for the case study. D: quantity of water demanded for a time period; x^*: quantity of water allocated for a time period; P: price of water.

transfer payments) is enough to reimburse the agents for the actual cost of water; however, as mentioned, USD 3 million are held back for annual operating expenses. Therefore the shortfall between the amount the RBA has to share and the claims of the agents is the total cost of cooperation for irrigation agents ($\sum CC_j$) plus operating expenses.

The situation in which the amount available to share between agents is less than the total claims of the agents is, by definition, a bankruptcy problem.

In this case study, the collected benefits are shared among the water use agents following a rule that was developed based on a number of well-defined properties in the bankruptcy literature (*feasibility, non-negativity, claims-boundedness*) as well as some that are specific to the problem (*solidarity, security of minimum benefits*).

It should be noted that, for this study, the properties of this rule were not developed with stakeholder input, as this was beyond the scope of this research project. Although stakeholder involvement is imperative in this institutional arrangement, in this case study, we are giving an objective viewpoint, and this analysis serves as a benchmark or reference point.

Benefits are shared in such a way as to ensure that each agent has the same proportion of final costs ($ENB_j - (FNB_j + tp_j)$) to benefits demanded (ENB_j) (where tp_j is the monetary transfer payment made to the agent) and that these are minimized. By extension, this rule also ensures that each agent receives an equal proportion of final benefits ($FNB_j + tp_j$) to benefits demanded (ENB_j) and that these are maximized. This rule also applies a *solidarity* property in which all agents take equal responsibility for the shortfall in benefits at certain nodes due to the efficient economic allocation of water over the basin, and a property of *security of minimum benefits* in which the benefits obtained from the use of water (FNB_j) are uncontested.

The compensation rule is defined as follows:

$$tp_j = ENB_j - (FNB_j + \gamma ENB_j), \qquad (4)$$

where γ is chosen such that

$$\sum tp_j \leq E. \qquad (5)$$

Equation (5) ensures the property of *feasibility*, which is the requirement that the sum of the transfer payments not exceed the amount available to share.

The following constraints also apply:

$$tp_j \geq 0, \qquad (6)$$
$$tp_j \leq c_j. \qquad (7)$$

Equation (6) ensures *non-negativity*, which requires that each agent receive a non-negative amount, and Eq. (7) ensures *claims boundedness*, which requires that each agent receive, at most, the amount of its claim.

Rewriting Eq. (4) to read

$$\gamma = (ENB_j - (FNB_j + tp_j)) / ENB_j \qquad (8)$$

shows that the property of *solidarity* is supported by ensuring that the final cost ($ENB_j - (FNB_j + tp_j)$) to expected benefit (ENB_j) ratio for all agents is the same.

In this final step, the transfer payments are calculated and the total final benefits ($FNB + tp$) for each agent are determined.

4 Results

The analysis of results was carried out on year 4 of the 10-year planning horizon. This ensures a steady-state condition that is not influenced by initial hydrological and storage conditions or by any end-effect distortion due to reservoir depletion that occurs as the end of the planning period approaches (Arjoon et al., 2014). As previously explained, the amount of water allocated to hydropower agents is equal to the amount demanded. This means that all hydropower agents receive 100 % of the water demanded. The efficient allocation of water results in most irrigation agents also receiving their unconstrained demand. The exceptions are agents I1, I4 and I14, who receive, on average, 1, 0 and 94 % of their unconstrained demand, respectively (see Fig. 8). This result is not unexpected because, from an economic standpoint, irrigation in the Eastern Nile River basin should take place downstream after water has been used for hydropower generation upstream (Whittington et al., 2005). These three irrigation agents have cooperation costs as well as, possibly, water costs. Looking at the cumulative distribution of the proportion of the allocated amount of water to the amount received for these agents (Fig. 9), we see that 95 % of the time, agent I1 does not receive any water. Agent I14, on the other hand, receives its full demand about 75 % of the time.

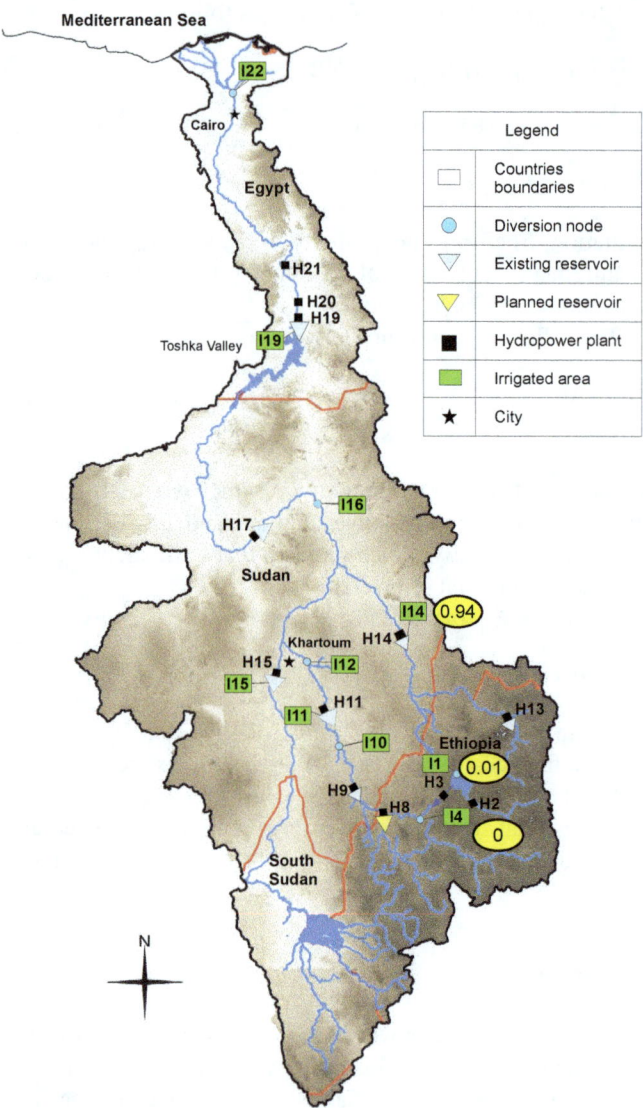

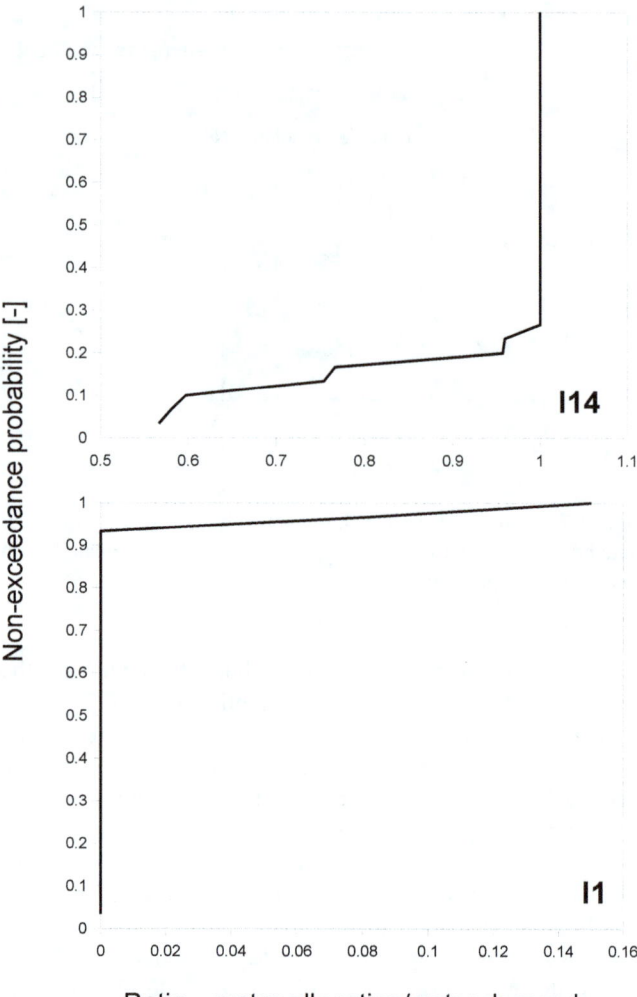

Figure 9. Cumulative distribution function for the proportion of water allocation to unconstrained demand for agents I1 and I14.

Figure 8. Average proportion of water allocation to unconstrained demand for all agents. Only the values for those agents in which the proportion is less than 1 are shown.

Agent I4 (not shown in Fig. 9) always receives 0 %. The rationing of water for upstream irrigation users is a result of the horizontal demand curve used for irrigation. If more detailed economic/agricultural data were available, a non-horizontal demand curve could be produced. This may result in irrigation schemes with high value crops having priority to water and those areas with low value crops not being irrigated. This means that the irrigation water users that are rationed may change and they may be more spread out over the basin.

Overall, the agents with the smallest claims are all hydropower agents in Sudan (H9, H11, H14, H15) with marginal values that are almost equal to marginal values at the downstream sites (see Fig. 10). This means that they sell water downstream at about the same price that they paid for

it, resulting in lower water costs. Figure 11 gives a basin-wide view of the percentage of the unconstrained benefits claimed by each agent, by agent type, on average. The irrigation agents upstream claim a larger percentage of their expected benefits because, first, they pay more for water and, second, they also have cooperation costs. With respect to hydropower agents, H8 and H19 (Grand Renaissance and High Aswan, respectively) claim the largest percentage of their expected benefits. In both cases, the cost of water at these sites is much greater than the cost of water at the respective downstream sites.

From the collection of bulk water charges for the period analyzed (year 4), the RBA ends up with USD 3894 million to allocate between the agents (after subtracting USD 3 million for operating costs). The total claims amount for all agents, for the year, is USD 4266 million, which means that there is a shortfall of USD 372 million between the

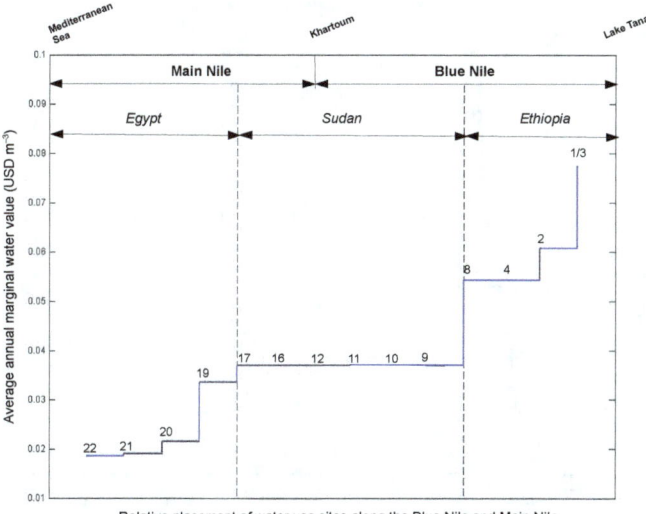

Figure 10. Marginal water value – Blue Nile and Main Nile

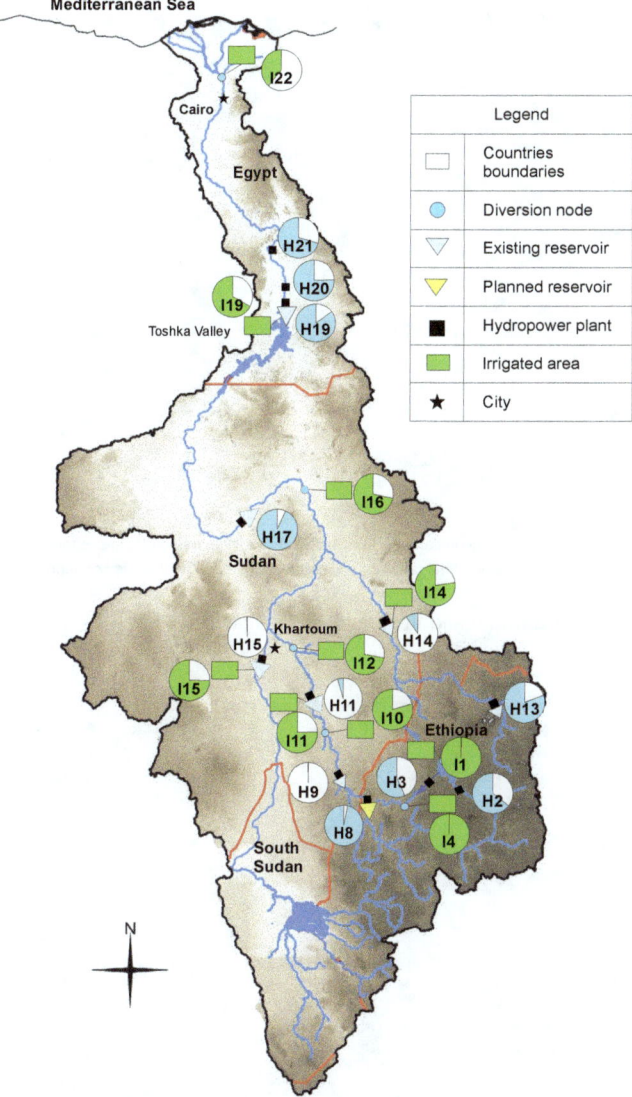

Figure 11. Percentage of unconstrained benefits claimed by agents.

amount available to share and the claims, or about 9 % of the total claims.

Using the bankruptcy rule developed for this example, the average amount of transfer payment is calculated for each agent. The ratio of FNB to ENB, referred to as the *initial ratio*, and final net benefits plus transfer payments (FNB+tp) to ENB, referred to as the *final ratio*, are determined and analyzed. These results were analyzed over the 30 different hydrologic sequences to assess how this rule performs under varying hydrologic conditions.

Figure 12 shows the mean values for initial ratios (shown as large filled squares) and final ratios (shown as large filled diamonds) for irrigation agents as well as the values for each of the hydrologic sequences. Agents I1 and I4 receive little or no irrigation water, on average, as discussed previously. Agent I14 initially receives about 23 % of its expected net benefits, on average. This agent is located at the Kashm El Girba dam, on the Tekeze-Atbara River. The flow of this river is highly seasonal, with annual flows entering Sudan from Ethiopia restricted to the flood period of July to October. The design storage capacity of the reservoir at this site is about 10 % of the inflow; however, high sedimentation in the reservoir dropped the storage capacity by 50 % as of 1977. This loss of storage capacity has resulted in severe water shortages during drought years and an associated decline in the crop area cultivated. As a result, the restriction of water for this irrigation agent is more probably due to the hydrology as opposed to being economic in nature. Due to flow variation, the marginal water values are highly variable at this site, resulting in a wide spread of initial ratios over the hydrologic sequences (as indicated by a large vertical spread of data points on the graph for this agent). All other agents always receive their full unconstrained demand. Variability

in the initial ratios of these agents is due to variability in the marginal water values over the hydrologic sequences.

Results for hydropower agents are shown in Fig. 13. Here we see more variation in the initial ratio than for the irrigation agents. The upstream hydropower agents (H2, H3), and those on the Tekeze-Atbara River (H13, H14), have large variations in initial ratios as a result of large inter- as well as intra-year variations in flow (and subsequently in marginal water values), which occurs because these sites are all upstream of flow regulating infrastructure. The agents with the smallest claims are the four smallest hydropower agents in Sudan (H9, H11, H14, H15). These agents have the largest initial ratios and, therefore, often do not receive monetary transfers. This also results in the final ratios for hydropower agents not being equal because the property of non-negativity, which is used to define the sharing rule, allows an agent to keep its initial

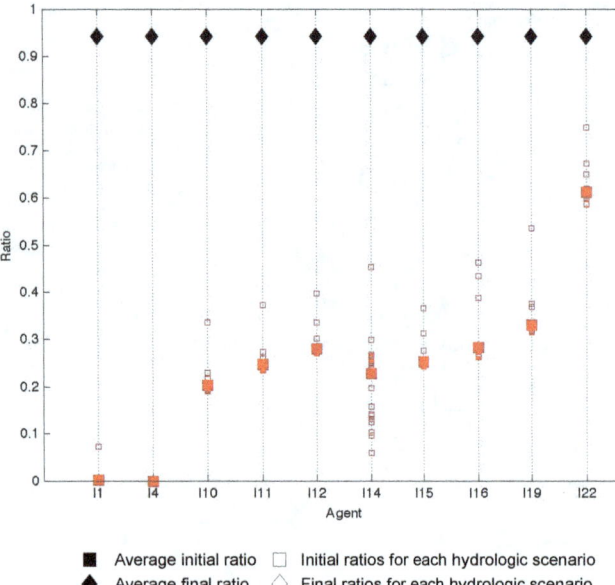

Figure 12. Initial and final ratios for irrigation agents.

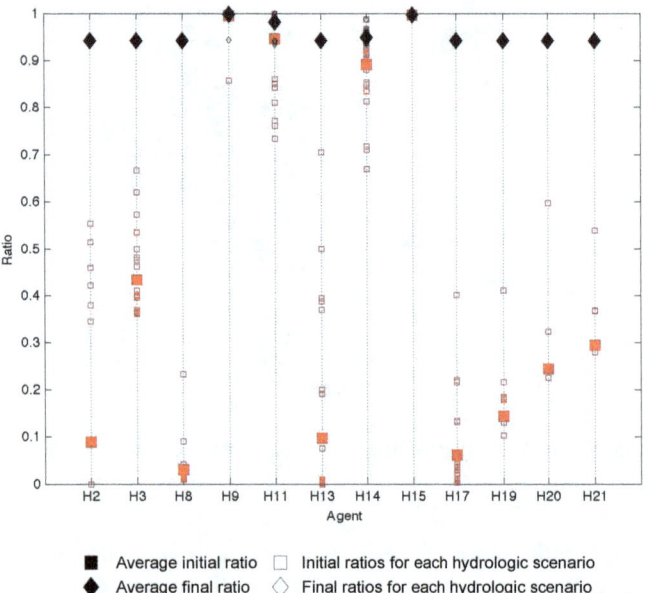

Figure 13. Initial and final ratios for hydropower agents.

benefits from water use even if this results in its final ratio being larger than those of the other agents.

Overall, the average final ratios for all agents (irrigation and hydropower) are equal, with the exception of agents H9, H11, H14 and H15, as mentioned above. There is also very little variation in final ratio values with respect to hydrologic sequence. The final ratio for irrigation agents varies from 93.5 to 95 % of their uncontested benefits. For hydropower agents the statistical distribution of final benefit ratios is shown in Fig. 14. We see that these final ratios also vary be-

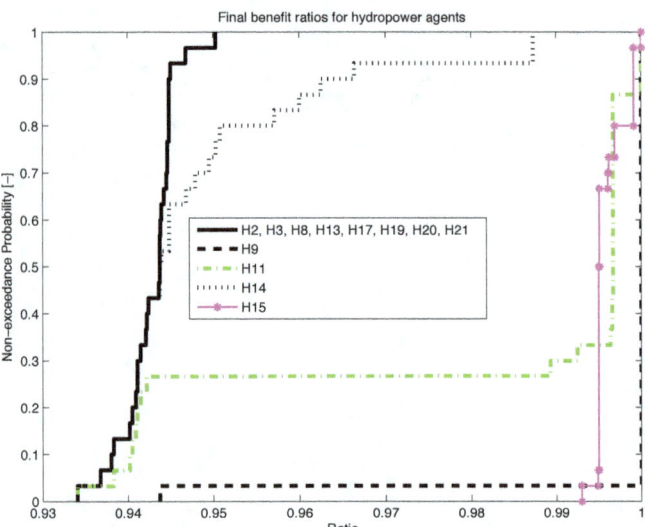

Figure 14. Final benefit ratio for hydropower agents.

tween 93.5 and 95 % with the exception, again, of agents H9, H11, H14 and H15, which have high initial ratios that vary with inter- and intra-annual variations in the marginal value of water. These results indicate that the sharing rule used is predictable in that agents can expect similar final benefits regardless of the hydrologic conditions.

Results that warrant a closer look are those for the upstream irrigation agents I1 and I4. We can conclude that, in this case study, given the economic information used in the model, it is economically inefficient to irrigate upstream in the basin regardless of the hydrologic sequence (meaning that even in situations of high flow years, there is no irrigation water allocated to these agents). However, these two irrigation agents consistently demand fairly substantial transfer payments even though they do not contribute economically to the basin. This becomes an obvious problem of fairness for the other agents. If these results persist over a number of years, the RBA could use this information for better management by ensuring that agriculture is developed downstream or that upstream agricultural sites have a high productivity value.

Finally, it should be noted that we make no attempt to compare the results of the case study with current water use in the basin. While the presented case study is hypothetical and is not consistent with the actual, current situation, it represents a possible long-term future scenario in the basin, and the results reflect these assumptions. In the case study, we assume complete cooperation; there is expanded irrigation in the basin and the Grand Ethiopian Renaissance Dam is online.

5 Conclusions

The sharing of benefits among agents in a transboundary river basin is based on three fundamental questions: (i) how can the benefits of water use be quantified and monetized, ii) what mechanism can be used to allocate benefits, and (iii) upon what criteria should the sharing of benefits be based to ensure efficiency and equitability. It should be noted that there is no unique response to these questions. In this paper, we propose one approach for distributing the benefits of co-operative management in a river basin system comprised of rival and non-rival uses. To illustrate the approach, we used the Eastern Nile River basin as a case study due to the important hydropower and agricultural sectors spread over three countries.

The methodology described in this paper is based on the welfare distribution for each agent being equal to the sum of its benefits from water use plus a monetary transfer. First, efficient water allocation is implemented through the application of a hydro-economic model in order to maximize the benefits in the river basin. Second, a charge for the use of water is established. The price that agents pay for the use of water is equal to the marginal value of water at the site at which the agent receives its allocation. The total of the water charges is equivalent to the overall value of water in the basin that is used in the sectors being studied. Finally, the total of the water charges are reallocated over the basin to ensure that all agents pay the same ratio of costs to benefits, using an axiomatic approach. The whole system is overseen by an RBA.

The two main goals of benefit sharing, efficiency and equitability, are the foundation of this methodology. The hydro-economic model results are the efficient water allocations for each agent. Efficiency is also inherent in the benefit-sharing rule used to implement the monetary transfers in that all of the available money is shared among the agents. The defined properties of fairness are embedded in the sharing rule through the axioms.

This methodology can be useful to policy-makers in that the solution is more likely to be perceived as equitable, resulting in water use agents being more open to cooperation. An additional advantage of this method is the predictability of the final results. These results, over varying hydrological sequences, are shown to be relatively constant.

The importance of this methodology is that it can be adopted for application in negotiations to cooperate in transboundary river basins. The methodology is flexible in that there is no set way to allocate the water over the basin. Any hydro-economic model (or another method) can be used as long as the amount of water allocated to each agent, as well as the marginal value of water for each agent, is available. Also, the development of the sharing rule can be based on stakeholder input and will depend on specific conditions in specific river basins.

One obvious constraint of this method is its dependence on the existence of a strong basin-wide authority to impose fees and that can enable negotiations between stakeholders for the development of a sharing rule. Allowing all stakeholders a place at the table might prove challenging, especially for large systems with diversified water use activities. In the irrigation sector, for instance, farmers could be represented by a water user association. For uses of water as a public good, such as for environmental flows, the representative could be the Ministry of Environment of the country of interest. For municipal uses, the system could be designed in such a way that a minimum amount of allocated water is guaranteed (a fixed constraint in the allocation system), while quantities beyond that minimum would be part of the pool for which municipalities would have to bid. Industrial and power companies are easier to handle. All users that can be rationed (mainly private water users) are allowed a place at the table for the purpose of defining fairness with respect to transfer payments. Another possibility is that the government (or at least a high level representative of the stakeholders) has the ultimate negotiation power, akin to negotiations on trade liberalizations. Clearly, different lobbies exist that would try to influence the government, implying, ultimately, some form of compensation (the analysis of which is outside the scope of this paper).

Another constraint is the availability of reliable data. Some information such as market prices, either national or international, can be observed and transportation costs can be estimated, allowing for an approximation of the mark-up that may accrue to farmers, for example. This paper describes a system in which it is assumed that there is cooperation over the whole basin and that water users have agreed to bid for water and to supply the information that is necessary to make the methodology work. Increasingly, the information required is becoming available through the use of remote sensing and monitoring of river basins.

Incentives for water users to cheat, with respect to the data they provide, will remain even if the river basin authority is able to audit the bids. For industrial uses, including hydropower generation, cheating might be more difficult because the market prices and production functions are often well characterized. The main challenge is to be found in the agricultural sector because (a) it is often the largest water use in a basin (and, hence, cheating might have serious basin-wide consequences), and (b) the heterogeneity in terms of cropping patterns and irrigation efficiency requires that significant data be collected and analyzed to audit the demands. We argue that the incentives to cheat might not be eliminated, but they can be suppressed, or at least kept within limits, through a robust monitoring system and a strong RBA to negotiate disputes. An example of how this has worked, with good success, is the Indus River basin. Zawahri (2009), in discussing the Permanent Indus Commission, states "The commission's ability to monitor development of the shared river system has permitted it to ease member states" fear of cheating and confirm the accuracy of all exchanged data. Finally, its conflict resolution mechanisms have permitted the commission to negotiate settlements to disputes and prevent defection from cooperation.

This paper adds to the analysis of the sharing of economic benefits in transboundary river basins by describing a methodology for efficient and equitable benefit sharing based on operating the river basin as a water pseudo-market with the advantages of resource use optimization, improved resource reliability and enhanced security of resource supply. Also, we impose specific axioms, based on a stakeholder vision of fairness, on the compensation scheme and derive a unique solution for the distribution of monetary payments. This technique may lead to a sharing solution that is more acceptable to shareholders because the definition of the sharing rule is not in question, as would be the case if we applied existing bankruptcy rules or other game theory solutions with their inherent definitions of fairness.

Appendix A

Hydro-economic modeling is a common tool used to analyze river basin systems and, specifically, water resources allocation problems. These models use a network representation of the system in order to physically connect various sources of supply with scarcity-sensitive water demands. Reviews of hydro-economic models can be found in Harou et al. (2009) and Brouwer and Hofkes (2008). Two classes of hydro-economic models exist: optimization-based and simulation-based. Both approaches have advantages and disadvantages, but the allocation decisions and the marginal costs of the binding constraints (the limiting resources or factors that prevent further improvement of the objective function) determined by an optimization model make this type of model attractive in the proposed methodology. In the system network, a water balance is evaluated at each node to determine the amount of water available for the demand sites connected to that node. The mass balance equation ensures that water is allocated to the connected water users to the extent permitted by water availability at the node. In the case of water scarcity, the marginal cost associated with the water balance indicates the shadow price of water or what the users would be willing to pay for an additional unit of water (Young, 2005).

In a hydro-economic water resource optimization problem, the objective function Z to be maximized includes the economic net benefits across all water uses over a given planning period.

$$Z^* = \max_{x_t} \left\{ \mathbf{E}_{q_t} \left[\sum_t^T \alpha_t b_t(\boldsymbol{w}_t, \mathbf{x}_t) + \alpha_{T+1} \nu(\boldsymbol{w}_{T+1}) \right] \right\}, \quad (A1)$$

where b_t are the basin-wide net benefits at time t, $\boldsymbol{x}_t$ the vector of allocation decisions, $\boldsymbol{w}_t$ the vector of state variables, α a discount factor, ν a terminal value function, $\mathbf{E}$ the expectation operator capturing he uncertainty that governs the hydrologic inflow q_t and Z the total benefit associated with the optimal allocations $(x_1^*, x_2^*, ..., x_T^*)$.

This function is maximized to the extent permitted by physical, institutional or economic constraints:

$$g_{t+1}(\boldsymbol{x}_{t+1}) \leq 0, \quad (A2)$$
$$h_{t+1}(\boldsymbol{w}_{t+1}) \leq 0, \quad (A3)$$
$$\boldsymbol{w}_{t+1} = f_t(\boldsymbol{w}_t, \boldsymbol{x}_t, \boldsymbol{q}_t), \quad (A4)$$

where g is a set of functions constraining the allocation decision, h a set of functions constraining the state of the system and f a set of functions describing the transition of the system from time t to time $t+1$.

Included in the functions in Eq. (A4) are the mass balance equations for the river basin:

$$s_{t+1} - \mathbf{R}(r_t + l_t) - \mathbf{I}(i_t) + e_t(s_t, s_{t+1}) = s_t + q_t, \quad (A5)$$

where s_t is the storage at time t, r_t the controlled outflows, l_t the uncontrolled outflows, i_t the water withdrawals, $\mathbf{R}$ and $\mathbf{I}$ the connectivity matrices representing the topology of the system (including return flows), and e_t the evaporation losses.

At the optimal solution of the problem (Eqs. A1 to A4), the solver provides the allocation decisions $(x_1^*, x_2^*,..., x_T^*)$ and the marginal values of water (shadow prices) $(\lambda_1, \lambda_2,...,\lambda_T)$ of the constraints. For the constrains in Eq. (A4), the shadow prices correspond to the marginal resource opportunity cost at the sites where water balances are computed.

Acknowledgements. The authors are grateful to Yasir Mohamed (HRC-Sudan) and Erik Ansink (Utrecht University) for valuable discussions early in the development of this work, and would like to thank the Institut Hydro-Québec en environnement, développement et société (EDS) for their financial support (grant 03605-FO101829).

Edited by: P. van der Zaag

References

Abed-Elmdoust, A. and Kerachian, R.: Water resources allocation using a cooperative game with fuzzy payoffs and fuzzy coalitions, Water Resour. Manage., 26, 3961–3976, 2012.

Ansink, E. and Houba, H.: The economics of transboundary river management, Discussion Paper TI 2014-132/VIII, Tinbergen Institute, 2014.

Ansink, E. and Weikard, H.: Sequential sharing rules for river sharing problems, Soc. Choice Welfare, 38, 187–210, 2012.

Arjoon, D., Mohamed, Y., Goor, Q., and Tilmant, A.: Hydro-economic risk assessment in the eastern Nile River basin, Water Resour. Econom., 8, 16–31, 2014.

Asmamaw, D. K.: A critical review of integrated river basin management in the upper Blue Nile river basin: the case of Ethiopia, Int. J. River Basin Manage., 1–14, doi:10.1080/15715124.2015.1013037, 2015.

Brunnée, J. and Toope, S.: The Nile Basin Regime: A Role For Law?, in: Water ResourcesPerspectives:Evaluation, Managementand Policy, edited by: Wood, A. A. W., 93–117, Elsevier Science, 2003.

de Clippel, G. and Rozen, K.: Fairness through the lens of cooperative game theory: an experimental approach, Discussion Paper 1925, Cowles Foundation for Research in Economics at Yale University, 2013.

Ding, N., Erfani, R., Mokhtar, H., and Erfani, T.: Agent Based Modelling for Water Resource Allocation in the Transboundary Nile River, Water, 8, 139, 2016.

Dombrowsky, I.: Revisiting the potential for benefit sharing in the management of trans-boundary rivers, Water Policy, 11, 125–140, 2009.

El-Kawy, O. A., Rød, J., Ismail, H., and Suliman, A.: Land use and land cover change detection in the western Nile delta of Egypt using remote sensing data, Appl. Geogr., 31, 483–494, 2011.

Gallego, F., Kussul, N., Skakun, S., Kravchenko, O., Shelestov, A., and Kussul, O.: Efficiency assessment of using satellite data for crop area estimation in Ukraine, Int J. Appl. Earth Obs., 29, 22–30, 2014.

Goor, Q., Halleux, C., Mohamed, Y., and Tilmant, A.: Optimal operation of a multipurpose multireservoir system in the Eastern Nile River Basin, Hydrol. Earth Syst. Sci., 14, 1895–1908, doi:10.5194/hess-14-1895-2010, 2010.

Grey, D. and Sadoff, C.: Sink or Swim? Water security for growth and development, Water Policy, 9, 545–571, 2007.

Hausman, D.: Fairness and trust in game theory, unpublished manuscript, 1999.

Herrero, C. and Villar, A.: The three musketeers: four classical solutions to bankruptcy problems, Math. Soc. Sci., 42, 307–328, 2001.

Jafarzadegan, K., Abed-Elmdoust, A., and Kerachian, R.: A fuzzy variable least core game for inter-basin water resources allocation under uncertainty, Water Resour. Manage., 27, 3247–3260, 2013.

Jeuland, M., Baker, J., Bartlett, R., and Lacombe, G.: The costs of uncoordinated infrastructure management in multi-reservoir river basins, Environ. Res. Lett., 9, 105006–105016, 2014.

Leb, C.: One step at a time: international law and the duty to cooperate in the management of shared water resources, Water Int., 40, 21–32, 2015.

Madani, K.: Game theory and water resources, J. Hydrol., 381, 225–238, 2010.

Madani, K., Zarezadeh, M., and Morid, S.: A new framework for resolving conflicts over transboundary rivers using bankruptcy methods, Hydrol. Earth Syst. Sci., 18, 3055–3068, doi:10.5194/hess-18-3055-2014, 2014.

Mekong River Commission: Operating expenses budget: Audited financial statements as at and for the year ended 31 December 2013, Mekong River Commission, 2013.

Mianabadi, H., Mostert, E., Zarghami, M., and van de Giesen, N.: A new bankruptcy method for conflict resolution in water resources allocation, J. Environ. Manage., 144, 152–159, 2014.

Mianabadi, H., Mostert, E., Pande, S., and van de Giesen, N.: Weighted Bankruptcy Rules and Transboundary Water Resources Allocation, Water Resour. Manage., 29, 2303–2321, 2015.

MRC Initiative on Sustainable Hydropower: Knowledge base on benefit sharing, Summary and guide to the knowledge base (KB) compendium, Volume 1 of 5, 2011.

Oesterle, M. and Hahn, M.: A Case Study for Updating Land Parcel Identification Systems by Means of Remote Sensing, in: Proceedings of the XXth ISPRS Congress 2003, Istanbul, Turkey, 2004.

Phillips, D., Daoudy, M., McCaffrey, S., Öjendal, J., and Turton, A.: Trans-boundary Water Cooperation as a Tool for Conflict Prevention and for Broader Benefit-sharing, Tech. rep., Ministry of Foreign Affairs, Sweden, 2006.

Pulido-Velazquez, M., Andreu, J., Sahuquillo, A., and Pulido-Velazquez, D.: Hydro- economic river basin modelling: The application of a holistic surface–groundwater model to assess opportunity costs of water use in Spain., Ecol. Econ., 66, 51–65, 2008.

Qaddumi, H.: Practical approaches to transboundary water benefit sharing, Working Paper 292, Overseas Development Institute, 2008.

Riegels, N., Pulido-Velazquez, M., Doulgeris, C., Sturm, V., Jensen, R., Møller, F., and auer Gottwein, B.: Systems Analysis Approach to the Design of Efficient Water Pricing Policies under the EU Water Framework Directive., J. Water Resour. Plan. Manag., 139, 574–582, 2013.

Rozenstein, O. and Karnieli, A.: Comparison of methods for land-use classification incorporating remote sensing and GIS inputs., Appl. Geogr., 31, 533–544, 2011.

Sadoff, C. and Grey, D.: Beyond the river: the benefits of cooperation on international rivers, Water Policy, 4, 389–403, 2002.

Sechi, G. and Zucca, R.: Water Resource Allocation in Critical Scarcity Conditions: A Bankruptcy Game Approach., Water Resour. Manage., 29, 541–555, 2015.

Skinner, J., Naisse, M., and Haas, L. (Eds.): Sharing the benefits of large dams in West Africa., Natural Resources Issues No. 19, International Institute for Environment and Development, London, UK, 2009.

Southern African Power Pool: 2009 Annual Report, Tech. rep., Southern African Power Pool, 2009.

Suhardiman, D., Wichelns, D., Lebel, L., and Sellamuttu, S.: Benefit sharing in Mekong Region hydropower: Whose benefits count?, Water Resour. Rural Develop., 4, 3–11, 2014.

Syme, G. J., Nancarrow, B. E., and McCreddin, J. A.: Defining the components of fairness in the allocation of water to environmental and human uses, J. Environ. Manage., 57, 51–70, 1999.

Tafesse, T.: Benefit-Sharing Framework in Transboundary River Basins: The Case of the Eastern Nile Subbasin, vol. CP 19 Project Workshop Proceedings, Addis Ababa, Ethiopia, 5–6 February, 2009.

Teasley, R. and McKinney, D.: Calculating the Benefits of Transboundary River Basin Cooperation: The Syr Darya Basin., J. Water Resour. Plan. Manage., 137, 1–12, 2011.

The World Bank: Zambezi River Basin Sustainable Agriculture Water Development: Angola, Botswana, Malawi, Mozambique, Namibia, Tanzania, Zambia, Zimbabwe, Tech. rep., The World Bank, 2008.

Thomson, W.: On the axiomatic method and its recent applications to game theory and resource allocation, Soc. Choice Welfare, 18, 327–386, 2001.

Thomson, W.: Axiomatic and game-theoretic analysis of bankruptcy and taxation problems: a survey, Math. Soc. Sci., 45, 249–297, 2003.

Thomson, W.: Axiomatic and game-theoretic analysis of bankruptcy and taxation problems: an update, Working Paper 578, Rochester Center for Economic Research, 2013.

Tilmant, A. and Kinzelbach, W.: The cost of non-cooperation in international river basins, Water Resour. Res., 48, W01503, doi:10.1029/2011WR011034, 2012.

Tilmant, A., Marques, G., and Mohamed, Y.: A dynamic water accounting framework based on marginal resource opportunity cost, Hydrol. Earth Syst. Sci., 19, 1457–1467, doi:10.5194/hess-19-1457-2015, 2015.

United Nations World Water Assessment Programme: The United Nations World Water Development Report 2014: Water and Energy, Tech. rep., UNESCO, Paris, 2014.

van der Zaag, P.: Asymmetry and equity in water resources management; critical governance issues for Southern Africa., Water Resour. Manage., 21, 1993–2004, 2007.

van der Zaag, P., Seyam, I., and Savenije, H.: Towards measurable criteria for the equitable sharing of international water resources, Water Policy, 4, 19–32, 2002.

Wang, L. Z., Fang, L., and Hipel, K. W.: Water Resources Allocation: A Cooperative Game Theoretic Approach, J. Environ. Inform., 2, 11–22, 2003.

Whittington, D., Wu, X., and Sadoff, C.: Water resources management in the Nile basin: the economic value of cooperation, Water Policy, 7, 227–252, 2005.

Young, R.: Determining the Economic Value of Water – Concepts and Methods, Resources of the Future, Washington, D.C., 2005.

Zawahri, N.: India, Pakistan and cooperation along the Indus River system, Water Policy, 11, 1–20, 2009.

14

Seasonal patterns of water storage as signatures of the climatological equilibrium between resource and demand

B. François[1,2], **B. Hingray**[1,2], **F. Hendrickx**[3], and **J. D. Creutin**[1,2]

[1]CNRS, LTHE – UMR5564, 38041 Grenoble, France
[2]University Grenoble Alpes, LTHE – UMR5564, 38041 Grenoble, France
[3]EDF, R&D, LNHE, 78400 Chatou, France

Correspondence to: B. François (benoit.hingray@ujf-grenoble.fr)

Abstract. Water is accumulated in reservoirs to adapt in time the availability of the resource to various demands like hydropower production, irrigation, water supply or ecological constraints. Deterministic dynamic programming retrospectively optimizes the use of the resource during a given time period. One of its by-products is the estimation of the marginal storage water value (MSWV), defined by the marginal value of the future goods and benefits obtained from an additional unit of storage water volume. Knowledge of the MSWV makes it possible to determine a posteriori the storage requirement scheme that would have led to the best equilibrium between the resource and the demand. The MSWV depends on the water level in the reservoir and shows seasonal as well as inter-annual variations. This study uses the inter-annual average of both the storage requirement scheme and the MSWV cycle as signatures of the best temporal equilibrium that is achievable in a given resource/demand context (the climatological equilibrium). For a simplified water resource system in a French mountainous region, we characterize how and why these signatures change should the climate and/or the demand change, mainly if changes are projected in the mean regional temperature (increase) and/or precipitation (decrease) as well as in the water demand for energy production and/or maintenance of a minimum reservoir level.

Results show that the temporal equilibrium between water resource and demand either improves or degrades depending on the considered future scenario. In all scenarios, the seasonality of MSWV changes when, for example, earlier water storage is required to efficiently satisfy increasing summer water demand. Finally, understanding how MSWV signatures change helps to understand changes in the storage requirement scheme.

1 Introduction

Mountain catchments yield most of the European hydroelectric production (Eurelectric gives ca. 140 TWh for Scandinavia and the Alps and refers to the "blue battery" of Europe). At high elevation (and/or latitude), spatial and temporal variations of the snowpack make the hydrological regime of rivers highly seasonal with low and high flows in the snow-accumulation and snowmelt seasons respectively. On the other hand, the electricity demand is also highly seasonal, with consumption peaks that mainly occur during the winter (e.g. Schaefli et al., 2007). The temporal deviations between the resource and the demand can be balanced with storage and release operations, transferring the resource in excess at a given time to times when it is insufficient. Most water storage reservoirs in Europe were designed and are managed to balance these two seasonal signals. Many of these reservoirs are not only dedicated to hydroelectricity production but are assigned other management objectives, related for instance to low flow maintenance, irrigation and drinking water supply (Loucks et al., 2005). In multi-purpose configurations, the time profile of the day-to-day storage levels resulting from storage and release operations aims at the best possible socio-economic equilibrium between water inflows and water demands. This optimal storage requirement scheme (for conciseness also denoted as storage scheme) is thus a signature of the best temporal equilibrium between the natural resource and the demand under a given climate, which we call climatological equilibrium.

Significant regional changes are expected worldwide for the next decades as a result of climate change. This will be especially the case for the hydrological regime of mountain rivers. Warmer temperatures are expected to reduce the ratio of snow to rainfall and shorten the snow accumulation period. The spring snowmelt is expected to reduce and to shift earlier in the year by two weeks to one month (Schneider et al., 2013; Lafaysse et al., 2014). Warmer temperatures are also expected to increase the demand for irrigation water (Rosenberg et al., 2003; Rosenzweig et al., 2004) and to modify the seasonal pattern of electricity demand, with lower consumption for heating during the winter and greater needs for cooling during the summer (Alcamo et al., 2007; Hekkenberg et al., 2009). As a result, climate change is expected to modify the seasonal equilibrium between water availability and demand (Raje and Mujumdar, 2010).

A number of recent studies have explored the potential impact of climate change on water systems (e.g. Gaudard et al., 2013). They are mostly based on the simulation of the management system over future periods and the statistical analysis of simulation outputs in terms of system performance. The simulation is classically based on day-to-day system operation scenarios obtained with either simple management models, based on rule curves or balance equations (Veijalainen et al., 2010; Ashofteh et al., 2013), or more sophisticated models mimicking a real operational context (e.g. Minville et al., 2009; Raje and Mujumdar, 2010; Vicuña et al., 2010). System performance is estimated using synthetic criteria such as the mean benefit from hydropower or agricultural production or the RRV criteria (Reliability, Resilience, and Vulnerability), a statistics of system failures such as day-to-day deviations between the effective supply and the demand (Hashimoto et al., 1982; Moy et al., 1986). Interpreting those performance criteria is not an easy task since (i) they may combine resource and demand modifications together with management adaptability issues and (ii) they summarize behind a single value quite complicated time patterns – namely, they cannot inform whether the tested management rules have to be modified or whether any better rules exist, nor can they describe the possible modification of the temporal resource/demand equilibrium over the considered period, even though understanding the time patterns behind such modification is likely to highlight the reasons for change in the system performance.

In the present work, we use the mean inter-annual pattern of the storage requirement as a first signature of the evolution of the climatological resource/demand equilibrium. We also consider the marginal value of storage water (MSWV) representing the future benefit that would be obtained at any given time from an additional unit of water volume stored in the reservoir. We estimate it as a by-product of deterministic dynamic programming (Masse, 1946; Bellman, 1957). The variations of MSWV with time for different levels in the reservoir drive the day-to-day storage scheme required to maximize a chosen benefit function coupling water inflows, demand and constraints. They provide a quite detailed description of the role played by the reservoir in redistributing the water throughout the year and from one year to another given the constraints. We propose the mean inter-annual pattern of MSWV as an alternative signature of the resource/demand disequilibrium. We also look at how these signatures are modified by changes in climate or demand. We compute both signatures under the present climate and a set of future climate scenarios, for a simplified water resource system with a single storage reservoir. This system is a catchment located in the Southern French Alps. We analyse the signature sensitivity to a mean regional temperature increase, a precipitation decrease and both together. We also explore the influence of the nature of water demand on both signatures (energy production and water level maintenance).

The paper is organized as follows. Section 2 briefly describes how the MSWV are estimated and how they are used for the determination of the storage scheme. Section 3 presents the simplified water resource system, the data and the simulation models considered in the application to the Upper Durance River (France). It also describes the future climate scenarios considered in this work. The storage scheme of this system is presented and discussed in Sect. 4. The inter-annual pattern of MSWV through the calendar year for the present and future climates are presented and discussed in Sects. 5 and 6, when they are interpreted as signatures of the climate change. Section 7 presents the conclusions.

2 Storage water values and storage requirement scheme

The optimal storage requirement scheme is the day-to-day storage level required over the analysis period $[t_0, t_N]$ to reach the best possible equilibrium between water resource and demand, given operational constraints. This scheme maximizes over the period the sum of the benefits at each time step t_i of the analysis period, plus the benefit expected from the water remaining in the reservoir at the end of the period. The benefit function for any time step, further referred to as the "current" benefit function, can be expressed as a weighted sum of the benefits and costs over different water uses or management objectives. This function thus reads as

$$g\left(u_{t_i}, s_{t_i}, t_i\right) = \sum_j c_j g_j\left(u_{t_i}, s_{t_i}, t_i\right), \qquad (1)$$

where g_j is a function representing the monetary benefits and costs associated with the different services by operation u_{t_i} at the storage level s_{t_i} during $[t_i, t_{i+1}]$ and c_j is a weighting constant defined according to the priority level assigned to use j.

For each time step t_i, an immediate use of water reduces the availability of stored water for all future water uses. The current benefits must therefore be balanced against losses in future benefits. Identifying the optimal storage variation at the current time step requires knowing the marginal value of storage water (MSWV) in the reservoir from the current

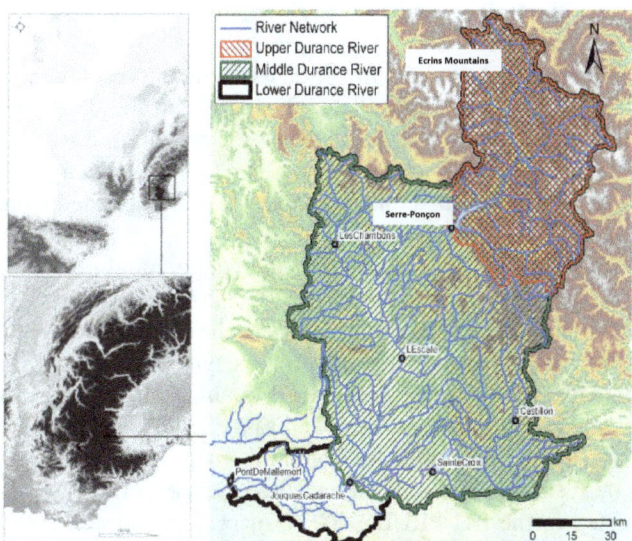

Figure 1. Map of the Durance River watershed. Serre-Ponçon reservoir is the outlet of the Upper Durance River watershed.

time step to the next. MSWV estimation is detailed within Appendix A.

As shown in Appendix A and discussed below, the MSWV is time and storage level dependent. It can be obtained a posteriori as a by-product of deterministic dynamic programming, an optimization method developed for multistage dynamic decision processes. In our case, MSWV values are estimated for the whole analysis period at a daily time step for 51 storage levels uniformly distributed between the minimum and maximum storage bounds $s_{\min}$ and $s_{\max}$. At any given day, these MSWV can be used in a second optimization stage to identify the optimal storage variation given the current water storage in the reservoir. For a given storage level at the beginning of the analysis period, the forward day-to-day optimization process therefore gives the optimal storage requirement scheme for the whole analysis period.

In the following, the MSWV is expressed in relative value units per cubic metre denoted as SWV m^{-3}.

3 Case study and data

3.1 Catchment characteristics and experimental setup

The Upper Durance River (UDR) basin at Serre-Ponçon is a meso-scale basin (3580 km^2) located in the southern French Alps (Fig. 1). Its outlet is the Serre-Ponçon reservoir, a storage reservoir that is part of a large hydroelectric system operated by Electricité de France (EDF). It plays a key role in the energy supply of the Provence region, which extends from the Alps to the Mediterranean shore. This region, which is connected to the rest of the French electric network by a unique line, is limited in terms of energy imports. Serre-Ponçon reservoir objectives and constraints are also related

to recreational activities on the lake, drinking and irrigation water supply and to the preservation of downstream ecological integrity. Contrary to most French mountain basins of this size, there are no significant reservoirs built along the UDR and its discharges are thus almost natural. The local climate is much drier than in the northern French Alps (Durand et al., 2009) due to the Mediterranean influence and to the protection from oceanic disturbances provided by the high Ecrins Mountains (Fig. 1). With elevations ranging from 700 to 4100 m a.s.l., the catchment presents highly seasonal flows due to snow accumulation and melt. Winter low flows can last 3 months or more. Long low flow sequences are also frequently observed in late summer and fall. During these seasons and when the precipitations are negligible, such a low flow episode can last several weeks after the end of the snow-covered period. Major floods are often observed in fall with intense rainfall events (Lafaysse et al., 2011).

We consider a simplified water resource system inspired by the real UDR system with two basic uses: hydroelectric production (HEP) and maintenance of a minimum water level in the reservoir lake during the summer season for recreational activities (Reservoir Level Maintenance denoted as RLM). We chose HEP and RLM because these two objectives present important differences in terms of adequacy with the water resource availability and are important for the real system of Serre-Ponçon.

The benefit function used in Eq. (1) for the determination of MSWV is the sum of the possible benefits from HEP as defined by Eq. (2) and benefits from RLM during a summer season as defined by Eq. (3):

$$g_{\mathrm{HEP}}\left(u_{t_i}, s_{t_i}, t\right) = \mathrm{HEPI}_{t_i} u_{t_i} r(s_{t_i}), \qquad (2)$$

where u_{t_i} in m^3 s^{-1} is the discharge released from the reservoir for HEP, HEPI is the daily interest of HEP in value units kWh^{-1} (see Sect. 3.4) and r is the hydropower production coefficient in kWh m^{-3} which depends on the water head in the reservoir:

$$\begin{cases} g_{\mathrm{RLM}}\left(s_{t_i}, t_i\right) = K\left[1 - b\left\{\max\left(s^* - s_{t_i}, 0\right)\right\}^2\right] \\ \quad \text{if } t_i \in \text{ summer season} \\ g_{\mathrm{RLM}}\left(s_{t_i}, t_i\right) = 0 \text{ if not} \end{cases} \qquad (3)$$

In Eq. (3), K is the maximal value of daily benefit (value units) that can be obtained during the summer period. It is achieved as soon as the storage is greater than a threshold $s^* = 85\%$ of the storage capacity, the volume below which recreational activities are expected to be reduced. The corresponding decrease in RLM benefit is assumed to be a quadratic function of the difference between the actual water storage and s^*. In Eq. (1), the values of the weighting parameters c_j are referred to as c_{HEP} and c_{RLM} for the HEP and RLM objectives respectively and set either to 1 when the objective is considered or to 0 when it is not.

In the water balance of the reservoir, the only input and output discharges are respectively the inflow from the upstream UDR basin and the optimized water release. Direct precipitations to the reservoir and evaporation from the reservoir are neglected. Their inter-annual means are actually of the same order, and the net balance between both terms is less than 1 % of the mean river discharge into the reservoir (Vachala, 2008).

In France, like in many countries where hydropower is not dominant, hydroelectric production is used to replace more expensive power generation facilities and the objective is to minimize the expected sum of other energy production costs for the national network as a whole. In this study, we consider a simplified daily interest of HEP estimated from a local daily temperature index (see Sect. 3.3) and the benefits are optimized for the system independently from other considerations of the energy production costs. In addition, summer RLM is a priority objective: an empirical guideline curve is used for reservoir operations (applied mostly in the spring season) and HEP optimization roughly applies to the water inflows that are not needed to satisfy the RLM objective.

However, it is expected that an increase of future energy costs will increase the interest of HEP and, as a consequence, benefits from recreational activities will be balanced with respect to benefits from HEP (or with respect to the reduction of other production costs allowed by the use of HEP). In this study, a benefit function (Eq. 3) is therefore used for RLM instead of a rule curve. This provides a rough estimate of the marginal value of storage water to satisfy the RLM objective. Recreational benefits are expressed as a function of water storage in the reservoir, similarly to Ward et al. (1996). However, our formulation does not include information about tourist affluence due to the lack of appropriate data in the region. The value of K in Eq. (3) is chosen so that the maximum benefit obtained from RLM is of the same order of magnitude as the one obtained from HEP, if they were considered separately. This makes it possible to analyse a double-objective configuration with objectives of equivalent economic value, a situation that could occur in the future.

The inflows to the reservoir are modelled with CEQUEAU (Morin et al., 1975), a semi-distributed hydrological model already applied by EDF for previous climate change impact studies on different mesoscale French basins. Snow accumulation and melt, effective rainfall, infiltration and evapotranspiration fluxes are estimated for each of the 99 sub-basins from daily series of mean areal precipitation and surface air temperature. The discharges produced by all hydrological units are routed through the river network to produce the total water inflow into the reservoir. The CEQUEAU model of UDR has been calibrated and validated by Bourqui et al. (2011) with a split sample test procedure. The Nash–Sutcliffe efficiency criterion (Nash and Sutcliffe, 1970) is 0.86 for the 1981–2005 calibration period and 0.83 for the 1959–1981 validation period.

3.2   Climate scenarios

The observed precipitation and temperature data for the 1970–1999 control period are obtained from the daily meteorological reanalyses developed by Gottardi et al. (2012) for French mountainous regions. The reference discharges to the reservoir for the control period are those obtained from CEQUEAU simulations.

The local-scale time series of temperature and precipitation for the future climate period 2070–2099 are obtained by perturbing the observed time series of the control period in a similar way to Horton et al. (2006). Six synthetic regional climate change scenarios are defined as an absolute change of the mean annual temperature and as a relative change of the mean annual precipitation. The magnitude of these changes is derived from a suite of climate modelling experiments conducted in the EU PRUDENCE project (Christensen, 2004) for SRES scenario A2 (Nakicenovic et al., 2001). It roughly corresponds to the 50th and 90th percentiles of changes estimated by the climate model experiments, representing respectively a 10 and 20 % decrease in precipitation and a 3 and 5 °C increase in temperature.

Control and future hydrological regimes obtained from CEQUEAU simulations for these scenarios are presented in Fig. 2. A temperature increase leads to reduced snow accumulation in winter and an earlier melting season. This in turn induces a higher winter low flow and a lower snowmelt flood peak (Fig. 2, left). The snowmelt flood peak shifts by one month for the warmest scenario (+5 °C). Besides this change in flow seasonality, an increase in temperature also leads to a slight reduction of the mean annual inflow to the reservoir due to increased evapotranspiration losses in summer (up to 22 % for the +5 °C scenario). Without temperature change, precipitation change scenarios modify the magnitude of the hydrological cycle (Fig. 2, middle). The mean inter-annual daily discharges decrease with the mean inter-annual precipitation, except for the winter period during which flows are sustained by deep underground storage. The large decrease of the snowmelt flood peak is the result of a smaller snowpack extent and thickness, induced by less winter to spring solid precipitation.

Scenarios with both precipitation and temperature changes lead to a modification of the hydrological regime that roughly combines the modifications previously discussed for temperature change (mainly modification in seasonality) or precipitation change alone (mainly modification in mean discharge).

3.3   Economic interest of hydroelectric production

A detailed representation of electricity prices is difficult to simulate because of the complex interaction with other energy production means and the high variability of the energy market. However, electricity prices in France tend to be higher for periods of high electricity consumption. Moreover, electricity consumption is higher during the cold season

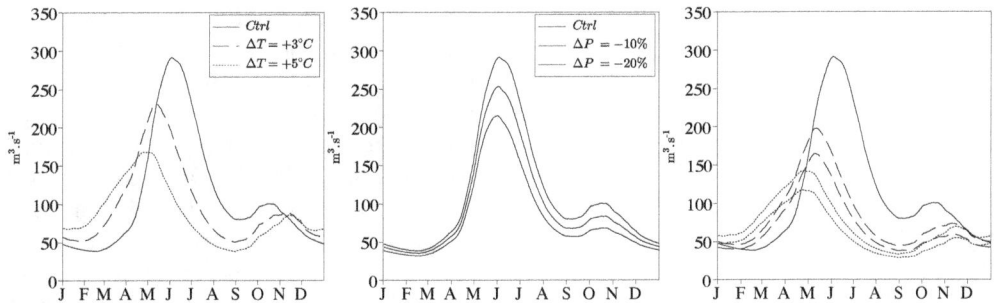

Figure 2. Mean inter-annual cycles of daily inflow to the reservoir for control data (black curve in all graphics, period 1970–1999) and two future meteorological scenarios (with prescribed changes of the mean annual temperature (ΔT) and precipitation (ΔP) over the period 2070–2099). Left: changes in mean annual temperature only. Middle: changes in mean annual precipitation only. Right: changes in both annual precipitation and temperature. The control hydrological regime is obtained from CEQUEAU simulations with the observed meteorological times series of the 1970–1999 period.

and highly correlated with the daily time variations of regional temperatures below an approximate heating threshold $T_{heat} = 15\,°C$ that governs heating demand. As a result, a convenient formulation for the daily interest of HEP (HEPI) can be based on daily regional temperatures (Paiva et al., 2010). The electricity consumption is assumed to linearly decrease with the temperature up to a given threshold and to remain constant above this threshold.

In a future climate with higher summer temperatures, an additional demand for hydroelectric production is expected for cooling purposes. The daily HEPI expected in the future during the hot season is assumed to linearly depend on regional temperatures above a cooling threshold $T_{cool} = 25\,°C$ (Buzoianu et al., 2005). In the following, the daily HEPI is therefore defined as a piece-wise linear function of daily temperature:

$$\begin{cases} \text{HEPI}_{t_i} = \text{HEPI}_0 + \text{HEPI}_h \cdot (T_{heat} - T_{t_i}) \text{ if } T_{t_i} < T_{heat} \\ \text{HEPI}_{t_i} = \text{HEPI}_0 \text{ if } T_{heat} < T_{t_i} < T_{cool} \\ \text{HEPI}_{t_i} = \text{HEPI}_0 + \text{HEPI}_c \cdot (T_{t_i} - T_{cool}) \text{ if } T_{t_i} > T_{cool}, \end{cases} \quad (4)$$

where HEPI_0 is the HEPI when temperatures are in between cooling and heating temperature thresholds, and HEPI_h and HEPI_c are the additional HEPI rates for each the heating and the cooling seasons respectively. The HEPI is expressed in value units per kWh denoted V hereafter. HEPI_0 and HEPI_h are set to unity ($=1\,\text{V}\,°\text{C}^{-1}$) in accordance to Paiva et al. (2010). A higher value was set for HEPI_c ($\text{HEPI}_c = 2.5\,\text{V}\,°\text{C}^{-1}$).

Time series of daily HEPI were obtained for each scenario of daily temperatures. The corresponding mean inter-annual values of daily HEPI are presented in Fig. 3 as characteristic seasonal HEPI patterns.

4 Storage signature

In order to briefly illustrate the kind of climate signature proposed in this work, we start the analysis of our results looking at the storage scheme obtained for the period 1970–1999

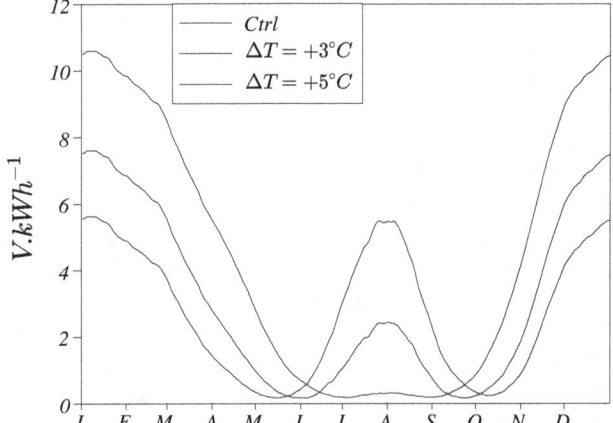

Figure 3. Mean inter-annual cycles of the interest hydroelectric production (HEPI) for the control period and two different future scenarios of annual temperature increase ΔT.

when both HEP and RLM objectives are taken into account (this configuration is denoted HEP + RLM in the following). The reservoir inflows and HEPI scenarios are produced as described in Sect. 3. Their optimal temporal balance is computed through dynamic programming as explained in Sect. 2. The constrained summer season for RLM runs from 15 June to 31 August and the minimum assigned storage level is $s^* = 85\,\%$ of s_{max} during this period, and $s^* = 0$ outside this period. As shown Fig. 4, the storage scheme presents a significant seasonality. The storage level continuously decreases during winter months, when HEPI is high and inflows are low. It then increases during spring-time with high spring snowmelt inflows and lower HEPI values. The inter-annual variability of the storage scheme is moderate (see dispersion between grey curves around the mean inter-annual pattern in Fig. 4), and much lower than the intra-annual variability that covers the full capacity range from 10 to 100 %. The lowest inter-annual variability of the scheme is obtained for the first days of November. Each year, the reservoir is roughly full at

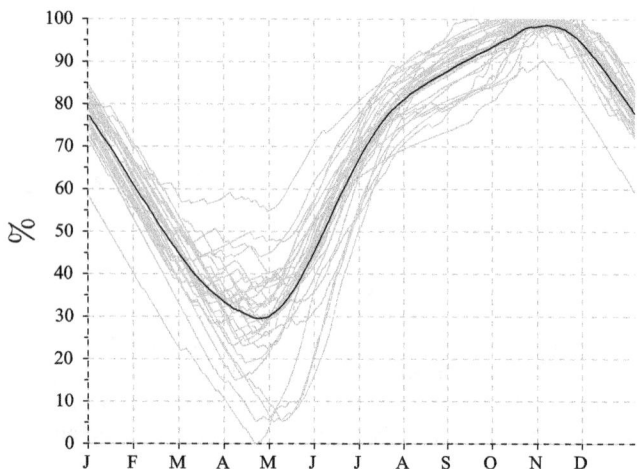

Figure 4. Storage requirement scheme for the period 1970–1999 (configuration HEP + RLM). Grey curves: day-to-day storage level trajectory required each year to reach the best possible resource/demand equilibrium, given the constraints; Black curve: mean inter-annual storage cycle.

this period. The highest inter-annual variability of the scheme is during spring period when storage levels vary from 10 to 60 % of the reservoir capacity. All storage curves converge next rapidly to a high storage level as required by the summer touristic level objective. Despite this, the summer level objective (i.e. 85 % of s_{max}) is never reached on time (i.e. the 15 June) but roughly one month later.

In the following sections, because the temporal variations of the storage scheme are mainly seasonal, we use its mean inter-annual pattern as a first signature of the disequilibrium between water resources and demand for the studied climatic and economical forcing. For brevity, we call this the storage signature.

5 Storage water value signature

The storage signature derives from temporal patterns of MSWV that we discuss now for various climate scenarios and various combinations of objectives. For a more comprehensive analysis, we consider in a preliminary step two objectives separately (HEP or RLM) and subsequently a double-objective configuration (HEP + RLM).

5.1 Hydroelectric production

The optimization of the HEP objective alone corresponds to $C_{HEP} = 1$ and $C_{RLM} = 0$ in Eq. (1). Note first that the efficiency of the hydroelectric production system is an increasing function of water head in the reservoir. If HEP were constant throughout the year, the storage scheme would be to maintain the water level at its highest possible value, which may be a bit lower than the storage capacity in order to avoid future spillage (see for example Turgeon, 2007). Except be-

fore large inflow periods such as the snowmelt season, this scheme would correspond to high MSWV for most reservoir levels, especially the lowest ones. In the studied configuration, MSWV is higher during the periods prior to the highest HEPI. The high seasonality of HEPI (Fig. 3) thus influences the seasonality of MSWV and modulates the storage scheme.

Figure 5 illustrates the variation of the HEPI and the water inflows to the reservoir with time over a 4-year period (1 January 1977 to 1 January 1981). It also presents the corresponding variations of the MSWV with time for different reservoir levels (corresponding to 10, 50 and 90 % of storage capacity) and the resulting optimal storage requirement scheme.

At any time, MSWV is lower at high storage levels (Fig. 5, top). At these levels, the increase of the future benefit related to an additional storage of water is very low. Indeed, additional stored water might be turbined during very low HEPI periods only, in order to avoid un-valorized spillages. At any storage level, MSWV fluctuates in time. At high storage levels (e.g. 90 %), MSWV is low except when a very high HEPI period is imminent (e.g. before winter periods). At low storage levels (e.g. 10 %), MSWV is conversely high to very high (up to 10 relative value units) except when a very high flow period is imminent (e.g. before spring flood periods). At all storage levels high MSWV prompts water storage for future use.

Periods of high HEPI alternate with periods of inflow discharge (Fig. 5, bottom). As a result, MSWV presents high seasonal variations for all reservoir levels (Fig. 5, top). During the late winter and early spring transition periods, the concomitant decrease of HEPI and rapid increase of snowmelt inflow diminishes the storage requirement. The following increase of MSWV is quite abrupt, as can be seen during the year 1979 in June for the storage level 50 % and in September for the storage level 90 %. It begins as soon as spillage is no longer required, given the known future inflows.

For any given storage level, MSWV varies with time reflecting the role of the reservoir in adjusting the adequacy between the future HEPI and the future availability of water from upstream catchment. Future resource abundance (respectively scarcity) decreases (respectively increases) the value of more storage water – like for example in May 1977 (respectively September 1977).

In addition to a marked seasonality, MSWV shows year-to-year variations related to the future ratio of HEPI and the inflow. MSWV is for instance higher in 1980 than in the previous 3 years. This inter-annual variability directly translates to the storage requirement scheme with a spring storage higher than 30 % of the capacity for 1980 whereas it roughly equals zero for previous years.

As for the storage requirement scheme, the variation of the MSWV in time reflects in a sophisticated way the temporal patterns of the climate variables governing the water demands and inflows. In the following sections we will use

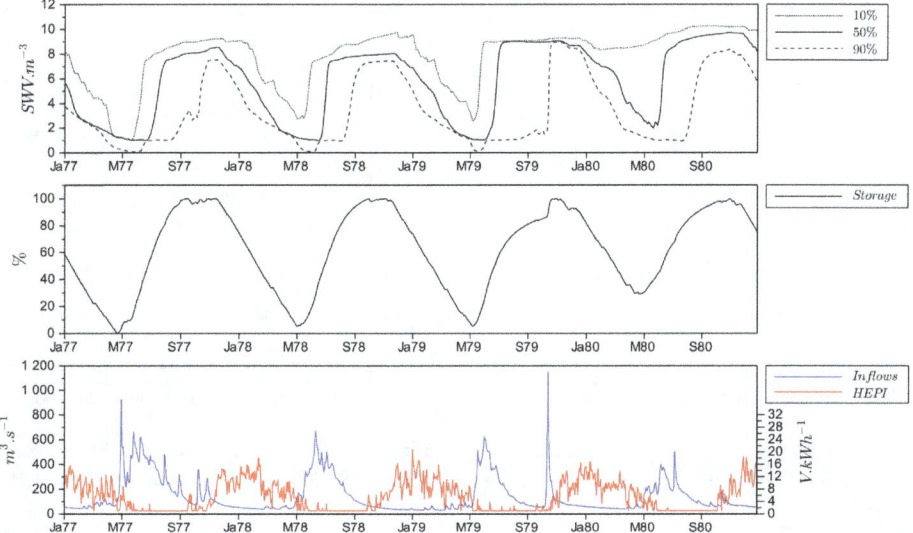

Figure 5. Variations of MSWV, reservoir level, inflows and interest for hydroelectric production (HEPI) from January 1977 to January 1981 for the meteorological control scenario (Ja: January, M: May, S: September). Top: marginal value of water (SWV m^{-3}) for different reservoir storage levels corresponding to 10, 50 and 90 % of the capacity. Middle: reservoir level (%) Bottom: water inflow to the reservoir (blue curve, m^3 s^{-1}) and interest of hydroelectric production (red curve, V kWh^{-1}).

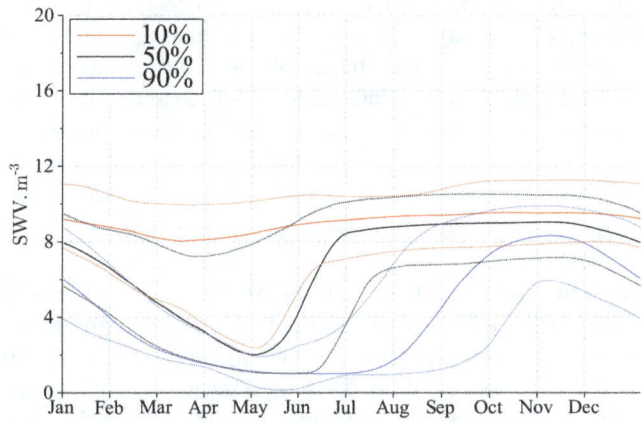

Figure 6. MSWV signature for the single hydroelectric production objective (HEP). The mean inter-annual MSWV variation obtained for the 1970–1999 period is plotted for three reservoir storage levels (10, 50 and 90 % of storage capacity). For each storage level, the upper, middle, and lower curves correspond respectively to the 95th percentile, the mean and the 5th percentile of MSWV calendar values obtained for the 30 years of the period.

the mean inter-annual patterns of MSWV for different reservoir levels as a second signature of the disequilibrium between water resource and demand under climatic and economic conditions. The MSWV signature obtained for the UDR system is presented in Fig. 6 for three storage levels (10, 50 and 90 % of storage capacity). In addition to the mean inter-annual value, Fig. 6 also shows the 5th or 95th percentiles of the MSWV calendar values. For the sake of conciseness, the expression "MSWV signature" will subsequently be used for this type of graph.

5.2 Summer reservoir level maintenance

We now consider a system for which the only objective would be to maintain a minimum water level in the reservoir during the summer months as explained in Sect. 3 (i.e. $C_{HEP} = 0$ and $C_{RLM} = 1$ in Eq. 1). Penalty costs are incurred in the event of failure to maintain the required level. The MSWV corresponds to the additional reduction of penalty costs that would be achieved by storing one more cubic metre of water at the current date. The MSWV signature is quite different from the one obtained for the HEP objective alone although it presents also a marked seasonality (see Fig. 7 compared to Fig. 6).

The possibility to achieve the objective depends on the current storage level and on the volume of inflow to the reservoir from the current date to the beginning of the next constrained period. At a given date, the higher the current storage level, the easier it is to achieve the objective.

For a given storage level, the longer the duration until the next constrained period, the larger the total future inflows to the reservoir and the easier it is to reach the objective. MSWV therefore slowly increases over the year to reach a maximum in early summer. According to Fig. 7 the MSWV maximum is nearly 1 month before the beginning of the constrained period for the most adverse situations (95th percentile envelope curve – corresponding to the driest spring years) or as late as mid-July for the most favourable situations (5th percentile envelope curve – corresponding to the wettest spring years). The lowest MSWV is zero, indicating that there is no interest to store water as forthcoming inflows will fill the reservoir to the required level on time (Fig. 7).

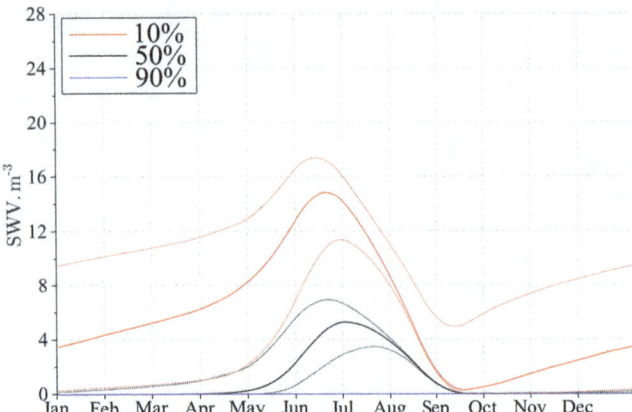

Figure 7. MSWV signature for the reservoir level maintenance objective (RLM). See Fig. 6 caption for details. The 90 % curves are confounded with the x-axis.

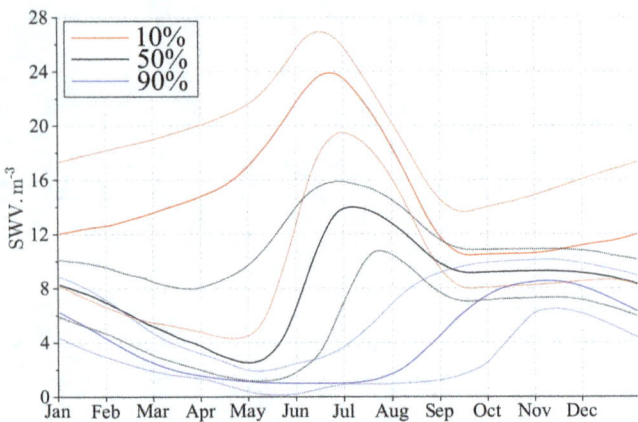

Figure 8. MSWV signature for the double-objective configuration (HEP + RLM). See Fig. 6 caption for details.

This is the case for almost all reservoir levels in September, after the end of the constrained period (an exception is for the driest years if the storage level is low). This applies also from mid-September to mid-April at more than 50 % of the storage capacity, when large inflows from the spring snowmelt flood are expected. In terms of seasonality the periods of high and low MSWV are roughly in phase opposition with those obtained previously for the HEP objective.

5.3 Double-objective configuration

Figure 8 presents the MSWV signature obtained when both HEP and RLM objectives must be fulfilled (i.e. $C_{HEP} = 1$ and $C_{RLM} = 1$ in Eq. 1). The storage signature for this configuration is the one discussed in Sect. 4 (Fig. 4).

For this configuration, MSWV is logically higher than those obtained for each single-objective configuration (Figs. 6 and 7). It is actually not possible to produce as much HEP and to fulfil the RLM objective as well as in the single-objective configurations. To limit the cost of RLM failure, water allocations previously determined for the single HEP objective configuration must be re-allocated to periods with lower HEPI thanks to higher MSWV at all reservoir levels, since high MSWV reduces the interest of immediate water use.

The MSWV signature for the double-objective configuration is not exactly an additive combination of the two single-objective signatures owing to the non-linearity of the optimization. The most significant difference between the HEP + RLM signature and the sum of the single-objective ones is during the winter season at low reservoir levels. The higher MSWV obtained for the double-objective configuration directly translates to the storage scheme. For instance, the minimum storage levels of the storage scheme are all greater than 10 % (see Fig. 4) whereas it can reach zero in the single HEP objective configuration (see spring storage level

for the year 1977 in Fig. 5). Similarly, the storage level in the early fall is always over 80–90 % in the double-objective configuration, whereas it may be lower than 80 % in the single HEP objective configuration (see year 1979 in Fig. 5).

In summary, the MSWV signature displays patterns of increasing complexity when the variety of assigned objectives increases. The seasonal shapes of the different objectives combine almost linearly and reflect with great detail the respective seasonality of the climate and the various demands.

6 Sensitivity of the signatures to climate change

We show now the sensitivity of the storage and MSWV signatures to a climate modification resulting from an annual temperature increase, an annual precipitation decrease and finally from both modifications simultaneously. This sensitivity analysis illustrates the interest of the presented results in terms of climate change signatures.

Figure 9 displays the storage signature for the double-objective configuration HEP + RLM. The signature is more sensitive to temperature warming than to precipitation decrease. For all scenarios, the average storage level increases and the magnitude of seasonal storage fluctuation is significantly lower which means that the resource–demand temporal equilibrium improves under the considered future climates. The temporal pattern of the storage signature is also modified: the late summer period for which high levels of storage were required is 2 months longer for a 3 °C warming. For the 5 °C warming scenario, a bimodal pattern is obtained and the period with the highest required storage levels is shifted to early summer.

Figure 10 shows the dependence of MSWV signatures to temperature for HEP and RLM objectives. For the HEP objective alone (first row Fig. 10), a temperature increase modifies the seasonality of the MSWV signature but does not

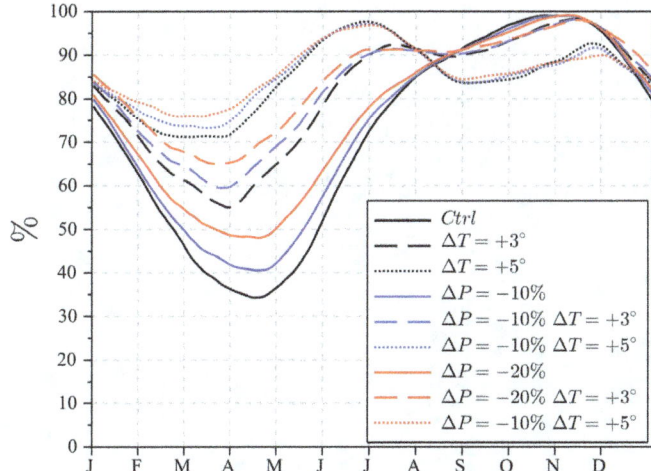

Figure 9. Sensitivity of storage requirement scheme to temperature increase or precipitation decrease or both together.

significantly change the average value of storage water. The MSWV seasonal peak is shifted from autumn to summer for high reservoir levels and disappears at low levels. At all levels the seasonality of MSWV is smoothed out; in particular for low and medium reservoir levels (10 and 50 %), MSWV becomes practically constant throughout the year. This observation corroborates the better temporal balance between resource and demand under a modified climate. At low and middle storage levels and compared to the control period, the increase of MSWV during the spring season is due to far less intense snowmelt floods (Fig. 2) and in turn to a large decrease of potential spillage risk. Potential spillage is also reduced because of a better temporal match between inflows and periods of high HEPI: for the control period, the main inflow period (spring) is almost 8 months before the highest HEPI (winter); for the increased-temperature scenarios, the snowmelt flood is up to 1 month earlier and a second period with high HEPI appears in the summer season only 3 to 4 months later. At high storage level, the MSWV signature modification is different but the reasons for these changes remain the same. The large MSWV values during the late spring and summer seasons increase the interest of raising the water head during this period without causing later spillage thanks to the new and greater interest of HEP in summer. The low MSWV values in winter result from the lower HEPI demand for this season.

For the RLM objective alone, lower mean inflow and earlier snowmelt increase MSWV earlier in the year for reservoir levels lower than the summer objective. The objective is therefore more difficult to meet on time than for the control period. For low reservoir storage levels the positive MSWV obtained in September even shows incapacity to meet this single objective.

Finally, the MSWV signature obtained for the double HEP + RLM configuration is approximately an additive combination of the two single-objective signatures, as for the present climate. For example, for the 50 % storage level, the large MSWV decrease observed in the control climate during the 6 months from December to May tends to disappear as a consequence of the smaller snowmelt flood and the increased HEP interest during the summer months.

Regarding now a precipitation decrease, Fig. 11 displays the MSWV signature for the HEP + RLM configuration. As changes in precipitation do not influence the seasonality of the inflow (Fig. 2) and the demand, the seasonality of MSWV is maintained, whatever the reservoir level. The decrease in precipitation leads to a reduced mean inflow to the reservoir and, in turn, to an increased MSWV mean value at all storage levels and all seasons (excepted during the summer season for the 90 % storage level where MSWV is zero). This means more severe conditions with a concentration of water allocations to HEP in the periods with the highest HEPI values. Similar results are obtained when considering HEP or RLM alone (not shown).

Finally, the MSWV signature resulting from a modification of both precipitation and temperature changes is shown for three storage levels in Fig. 12. Seasonality and mean value of MSWV are modified. Changes of MSWV for this combined change are approximately an additive combination of the partial ones, and directly translate to modifications of the storage scheme described previously. They lead for instance to building the storage earlier in order to better use the earlier spring snowmelt flood. They also lead to reducing the magnitude of storage fluctuations and thus to increase the water head, especially before the period of high HEPI in summer due to cooling needs.

7 Conclusions

In this study we formalized the central role of water storage management in balancing seasonal fluctuations of the water resource/demand equilibrium using an elementary optimization technique. The representation of the water system is reduced to a small set of objectives and free of any hypothesis on the constraints and uncertainties of the real-time management. Derived storage water values and reservoir levels exhibit seasonal patterns that we propose to read as signatures of this climatological equilibrium and its potential modification under changing hydro-climatic conditions. We consider such signatures as attractive alternatives to performance indicators like statistics of a system's failures in the sense that they preserve quite complicated seasonal patterns giving more insight into the socio-technical system behaviour.

The presented case study illustrates how the proposed signatures contain, under a synthetic set of graphs, much information on the seasonality of the governing processes and their eventual shifts in time. The multi-purpose system taken

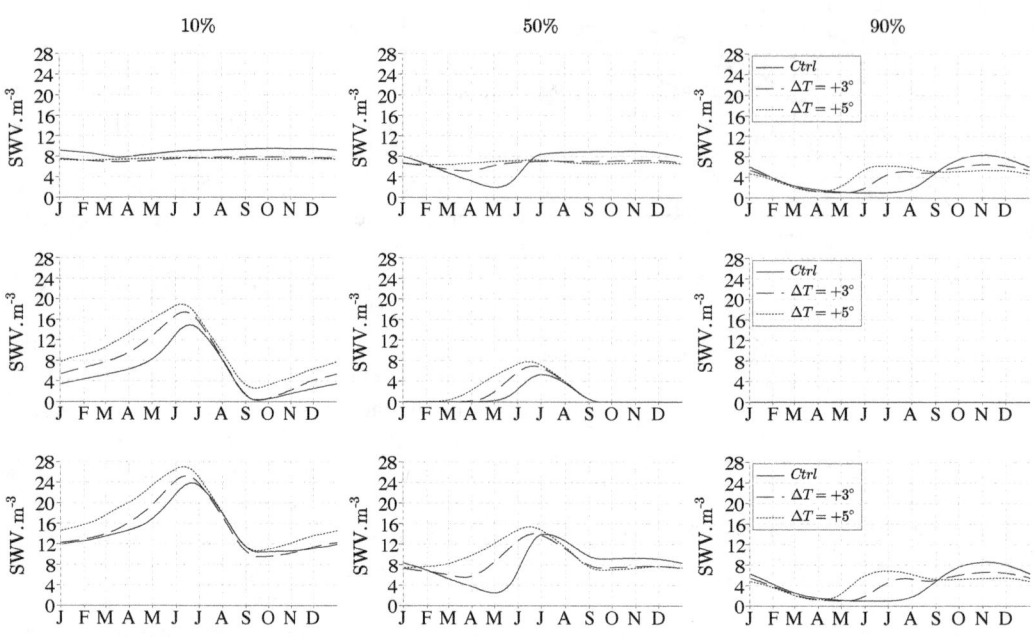

Figure 10. Sensitivity of MSWV signatures to temperature. The different curves correspond to the control data set and to two scenarios of warming. The different columns correspond to storage levels of 10 % (left), 50 % (middle) and 90 % (right) of storage capacity. The objectives considered are the HEP (top graphs), the RLM (middle) and a combination of the two (bottom).

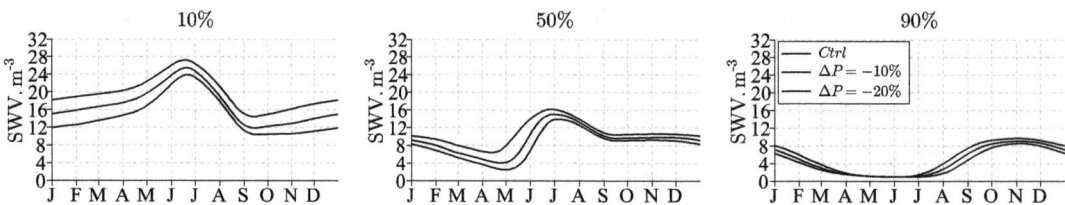

Figure 11. Sensitivity of MSWV signatures to precipitation changes in the case of the double-objective configuration (HEP + RLM). The different columns correspond to storage levels of 10 % (left), 50 % (middle) and 90 % (right) of storage capacity.

in the French Alps is reduced to the management of a single reservoir responding to a demand for hydroelectricity and reservoir level maintenance during a touristic period in a climate change context. This case study led to the following considerations:

- When considering several management objectives, each individual objective signature sheds light on its specific role and the multi-purpose signature is not the mere linear combination of the individual signatures, which reveals the potentially non-linear interaction or competition between objectives.

- When analysing signatures one by one, the smoothness of their shape and their amplitude seems to be informative. Both for MSWV and storage signatures, a smoother shape shows a better seasonal fit between resource and demand and thus an easier manageability or lower storage fluctuation needs.

- When comparing signatures under different climatic conditions, changes in shape reveal changes of the governing processes. For instance, the studied water system seems to be more sensitive to warmer conditions than to drier ones. Warmer conditions deeply modify the different signatures (MSWV and storage) in relation with the behaviour of the snow-pack and the electricity demand. Drier conditions provide more homothetic shape modifications, revealing less impact on the management and the storage signatures.

- As a last consideration, we can note that the storage signature is more straightforward to interpret, both in terms of shape (management difficulty) and amplitude (reservoir relevance). Nevertheless, this signature only reflects the satisfaction of the objective. Its shape can be weakly informative when this objective is simple like in the case of the RLM alone – the storage signature is then almost flat throughout the year. Interpreting MSWV signatures requires a more economical reasoning about the

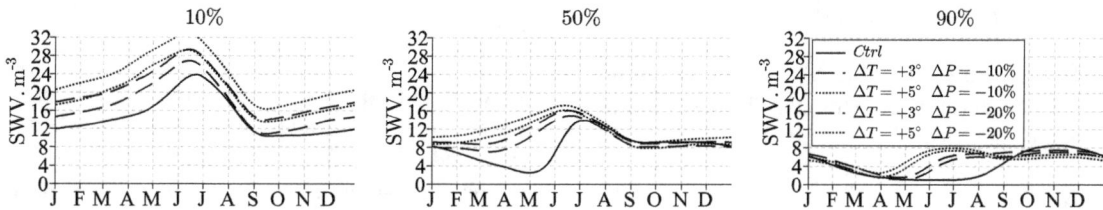

Figure 12. Sensitivity of MSWV signatures to changes of both precipitation and temperature in the case of the double-objective configuration (HEP + RLM). The different columns correspond to storage capacity levels of 10 % (left), 50 % (middle) and 90 % (right).

interest of water allocation in time. They express in more detail the full set of mechanisms behind the satisfaction of the assigned objective. For instance, in the case of the RLM objective alone, the MSWV signature will display the rather marked seasonality of the needed management, and not only its mere result expressed by a single reliability value, for example. In that sense, we suggest that both signatures are useful.

This study has shown some limitations, opening perspectives for future studies. There is for instance the relative complexity of the system used for the demonstration. Real water resource systems generally deal with more objectives and constraints and with a number of interconnected reservoirs. With an optimization algorithm such as dynamic programming, additional constraints and requirements can be integrated quite easily (e.g. irrigation water demand, dam safety management during floods or minimum flow maintenance for ecosystem integrity). In the case of multi-reservoir systems, MSWV will be site dependent in addition to being time and storage level dependent (Tilmant et al., 2008, 2009; Wolfgang et al., 2009).

The simulation of future hydrological scenarios was here driven by observed precipitation and temperature time series modified according to synthetic climate change scenarios using a classical perturbation methodology. The temporal variability of future meteorological variables is therefore the same as that of the historical period. In particular, no changes in the sequences of wet and dry periods are considered from seasonal to pluri-annual timescales. Such changes are however expected to be potentially as critical as changes in the means of meteorological driving variables. They at least fully determine changes in the temporal variability of natural inflows into a reservoir, a determinant factor in the analysis of the performance of the system (McMahon et al., 2006). Changes in precipitation seasonality are expected to modify the seasonality of inflows. A higher variability of annual or pluri-annual inflows to the reservoir is also expected to lead to longer and more frequent periods of resource scarcity. The influence of such regional climate modifications will be analysed considering a large set of scenarios recently developed within the RIWER2030 research project (http://www.lthe.fr/RIWER2030/). For the studied region, those scenarios are obtained using different statistical downscaling models from a suite of GCM (global circulation model) experiments (Lafaysse et al., 2014).

Finally, we note that MSWV is also frequently estimated to determine an operating strategy for the real-time management of a water system. In such a case, the MSWV can be obtained using stochastic dynamic programming in a configuration in which future inflows and water demands are unknown (e.g. Wolfgang et al., 2009). As a result of inflow variability and uncertain predictability, the MSWV is expected to increase when compared to the MSWV obtained in the configuration of the present work (Draper et al., 2003; François, 2013). MSWV signatures obtained for an uncertain future are also potentially very informative with regard to how an operational strategy is organized, what its key features are and how it could change should the climate or demand change, or both together. When they are conversely obtained for a known sequence of inflow and demand, as in the present work, MSWV signatures define the best possible manageability of the system. They are, therefore, not influenced by possible changes in the predictability of future inflows and demand. They furthermore separate in a sense the socioclimatic and the management components of the equilibrium. In that regard, analysing changes in this signature is expected to improve our understanding of modifications of the optimal storage requirement scheme for this socio-climatic context as well as modifications of system performance classically reported on the basis of a variety of performance criteria in climate change impact analyses.

Appendix A

In deterministic dynamic programming, the optimal storage variation for each time step t_i of the considered simulation period $[t_0, t_N]$ is identified in order to maximize the sum, over the simulation period $[t_i, t_N]$, of the current benefits, i.e. the benefits that would result from an immediate use of water at time step t_i, and of the optimal future benefits, i.e. the benefits that would result from optimal storage variations over the future simulation period $[t_{i+1}, t_N]$. The optimal future benefit $F_{t_i}(s_{t_i})$ obtainable from a hypothetical reservoir level s_{t_i} at time t_i is often referred to as the Bellman Value for this storage and time configuration (Bellman, 1957). It is obtained from a backward recursive calculation from the future benefits estimated for time t_{i+1}:

210   Water Resources Management: Principles and Practice

$$F_{t_i}\left(s_{t_i}\right) = \left\{g\left(u_{t_i}, s_{t_i}, t_i\right) + F_{t_{i+1}}\left(s_{t_{i+1}}\right)\right\}, \tag{A1}$$

where the different terms are subject to upper and lower bounds and mass conservation constraints. The state and decision variables are such that

$$s_{\min} \leq s_{t_i} \leq s_{\max} \tag{A2}$$

and

$$u_{\min} \leq u_{t_i} \leq u_{\max}, \tag{A3}$$

where $s_{\min}$ and $s_{\max}$ are minimum and maximum bounds for water storage volumes in the reservoir and $u_{\min}$ and $u_{\max}$ the minimum and maximum bounds for release discharges. The mass conservation equation is

$$s_{t_{i+1}} = s_{t_i} + q_{t_i} - u_{t_i} - o_{t_i}, \tag{A4}$$

where q_{t_i} is the inflow to the reservoir during the period $[t_i, t_{i+1}]$, and o_{t_i} the losses (evaporation above the reservoir, controlled and uncontrolled withdrawals from the reservoir for irrigation, drinking water and other uses).

A discrete approach can be used to estimate the benefit function $F_t(s)$ when the dimension of the state vector is not too large. An extensive discussion about the dimensionality issue is presented in Yakowitz (1982). The final result is a table that gives the future benefits for different water levels and each time step of the simulation period. For storage levels in-between the a priori selected states, $F_t(s)$ can be obtained via interpolation. In our case, $F_t(s)$ is estimated at a daily time step and at 51 storage levels uniformly distributed between the minimum and maximum storage bounds $s_{\min}$ and $s_{\max}$. A cubic spline interpolation method is used when needed (Foufoula-Georgiou and Kitanidis, 1988).

In the present study, end values are estimated as proposed by Wolfgang et al. (2009). The duration of the simulation period is artificially increased with a fictitious n-year initialization period, added at the end of the simulation period. The initialization period is composed from several duplications of the final year so that the storage water values at t_N are no longer influenced by the boundary conditions chosen at the end of the extended planning period. The storage water values at t_N are next used to estimate the corresponding Bellman value $F_{t_N}(s)$ from the reciprocal function of Eq. (A1).

The derivative of the future benefit function $F_t(s)$ for a given storage level s in the reservoir gives the optimal benefit for a future use of one additional unit of water stored at this storage level (Eq. 1). It corresponds to the marginal value of storage water for this storage level s and time t:

$$V_t(s) = \frac{\partial F_t(s)}{\partial s}. \tag{A5}$$

As shown in Eq. (A5), the marginal value of storage water V is time and storage level dependent. The MSWV signatures proposed in Sect. 5 are derived from this computation.

The above-mentioned optimization stage provides the optimal future benefit $F_t(s)$ for all storage levels s of the state-time table. This table can be used to derive the storage water values V for the same state–time grid. In a discrete approach, the derivatives are calculated with finite differences from neighbouring water level states in the table.

The storage water values can be used in a second optimization stage to identify the optimal operation decision for the current time t_i, given the water level in the reservoir s_{t_i}. This operation maximizes the following equation:

$$\left\{g\left(u_{t_i}, s_{t_i}, t_i\right) + \left(s_{t_{i+1}} - s_{t_i}\right).V_{t_{i+1}}(s_{t_{i+1}})\right\}. \tag{A6}$$

The forward iterative optimization of Eq. (A6) can therefore give the optimal sequence of storage variations, resulting reservoir water levels, benefits and penalty costs over the whole simulation horizon $[t_0, t_N]$. This simulation method is usually referred to as the water value method (e.g. Hveding, 1968). The storage signature proposed in Sect. 4 is derived from this computation.

Acknowledgements. This work was initiated within the RI-WER2030 research project on Regional Climate, Water, Energy Resources and Uncertainties from 1960 to 2030 (grant number ANR-08-VULN-014-01 funded by the VMCS Program of the French National Research Agency; http://www.lthe.fr/RIWER2030/). It also contributes to the COMPLEX Project on Knowledge Based Climate Mitigation Systems for a Low Carbon Economy (European Collaborative Project FP7-ENV-2012 number: 308601; http://www.complex.ac.uk/).

References

Alcamo, J., Moreno, J. M., Novaky, B., Binidi, M., Corobov, R., Devoy, R. J., Giannakopoulos, C., Martin, E., Olesen, J., and Shvidenko, A.: Europe, in: Climate Change 2007: Impacts, Adaptations and Vulnerability. Contribution of Working Group II to the Fourth Assessment Report of the Intergovernmental Panel on Climate Change, 541–580, Cambridge, 2007.

Ashofteh, P. S., Haddad, O. B., and Mariño, M. A.: Climate Change Impact on Reservoir Performance Indexes in Agricultural Water Supply, J. Irrig. Drainage Eng., 139, 85–97, doi:10.1061/(ASCE)IR.1943-4774.0000496, 2013.

Bellman, R.: Dynamic Programming, Princeton University Press., Defense Technical Information Center, New Jersey, 366 pp., 1957.

Bourqui, M., Mathevet, T., Gailhard, J., and Hendrickx, F.: Hydrological validation of statistical downscaling methods applied to climate model projections, IAHS-AISH publication, 344, 32–38, 2011.

Buzoianu, M., Brockwell, A. E., and Seppi, D. J.: A Dynamic Supply-Demand Model for Electricity Prices, Carnegie Mellon University,Department of statistics, available at: http://search.stat.cmu.edu/tr/tr817/tr817.pdf (last access: 11 October 2012), 2005.

Christensen, N.: Prediction of Regional scenarios and Uncertainties for Defining EuropeaN Climate change risks and Effects – Prudence Final Report, PREUDENCE EVK2-CT2001-00132, available at: http://prudence.dmi.dk (last access: October 2013), 2004.

Draper, A. J., Jenkins, M. W., Kirby, K. W., Lund, J. R., and Howitt, R. E.: Economic-engineering optimization for California water management, J. Water Resour. Plann. Manage., 129, 155–164, 2003.

Durand, Y., Laternser, M., Giraud, G., Etchevers, P., Lesaffre, B., and Merindol, L.: Reanalysis of 44 Yr of Climate in the French Alps (1958-2002): Methodology, Model Validation, Climatology, and Trends for Air Temperature and Precipitation, J. Appl. Meteorol. Climatol., 48, 429–449, doi:10.1175/2008JAMC1808.1, 2009.

Foufoula-Georgiou, E. and Kitanidis, P. K.: Gradient dynamic programming for stochastic optimal control of multidimensional water resources systems, Water Resour. Res., 24, 1345–1359, doi:198810.1029/WR024i008p01345, 1988.

François, B.: Gestion optimale d'un réservoir hydraulique multiusages et changement climatique. Modèles, projections et incertitudes, Université de Grenoble, Grenoble, available at: http://tel.archives-ouvertes.fr/docs/00/99/70/12/PDF/33716_FRANCOIS_2013_archivage.pdf (last access: 5 June 2014), 2013.

Gaudard, L., Gilli, M., and Romerio, F.: Climate Change Impacts on Hydropower Management, Water Resour. Manage., 27, 5143–5156, doi:10.1007/s11269-013-0458-1, 2013.

Gottardi, F., Obled, C., Gailhard, J., and Paquet, E.: Statistical reanalysis of precipitation fields based on ground network data and weather patterns: Application over French mountains, J. Hydrol., 432–433, 154–167, doi:10.1016/j.jhydrol.2012.02.014, 2012.

Hashimoto, T., Stedinger, J. R., and Loucks, D. P.: Reliability, resiliency, and vulnerability criteria for water resource system performance evaluation, Water Resour. Res., 18, 14–20, doi:10.1029/WR018i001p00014, 1982.

Hekkenberg, M., Benders, R. M. J., Moll, H. C., and Schoot Uiterkamp, A. J. M.: Indications for a changing electricity demand pattern: The temperature dependence of electricity demand in the Netherlands, Energ. Pol., 37, 1542–1551, doi:10.1016/j.enpol.2008.12.030, 2009.

Horton, P., Schaefli, B., Mezghani, A., Hingray, B., and Musy, A.: Assessment of climate-change impacts on alpine discharge regimes with climate model uncertainty, Hydrol. Process., 20, 2091–2109, doi:10.1002/hyp.6197, 2006.

Hveding, V.: Digital simulation techniques in power system planning, Econom. Plann., 8, 118–139, doi:10.1007/BF02481379, 1968.

Lafaysse, M., Hingray, B., Etchevers, P., Martin, E., and Obled, C.: Influence of spatial discretization, underground water storage and glacier melt on a physically-based hydrological model of the Upper Durance River basin, J. Hydrol., 403, 116–129, doi:10.1016/j.jhydrol.2011.03.046, 2011.

Lafaysse, M., Hingray, B., Mezghani, A., Gailhard, J., and Terray, L.: Internal variability and model uncertainty components in future hydrometeorological projections: The Alpine Durance basin, Water Resour. Res., 50, 3317–3341, doi:10.1002/2013WR014897, 2014.

Loucks, D. P., van Beek, E., Stedinger, J. R., Dijkman, J. P. M., and Villars, M. T.: Water Resources Systems Planning and Management: An Introduction to Methods, Models and Applications, Paris?: UNESCO, available at: http://ecommons.library.cornell.edu/handle/1813/2804 (last access: 23 May 2013), 2005.

Marnezy, A.: Les barrages alpins. De l'énergie hydraulique à la neige de culture, Revue de géographie alpine, J. Alpine Res., 1, 92–102, doi:10.4000/rga.422, 2008.

Masse, P.: Les réserves et la régulation de l'avenir dans la vie économique, Actualités scientifiques et industrielles 1008, Hermann, Paris, 1946.

McMahon, T., Adeloye, A., and Zhou, S.: Understanding performance measures of reservoirs, J. Hydrol., 324, 359–382, doi:10.1016/j.jhydrol.2005.09.030, 2006.

Minville, M., Brissette, F., Krau, S., and Leconte, R.: Adaptation to Climate Change in the Management of a Canadian Water-Resources System Exploited for Hydropower, Water Resour. Manage, 23, 2965–2986, doi:10.1007/s11269-009-9418-1, 2009.

Morin, G., Fortin, J. P., and Charbonneau, R.: Utilisation du modèle hydrophysiographique CEQUEAU pour l'exploitation des réservoirs artificiels, IAHS Publication no 115, 176–184, 1975.

Moy, W.-S., Cohon, J. L., and ReVelle, C. S.: A Programming Model for Analysis of the Reliability, Resilience, and Vulnerability of a Water Supply Reservoir, Water Resour. Res., 22, 489–498, doi:198610.1029/WR022i004p00489, 1986.

Nakicenovic, N., Alcamo, J., Davis, G., de Vries, B., Fenhann, J., Gaffin, S., Gregory, K., Grubler, A., Jung, T., Kram, T., La Rovere, E. L., Michaelis, L., Mori, S., Morita, T., Pepper, W., Pitcher, H. M., Price, L., Riahi, K., Roehrl, A., Rogner, H. H., Sankovski, A., Sclesinger, M., Shukla, P., Smith, S. J., Swart, R., van Rooijen, S., Victor, N., and Dadi, Z.: Special Report on Emissions Scenarios?: a special report of Working Group III of the Intergovernmental Panel on Climate Change, available at: http://www.osti.gov/energycitations/servlets/purl/15009867-Kv00FB/native/ (last access: 20 April 2012), 2001.

Nash, J. E. and Sutcliffe, J. V.: River flow forecasting through conceptual models part I – A discussion of principles, J. Hydrol., 10, 282–290, doi:10.1016/0022-1694(70)90255-6, 1970.

Paiva, R. C. D., Collischonn, W., Schettini, E. B. C., Vidal, J.-P., Hendrickx, F., and Lopez, A.: The Case Studies, in: Modelling the Impact of Climate Change on Water Resources, edited by: Fung, F., Lopez, A., and New, M., 136–182, John Wiley & Sons, Ltd., available at: http://onlinelibrary.wiley.com/doi/10.1002/9781444324921.ch6/summary (last access: 18 May 2012), 2010.

Raje, D. and Mujumdar, P. P.: Reservoir performance under uncertainty in hydrologic impacts of climate change, Adv. Water Resour., 33, 312–326, doi:10.1016/j.advwatres.2009.12.008, 2010.

Rosenberg, N. J., Brown, R. A., Izaurralde, R. C., and Thomson, A. M.: Integrated assessment of Hadley Centre (HadCM2) climate change projections on agricultural productivity and irrigation water supply in the conterminous United States, Agr. Forest Meteorol., 117, 73–96, doi:10.1016/S0168-1923(03)00025-X, 2003.

Rosenzweig, C., Strzepek, K. M., Major, D. C., Iglesias, A., Yates, D. N., McCluskey, A., and Hillel, D.: Water resources for agriculture in a changing climate: international case studies, Global Environ. Change, 14, 345–360, doi:10.1016/j.gloenvcha.2004.09.003, 2004.

Schaefli, B., Hingray, B., and Musy, A.: Climate change and hydropower production in the Swiss Alps: quantification of potential impacts and related modelling uncertainties, Hydrol. Earth Syst. Sci., 11, 1191–1205, doi:10.5194/hess-11-1191-2007, 2007.

Schneider, C., Laizé, C. L. R., Acreman, M. C., and Flörke, M.: How will climate change modify river flow regimes in Europe?, Hydrol. Earth Syst. Sci., 17, 325–339, doi:10.5194/hess-17-325-2013, 2013.

Tilmant, A., Pinte, D., and Goor, Q.: Assessing marginal water values in multipurpose multireservoir systems via stochastic programming, Water Resour. Res., 44, W12431, doi:10.1029/2008WR007024, 2008.

Tilmant, A., Goor, Q., and Pinte, D.: Agricultural-to-hydropower water transfers: sharing water and benefits in hydropower-irrigation systems, Hydrol. Earth Syst. Sci., 13, 1091–1101, doi:10.5194/hess-13-1091-2009, 2009.

Turgeon, A.: Stochastic optimization of multireservoir operation: The optimal reservoir trajectory approach, Water Resour. Res., 43, W05420, doi:10.1029/2005WR004619, 2007.

Vachala, S.: Evaporation sur les retenues du Sud de la France (In French), Master Thesis, Université Pierre et Marie Curie, Ecole des Mines de Paris & Ecole Nationale du Génie Rural des Eaux et des Forêts, available at: http://www.sisyphe.upmc.fr/~m2hh/arch/memoires2008/Vachala.pdf (last access: 29 October 2013), 2008.

Veijalainen, N., Dubrovin, T., Marttunen, M., and Vehviläinen, B.: Climate Change Impacts on Water Resources and Lake Regulation in the Vuoksi Watershed in Finland, Water Resour. Manage., 24, 3437–3459, doi:10.1007/s11269-010-9614-z, 2010.

Vicuña, S., Dracup, J. A., Lund, J. R., Dale, L. L., and Maurer, E. P.: Basin-scale water system operations with uncertain future climate conditions: Methodology and case studies, Water Resour. Res., 46, W04505, doi:10.1029/2009WR007838, 2010.

Ward, F. A., Roach, B. A., and Henderson, J. E.: The Economic Value of Water in Recreation: Evidence from the California Drought, Water Resour. Res., 32, 1075–1081, doi:10.1029/96WR00076, 1996.

Wolfgang, O., Haugstad, A., Mo, B., Gjelsvik, A., Wangensteen, I., and Doorman, G.: Hydro reservoir handling in Norway before and after deregulation, Energy, 34, 1642–1651, doi:10.1016/j.energy.2009.07.025, 2009.

Yakowitz, S.: Dynamic programming applications in water resources, Water Resour. Res., 18, 673–696, doi:10.1029/WR018i004p00673, 1982.

15

A coupled modeling framework for sustainable watershed management in transboundary river basins

Hassaan Furqan Khan[1]**, Y. C. Ethan Yang**[2]**, Hua Xie**[3]**, and Claudia Ringler**[3]

[1]Department of Civil and Environmental Engineering, University of Massachusetts, Amherst, MA 01003, USA
[2]Department of Civil and Environmental Engineering, Lehigh University, Bethlehem, PA 18015, USA
[3]International Food Policy Research Institute, Washington, DC, USA

Correspondence: Y. C. Ethan Yang (yey217@lehigh.edu)

Abstract. There is a growing recognition among water resource managers that sustainable watershed management needs to not only account for the diverse ways humans benefit from the environment, but also incorporate the impact of human actions on the natural system. Coupled natural–human system modeling through explicit modeling of both natural and human behavior can help reveal the reciprocal interactions and co-evolution of the natural and human systems. This study develops a spatially scalable, generalized agent-based modeling (ABM) framework consisting of a process-based semi-distributed hydrologic model (SWAT) and a decentralized water system model to simulate the impacts of water resource management decisions that affect the food–water–energy–environment (FWEE) nexus at a watershed scale. Agents within a river basin are geographically delineated based on both political and watershed boundaries and represent key stakeholders of ecosystem services. Agents decide about the priority across three primary water uses: food production, hydropower generation and ecosystem health within their geographical domains. Agents interact with the environment (streamflow) through the SWAT model and interact with other agents through a parameter representing willingness to cooperate. The innovative two-way coupling between the water system model and SWAT enables this framework to fully explore the feedback of human decisions on the environmental dynamics and vice versa. To support non-technical stakeholder interactions, a web-based user interface has been developed that allows for role-play and participatory modeling. The generalized ABM framework is also tested in two key transboundary river basins, the Mekong River basin in Southeast Asia and the Niger River basin in West Africa, where water uses for

ecosystem health compete with growing human demands on food and energy resources. We present modeling results for crop production, energy generation and violation of eco-hydrological indicators at both the agent and basin-wide levels to shed light on holistic FWEE management policies in these two basins.

1 Introduction

Comprehensive watershed management is a challenging task that requires multidisciplinary knowledge. An emerging research area highlights the importance of using watershed management to sustain various ecosystem services for human society (Jewitt, 2002; Lundy and Wade, 2011). While the various services provided by a river are primarily viewed through the prism of human benefits, maintaining a healthy ecosystem can be mutually beneficial to both human society and ecological systems. A failure to maintain adequate levels of riverine ecosystem health may result in compromised human benefits for future generations (Baron et al., 2004). There is therefore a growing recognition among water resource managers that sustainable watershed management needs to not only account for the diverse ways humans benefit from the environment, but also incorporate the impact of human actions on the natural system (Vogel et al., 2015). This is perhaps most prominently advocated in the emerging science of socio-hydrology, which calls for an understanding of the two-way interactions and co-evolution of coupled human–water systems (Sivapalan et al., 2012). This two-way coupling, then, needs to be integrated into computational tools used to aid watershed management.

A coupled human natural systems modeling approach, where the stochastic interactions between agents are represented, also facilitates stakeholder involvement. It can be used as a communication tool to organize information between hydrologists, systems analysts, policy makers and other stakeholders to inform the model and provide meaning to its results. The process of involving stakeholders in the modeling process allows them to observe how their actions affect other agents and observe the system-wide trends that emerge based on low-level agent interactions (Lund and Palmer, 1997).

Traditional watershed modeling does not effectively capture system heterogeneity, limiting its ability to effectively represent the two-way interaction between human and natural systems. Conventional models of water resource systems developed for assisting decision-making treat human benefits as a single objective using a centralized optimization approach, which ignores the heterogeneity among water users and uses (e.g., priority of different water uses along a river system based on socioeconomic differences) (Yang et al., 2009). The decision-maker is usually assumed to possess perfect information with respect to demand and supply of water and other resources in the watershed. If they are considered at all, most ecological functions are considered as constraints in the system, often for numerical convenience and frequently leading to oversimplification (Stone-Jovicich, 2015).

In this paper, we develop a modeling framework that can effectively address both system heterogeneity and the linkage between human society and hydrology that influences water cycling in the watershed. We do so by differentiating key stakeholders of ecosystem services as active agents based on their characteristics such as location and water use preferences, and tightly couple the human system with a process-based watershed model that simulates the stock and flow of environmental variables needed by the stakeholders.

In this two-way coupled natural–human systems modeling framework, the human system is modeled as a decentralized water systems model and is linked to a process-based, semi-distributed hydrologic model. Empirical data obtained from surveys of water practitioners are used to develop behavior rules for water use, providing a realistic representation of human behaviors in water resource modeling. In addition to incorporating indirect interaction between the agents through the environment, i.e., surface water flows, a novel advancement offered in this framework is the ability of agents to *directly* interact by requesting assistance from other agents based on their level of cooperation. A web-based user interface for this coupled model has been developed which enables non-technical stakeholders to use this modeling platform online. The online portal allows for role-play and participatory modeling. We apply this modeling framework to two different transboundary basins where ecological needs are competing with growing human demands on the water

resources: the Mekong River basin in Southeast Asia and the Niger River basin in West Africa.

2 Previous studies of coupled natural–human system modeling

Coupled natural–human system modeling through explicit modeling of both natural processes (e.g., rainfall–runoff for water supply) and human behavior (e.g., services that humans derive from natural systems, such as water resources) helps reveal the reciprocal interactions and coevolution of the natural and human systems. Modeling efforts coupling the natural and human systems have increased in recent years (Liu et al., 2007), evolving from an approach that focused mostly on understanding the natural processes and that treated human actions as fixed boundary conditions (Sivakumar et al., 2005). The human system coupled with the natural system can be simulation (descriptive) or optimization (prescriptive) based, depending on the modeling objective (Giuliani et al., 2016).

A watershed is a self-organizing system characterized by distributed albeit interactive decision processes. If a coordination mechanism exists, it will guide the interactions among individual decision processes. The agent-based modeling (ABM) framework provides such a mechanism for integrating knowledge and understanding across diverse domains (Berglund, 2015; Yang et al., 2009). In an ABM, individual actors are represented as unique and autonomous "agents" with their own interests. Agents follow certain behavioral rules and interact with each other in a shared environment allowing for a natural representation of real-world, "bottom–up" watershed management processes. A (semi-)distributed hydrological model that can simulate the environment, and which provides ecosystem services, can then be linked with the agent-based model that represents decentralized decision-making processes. This linkage allows us to utilize the strength from both models and better represent a watershed as a coupled natural–human complex system.

Distributed process-based hydrologic models are well suited for linkage with ABMs. Compared to statistical or data driven models, process-based models are more robust for extrapolation or in simulating conditions under changing management practices. Distributed and semi-distributed models have the capacity to reflect the spatial heterogeneity of hydrologic and water quality processes within a river basin. This capacity also facilitates the evaluation of spatially variable user demands for ecosystem services. Open-source hydrologic models, where it is possible for third-party users to incorporate region-specific knowledge into the models to improve performance or extend model capability, are especially suitable for coupling with decentralized water system models. The spatial structure of the hydrologic model and its consistency with the model structure of the ABM it is being coupled to are additional important considerations.

SWAT (Soil and Water Assessment Tool) is one such hydrologic modeling platform with many of the features described above that has been used previously to explore effects of human intervention on basin water resources. It provides built-in functions to simulate reservoir operations, irrigation and a variety of best management practices (BMPs) for nutrient pollution control (Bracmort et al., 2006; Strauch et al., 2013). Its open-source nature allows users to incorporate locale-specific knowledge into the model to improve model performance or extend a model's capabilities. SWAT conducts simulations at the level of the sub-watershed, or hydrological response unit. When the modeling domain of an agent-based model is delineated following the boundaries of a sub-watershed, it has the advantage of spatial unit consistency with agent-based models. Furthermore, it has been coupled with (non-ABM) decision modeling tools to identify cost-effective solutions to basin water resource management challenges (Ciou et al., 2012; Karamouz et al., 2010). We therefore choose SWAT as the hydrologic model for this study.

A fully coupled modeling framework involves continuous information exchange between the agent-based and hydrologic models such that the two models are solved simultaneously or iteratively in each time step. Relevant existing studies that link agent-based models with other simulation models are summarized in Table S1 in the Supplement. A review of the existing literature shows that most coupled natural–human systems models, especially in the context of surface-water management, are only loosely linked and thus do not fully capture the impact of human actions on hydrology (Berger et al., 2007; Giacomoni et al., 2013; Ng et al., 2011; Yang et al., 2012). "Fully coupled" models can be found for groundwater analysis (e.g., Reeves and Zellner, 2010). This is because the common outputs from groundwater models are "stock variables" such as groundwater head, and it is relatively easy to restart the simulation model from the previous step. Surface hydrologic models, on the other hand, usually output flux (i.e., streamflow) and not stock variables (e.g., lake storage and soil moisture). To be "fully coupled" with an agent-based model, a modification of the programming code of the watershed model is usually necessary to output state variables and allow the agent-based model to interact with the watershed model at monthly or daily time steps (Mishra, 2013).

The methodology proposed here is designed primarily to help improve stakeholder understanding of a complex system as well as recognition of various, alternative development pathways for the basin in question. A linkage between an agent-based model and a process-based watershed model, incorporating direct interactions between agents, is a promising method to accurately represent complex coupled natural–human systems as well as to appropriately involve non-technical stakeholders in the assessment.

3 Methodology

The generalized framework for the two-way coupling between an agent-based model and a process-based watershed model is described here in greater detail. In this framework, the river basin is divided into politically and hydrologically similar sub-regions, where water management is primarily carried out under the ambit of a single administrative unit, which represents an autonomous agent. This approach to delineating regions is also found in other studies, e.g., the Food Production Unit in the International Model for Policy Analysis of Agricultural Commodities and Trade (Robinson et al., 2015).

In this framework, agents follow prescribed rules, based on which their benefits are calculated. Agents make water management decisions, on an annual time step, for agricultural production, hydropower generation and ecological management based on targets set using long-term historical data. They update their actions every year based on their experience from previous years; this behavior can be classified as a hybrid between reactive and deliberative approaches (Akhbari and Grigg, 2013). In this modeling framework, agents can interact both directly and indirectly. Agents interact indirectly through their water usage for agriculture, and changes in streamflow in response to hydropower production. For direct communication between agents, we include a level of cooperation (LOC) parameter that signifies the willingness of an agent to alter their own water management actions to benefit a downstream agent. This setting allows for the incorporation of stochasticity into the agent decision-making process.

Figure 1 shows the higher-level coupled modeling framework. First, user-defined preferences and level of cooperation are defined based on stakeholder input. These input parameters can either be defined by individual users according to specific scenarios of interest, or be determined by directly eliciting the information from the various water-using stakeholders, for example, through surveys. As part of this project, we conducted comprehensive surveys across three transboundary river basins (Indus, Mekong and Niger) to identify water use preferences (Khan et al., 2017). A sample survey questionnaire is provided in the Supplement. The surveys were developed to elicit the perceived importance of various ecosystem services across each basin under a variety of economic and hydrologic future conditions. One of the questions in the survey asked respondents to rank different ecosystem services in order of importance for each agent. These responses were then averaged across all the respondents for each agent to obtain a ranking of the importance of the different ecosystem services. These rankings were used in the decision algorithm for the case study models developed and presented in Sect. 4. Second, other initial input parameters are incorporated into the ABM framework. These include reservoir characteristics, such as storage, release capacity, efficiency and operational rules for each reser-

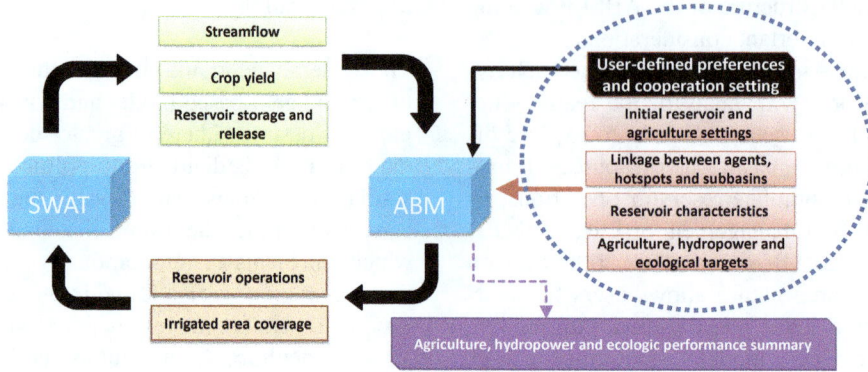

Figure 1. Overview of the modeling framework coupling ABM with SWAT.

voir. The geographic linkages between subbasins, ecosystem hotspots and agents across the entire river basin are defined in the ABM as well. For each subbasin, agricultural parameters are defined, including the type of land cover, total cropped area and type of crop produced. For each agent, targets are defined for each of the three water uses based on historical flow conditions. These targets form the basis relative to which the agents make their water management decisions.

The ABM, built using the *R* statistical language, reports agent decisions concerning reservoir operation and irrigated area that are then used as input for the calibrated SWAT model that simulates the hydrology for the next time step. The crop production and reservoir modules in the SWAT model are driven using water management decisions from the ABM and hydroclimatologic conditions. Upon completion, the SWAT model generates three primary output files that are used as input for the agent-based model. These files include the following.

- Proportion of cropped area and crop yield for each hydrologic-response unit (HRU) in each subbasin in each agent.

- Daily storage volume and releases from each reservoir.

- Daily streamflow at the outlet of each of the subbasins across the basin.

The output from the SWAT model is then fed back into the ABM, based on which the agents make water management decisions for the next time step. In the last time step of the modeling run, the ABM provides a summary file summarizing the performances for each of the three water uses: agriculture, hydropower and ecology.

Figure 2 shows the algorithm through which the ABM and the hydrologic model interact, and the process through which various agents make their water management decisions, in two distinct parts. In the first part, the agent's water management decision is made based on its preferences of water use, while in the second part the decisions are made based on its willingness to cooperate. In the first part, the algorithm uses the water use preferences for each agent, and compares the target value with the output from the SWAT model for each of the water uses to make the water management decision for each agent. Under the current setting, the agent is allowed to only make one water management decision every year. However, this can be modified in future studies to allow multiple decisions to be made in a year. Additional information from stakeholders (such as rules of tiebreak) would be needed for this.

For instance, consider an agent that ranks agricultural production higher than other water uses. In this case, the ABM checks to see whether crop production meets the target crop production. If crop production is significantly lower than the target crop production, then the agent decides to increase the irrigated area. If crop production meets the target production, then the ABM checks to see whether hydropower generation for the current time step meets the hydropower generation target. If the hydropower generation target is not met, the agent decides to decrease the number of days actual storage needs to meet the target storage. This allows for greater releases and increased hydropower generation. If the hydropower generation target has also been satisfied, then the ABM moves to the second part of the decision-making algorithm.

An important input to the ABM is the identification of ecosystem hotspots. Ecosystem hotspots are specific regions in the river basin that are especially critical to or indicative of the health of the ecosystem in the entire basin. Ecosystem hotspots can be identified in a variety of ways including through a literature review of critical ecological concerns in a basin and/or input from local ecological experts. For this analysis, for each ecosystem hotspot, relevant Indicators of Hydrologic Alteration (IHA) and Environmental Flow Component (EFC) parameters are selected based on expert opinion to measure ecosystem health (Richter et al., 1997, 1996). Baseline values for relevant IHA and EFC parameters, which are streamflow-based indicators, are calculated from daily streamflow of the calibrated SWAT model. The IHA and EFC parameters included for the case study applications described

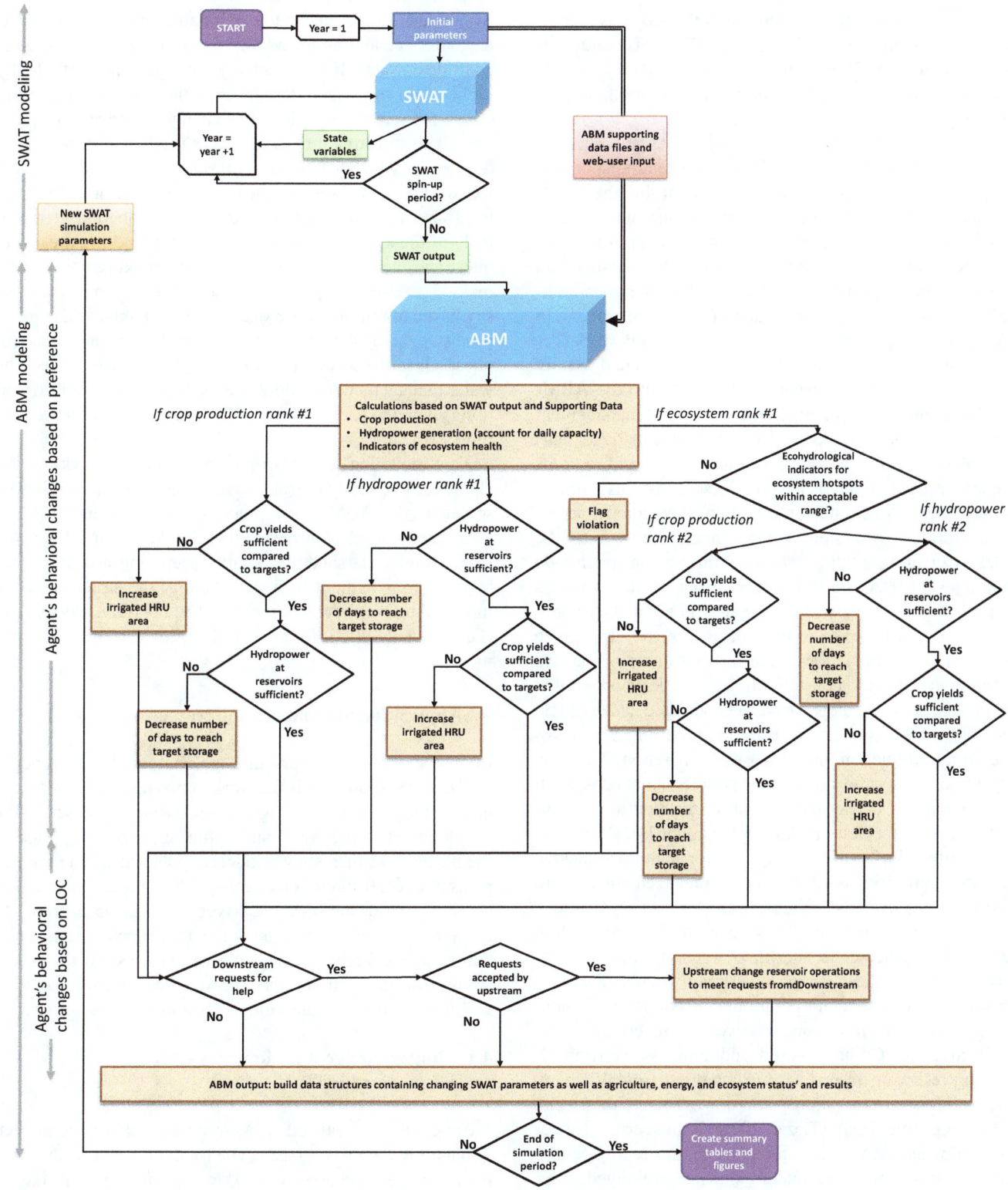

Figure 2. Modeling workflow including the two-part algorithm through which agents make water management decisions.

in Sect. 4 include monthly median flows, 7-day annual maximum flow, small and large flood event duration, timing and duration of extreme low flows, etc. We use $\pm 10\%$ from the baseline value as a decision threshold in the ABM as recommended by research consortium partner WorldFish and Wetlands International. This means the modeled IHA and EFC values deviating from the baseline value by more than 10 % would require an agent to take action.

Water management to satisfy ecological targets depends on the specific hydro-ecology of the ecosystem hotspot. For example, a river reach may need low flows during the breeding season, while a downstream wetland may need higher flows to avoid eutrophic conditions. Satisfying multiple ecologic needs, as is often the case in large river basins, can require contrasting interventions and add tremendous complexity to the water management decision-making process. In the case study applications for this modeling framework (detailed in Sect. 4), we find that the information needed to fully incorporate ecosystem hotspot management into the ABM-SWAT framework is limited. The link between management actions (e.g., reservoir operations, crop land management) and ecological concerns is not well understood and requires further investigation that is beyond the scope of this work.

In the absence of detailed information on ecological needs, we incorporate ecosystem hotspot management into the model by creating a "flag" when the timing and magnitude of the relevant IHA and EFC deviate from the target values in each hotspot. Thus, while the agents do not actively consider ecosystem hotspots in their decisions, they recognize when violations (deviations from target values) occur. We use these violations to constrain the agent's decision, so that if any of the ecologic targets have been violated and ecologic needs are ranked highest, no action can be undertaken for agricultural production or hydropower generation. This current setting mimics most real-world policies about ecosystem conservation that do not have an active reaction to environmental issues, especially in developing countries. Of course, this algorithm is flexible and allows for a more proactive decision-making process for ecologic management if more information regarding stakeholder perceptions is available.

In the second part of the decision-making algorithm, agents decide whether to alter their water management actions based on requests from downstream agents. This feature aims to represent the possibility of cooperative water management in a transboundary river basin. For instance, in March 2016, China released additional water from its Jinghong reservoir, in response to a request from Vietnam, to help alleviate water shortages in downstream countries in the Mekong River basin (Tiezzi, 2016). In the current framework, a downstream agent can request an upstream agent to change its reservoir operations to alleviate prolonged water scarcity (at least two time steps). For instance, if a downstream agent has been unable to meet its agricultural production target for 2 years, then it can request an upstream agent

to increase releases. Wherever available, one upstream reservoir is identified for each agent.

Once a request is made by a downstream agent, the upstream agent first checks to see whether it has surplus storage, after accounting for its own needs, to consider releasing additional water. If the available storage is not sufficiently higher than the target storage, then the upstream agent declines the request and does not change its reservoir operations. If the upstream reservoir has sufficient storage, then it decides on whether to respond favorably to the downstream request based on its willingness to cooperate. In this modeling framework, the LOC represents the probability (from 0 to 1) of the agent responding favorably to a downstream request and incorporates human decision-making uncertainty, making the second part of the decision-making algorithm stochastic to mimic human decision uncertainty. In any given time step, an upstream reservoir can only respond to one request. Once the second part of the algorithm is executed, the water management decisions are made and relevant information is then fed back to the SWAT model as input for the next time step.

This modeling framework is generalizable, tackling the challenge of paucity of transparency and reusability often associated with ABM development (O'Sullivan et al., 2016). The framework design means that the ABM can be adapted to different watersheds by simply preparing a different set of input files without having to modify the structure of the model. An Overview, Design, and Details (ODD) document (Grimm et al., 2010) for the ABM is provided in the Supplement.

4 Application of the modeling framework

In this section, we show the application of this generalized coupled modeling framework to two transboundary river basins: the Mekong and Niger River basins. We describe the development of the ABM and hydrology model for each of the basins, and then show model outputs illustrating the impacts of agent behavior on agent-specific and basin-wide outcomes. We use the Mekong River basin as an example to show how agents' preferences impact different water uses, while the Niger River basin is used as a case study to demonstrate how interactions between different agents and their willingness to cooperate influence basin-wide outcomes.

4.1 Impact of agent preferences – Mekong demonstration

We apply the generalized ABM framework described in Sect. 3 to the Mekong River basin. The Mekong River, with an annual average discharge of $450\,\mathrm{km}^3$, drains the sixth largest river basin in the world in terms of runoff (Kite, 2001). It is a transboundary river originating in China and flows through or borders Myanmar, Thailand, Laos and Cambodia before

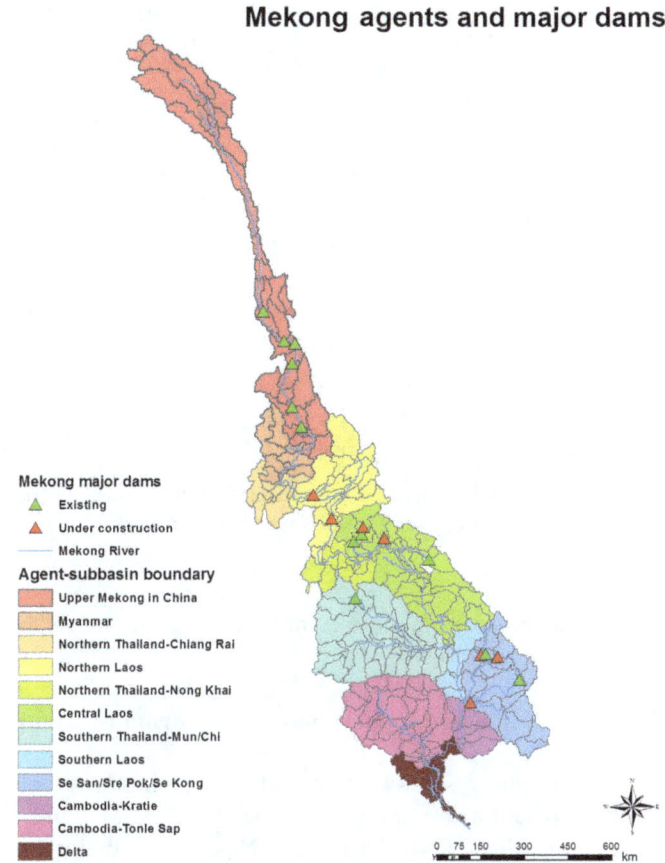

Figure 3. Basin map for the Mekong River basin showing agent boundaries and major dams included in the model.

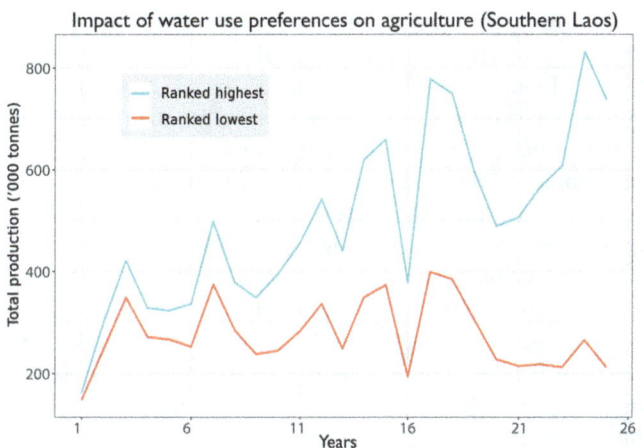

Figure 4. Difference in crop production caused by the differing prioritization of agriculture for the southern Laos agent.

finally draining in the Mekong delta in Vietnam. Flow in the upper Mekong in China is mainly comprised of snowmelt, while precipitation from the two monsoon systems provides the bulk of the flow in the lower Mekong (Ringler, 2001). Around 70 million people depend upon the Mekong River for food, water and economic sustenance, and the basin is home to several diverse and productive ecosystems. The Tonle Sap lake, among the most productive ecosystems in the world (Bakker, 1999), is an example of the unique ecology and biodiversity in the basin. Agriculture accounts for about 80–90 % of total freshwater consumption in the Mekong (MRC, 2002), with rice being the most widely grown crop. The Mekong delta is another hotspot of economic activity and produces approximately half of Vietnam's annual rice harvest and over half of Vietnam's fish exports (Kite, 2001). The Mekong is currently in a phase of rapid infrastructure development (storage and hydropower), raising concerns regarding the downstream ecological impact (Urban et al., 2013).

The Mekong was spatially delineated into 12 distinct hydrologically similar agents who make water management decisions to satisfy their own targets. Figure 3 shows the distribution of the agents across the basin and the locations of major existing and planned water infrastructure facilities, and

important ecological hotspots identified by local ecological experts. In total, there are 19 major dams (7 existing and 12 planned) and 23 ecological hotspots identified by local ecological experts using the existing literature (Baran et al., 2012). To allow for a more intuitive interpretation of results, here we only model crop production for irrigated rice, but the modeling framework allows for incorporation of any number of crop types. The modeling structure allows for simulations under either existing water infrastructure or future conditions that also include dams under construction. For demonstration purposes, we present results under future water infrastructure.

A SWAT hydrology model was developed, calibrated and validated with streamflow data from 1978 to 2007. Details on model setup and calibration and validation results for the hydrology model are provided in the Supplement. In addition, Fig. S4 in the Supplement shows simulated average hydropower generation under historic streamflow conditions and compares it with the observed hydropower generation for five existing reservoirs during the period of comparison as validation for the ABM.

Figure 4 shows an example of how total crop production (of irrigated rice) changes over the simulation period with a different assigned priority (lowest vs. highest) for agriculture for the agent representing southern Laos. Both these simulated crop production time series are run with the same hydrologic time series, so the differences between the levels of crop production are caused by different water management actions. Over the simulation period of 25 years, there is a significant cumulative difference in agricultural production largely because of the compounding effect of increasing irrigated area whenever the crop production target is not met. When agriculture is assigned a lower priority, the agent prioritizes either hydropower generation or ecosystem health and is less likely to make decisions to increase agricultural production.

Different ecosystem services respond differently to changes in external drivers, depending on the nature of water use. Figure 5 shows a comparison of the effect of different priorities on hydropower generation for the Nam Theun 2 dam in the agent representing central Laos. As in the previous example, both the simulated time series are run with similar hydrology to isolate the difference in hydropower generation due only to different agent behavior. For this model, if simulated hydropower generation is less than 90 % of historic (for existing dams) or expected (for future dams) mean annual energy, the agent can decide to change its operation rules for the dam to increase hydropower generation. In this model specifically, agents do so by increasing the minimum monthly releases from their reservoirs.

The fluctuations in HP generation from year to year are caused by changes in hydrology, while the differences between the blue and red lines represent the agent preference regarding the relative importance of hydropower. We observe that the annual fluctuations in hydropower generation (due to hydrology) are significantly greater than the slight changes in generation stemming from modified reservoir operations. Time steps with high streamflow conditions lead to very similar outcomes regardless of preference. The difference is more prominent in low-flow conditions, where a higher prioritization of hydropower leads to an increased "minimum" level of hydropower. Despite the fact that the difference between hydropower generation due to a change in prioritization is not as significant as that for the agricultural production, annual differences in hydropower generation can be as high as 8 % (210 GWh). In the context of energy shortages in the Mekong, this difference is non-trivial. Another interesting feature to note in Fig. 5 is that when the agent decides to increase releases in a time step for larger hydropower generation, generation in the next time step is reduced because of reduced storage. The emergence of this myopic behavior pattern also gives us confidence in the model as it replicates how hydropower generation decisions are made in the real world.

Finally, we also investigate the impact of changing priorities on ecologic performance. For each of the 23 hotspots, relevant indicators of ecologic health using the IHA and EFC framework are identified. As explained in Sect. 3, agents can protect ecological health by choosing to limit water management actions for other water uses (agriculture and hydropower). Simulation results for this model showed that different agent preferences do not have a significant impact on ecological violations. The amount of water available (hydrology) has a much more pronounced impact. A reason for the lack of a negative impact of changes in reservoir operations on ecological performance is that reservoir capacities are low relative to streamflow. It is important to note here that the eco-hydrological indicators we used in the current modeling framework do not account for fish migration patterns and sediment transport, which are among the biggest concerns about hydropower in the Mekong. Future studies can link the

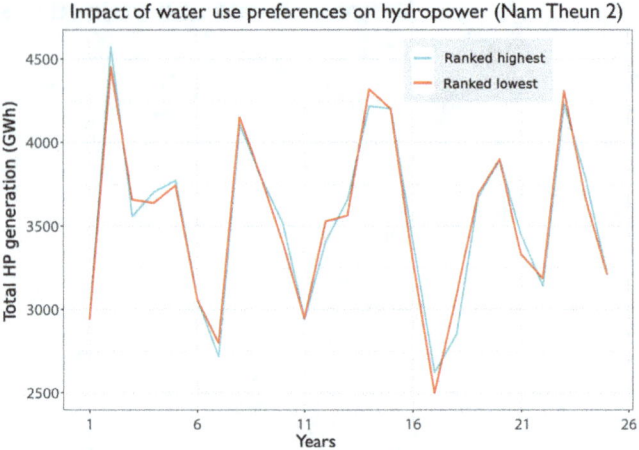

Figure 5. Difference in hydropower generation due to changes in prioritization of hydropower for the Nam Theun 2 reservoir.

current framework with more complex ecological models to address these concerns.

4.2 Impact of agent cooperation – Niger demonstration

To illustrate the system-wide impacts of varying levels of agent cooperation, we apply this generalized ABM framework to the Niger River basin. The Niger River drains an area of over 2 million km^2 spanning nine riparian countries in West Africa, making it the ninth largest river basin globally in terms of area. The Niger River is spread across a wide range of ecosystem zones, and the basin is thus notable for its high spatial and temporal hydrologic variability on interannual and decadal scales (Ghile et al., 2014). Based on GDP, all nine countries of the Niger basin fall in the bottom quartile of national incomes (Ogilvie et al., 2010). Agriculture constitutes a large part of the economic output for the region (approximately 33 %), with livestock and fisheries also contributing substantially in some areas (Welcomme, 1986). Owing to the lack of a well-developed irrigation system, most of the agriculture in the Niger is rainfed, with only 20 % of available arable land under cultivation. Investment in water resource infrastructure and institutions offers a potential pathway to economic development for the basin population and several large dams are slated for construction under the existing Niger Basin Authority investment plan. However, the downstream impacts of upstream infrastructure have become a contentious issue.

For the Niger basin, 15 agents were identified based on hydrologic characteristics and administrative boundaries. A map of the system showing the agent and subbasin boundaries, and existing and planned water infrastructure, is provided in Fig. 6. Nineteen ecologic hotspots identified by local ecological experts using the Niger Basin Atlas (Aboubacar, 2007) and 10 dams (6 existing + 4 planned) are included in the model. For the agricultural module, we simulate irrigated

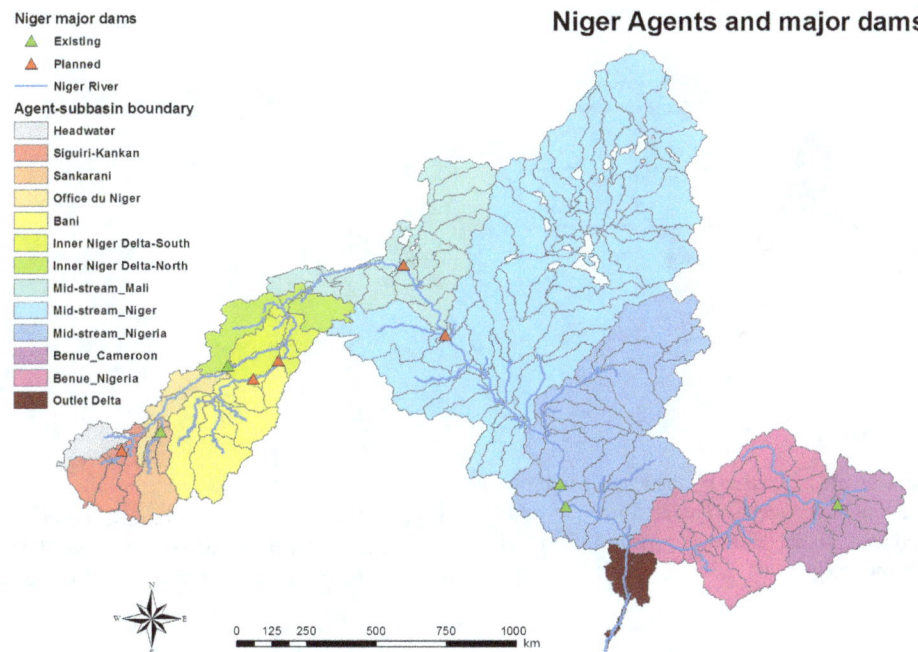

Figure 6. Basin map for the Niger River basin showing agent boundaries and major dams included in the model.

rice and upland crops. A SWAT hydrology model was developed, calibrated and validated with streamflow data from 1985 to 2010. Details on model setup and calibration and validation results for the hydrology model are provided in the Supplement.

We run this model under two different settings and then compare the results to evaluate the basin-wide impacts of cooperation between agents. In the first setting, agents make water management decision solely to satisfy their own objectives without interacting directly with other agents. In the second setting, agents' decisions are driven by both their own objectives, and their willingness to cooperate with other agents. Willingness to cooperate, represented in the model with the level of cooperation parameter (LOC), can be set on a scale of 0 to 1 and signifies the probability of an agent responding favorably to a request from another agent to alter its water management decisions. In this model, agents with reservoirs respond to a downstream request by increasing the minimum flow if storage in the reservoir is above the target storage. For the purposes of demonstration, we set the LOC for agents to 1 to simulate a fully cooperative environment. Both model runs are made with the same set of agent preferences. To illustrate impacts of future infrastructure development, we run both the simulations under the future state of water infrastructure.

Over the course of the 26-year simulation period, we observe 73 instances of agents requesting help successfully, with many of these requests made during low-flow years. We see that additional releases from an upstream agent willing to cooperate can often, but not always, result in an appreciable increase in crop production compared to when the agents are

solely interested in satisfying their own objectives. For example, in year 20 of the simulation, the Outlet Delta agent successfully requests the upstream Jebba reservoir for additional water releases, and experiences an increase in food production of almost 50 000 tons without any decrease in production in the upstream agent.

Figures 7 and 8 illustrate the changes in reservoir operation and its impact on streamflow downstream when an upstream agent decides to cooperate. For Jebba reservoir, Fig. 7 shows the difference in reservoir releases between the "cooperation" and "no cooperation" runs, the blue region representing the additional volume that is released based on the decision of the agent to cooperate. Figure 8 shows the available streamflow downstream of the dam under both the simulation scenarios: the red line indicates releases when the agent alters its reservoir operations in response to the request while the blue line shows releases in the model where the agents do not cooperate. It is interesting, but not surprising to note, that additional water released leads to reduced releases in subsequent time steps due to reduced storage.

This change in the timing of water availability has the potential to both negatively and positively affect all downstream users, including those that were not part of the negotiation that led to the altered water management action (i.e., "third-party impacts"). The occurrence of third-party impacts is dependent on the context; they do not necessarily occur every time, and if they do occur, they can be either positive or negative. In these modeling runs, we observe many instances of varying third-party impacts. For example, in response to consecutive years of reduced agricultural production, the Niger Inner Delta (South) Agent requests the upstream Fomi Dam

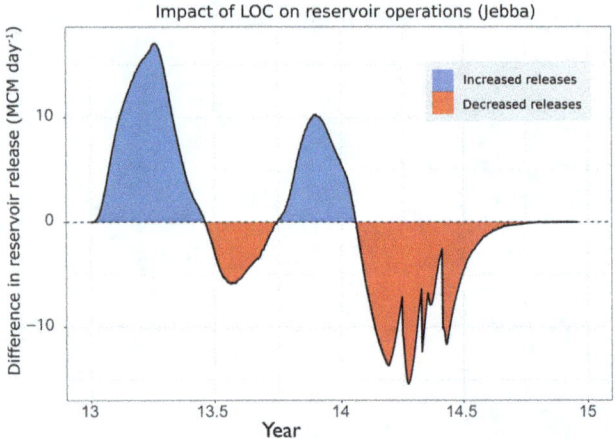

Figure 7. Change in reservoir release caused by the agent's willingness to cooperate with downstream agents. The area in blue (red) represents additional (reduced) water released compared to model runs where the agent does not cooperate.

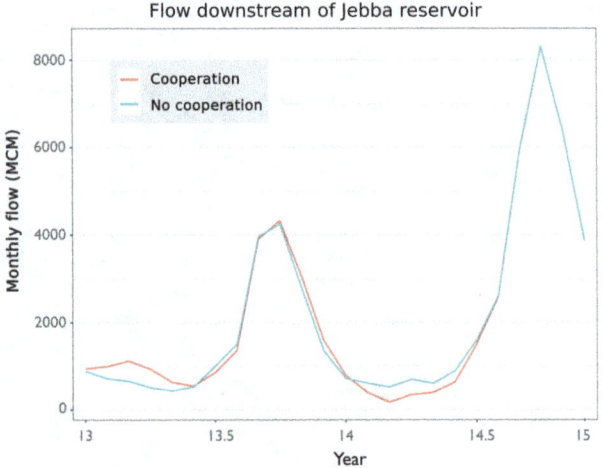

Figure 8. Comparison of monthly streamflow immediately downstream of Jebba reservoir between model runs when an agent decides to cooperate and when it does not cooperate.

for additional releases in year 13 of the simulation. The agent managing Fomi Dam, Siguiri-Kankan, agrees to the request and increases its minimum releases. Not only does crop production in Niger Inner Delta (South) increase as a result, but crop production in Niger Inner Delta (North) is also positively impacted. However, the Office Du Niger Agent suffers from a decrease in food production.

It is pertinent to note here that additional releases do not necessarily increase crop production; it is possible that there are constraints other than water availability that are limiting crop production. In the same year of the simulation as the previous example, the agent representing Mid-stream Niger requests additional releases from Touassa Dam and experiences an increase in crop production. Crop production in the mid-stream does not change appreciably as a result; however, production in another downstream agent, Mid-Stream Nigeria, is increased. In the current model, agents make requests when they are unable to meet crop production targets. However, the modeling framework allows for making requests dependent on other factors (e.g., ecological needs).

These third-party impacts, also referred to as *externalities* in the natural resource economics literature, are also seen in ecologic performance. The nature and magnitude of third-party impacts on ecologic performance are dependent on the specific ecosystem. Arguably, ecologic health is even more sensitive than agricultural production to changes in the timing and magnitude of streamflow. In these simulations, we see evidence of this impact. In year 9, in response to a request from Mid-Stream Nigeria, Kandaji reservoir releases additional water that (compared to the no cooperation setting) positively affects the ecosystem hotspots in Mid-Stream Niger and Mid-Stream Nigeria, but results in increased violations of ecological targets in the downstream Outlet Delta. In particular, the ecological parameter seen to be violated is the

IHA parameter for minimum average 7-day flow. Despite the increase in total annual flow due to the additional releases, the change in the flow timing leads to an ecologically inferior outcome for the Outlet Delta. This finding supports the argument that evaluations of ecological health performed at coarse timescales (e.g., annual) may overlook finer timescale flow parameters that are critical to ecosystems (Palmer et al., 2005). In the absence of detailed data relating flow conditions to aquatic health in the Niger Outlet Delta, it is difficult to ascertain the exact impact that the violation of this target would have on the delta's ecosystem.

5 Discussion

5.1 Dynamic coupled natural–human systems modeling

The generalized coupled modeling framework presented in this paper adopts many of the principles from the Shared Vision Modeling (SVM) approach (Palmer et al., 2013). To improve allocation of scarce resources across competing uses, it is crucial to understand the values placed on various water uses by stakeholders in the watershed. For the case study applications, model development was preceded and followed by extensive stakeholder engagements. Before the model development began, an electronic survey of water users in each of the river basins was conducted to analyze perceptions of the relative importance of different water uses. Rules derived from these surveys improve representation of the interactions between heterogeneous subsystems. Moreover, to make this modeling framework more accessible to users, a web-based interface has been developed where users can perform model simulations with differently specified agent behavior rules.

The online interface allows users to visualize and save results from several modeling runs. Information from the modeling runs made on the online platform can be used to further develop agent behavior rules and have stakeholders evaluate the results to gain insight into emerging development pathways in the basin. In addition to the utility provided by the visualization of the outcomes, the exercise of tailoring the modeling framework to a specific basin requires stakeholders to conceptualize the water system better. A beta version of the website with the model for the Mekong River basin has been developed and tested with stakeholders in the Mekong.

Third-party impacts, which are costs or benefits borne by a party due to the actions of others, have been recognized as an obstacle to promoting cooperative water management practices in a water system with many heterogeneous users (Petersen-Perlman et al., 2017). While the existence and importance of third-party impacts are widely acknowledged, they are not easily quantified, making them difficult to incorporate into stakeholder discussions on water management in transboundary settings. The case study results for the Niger River basin presented here quantify these third-party impacts on agricultural production, hydropower generation and ecological performance. Quantification of the impacts, both positive and negative, of the actions of water users can help develop a shared understanding of the water system dynamics among stakeholders (Skurray et al., 2012). By offering a way to fully couple human and natural systems with several ecosystem services, with flexibility to incorporate varying levels of importance for heterogeneous users, the modeling framework presented here can be useful as a tool to stimulate cooperative water management in transboundary settings.

5.2   Limitations and future work

The case study models developed use observed climate data to develop hydrologic time series for model simulations. Observed streamflow data are used for model simulations under the future infrastructure setting as well. However, significant uncertainty exists regarding future hydroclimatology and its impact on water resources in these basins (Lauri et al., 2012). A climate stress-test approach where the agent's response to varying hydroclimatological conditions is evaluated can provide insight into sensitivity to climate variables (Brown et al., 2012).

Another useful extension of this modeling framework would be to incorporate seasonal forecasts of water availability into the decision-making process of agents. Water managers often perceive the advantages offered by seasonal forecasts as being low (Pagano et al., 2002), even though the economy-wide benefits of seasonal forecasts can be substantial (Rodrigues et al., 2016). This modeling framework can be used to highlight the potential benefits of short-term seasonal forecasts for agents' decisions on water allocation and willingness to cooperate with other agents, and introduce another dimension of stochasticity to the agent decision-making process. The seasonal forecasts used, however, would need to be geographically suitable and temporally appropriate for each agent's operations.

The development of coupled river basin models needs to carefully address several tradeoffs to ensure that the models are scientifically sound and computationally tractable. The focus of this work is to develop a generalized ABM framework that addresses model transparency and model/module reusability (An, 2012; Parker et al., 2003). To address this, the geographic delineations of our agents are relatively larger than traditional agent-based models (which define individual water users as agents). This is a necessary simplification in order to balance model complexity (or the level of detail of simulated decision processes) and computational resource and data availability. Furthermore, it is pertinent to recognize that agent-based models are best used to explain existing relationships or phenomena, rather than as prediction tools. Another related limitation associated with large-scale agent-based models is reliance on informal validation. For the case studies presented here, we validate the ABM with internal checks, for instance by comparing modeled and observed hydropower generated (Fig. S4). We also address this limitation through the use of surveys to inform agent behavior rules.

To further improve the agent decision module, Bayesian decision theory would be a useful avenue of future research to better address uncertainty of human decisions (Kocabas and Dragicevic, 2013; Van Oijen et al., 2011). However, this approach is computationally costly, especially in our setting with a variety of different agents, water use preferences and willingness to cooperate. High performance computing technology might become necessary for this purpose.

The coupled modeling framework described in this paper operates on an annual time step. This means that exchange of information between the ABM and SWAT takes place at the start of every year. The framework can be made more realistic by configuring the models to interact at the finer timescale at which water management decisions are made, i.e., monthly or weekly. While the modeling framework is sufficiently flexible to allow for a range of water management actions, in the modeling framework described here, we model ecological health management in a passive rather than active manner. Active ecologic health management, where the agents make specific decisions (especially with regards to reservoir operations), requires a more in-depth understanding of the basin ecology than was available for either of the two transboundary rivers used as case studies for this paper.

6　Conclusion

Sustainable watershed management requires water managers and policy makers to have a clear understanding of their water system and its interactions with the natural environment. This study develops a spatially scalable, generalized agent-based modeling (ABM) framework consisting of a process-based semi-distributed hydrologic model, SWAT and a decentralized water system model to simulate the impacts of water resource management decisions on the food–water–energy–environment nexus (FWEE) at the watershed scale. The two-way coupling provides a holistic understanding of the FWEE nexus. A novel advancement offered in this framework is the ability of agents to *directly* interact by requesting assistance from other agents based on their level of cooperation (LOC). Quantification of the LOC is especially useful for transboundary river basins with several unique actors with different water management objectives. Among various other future uses, this modeling system has been developed for the CGIAR Research Program on Water, Land and Ecosystems to assess tradeoffs between agricultural production, productivity, other water-based ecosystem services and ecosystem health. To support non-technical stakeholder interactions in developing country settings, where CGIAR operates, a web-based user interface has been developed. This online portal allows for end-user role-play, participatory modeling and inference of prioritized ecosystem services and ecosystem health.

We show the flexibility of this modeling framework by applying it to two large transboundary rivers as case studies and demonstrate its ability to reveal the impact of water use preferences and willingness to cooperate on region-specific and basin-wide outcomes. In the case studies, we see that agent preferences have a more pronounced effect on crop production compared to hydropower generation. Changing preferences has a relatively smaller impact on ecological health, but that is heavily dependent on the river basin, ecological health indicators and water management actions. The impact of agent cooperation revealed the presence of both positive and negative third-party impacts that need to be acknowledged and accounted for when considering cooperative river management in transboundary settings, especially at finer timescales.

Author contributions. HFK and YCEY developed the ABM. HX developed the SWAT hydrologic models. CR provided guidance on project direction and manuscript preparation. HFK prepared the manuscript with contributions from all co-authors.

Competing interests. The authors declare that they have no conflict of interest.

Special issue statement. This article is part of the special issue "Coupled terrestrial-aquatic approaches to watershed-scale water resource sustainability". It is not connected with a conference.

Acknowledgements. This paper was developed under the Innovation Fund modus of the CGIAR Research Program on Water, Land and Ecosystems, which receives support from CGIAR fund donors, including the Australian Department of Foreign Affairs and Trade (DFAT), the Bill and Melinda Gates Foundation, the Netherlands Directorate-General for International Cooperation (DGIS), the Swedish International Development Cooperation Agency (Sida) and Switzerland: Swiss Agency for Development Cooperation (SDC).

Edited by: Xuesong Zhang

References

Aboubacar, A.: Niger River Basin Atlas, Niger Basin Authority, Niamey, 2007.

Akhbari, M. and Grigg, N. S.: A Framework for an Agent-Based Model to Manage Water Resources Conflicts, Water Resour. Manag., 27, 4039–4052, https://doi.org/10.1007/s11269-013-0394-0, 2013.

An, L.: Modeling human decisions in coupled human and natural systems: Review of agent-based models, Ecol. Model., 229, 25–36, https://doi.org/10.1016/j.ecolmodel.2011.07.010, 2012.

Arnold, J. G., Srinivasan, R., Muttiah, R. S., and Williams, J. R.: Large area hydrologic modeling and assesment Part I: Model development, J. Am. Water Resour. As., 34, 73–89, https://doi.org/10.1111/j.1752-1688.1998.tb05961.x, 1998.

Bakker, K.: The politics of hydropower: Developing the Mekong, Polit. Geogr., 18, 209–232, https://doi.org/10.1016/S0962-6298(98)00085-7, 1999.

Baran, E., Chum, N., Fukushima, M., Hand, T., Hortle, K. G., Jutagate, T., and Kang, B.: Fish Biodiversity Research in the Mekong Basin, in: The Biodiversity Observation Network in the Asia-Pacific Region: Toward Further Development of Monitoring, edited by: Nakano, S., Yahara, T., and Nakashizuka, T., Ecological Research Monographs, Tokyo, Springer, 149–164, 2012.

Baron, J., Poff, N. L., Angermeier, P. L., Dahm, C., Gleick, P. H., Hairston, N. G., Jackson, R. B., Johnston, C. A., Richter, B. D., and Steinman, A. D.: Sustaining healthy freshwater ecosystems, Water Resour., 127, 25–58, 2004.

Berger, T., Birner, R., Diaz, J., McCarthy, N., and Wittmer, H.: Capturing the complexity of water uses and water users within a multi-agent framework, Integr. Assess. Water Resour. Glob. Chang. A North-South Anal., 129–148, https://doi.org/10.1007/978-1-4020-5591-1_9, 2007.

Berglund, E. Z.: Using Agent-Based Modeling for Water Resources Planning and Management, J. Water Res. Pl., 141, 1–17, https://doi.org/10.1061/(ASCE)WR.1943-5452.0000544, 2015.

Bracmort, K. S., Arabi, M., Frankenberger, J. R., Engel, B. A., and Arnold, J. G.: Modeling Long-Term Water Quality Impact of Structural BMPs, Trans. Am. Soc. Agric. Biol. Eng., 49, 367–374, https://doi.org/10.13031/2013.20411, 2006.

Brown, C., Ghile, Y., Laverty, M., and Li, K.: Decision scaling: Linking bottom-up vulnerability analysis with climate projections in the water sector, Water Resour. Res., 48, W09537, https://doi.org/10.1029/2011WR011212, 2012.

Ciou, S.-K., Kuo, J.-T., Hsieh, P.-H., and Yu, G.-H.: Optimization Model for BMP Placement in a Reservoir Watershed, J. Irrig. Drain. Eng., 138, 736–747, https://doi.org/10.1061/(ASCE)IR.1943-4774.0000458, 2012.

Ghile, Y. B., Taner, M., Brown, C., Grijsen, J. G., and Talbi, A.: Bottom-up climate risk assessment of infrastructure investment in the Niger River Basin, Climatic Change, 122, 97–110, https://doi.org/10.1007/s10584-013-1008-9, 2014.

Giacomoni, M. H., Kanta, L., and Zechman, E. M.: Complex Adaptive Systems Approach to Simulate the Sustainability of Water Resources and Urbanization, J. Water Res. Pl., 139, 554–564, https://doi.org/10.1061/(ASCE)WR.1943-5452.0000302, 2013.

Giuliani, M., Li, Y., Castelletti, A., and Gandolfi, C.: A coupled human-natural systems analysis of irrigated agriculture under changing climate, Water Resour. Res., 52, 6928–6947, https://doi.org/10.1002/2016WR019363, 2016.

Grimm, V., Berger, U., DeAngelis, D. L., Polhill, J. G., Giske, J., and Railsback, S. F.: The ODD protocol: A review and first update, Ecol. Model., 221, 2760–2768, https://doi.org/10.1016/j.ecolmodel.2010.08.019, 2010.

Jewitt, G.: Can Integrated Water Resources Management sustain the provision of ecosystem goods and services?, Phys. Chem. Earth, 27, 887–895, https://doi.org/10.1016/S1474-7065(02)00091-8, 2002.

Karamouz, M., Taheriyoun, M., Baghvand, A., Tavakolifar, H., and Emami, F.: Optimization of watershed control strategies for reservoir eutrophication management, J. Irrig. Drain. Eng., 136, 847–861, https://doi.org/10.1061/(ASCE)IR.1943-4774.0000261, 2010.

Khan, H. F., Yang, Y.-C. E., and Ringler, C.: Heterogeneity in Riverine Ecosystem Service Perceptions?: Insights for Water-decision Processes in Transboundary Rivers, IFPRI Discussion Paper 1668, Washington, DC, 2017.

Kite, G.: Modelling the mekong: Hydrological simulation for environmental impact studies, J. Hydrol., 253, 1–13, https://doi.org/10.1016/S0022-1694(01)00396-1, 2001.

Kocabas, V. and Dragicevic, S.: Bayesian networks and agent-based modeling approach for urban land-use and population density change: A BNAS model, J. Geogr. Syst., 15, 403–426, https://doi.org/10.1007/s10109-012-0171-2, 2013.

Lauri, H., de Moel, H., Ward, P. J., Räsänen, T. A., Keskinen, M., and Kummu, M.: Future changes in Mekong River hydrology: impact of climate change and reservoir operation on discharge, Hydrol. Earth Syst. Sci., 16, 4603–4619, https://doi.org/10.5194/hess-16-4603-2012, 2012.

Liu, J., Dietz, T., Carpenter, S. R., Alberti, M., Folke, C., Moran, E., Pell, A. N., Deadman, P., Kratz, T., Lubchenco, J., Ostrom, E., Ouyang, Z., Provencher, W., Redman, C. L., Schneider, S. H., and Taylor, W. W.: Complexity of Coupled Human and Natural Systems, Science, 317, 1513–1516, https://doi.org/10.1126/science.1144004, 2007.

Lund, J. R. and Palmer, R. N.: Water Resource System Modeling for Conflict Resolution, Water Resour., 3, 70–82, 1997.

Lundy, L. and Wade, R.: Integrating sciences to sustain urban ecosystem services, Prog. Phys. Geogr., 35, 653–669, https://doi.org/10.1177/0309133311422464, 2011.

Mishra, S. K.: Modeling water quantity and quality in an agricultural watershed in the midwestern US using SWAT?: assessing implications due to an expansion in biofuel production and climate change, University of Iowa, 2013.

MRC: Annual Report, Phnom Penh., 2002.

Ng, T. L., Eheart, J. W., Cai, X., and Braden, J. B.: An agent-based model of farmer decision-making and water quality impacts at the watershed scale under markets for carbon allowances and a second-generation biofuel crop, Water Resour. Res., 47, 1–17, https://doi.org/10.1029/2011WR010399, 2011.

Ogilvie, A., Mahé, G., Ward, J., Serpantié, G., Lemoalle, J., Morand, P., Barbier, B., Tamsir Diop, A., Caron, A., Namarra, R., Kaczan, D., Lukasiewicz, A., Paturel, J.-E., Liénou, G., and Charles Clanet, J.: Water, agriculture and poverty in the Niger River basin, Water Int., 35, 594–622, https://doi.org/10.1080/02508060.2010.515545, 2010.

O'Sullivan, D., Evans, T., Manson, S., Metcalf, S., Ligmann-Zielinska, A., and Bone, C.: Strategic directions for agent-based modeling: avoiding the YAAWN syndrome, J. Land Use Sci., 11, 177–187, https://doi.org/10.1080/1747423X.2015.1030463, 2016.

Pagano, T. C., Hartmann, H. C., and Sorooshian, S.: Factors affecting seasonal forecast use in Arizona water management: A case study of the 1997–98 El Niño, Clim. Res., 21, 259–269, https://doi.org/10.3354/cr021259, 2002.

Palmer, M. A., Bernhardt, E. S., Allan, J. D., Lake, P. S., Alexander, G., Brooks, S., Carr, J., Clayton, S., Dahm, C. N., Follstad Shah, J., Galat, D. L., Loss, S. G., Goodwin, P., Hart, D. D., Hassett, B., Jenkinson, R., Kondolf, G. M., Lave, R., Meyer, J. L., O'Donnell, T. K., Pagano, L., and Sudduth, E.: Standards for ecologically successful river restoration, J. Appl. Ecol., 42, 208–217, https://doi.org/10.1111/j.1365-2664.2005.01004.x, 2005

Palmer, R. N., Cardwell, H. E., Lorie, M. A., and Werick, W.: Disciplined planning, structured participation, and collaborative modeling – applying shared vision planning to water resources, J. Am. Water Resour. As., 49, 614–628, https://doi.org/10.1111/jawr.12067, 2013.

Parker, D. C., Manson, S. M., Janssen, M. A., Hoffmann, M. J., and Deadman, P.: Multi-agent systems for the simulation of land-use and land-cover change: A review, Ann. Assoc. Am. Geogr., 93, 314–337, https://doi.org/10.1111/1467-8306.9302004, 2003.

Petersen-Perlman, J. D., Veilleux, J. C., and Wolf, A. T.: International water conflict and cooperation: challenges and opportunities, Water Int., 42, 105–120, https://doi.org/10.1080/02508060.2017.1276041, 2017.

Reeves, H. W. and Zellner, M. L.: Linking MODFLOW with an agent-based land-use model to support decision making, Ground Water, 48, 649–660, https://doi.org/10.1111/j.1745-6584.2010.00677.x, 2010.

Richter, B., Baumgartner, J. V, Wigington, R., and Braun, D. P.: How much water does a river need?, Freshwater Biol., 37, 231–249, https://doi.org/10.1046/j.1365-2427.1997.00153.x, 1997.

Richter, B. D., Baumgartner, J. V., Powell, J., and Braun, D. P.: A Method for Assessing Hydrologic Alteration within Ecosystems,

Conserv. Biol., 10, 1163–1174, https://doi.org/10.1046/j.1523-1739.1996.10041163.x, 1996.

Ringler, C.: Optimal Water Allocation in the Mekong River Basin, Center for Development Research, Bonn, 2001.

Robinson, S., Mason-D'Croz, D., Islam, S., Sulser, T. B., Robertson, R., Zhu, T., Gueneau, A., Pitois, G., and Rosegrant, M.: The International Model for Policy Analysis of Agricultural Commodities and Trade (IMPACT): model description for Version 3, Washington, DC, available at: http://ebrary.ifpri.org/cdm/ref/collection/p15738coll2/id/129825 (30 July 2017), 2015.

Rodrigues, J., Thurlow, J., Landman, W., Ringler, C., Robertson, R., and Zhu, T.: The economic value of seasonal forecasts: Stochastic Economy-Wide Analysis for East Africa, IFPRI Discussion Paper 1546, Washington, DC, 2016.

Sivakumar, M., Das, H., and Brunini, O.: Impacts of present and future climate variability and change on agriculture and forestry in the arid and semi-arid tropics, Climatic Change, 70, 31–72, https://doi.org/10.1007/s10584-005-5937-9, 2005.

Sivapalan, M., Savenije, H. H. G., and Blöschl, G.: Socio-hydrology: A new science of people and water, Hydrol. Process., 26, 1270–1276, https://doi.org/10.1002/hyp.8426, 2012.

Skurray, J. H., Roberts, E. J., and Pannell, D. J.: Hydrological challenges to groundwater trading: Lessons from south-west Western Australia, J. Hydrol., 412–413, 256–268, https://doi.org/10.1016/j.jhydrol.2011.05.034, 2012.

Stone-Jovicich, S.: Probing the interfaces between the social sciences and social-ecological resilience: Insights from integrative and hybrid perspectives in the social sciences, Ecol. Soc., 20, 25, https://doi.org/10.5751/ES-07347-200225, 2015.

Strauch, M., Lima, J. E. F. W., Volk, M., Lorz, C., and Makeschin, F.: The impact of BMPs on simulated streamflow and sediment loads in a Central Brazilian catchment, 127, 1–18, 2013.

Tiezzi, S.: Facing Mekong Drought, China to Release Water from Yunan Dam, Dipl., 2–3, available at: http://thediplomat.com/2016/03/facing-mekong-drought-china-to-release-water-from-yunnan-dam/ (last access: 30 July 2017), 2016.

Urban, F., Nordensvärd, J., Khatri, D., and Wang, Y.: An analysis of China's investment in the hydropower sector in the Greater Mekong Sub-Region, Environ. Dev. Sustain., 15, 301–324, https://doi.org/10.1007/s10668-012-9415-z, 2013.

Van Oijen, M., Cameron, D. R., Butterbach-Bahl, K., Farahbakhshazad, N., Jansson, P. E., Kiese, R., Rahn, K. H., Werner, C., and Yeluripati, J. B.: A Bayesian framework for model calibration, comparison and analysis: Application to four models for the biogeochemistry of a Norway spruce forest, Agr. Forest Meteorol., 151, 1609–1621, https://doi.org/10.1016/j.agrformet.2011.06.017, 2011.

Vogel, R. M., Lall, U., Cai, X., Rajagopalan, B., Weiskel, P. K., Hooper, R. P., and Matalas, N. C.: Hydrology: The interdisciplinary science of water, Water Resour. Res., 51, 4409–4430, https://doi.org/10.1002/2015WR017049, 2015.

Welcomme, R. L.: The effects of the Sahelian drought on the fishery of the central delta of the Niger River, Aquac. Res., 17, 147–154, 1986.

Yang, Y. C. E., Cai, X., and Stipanović, D. M.: A decentralized optimization algorithm for multiagent system-based watershed management, Water Resour. Res., 45, 1–18, https://doi.org/10.1029/2008WR007634, 2009.

Yang, Y. E., Zhao, J., and Cai, X.: Decentralized optimization method for water allocation management in the Yellow River basin, J. Water Resour. Pl., 138, 313–325, https://doi.org/10.1061/(ASCE)WR.1943-5452.0000199, 2012.

PERMISSIONS

LIST OF CONTRIBUTORS

C. Dong and G.-H. Huang
MOE Key Laboratory of Regional Energy and Environmental Systems Optimization, Resources and Environmental Research Academy, North China Electric Power University, Beijing 102206, China

Y.-P. Cai
State Key Laboratory of Water Environment Simulation, School of Environment, Beijing Normal University, Beijing 100875, China
Institute for Energy, Environment and Sustainable Communities, University of Regina, Regina S4S 7H9, Canada

Q. Tan
MOE Key Laboratory of Regional Energy and Environmental Systems Optimization, Resources and Environmental Research Academy, North China Electric Power University, Beijing 102206, China
Institute for Energy, Environment and Sustainable Communities, University of Regina, Regina S4S 7H9, Canada

Suria Tarigan
Department of Soil Sciences and Natural Resource Management, Bogor Agricultural University, Bogor, Indonesia

Kerstin Wiegand
Department of Ecosystem Modeling, University of Göttingen, Büsgenweg 4, 37077 Göttingen, Germany

Sunarti
Faculty of Agriculture, University of Jambi, Jambi, Indonesia

Bejo Slamet
Faculty of Agriculture, North Sumatra University, Medan, Indonesia

A. Nazemi and H. S. Wheater
Global Institute for Water Security, University of Saskatchewan, 11 Innovation Boulevard, Saskatoon, SK, S7N 3H5, Canada

Tracy Ewen
Department of Geography, University of Zurich, Zurich, Switzerland
Center for Climate Systems Modeling, ETH Zurich, Zurich, Switzerland

Jan Seibert
Department of Physical Geography and Quaternary Geology, Stockholm University, Stockholm, Sweden
Department of Earth Sciences, Uppsala University, Uppsala, Sweden

Morten Grum and Thomas Munk-Nielsen
Krüger Veolia, Søborg, 2860, Denmark

Vianney Courdent
Krüger Veolia, Søborg, 2860, Denmark
Department of Environmental Engineering, Technical University of Denmark, Kgs. Lyngby, 2800, Denmark

Peter S. Mikkelsen
Department of Environmental Engineering, Technical University of Denmark, Kgs. Lyngby, 2800, Denmark

L. J. M. Peeters
CSIRO Land and Water, Water for a Healthy Country Flagship, Adelaide, Australia

G. M. Podger, S. M. Cuddy and T. Smith
CSIRO Land and Water, Water for a Healthy Country Flagship, Canberra, Australia

T. Pickett
CSIRO Land and Water, Water for a Healthy Country Flagship, Brisbane, Australia

R. H. Bark
CSIRO Ecosystem Sciences, Water for a Healthy Country Flagship, Brisbane, Australia

Juan Fernando Salazar, Juan Camilo Villegas, Angela María Rendón, Estiven Rodríguez and Daniel Mercado-Bettín
GIGA, Escuela Ambiental, Facultad de Ingeniería, Universidad de Antioquia, Medellín, Colombia

Peter S. Mikkelsen
Department of Environmental Engineering, Technical University of Denmark, Kgs. Lyngby, 2800, Denmark

L. J. M. Peeters
CSIRO Land and Water, Water for a Healthy Country Flagship, Adelaide, Australia

G. M. Podger, S. M. Cuddy and T. Smith
CSIRO Land and Water, Water for a Healthy Country Flagship, Canberra, Australia

T. Pickett
CSIRO Land and Water, Water for a Healthy Country Flagship, Brisbane, Australia

R. H. Bark
CSIRO Ecosystem Sciences, Water for a Healthy Country Flagship, Brisbane, Australia

Juan Fernando Salazar, Angela María Rendón, Estiven Rodríguez and Daniel Mercado-Bettín
GIGA, Escuela Ambiental, Facultad de Ingeniería, Universidad de Antioquia, Medellín, Colombia

Juan Camilo Villegas
GIGA, Escuela Ambiental, Facultad de Ingeniería, Universidad de Antioquia, Medellín, Colombia
School of Natural Resources and the Environment, University of Arizona, Tucson, USA

Isabel Hoyos
GAIA, Escuela Ambiental, Facultad de Ingeniería, Universidad de Antioquia, Medellín, Colombia
Instituto de Física, Universidad de Antioquia, Medellín, Colombia

Germán Poveda
Universidad Nacional de Colombia, Sede Medellín, Departamento de Geociencias y Medio Ambiente, Facultad de Minas, Medellín, Colombia

Martine Rutten
Department of Water Resource Management, Delft University of Technology, Delft, 2628 CN, the Netherlands

Maarten van der Sanden
Department of Science Education and Communication, Delft University of Technology, Delft, 2628 CJ, the Netherlands

Ellen Minkman
Department of Water Resource Management, Delft University of Technology, Delft, 2628 CN, the Netherlands
Department of Science Education and Communication, Delft University of Technology, Delft, 2628 CJ, the Netherlands
Presently at Department of Public Administration and Sociology, Erasmus University Rotterdam, Burgemeester Oudlaan 50, 3062 PA Rotterdam, the Netherlands

David Eschbach, Laurent Schmitt and Grzegorz Skupinski
Laboratoire Image, Ville, Environnement (LIVE UMR 7362), Université de Strasbourg, CNRS, ENGEES, ZAEU LTER, Strasbourg, France

Gwenaël Imfeld and Sylvain Payraudeau
Laboratoire d'Hydrologie et de Géochimie de Strasbourg (LHyGeS UMR 7517), Université de Strasbourg, CNRS, ENGEES, Strasbourg, France

Jan-Hendrik May and Frank Preusser
Institute of Earth and Environmental Sciences, University of Freiburg, Freiburg, Germany

Mareike Trauerstein
Institute of Geography, University of Bern, Bern, Switzerland

H. Macian-Sorribes and M. Pulido-Velazquez
Research Institute of Water and Environmental Engineering (IIAMA), Universitat Politècnica de València, Valencia, Spain

A. Tilmant
Department of Civil and Water Engineering, Université Laval, Québec City, Québec, Canada

D. C. Verdon-Kidd
Environmental and Climate Change Research Group, School of Environmental and Life Sciences, University of Newcastle, Callaghan, Australia

A. S. Kiem and R. Moran
Water Group, Department of Environment and Primary Industries, Victoria, Australia

Dingbao Wang
Department of Civil, Environmental, and Construction Engineering, University of Central Florida, Orlando, Florida, USA

Diane Arjoon and Amaury Tilmant
Department of Civil Engineering and Water Engineering, Université Laval, Québec, Canada

Markus Herrmann
Department of Economics, Université Laval, Québec, Canada

B. François, B. Hingray and J. D. Creutin
CNRS, LTHE – UMR5564, 38041 Grenoble, France
University Grenoble Alpes, LTHE – UMR5564, 38041 Grenoble, France

F. Hendrickx
EDF, R&D, LNHE, 78400 Chatou, France

Hassaan Furqan Khan
Department of Civil and Environmental Engineering,
University of Massachusetts, Amherst, MA 01003,
USA

Y. C. Ethan Yang
Department of Civil and Environmental Engineering,
Lehigh University, Bethlehem, PA 18015, USA

Hua Xie and Claudia Ringler
International Food Policy Research Institute,
Washington, DC, USA

Index